Multigrid Finite Element Methods for Electromagnetic Field Modeling

Multigrid Finite Element Methods for Electromagnetic Field Modeling

Yu Zhu
Cadence Design Systems, Inc.

Andreas C. Cangellaris
University of Illinois at Urbana-Champaign

IEEE Antennas and Propagation Society, *Sponsor*

 IEEE Press Series on Electromagnetic Wave Theory
Donald G. Dudley, Series Editor

IEEE

IEEE PRESS

A JOHN WILEY & SONS, INC., PUBLICATION

Published by John Wiley & Sons, Inc., Hoboken, New Jersey.
Published simultaneously in Canada.

For general information on our other products and services or for technical support, please contact our Customer Care Department within the United States at (800) 762-2974, outside the United States at (317) 572-3993 or fax (317) 572-4002.

Wiley also publishes its books in a variety of electronic formats. Some content that appears in print may not be available in electronic formats. For more information about Wiley products, visit our web site at www.wiley.com.

Library of Congress Cataloging-in-Publication Data is available.

ISBN-13 978-0-471-74110-7
ISBN-10 0-471-74110-8

Printed in the United States of America.

10 9 8 7 6 5 4 3 2 1

To Jie and Helen, for being there when no one else is.

CONTENTS

LIST OF FIGURES

LIST OF TABLES

PREFACE

The central theme of this book is the development of robust preconditioners for the iterative solution of electromagnetic field boundary value problems (BVPs) discretized by means of finite methods. More precisely, the book provides a detailed presentation of our successful attempts to utilize concepts from multigrid and multilevel methods for the preconditioning of matrices resulting from the approximation of electromagnetic BVPs using finite elements.

Solution by iteration is the reluctant compromise of modern scientific computing. A brief review of the young history of numerical computation offers unequivocal support for Edward Teller's bleak conjecture that "*A state-of-the-art calculation requires 100 hours of CPU time on the state-of-the-art computer, independent of the decade.*" Hidden in this statement is recognition and praise of our innate drive to advance our understanding of the physical world and, eventually, through its predictive modeling, establish our dominance over it. Yet, the deeper our understanding becomes the higher the complexity of the mysteries we need to unravel. And this complexity pushes our computing technology to its current limits.

In our desire to stay one step ahead of this seemingly perpetual complexity barrier, iteration seems to be an invaluable ally. Loosely speaking an iterative process should be interpreted as the attempt to solve a given problem by trial and error. In the context of the theme of this book, the pertinent problem is the solution of the system of linear equations resulting from the discretization of the electromagnetic BVP using finite elements. A guess for the unknown vector is made and the error (residual) is calculated between the forcing vector and the vector resulting from the multiplication of the trial vector by the matrix. A convergent iterative process is one through which the residual is iteratively reduced below some a-priori defined threshold.

Thanks to continuing advances in iterative matrix-solving methods, we are now equipped with solid knowledge about the right ingredients for the construction of an effective iterative solver. Among them, the availability of a good preconditioner stands out. Effective preconditioning of the iteration matrix expedites convergence and can enhance solution accuracy. Multigrid and multilevel methods have been shown to be most effective as preconditioners. Their success stems from the fact that the convergence of standard relaxation techniques can be accelerated by utilizing different spatial samplings for the damping of different components in the decomposition of the error in terms of the eigenvectors of the iteration matrix. In the context of finite methods spatial sampling is most typically effected through the size of the element in the numerical grid or the order of the interpolating polynomial.

The construction of an effective multigrid/multilevel preconditioner requires the understanding of the properties of the eigenvectors and eigenvalues of the iteration matrix. This, in turn, calls for a thorough insight in the way the development of the finite element approximation of the boundary value problem maps the properties of the eigenpairs of the governing mathematical operator onto those of its finite element matrix approximation.

It was primarily due to lack of such insight and understanding that the application of multigrid/multilevel ideas to the development of effective preconditioners for finite element matrix approximations of vector electromagnetic BVPs was slow in coming. However, over the past ten years significant progress has been made toward filling this void. This book attempts to provide the reader with an elucidating description of the path that led us to today's state-of- the-art in multigrid/multilevel preconditioners for finite element-based iterative electromagnetic field solvers.

Chapter 1 begins with a brief overview of the finite element method and its application to the numerical solution of the electromagnetic (BVP). This overview is followed by a discussion of current and future challenges in finite element-based electromagnetic field modeling, which helps motivate the methodologies and algorithms presented in the following chapters.

Chapter 2 presents a complete and systematic development of spaces of functions suitable for the expansion of scalar potential fields, tangentially continuous vector fields, normally continuous vector fields, and scalar charge densities on the triangles and tetrahedra used for the discretization of the computational domain. The corresponding spaces for these four classes of quantities are often referred to as Whitney-0, Whitney-1, Whitney-2, and Whitney-3 forms. They are the natural choices for the representation of the scalar potential, the field intensities ($\vec{E}$ and $\vec{H}$), the flux densities ($\vec{D}$ and $\vec{B}$), and the scalar charge density, respectively. The reason for this is that their development is guided by their physical attributes, as described mathematically by the governing equations and associated boundary conditions. It is shown in Chapter 2 that explicit, hierarchical basis functions of arbitrary order can be developed for all four spaces. In addition, the mathematical relations of the four spaces are presented and discussed. The proposed basis functions can be used, either directly or after a further partial orthogonalization post-processing, to improve the condition number of the finite element matrix.

Chapter 3 is devoted to the development of the finite element formulations of various classes of electromagnetic BVPs. In particular, formulations pertinent to static, quasi-static, and dynamic problems are presented, along with their subsequent reduction into matrix equation statements of their discrete approximation using the Galerkin process. The emphasis of the chapter is on the appropriate choice of basis functions for the expansion of pertinent unknown field quantities and their sources, in a manner such that the requirements for solvability and uniqueness of the solution of the continuous problem are mapped correctly onto the finite element matrix approximation of the BVP. The chapter concludes with an investigation of the source of the so-called low-frequency numerical instability of integral equation-based approximations of electrodynamic problems, in the context of the properties of the basis functions used for the approximation of the unknown electric current and charge densities.

In Chapter 4 an overview of iterative methods and preconditioning techniques for sparse linear systems of equations is provided. The discussion leads to the motivation for multigrid methods and a brief overview of the fundamental steps in the development of a multigrid algorithm. This is followed, in Chapter 5, by the demonstration of the application of the nested multigrid process as an effective preconditioner for enhancing the convergence of the iterative solution of finite element approximations of the scalar wave equation.

Chapter 6 extends the application of multigrid processes as preconditioners for enhancing the convergence of the iterative solution of finite element approximations of vector electromagnetic BVPs. Applications of the proposed nested multigrid preconditioners to the solution of numerous electromagnetic BVPs are used to demonstrate the superior numerical convergence and efficient memory use of these algorithms.

The nested multigrid algorithms discussed in Chapters 5 and 6 can be viewed as an h-adaptive finite element method, since use is made of the lowest order basis functions, while the finite element mesh is refined in a nested fashion. However, p-adaptive techniques, where the order of the basis functions is increased while the finite element mesh is kept unchanged, are oftentimes more effective in reducing the discretization error, especially when used in regions in the computational domain where the field quantities exhibit smooth spatial variation. Thus, in Chapter 7 we present a robust, hierarchical, multilevel preconditioning technique for the fast finite element analysis of electromagnetic BVPs.

As a precursor to the discussion of electromagnetic eigenvalue problems, Chapter 8 briefly introduces various Krylov-based techniques for the solution of large matrix eigenvalue problems. In Chapter 9 the Krylov-based matrix eigenvalue solution machinery is streamlined for application to the solution of matrix eigenvalue problems resulting from the finite element approximation of two-dimensional electromagnetic eigenvalue problems. In particular, the emphasis is on the finite element-based eigenvalue analysis of uniform, inhomogeneous, anisotropic electromagnetic waveguides.

In Chapter 10 an efficient algorithm is developed for the robust solution of sparse matrix eigenvalue problems resulting from the finite element approximation of three-dimensional electromagnetic cavities, as well as unbounded and lossy electromagnetic resonators. The proposed algorithm is based on a field-flux formulation of the finite element approximation of Maxwell's equations. The special relationship between the vector bases used for the expansion of the electric field vector $\vec{E}$ and the magnetic flux density vector $\vec{B}$ is used to reduce the computational complexity of the numerical solution of the finite element matrices resulting from the proposed formulation.

The formulation introduced in Chapter 10 is further exploited in Chapter 11 for the establishment of a systematic methodology and associated computer algorithms for the generation of reduced-order matrix transfer function representations of multi-port electromagnetic systems. The emphasis of the presentation in Chapter 10 is on two topics. First, it is shown that the field-flux formulation of the electromagnetic system leads to a passive discrete, state-space model, which is compatible with the well-established Krylov-subspace model order reduction techniques for reduced-order macromodeling of linear systems. Second, it is shown that through the proper selection of the expansion functions for the finite element approximation of the electric field and the magnetic flux density, a computationally efficient algorithm can be devised for the calculation of the broadband reduced-order matrix transfer function of the electromagnetic system.

Finally, in Chapter 12, we consider the application of finite elements to the modeling of periodic structures. In addition to the discussion of how the finite element formulation of the electromagnetic BVP must be modified to incorporate the results of Floquet's theory, the way the convergence of the iterative solution of the resulting finite element matrix can be enhanced via a multigrid/multilevel preconditioner is presented also.

To facilitate the reader with the computer implementation of the various methodologies presented in this book we have provided their algorithmic description in pseudo-code form. It is hoped that this feature will help expedite the incorporation of these algorithms in existing finite element-based solvers for wave problems.

The numerical examples presented in the applications section of each chapter are used to validate the presented methodologies and demonstrate their computational efficiency and robustness. As such, they tend to involve rather simple electromagnetic problems for which the answer is already known or can be obtained by other means. We hope the reader will appreciate the important role that such simple exercises play in building confidence in the tools we develop to tackle harder problems with yet unknown answers.

In all, this is a book about making iterative methods work well when applied to solving wave problems using finite methods. Our hope is that its readers will benefit from using some of the presented ideas to enrich their bag of tricks with new ones, and advance their wave field solvers to a new plateau of modeling versatility, computational efficiency and numerical robustness.

Y. Zhu

A. C. Cangellaris

CHAPTER 1

INTRODUCTION

The finite element method (FEM) is a numerical technique for the approximate solution of boundary value problems arising from the differential equation-based mathematical modeling of physical phenomena. Early developments and applications of the method were prompted by problems in structural analysis in the 1960s and 1970s [1], [2]. Its introduction in the discipline of computational electromagnetics as a tool for the numerical solution of electromagnetic boundary value problems occurred in the 1970s. For a historical overview of the application of the method to the solution of electromagnetic boundary value problems the reader is referred to [3].

While the emphasis of early applications of finite elements to electromagnetics was on static, quasi-static, and guided-wave eigenvalue problems [4], the modeling versatility of the method led several researchers to pursue ways in which it could be used for the solution of electrodynamic boundary value problems, particularly those concerned with electromagnetic radiation and scattering phenomena in unbounded regions. However, it was not until the mid 1980s that FEM started becoming widely accepted by electromagnetic researchers and designers alike as an important computer-aided modeling tool for electromagnetic analysis.

Since then, advances in computing technology, combined with aggressive research into sparse matrix solution, mesh generation technology and the mathematical attributes of finite element approximations of electromagnetic boundary value problems, have helped establish FEM as one of the most versatile and effective numerical techniques for computer-aided electromagnetic analysis and design.

The maturity of the finite element method as a numerical tool for electromagnetic analysis becomes evident from a brief overview of its very rich literature that includes its application to quantitative electromagnetic waveguide analysis for the design and optimization of microwave, millimeter-wave and optical devices, components and systems, to electromagnetic scattering for target identification, to electromagnetic radiation for antenna design [3]-[6].

The purpose of this introductory chapter is three-fold. First, a brief overview is provided of the two most commonly used approaches in practice for the development of finite element approximations of a differential equation-based boundary value problem. Second, the electromagnetic boundary value problem of interest in this book is defined and discussed briefly. This is followed by the discussion of current and future challenges in finite element-based electromagnetic field modeling. This discussion helps motivate the methodologies and algorithms presented in the following chapters.

1.1   STATEMENT OF THE BOUNDARY VALUE PROBLEM

A typical boundary value problem (BVP) is stated in terms of its governing differential equation over a domain Ω,

$$L\phi = f \tag{1.1}$$

together with appropriate boundary conditions imposed on the boundary Γ that encloses the domain Ω. In (1.1), L is the governing differential operator, ϕ is the physical field quantity that is solved for, and f is the forcing term (or excitation).

The analytic solution of (1.1) is not possible in the general case where the domain Ω involves nonseparable geometries and materials exhibiting arbitrary position dependence and anisotropy. More specifically, in the case of electromagnetic BVPs, analytic solutions are available only for a few classes of BVPs involving separable geometries with mostly homogeneous materials [7]-[11].

For those cases for which an analytic solution of (1.1) is not possible, a numerical method must be used. Among the various techniques possible [12] [13], the method of finite elements is most attractive in the presence of substantial geometric and material complexity. The basic steps of two of the most commonly used methodologies for the development of a finite element approximation of (1.1), namely, the Ritz and the Petrov-Galerkin's method, are discussed next.

1.2   RITZ FINITE ELEMENT METHOD

The Ritz method is a variational method where the solution to (1.1) is formulated in terms of the minimization of a functional. More specifically, a functional is derived, the minimum of which corresponds to the solution of (1.1), subject to the given boundary conditions [13]. For the purpose of keeping the mathematical development simple, we refrain from using the Hilbert space and formal linear operator theory nomenclature in this brief overview of machinery of variational methods. The reader is referred to [13] and [14] for an exposure to the rigorous mathematical framework in which variational methods are founded and developed.

For the purposes of this introductory discussion we consider only the case where the operator and the unknown and forcing functions in (1.1) are scalar. Furthermore, the properties of L are further constrained in a manner best described with the aid of the *inner product* of two functions, ϕ and ψ, defined as follows:

$$\langle \phi, \psi \rangle = \iiint_{\Omega} \phi \, \psi \, dv. \tag{1.2}$$

The operator L is said to be *self-adjoint* if the following equality holds,

$$\langle L\phi, \psi \rangle = \langle \phi, L\psi \rangle. \tag{1.3}$$

The operator L is said to be *positive definite* if

$$\langle L\phi, \phi \rangle = \begin{cases} > 0, & \phi \neq 0, \\ = 0, & \phi = 0. \end{cases} \tag{1.4}$$

If L in (1.1) is self-adjoint and positive definite then the solution to (1.1) can be obtained through the minimization of the functional [3]

$$F(\phi) = \frac{1}{2}\langle L\phi, \phi \rangle - \frac{1}{2}\langle \phi, f \rangle - \frac{1}{2}\langle f, \phi \rangle. \tag{1.5}$$

To show that the function, ϕ, that minimizes (1.5) is also the solution to (1.1), let us consider the change, δF, in F, caused by an arbitrary variation, $\delta\phi$, in ϕ,

$$\begin{aligned} \delta F &= F(\phi + \delta\phi) - F(\phi) \\ &= \frac{1}{2}\langle L\delta\phi, \phi \rangle + \frac{1}{2}\langle L\phi, \delta\phi \rangle - \frac{1}{2}\langle \delta\phi, f \rangle - \frac{1}{2}\langle f, \delta\phi \rangle \\ &= \frac{1}{2}\langle \delta\phi, L\phi \rangle - \frac{1}{2}\langle \delta\phi, f \rangle + \frac{1}{2}\langle L\phi, \delta\phi \rangle - \frac{1}{2}\langle f, \delta\phi \rangle \\ &= \langle \delta\phi, (L\phi - f) \rangle. \end{aligned} \tag{1.6}$$

Since $\delta\phi$ is arbitrary, it is evident from the above result that the extreme point of the functional $F(\phi)$ corresponds to the solution of (1.1).

The FEM approximation and its solution are obtained from (1.5) in the following manner. The domain Ω is discretized into a number of smaller sub-domains, which are referred to as *elements*. For example, for a one-dimensional (1D) domain, short line segments, interconnected to cover the entire domain, may be used as elements. For a two-dimensional (2D) domain, the most commonly used elements are the triangle and the quadrilateral. Finally, for a three-dimensional (3D) domain, tetrahedra, triangular prisms, or hexahedron bricks are used as elements. In all cases, connectivity of the elements in a manner such that the entire domain is covered by the resulting *finite-element mesh* or *finite-element grid*, must be ensured. Inside each element the unknown physical field quantity ϕ is expanded in terms of known polynomial functions, w_j. Hence, an approximation is obtained for the unknown field over Ω,

$$\hat{\phi} = \sum_{j=1}^{N} w_j c_j = w^T c, \tag{1.7}$$

where the vector c contains the coefficients c_j, $j = 1, 2, \ldots, N$, which are the unknown degrees of freedom in the approximation, while the vector w contains the expansion functions w_j, $j = 1, 2, \ldots, N$. Their number, N, is dictated by various factors, among which geometric/material complexity and accuracy of the approximation are the most important. Substitution of (1.7) into (1.5) yields

$$F(c) = \frac{1}{2}c^T \left(\iiint_{\Omega} wLw^T dv \right) c - \left(\iiint_{\Omega} fw^T dv \right) c. \tag{1.8}$$

The minimization of $F(c)$ and, hence, the calculation of the expansion coefficients in (1.7), reduces to the solution of the linear system of equations that is obtained by setting all first derivatives of $F(c)$ with respect to c_j, $j = 1, 2, \ldots, N$, equal to zero,

$$\left(\iiint_{\Omega} w L w^T dv \right) c = \iiint_{\Omega} w f dv. \tag{1.9}$$

A more in-depth discussion of the mathematical attributes of the Ritz method can be found in [13]. Numerous examples from its utilization for the development of finite element approximations of both scalar and vector electromagnetic BVPs are given in [3].

1.3   PETROV-GALERKIN'S FINITE ELEMENT METHOD

The Petrov-Galerkin's method relies upon the idea that for $\hat{\phi}$ of (1.7) to be a good approximation to the solution of (1.1) the integration of the *residual* obtained by substituting $\hat{\phi}$ in (1.1) over the domain Ω, weighted by a set of appropriate *weighting* (or *testing*) functions, is zero. Because of this weighting process, the method is also called the *method of weighted residuals*, since it seeks an approximate solution to (1.1) by enforcing its weighted residual to be, on the average, zero over the domain of interest.

Substitution of (1.7) in (1.1) yields to the following residual error,

$$r = L w^T c - f. \tag{1.10}$$

Let u_j, $j = 1, 2, \ldots, N$, be a set of testing functions. The linear system of equations for the calculation of the expansion coefficients in (1.7) is obtained by requiring that

$$\iiint_{\Omega} u_j r \, dv = 0, \quad j = 1, 2, \ldots, N. \tag{1.11}$$

The resulting system has the form

$$\left(\iiint_{\Omega} u L w^T dv \right) c = \iiint_{\Omega} u f dv, \tag{1.12}$$

where the vector u contains the N testing functions, u_j, $j = 1, 2, \ldots, N$.

The special case where the testing functions are taken to be the same with the expansion functions is most commonly used in practice and is referred to as *Galerkin's method*. Clearly, in this case (1.12) is identical to (1.9).

It is apparent that the development of the finite element approximation via the Petrov-Galerkin procedure is more straightforward than the Ritz process. The simplicity of the process is most attractive when the development of a functional for (1.1) is hindered by the complexity of the operator L. The Galerkin's method will be used throughout the book for the development of the finite element approximation to the electromagnetic BVPs of interest.

1.4   TIME-HARMONIC MAXWELL'S EQUATIONS AND BOUNDARY CONDITIONS

Of interest to this work is the solution of linear electromagnetic BVPs under the assumption of time-harmonic, sinusoidal excitation of angular frequency, $\omega = 2\pi f$. Thus, with the time convention $e^{j\omega t}$, $j = \sqrt{-1}$, assumed and suppressed for simplicity, the complex phasor form of the governing system of Maxwell's equations becomes

$$
\begin{aligned}
\nabla \times \vec{E} &= -j\omega \vec{B} & \text{(Faraday's law)} \\
\nabla \times \vec{H} &= j\omega \vec{D} + \vec{J}_c + \vec{J}_v & \text{(Ampere's law)} \\
\nabla \cdot \vec{D} &= \rho_v & \text{(Gauss' electric field law)} \\
\nabla \cdot \vec{B} &= 0 & \text{(Gauss' magnetic field law).}
\end{aligned}
\tag{1.13}
$$

In the above equation, the familiar notation $\vec{E}$, $\vec{H}$, $\vec{D}$, $\vec{B}$ has been used for the electric field intensity, magnetic field intensity, electric flux density, and magnetic flux density, respectively. The *conduction* electric current density, $\vec{J}_c$, accounts for induced current flow in the medium due to the presence of conductor and/or dielectric loss, while the electric current density $\vec{J}_v$ represents the impressed current sources in the domain of interest. Similarly, electric charge density ρ_v represents the impressed electric charge in the domain.

Maxwell's equations are consistent with the conservation of charge statement

$$
\nabla \cdot \left(\vec{J}_c + \vec{J}_v \right) = -j\omega \rho_v.
\tag{1.14}
$$

This is easily seen by observing that this result is readily obtained from Ampere's law by taking the divergence of both sides and making use of Gauss' law for the electric field.

The properties of the media inside the domain in which a solution of (1.13) is sought enter the BVP through the *constitutive relations* between electromagnetic field quantities and their associated flux densities. The constitutive relations are dictated by the macroscopic electromagnetic properties of the media of interest. For the case of simple, linear media, these relations are simply given by

$$
\vec{D} = \bar{\epsilon} \cdot \vec{E}, \quad \vec{B} = \bar{\mu} \cdot \vec{H}, \quad \vec{J}_c = \bar{\sigma} \cdot \vec{E},
\tag{1.15}
$$

where the tensors $\bar{\epsilon}$, $\bar{\mu}$, and $\bar{\sigma}$ are, respectively, the electric permittivity, magnetic permeability, and electric conductivity tensors of the media. Their linearity is understood to mean that they are independent of the field intensities. However, they are allowed to exhibit frequency and position dependence.

1.4.1 Boundary conditions at material interfaces

As it will be discussed in detail in Chapter 3, all four equations in (1.13) are needed for the well-posed definition of a uniquely solvable electromagnetic BVP. In addition, for the electromagnetic BVP to be well-posed (1.13) must be complemented by appropriate boundary conditions at material interfaces, across which the constitutive parameters exhibit discontinuities, and at the enclosing boundary, Γ, of the domain Ω.

At material interfaces the following set of boundary conditions hold

$$
\begin{aligned}
\hat{n} \times \left(\vec{E}_1 - \vec{E}_2 \right) &= 0 \\
\hat{n} \cdot \left(\vec{D}_1 - \vec{D}_2 \right) &= \rho_s \\
\hat{n} \times \left(\vec{H}_1 - \vec{H}_2 \right) &= \vec{J}_s \\
\hat{n} \cdot \left(\vec{B}_1 - \vec{B}_2 \right) &= 0,
\end{aligned}
\tag{1.16}
$$

where $\hat{n}$ is the unit normal on the interface (taken to be in the direction from Medium 2 toward Medium 1), and $\vec{J}_s$ and ρ_s are, respectively, any surface electric current and charge densities present on the interface. These equations simplify for the case where one of the media is a *perfect* electric conductor. A perfect electric conductor is an idealization of a highly conducting medium, obtained in the limit where the electric conductivity is assumed to become infinite. In this limit, and under the assumption of zero-field initial conditions, both electric and magnetic fields are identically zero inside the conductor. Hence, with Medium 2 assumed to be a perfect electric conductor, (1.16) simplify as follows:

$$\hat{n} \times \vec{E}_1 = 0$$
$$\hat{n} \cdot \vec{D}_1 = \rho_s$$
$$\hat{n} \times \vec{H}_1 = \vec{J}_s$$
$$\hat{n} \cdot \vec{B}_1 = 0.$$

(1.17)

It is noted that in this case the electric current and charge densities on the perfectly conducting surface are induced by the presence of non-zero fields in the exterior of the conductor (i.e., in Medium 1).

1.4.2 Boundary conditions at the enclosing boundary

With regards to boundary conditions at the enclosing boundary Γ, the assignment of appropriate boundary conditions is dictated by the requirement that the solution of the resulting BVP is unique. The pertinent conditions are well-documented in the electromagnetic literature (e.g., [7], [8]), and require that, in the presence of some loss in Ω, the solution of (1.13) subject to (1.16) is uniquely defined provided that either the tangential components of the electric field are specified over Γ, or the tangential components of the magnetic field are specified over Γ, or the tangential components of the electric field are specified over part of Γ and the tangential components of the magnetic field are specified over the remaining part.

The final case of a boundary condition of relevance to the finite element-based solution of electromagnetic boundary value problems is the case of a *surface impedance condition*, where a relationship is imposed between the tangential components of the electric and the magnetic fields on the enclosing boundary, Γ, as follows:

$$\vec{E}_t = \hat{t}_1 E_{t_1} + \hat{t}_2 E_{t_2} = Z_s \vec{H} \times \hat{n},$$

(1.18)

where, $\vec{E}_t$ denotes the tangential components of $\vec{E}$ on the conductor surface, $\hat{n}$ is the inward-pointing unit normal on the conductor surface, and the tangential unit vectors $\hat{t}_1$, $\hat{t}_2$, are such that $\hat{t}_1 \times \hat{t}_2 = \hat{n}$. The linear surface impedance Z_s is, in general, frequency and position dependent.

Surface impedance conditions are extremely useful in finite element applications since they serve as effective means for containing the complexity of the solution domain Ω. For example, consider the case where a portion of the domain of interest is occupied by a good conductor of conductivity high enough for magnitude of the displacement current density, $j\omega\epsilon\vec{E}$, to be negligible compared to that of the conduction current density, $\sigma\vec{E}$. This is recognized as the good conductor approximation [15], under which electromagnetic field penetration inside the good conductor decays exponentially from the surface to the interior with attenuation constant, α, given by

$$\alpha = \sqrt{\pi f \mu \sigma}.$$

(1.19)

For example, for the case of aluminum, with conductivity $\sigma = 4 \times 10^7$ S/m, and permeability, $\mu = 4\pi \times 10^{-7}$ H/m, the attenuation constant is 40π Nepers/mm at 100 MHz. Clearly, under such conditions, field penetration inside the conductor is restricted to a thin layer below the conductor surface; hence, the term *skin effect* is used to refer to this situation. As elaborated in [15], the interior of the good conductor can be removed from the domain of solution, with its presence taken into account by imposing the surface impedance condition of 1.18 with

$$Z_s = (1 + j) \sqrt{\frac{\pi f \mu}{\sigma}}. \tag{1.20}$$

A second example of a surface impedance boundary condition used frequently in practice is associated with the case where either a portion of or the entire enclosing boundary, Γ, is at infinity. Clearly, this is the case of an unbounded domain, characteristic of electromagnetic scattering and radiation problems. One approximate way for truncating the domain Ω such that a finite element solution becomes feasible is through the introduction of a mathematical (non-physical) boundary, placed at a finite distance from the electromagnetic device/structure under analysis, on which an *absorbing* or *radiation* surface boundary condition is imposed. The name reflects the fact that the boundary condition is constructed in a manner such that it supports one-way propagation of electromagnetic waves away from the domain Ω, with minimum spurious (non-physical) reflection.

From the numerous absorbing boundary conditions for time-harmonic electromagnetic BVPs [3], the simplest one is known as the first-order radiation boundary condition, which, under the assumption of lossless media of permittivity ϵ and permeability μ occupying the unbounded region, is given by

$$\vec{E}_t = \hat{t}_1 E_{t_1} + \hat{t}_2 E_{t_2} = \sqrt{\frac{\mu}{\epsilon}} \vec{H} \times \hat{n}. \tag{1.21}$$

In the above expression the unit normal, $\hat{n}$, on the truncation boundary is taken to be pointing in the outward direction, away from the domain Ω, and, as before, the tangential unit vectors $\hat{t}_1, \hat{t}_2$, are such that $\hat{t}_1 \times \hat{t}_2 = \hat{n}$.

1.4.3 Uniqueness in the presence of impedance boundaries

In the presence of surface impedance conditions on the boundary enclosing the domain of interest, uniqueness of the solution of the electromagnetic BVP is guaranteed only when the real part of the surface impedance is non-negative. Prior to giving a proof of this result, we note that this requirement is satisfied by both the good-conductor surface impedance given in (1.20) and the impedance coefficient in the first-order radiation boundary condition (1.21).

Without loss of generality, our proof is given for the case where an impedance boundary condition is imposed over the entire surface of the enclosing boundary Γ. Consider two possible solutions, $(\vec{E}_a, \vec{H}_a)$ and $(\vec{E}_b, \vec{H}_b)$. Both solutions satisfy Maxwell's equations for the given set of sources. Also, both satisfy the surface impedance boundary condition on Γ. Thus it follows that their difference, $(\delta\vec{E}, \delta\vec{H})$, satisfies the source-free equations,

$$\begin{aligned} -\nabla \times \delta\vec{E} &= \hat{z}\delta\vec{H}, \\ \nabla \times \delta\vec{H} &= \hat{y}\delta\vec{E}, \end{aligned} \tag{1.22}$$

where the short-hand notation $\hat{z} = j\omega\mu$, $\hat{y} = \sigma + j\omega\epsilon$ has been adopted. Furthermore, it is on Γ,

$$\delta\vec{E}_t = Z_s\delta\vec{H} \times \hat{n}. \tag{1.23}$$

Taking the inner product of the first of (1.22) with the complex conjugate of $\delta\vec{H}$, and adding to it the inner product of the complex conjugate of the second with $\delta\vec{E}$ yields

$$\nabla \cdot \left(\delta\vec{E} \times \delta\vec{H}^*\right) + \hat{z}|\delta\vec{H}|^2 + \hat{y}^*|\delta\vec{E}|^2 = 0, \tag{1.24}$$

where the superscript $*$ denotes complex conjugation and use was made of the vector identity (A.6) from Appendix A. Integration of the above equation over Ω, followed by the application of the divergence theorem (A.14) yields

$$\oiint_\Gamma \left(\delta\vec{E} \times \delta\vec{H}^*\right) \cdot \hat{n}ds + \iiint_\Omega \left(\hat{z}|\delta\vec{H}|^2 + \hat{y}^*|\delta\vec{E}|^2\right) dv = 0. \tag{1.25}$$

Use of (1.23) in the integrand of the surface integral allows us to recast the above equation in the following form:

$$\oiint_\Gamma Z_s|\delta\vec{H} \times \hat{n}|^2ds + \iiint_\Omega \left(\hat{z}|\delta\vec{H}|^2 + \hat{y}^*|\delta\vec{E}|^2\right) dv = 0. \tag{1.26}$$

Considering the real and imaginary parts of this equation separately, we have

$$\begin{aligned}
\oiint_\Gamma \mathrm{Re}(Z_s)|\delta\vec{H} \times \hat{n}|^2ds + \iiint_\Omega \left(\mathrm{Re}(\hat{z})|\delta\vec{H}|^2 + \mathrm{Re}(\hat{y})|\delta\vec{E}|^2\right) dv = 0, \\
\oiint_\Gamma \mathrm{Im}(Z_s)|\delta\vec{H} \times \hat{n}|^2ds + \iiint_\Omega \left(\mathrm{Im}(\hat{z})|\delta\vec{H}|^2 - \mathrm{Im}(\hat{y})|\delta\vec{E}|^2\right) dv = 0.
\end{aligned} \tag{1.27}$$

Since for lossy media $\mathrm{Re}(\hat{z})$ and $\mathrm{Re}(\hat{y})$ are non-negative, it is immediately evident that, with $\mathrm{Re}(Z_s) \geq 0$ everywhere on Γ, the equations above are satisfied only if $\delta\vec{E} = \delta\vec{H} = 0$ everywhere inside Ω. Hence, we conclude that uniqueness of the solution to Maxwell's equations in a domain involving surface impedance boundary conditions requires the real part of the surface impedance to be non-negative.

This concludes the discussion of the governing equations for the linear electromagnetic problem. A more in-depth discussion of the electromagnetic system, along with possible simplifications (or reductions) of the governing equations for the special cases of static and quasi-static conditions, can be found in Chapter 3.

1.5   PRESENT AND FUTURE CHALLENGES IN FINITE ELEMENT MODELING

To date, the application of FEM to electromagnetic field modeling has reached significant maturity. Evidence of this maturity is the availability of several commercially available finite element solvers, which are used extensively for the design of state-of-the-art electromagnetic devices of relevance to static, quasi-static and dynamic electromagnetic applications. However, as the community of finite element developers and users continues to grow and as the sophistication of the computational electromagnetics practitioner continues to advance, the expectations for the capabilities of electromagnetic computer-aided design (CAD) tools also continue to increase. In particular, irrespective of the engineering application of interest, ease of use, modeling versatility, robustness, solution accuracy, and computation expediency are all attributes that an increasing pool of users demands from state-of-the-art and future electromagnetic CAD tools.

Improvements in ease of use and modeling versatility are both driven by continuing advances in the sophistication of computer-aided design frameworks, geometric modeling and visualization software, and mesh generation algorithms. However, numerical robustness, solution accuracy and computation expediency, are all intimately related to the numerical method used for the solution of the BVP of interest. More specifically, in the context of finite element-based solution of electromagnetic BVPs, the primary issue that impacts numerical robustness, solution accuracy and computation expediency, is the large disparity in electrical size of geometric features in the computational domain. This is a pressing issue, already encountered in today's electronic/electromagnetic engineering applications, and stands out as one of the primary impeding factors in the routine application of state-of-the-art FEM technology to the solution of challenging electromagnetic BVPs at the system level. As such, it deserves some more discussion.

Finite element methods are best suited for the electromagnetic analysis of highly inhomogeneous structures that exhibit significant geometric complexity. Among the several application areas where such structures are encountered (e.g., scattering by large, multi-body structures, analysis of antenna arrays on platforms, analysis of integrated optical circuits, analysis of integrated electronic systems, system-level electromagnetic interference and electromagnetic compatibility analysis), the analysis of high-speed/high-frequency, mixed-signal integrated circuits (ICs) will be used to illustrate and discuss the impact of the large disparity in size in geometry features on the numerical robustness, accuracy and computation expediency of electrodynamic field solvers.

To begin with, *electrical length* is understood to mean length measured in terms of the wavelength, λ, at the operating frequency at which the electromagnetic analysis is carried out. For example, a thin wire, designed to operate as a half-wavelength dipole at 100 MHz, is (approximately) 1.5 m long. Its electrical length is $\lambda/2$ at 100 MHz, $\lambda/200$ at 1 MHz, 5λ at 1 GHz, and 50λ at 10 GHz.

A first indication of how electrical size impacts model complexity is provided by recalling that numerical solution accuracy requires the spatial sampling of the structure by the finite element mesh to be in the order of ten elements per wavelength [3]. Consequently, while a finite element mesh with ~ 5 elements along the wire length would suffice for an accurate finite element solution at 100 MHz, a mesh with ~ 500 elements along the wire length will be required for an accurate finite element solution at 10 GHz.

Furthermore, if the finite element analysis of the same wire at 1 MHz is required, it should be clear that, even though its electrical size at this frequency is $\lambda/200$, a finite element mesh of element size smaller than the wire length must be used for the accurate resolution of the electromagnetic field distribution in the immediate neighborhood of the wire, especially in view of the singular behavior of the electric field at the wire end points. Actually, one could argue that, instead of an electrodynamic model, a quasi-static model could be used in this case.

This observation suggests that, depending on the electrical size of the structure under analysis, a different finite element-based field solver may be used at different frequencies. Thus a static or quasi-static field solver may be used when the electrical size of the structure is small enough for electromagnetic retardation to have negligible effect on the field distribution and, thus the fields exhibit a predominantly static behavior. On the other hand, an electromagnetic field solver becomes necessary at higher frequencies at which geometric features are in the order of (or larger) than the wavelength and, thus electromagnetic retardation must be taken into account to ensure solution accuracy.

Next, we extend these preliminary observations to the case of electromagnetic field modeling in the context of high-speed/high-frequency IC analysis. We begin with the observation that the complexity of integrated electronics systems continues to escalate rapidly as the semiconductor industry aggressively moves toward reduced device feature size, in support of enhanced functionality integration and higher device speed and system operating bandwidths. The floor-planning of such systems and the design of the interconnection and power distribution networks for their functional blocks cannot be effected without the support of electromagnetic modeling for signal integrity assessment, interference prevention or mitigation, and electrical performance verification.

While the strongly heterogeneous environment of such a system calls for a finite element-based model, the large disparity in geometric feature size makes the use of such a numerical model computationally cumbersome if not prohibitive. More specifically, the several-orders-of magnitude variation in geometric feature size at all levels of circuit integration (e.g., at the chip level geometric features vary in size from sub-microns to centimeters; at the package level this variation is from several microns to centimeters; at the board and system level the variation in size is from the order of millimeters to tens of centimeters of even meters) would require an electromagnetic finite element model involving tens or even hundreds of millions of unknowns.

While the actual size of the model would depend on the portion of the IC under analysis and the frequency at which such analysis is required, in all cases of practical interest the dimension of the finite element matrix is large enough that the direct solution, despite the sparsity of the matrix, is, at least for today's computer technology, computationally overwhelming.

The alternative to a direct solution is an iterative solution. Over the past twenty years major advances have been made in the robustness and sophistication of iterative methods for the solution of sparse linear systems [16]. A prerequisite for a matrix to be suitable for solution through an iterative process is that the matrix is not ill-conditioned. The reason for this is that the convergence of the iterative solution process is critically dependent on the *condition number* of the matrix. We will discuss in detail the condition number of a matrix and its relevance to the convergence of the iterative solution in Chapter 4. For the purposes of this discussion it suffices to say that the condition number of a matrix is roughly given by the magnitude of the ratio of its largest to its smallest eigenvalue. The larger the condition number, the more ill-conditioned the matrix becomes. Severe ill-conditioning leads to numerical instability of the iterative solver, which manifests itself either in terms of stalled convergence or (if convergence is achieved) an inaccurate solution. Clearly, such lack of numerical robustness is unacceptable for a solver aimed for use as a computer-aided analysis and design tool.

The ill-conditioning of matrices resulting from the discretization of a structure exhibiting several-orders-of-magnitude variation in geometric feature size, is easily argued by considering the simple case of a one-dimensional, homogeneous domain of length d, discretized by means of a uniform grid of size h such that $h \ll d$; hence, the number of elements, $N = d/h$, in the domain, is very large. For simplicity, a finite-difference approximation to the one-dimensional Helmholtz operator, $d^2/dx^2 + k^2$, which governs electromagnetic wave propagation in the one-dimensional domain, will be considered. The wavenumber, k, depends on the angular frequency and the speed of light, $v_p = 1/\sqrt{\mu\epsilon}$, in the medium, and is given by

$$k = \frac{\omega}{v_p} = \frac{2\pi}{\lambda}. \tag{1.28}$$

Assuming zero field values at the end points of the domain, the finite difference matrix of the discrete problem is of dimension $(N - 1) \times (N - 1)$, and is given by

$$
\begin{pmatrix}
2 - (kh)^2 & -1 & & & & \\
-1 & 2 - (kh)^2 & -1 & & & \\
& \ddots & \ddots & \ddots & & \\
& & -1 & 2 - (kh)^2 & -1 & \\
& & & -1 & 2 - (kh)^2 &
\end{pmatrix}. \tag{1.29}
$$

The availability of the elements of the matrix allows us to obtain rough estimates for the eigenvalues by making use of Gershgorin's theorem which states that any eigenvalue of a matrix A of dimension N is located inside one of the closed discs of the complex plane centered at A_{ii}, $i = 1, 2, \ldots, N$, and having radius

$$
r_i = \sum_{j=1, j \neq i}^{j=N} |A_{ij}|. \tag{1.30}
$$

Thus the following bounds can be obtained for the eigenvalues, s, of (1.29),

$$
|s - (2 - (kh)^2)| \leq 2 \Rightarrow -(kh)^2 \leq s \leq 4 - (kh)^2 \tag{1.31}
$$

or, written in terms of the number of elements, N, in the grid,

$$
-\left(\frac{2\pi d_\lambda}{N}\right)^2 \leq s \leq 4 - \left(\frac{2\pi d_\lambda}{N}\right)^2, \tag{1.32}
$$

where $d_\lambda = d/\lambda$ is the electrical length of the domain.

With $d_\lambda \sim \mathcal{O}(1)$ and $N \gg 1$, it is immediately evident from (1.32) that the smallest eigenvalue is of order N^{-2}, while the largest eigenvalue is about 4. Hence, the condition number grows, roughly, as $\mathcal{O}(N^2)$, making evident the convergence difficulties associated with the application of iterative solvers for the solution of finite element matrices of dimension in the order of tens of millions of unknowns.

One class of methods aimed at tackling this hurdle are the so-called multigrid and multi-level methods. It is the primary objective of this book to establish the necessary mathematical infrastructure and develop effective methodologies and computer algorithms for the utilization of multigrid and multi-level methods for robust iterative solution of finite element approximations of electromagnetic BVPs.

However, multigrid methods alone are not capable of tackling the escalating complexity of the electromagnetic BVP associated with the electrical analysis of state-of-the-art and future-generation ICs. To alleviate the complexity, *domain decomposition* strategies can be employed. As the name suggests, this class of methods calls for the decomposition of the structure into sub-domains. Each sub-domain is modeled independently of the others, possibly utilizing different models or even different numerical methods. The generated solutions for all sub-domains must be made consistent with each other through the enforcement of the appropriate field continuity conditions at the sub-domain interfaces.

The process used for stitching together the sub-domain models in a consistent manner can follow one of several alternatives within the framework of iterative methods (e.g., [16]-[19]). In the context of linear system analysis, one can view the sub-domain models as *multi-port networks*, the port variables of which are the shared degrees of freedom between sub-domain models. In this context, one can consider the modeling of each individual sub-domain as a process for generating the *transfer function matrix* for the sub-domain. Once the transfer function matrix has been constructed, general-purpose, network-analysis oriented simulation techniques can be utilized for the analysis of the entire structure.

Recognizing the importance of such capability, a systematic methodology is presented in this book for the expedient, broadband generation of matrix transfer function representations of portions of an electromagnetic structure. Combined with the multigrid/multi-level machinery, streamlined and optimized for electromagnetic BVPs, the proposed algorithms constitute a major step toward the establishment of a robust and computationally efficient finite element-based modeling framework for the analysis of any electromagnetic system that exhibits large disparity in geometry feature size over its spatial extent.

REFERENCES

1. H. C. Martin and G. F. Carey, *Introduction to Finite Element Analysis: Theory and Application*, New York: McGraw-Hill, 1973.

2. R. K. Livesley, *Finite Elements: An Introduction for Engineers*, Cambridge: Cambridge University Press, 1983.

3. J. M. Jin, *The Finite Element Method in Electromagnetics*, New York: John Wiley & Sons, Inc., 1993.

4. P. P. Silvester and R.L. Ferrari. *Finite Elements for Electrical Engineering*, Cambridge: Cambridge University Press, 1983.

5. J. L. Volakis, A. Chatterjee, L. C. Kempel, *Finite Element Method for Electromagnetics: Antennas, Microwave Circuits, and Scattering Applications*, Piscataway, NJ: Wiley-IEEE Press Series on Electromagnetic Wave Theory, 1998.

6. M. S. Palma, T. K. Sarksar, L.E. García-Castillo, T. Roy, and A. Djordjević. *Iterative and Self-Adaptive Finite-Elements in Electromagnetic Modeling*, Norwood, MA: Artech House, 1998.

7. C. A. Balanis, *Advanced Engineering Electromagnetics*, New York: John Wiley & Sons, Inc., 1992.

8. R. F. Harrington, *Time-Harmonic Electromagnetic Field*, New York: McGraw-Hill, 1961.

9. J. Van Bladel, *Electromagnetic Fields*, New York: Hemisphere Publishing Corporation, 1985.

10. J. R. Wait, *Geo-Electromagnetism*, New York: Academic Press, 1982.

11. R. E. Collin, *Field Theory of Guided Waves*, Piscataway, NJ: IEEE Press, 1991.

12. A. F. Peterson, S. L. Ray, and R. Mittra, *Computational Methods for Engineering*, New York: IEEE Press, 1998.

13. D. S. Jones, *Methods in Electromagnetic Wave Propagation*, 2nd ed., Piscataway, NJ: IEEE Press, 1994.

14. D. G. Dudley, *Mathematical Foundations for Electromagnetic Theory*, Piscataway, NJ: IEEE Press, 1994.

15. J. D. Jackson, *Classical Electrodynamics*, 3rd ed., New York: John Wiley & Sons, Inc., 1999.

16. Y. Saad, *Iterative Methods for Sparse Linear Systems*, 2nd ed., Philadelphia: SIAM, 2003.

17. B. I. Wohlmuth, *Discretization Methods for Iterative Solvers Based on Domain Decomposition*, Berlin: Springer-Verlag, 2001.

18. M. Gander, F. Magoulès, and F. Nataf, "Optimal Schwartz methods without overlap for the Helmholtz equation," *SIAM J. Sci. Comput.*, vol. 24, no. 1, pp. 38-60, 2002.

19. S.-C. Lee, M.N. Vouvakis, and J.-F. Lee, "Domain decomposition method for large finite antenna arrays," *Proc. IEEE Antennas and Propagation Society Symposium*, vol. 4, pp. 3501 - 3504, Monterey, CA, Jun. 2004.

CHAPTER 2

HIERARCHICAL BASIS FUNCTIONS FOR TRIANGLES AND TETRAHEDRA

The objective of the finite element method is to obtain an approximate solution to the boundary value problem of interest. This is done through the weighted minimization of the error that results from the approximation of the governing partial differential equations over the set of finite volumes, the so-called elements, the union of which is the entire volume of the computational domain. The most common elements are triangles for two-dimensional problems and tetrahedra in three dimensions. Their popularity stems from their versatility in modeling very complex geometries, especially those exhibiting a complicated blend of fine and coarse geometric features and abrupt changes in material properties. It is because of this superior modeling versatility that triangles and tetrahedra have become the elements of choice in the development of design-driven electromagnetic FEM solvers, even though the geometric attributes of specific classes of electromagnetic devices (e.g., layered planar circuits) may favor other types of elements (e.g., hexahedrons). Thus triangles and tetrahedra are the ones that have received the most attention in the advancement of general-purpose finite element mesh generators and adaptive meshing algorithms.

We will not deviate from this trend. In addition to the mentioned advantages, triangles and tetrahedra turn out to be the most convenient sets of element families for the systematic development of *hierarchical*, polynomial interpolation functions for the representation of the various electromagnetic quantities of interest. What is meant by the word *hierarchical* in the context of multigrid methods is revealed through the discussion in this and later chapters. At this point it suffices to recall the standard definition of *hierarchical* sets of polynomial interpolation functions in the context of finite element methods. Prior to doing so, let us adopt some nomenclature. To be consistent with standard literature [1], given a *space* of functions we will make use of the term *basis* to refer to a given set of functions in this space that will be used for the finite-element interpolation of the field quantities of interest. Also, the term *expansion functions* or *shape functions* will be used often in place of interpolation functions. However, we will refrain from using strict mathematical rigor in our use of these concepts and results from the theory of linear spaces, and we will assume that the mathematically inclined reader will consult more appropriate textbooks like [1] and [2] for a more in-depth mathematical discussion of these concepts.

A given set of polynomial basis functions will be called *hierarchical* if they are constructed in such a way that the basis space of degree p contains polynomials of degrees $1, 2, \cdots, p-1$. Furthermore, these lower-degree polynomials should be present explicitly in the base of degree p, such that the construction of higher-accuracy finite element approximations can be done efficiently through the calculation and addition of only the contributions of the higher-degree terms to the finite-element approximation on the basis space of degree $p-1$.

A methodology for the development of such a hierarchical set of interpolation functions for triangles and tetrahedra is presented in this Chapter. Our approach is inspired by recent works by many researchers in the areas of computational engineering and applied mathematics. Furthermore, it is driven by the conviction that numerical solution accuracy is greatly enhanced through the direct imposition, to the degree possible, of all known physical attributes of the approximated field quantity in the expansion functions used for its approximation. In the context of electromagnetic field modeling, several examples come to mind. Approximations of electric or magnetic field intensities must exhibit tangential field continuity at material interfaces in the absence of surface electric current and surface magnetic current densities, respectively. The electrostatic potential function must be continuous across material interfaces. Approximations for time-varying electric and magnetic fields must be of high enough order to allow for the representation of non-zero rotation over each element.

These examples indicate that the development of the interpolation functions will be impacted by both the governing equations and the boundary conditions for the field quantities. There is one more remark we would like to make at this point. In our discussion we assume that the reader is familiar with the basic machinery of finite element interpolation, at the level found in texts such as [3] and [4]. Thus terms such as *node-based function* or *nodal elements*, understood to refer to expansion functions that assume a specific value at a set of pre-determined points (nodes) on an element, will be used hereafter without any further explanations. The same holds true for *edge-based functions* or *edge elements*.

2.1   THE IMPORTANCE OF PROPER CHOICE OF FINITE ELEMENT BASES

A survey of the computational electromagnetic literature reveals that the first space of expansion functions used in the finite element solution of electromagnetic boundary value problems was the space of node-based scalar functions. Intuitively one expects that such a space will be suitable for the approximation of the scalar potentials associated with the mathematical formulation of static boundary value problems [3], as well as the approximation of the scalar potential function used in the *potential formulation* statement of electrodynamic boundary value problems [5]. However, its subsequent use for the expansion of the electric field intensity $\vec{E}$ or magnetic field intensity $\vec{H}$, where each Cartesian component of the vector fields is expanded in terms of the node-based scalar functions, was soon found to be problematic. The obstacles were several. The ones that turned out to be the most cumbersome in the context of electromagnetic field modeling are discussed next.

The first difficulty is encountered in the enforcement of the tangential electric and magnetic field intensity continuity at material interfaces. The difficulty is easily recognized if one considers the case of a finite element node present at an arbitrarily shaped material interface. With three components of a field quantity assigned at such a node, a constraint equation must be introduced to enforce the appropriate boundary condition. However, a node-by-node enforcement of the boundary condition cannot guarantee that the interpolated field over the element side defined by a set of nodes on the material interface complies with the boundary condition that is imposed at each node. Particularly troublesome is the enforcement of physically consistent boundary conditions at perfectly conducting wedges and corners, the reason being the singular behavior of the fields in the vicinity of such features.

The consequences of such boundary condition violations on numerical solution inaccuracy can be significant, as easily seen by the following argument. For the electrodynamic case it is well known that enforcing the continuity of the tangential components of the electric and magnetic field intensities at source-free, material interfaces suffices for ensuring the continuity of the normal components of the magnetic flux and electric flux densities, respectively. However, as pointed out in the previous paragraph, in the finite element approximation of the problem, despite the enforcement of the continuity of the tangential components of the field intensities at each node on a source-free, material interface, continuity of the normal component of the fluxes cannot be guaranteed. The consequence of such a violation will be the occurrence of spurious (non-physical) charge densities, which in turn may give rise to a non-physical parasite component in the numerical solution of the problem. Considering its source, such a parasite solution is expected to manifest itself as a static field. Examples from the occurrence of such non-physical solutions in conjunction with the use of node-based scalar bases for vector field approximations abound in the literature, along with numerous, often tedious, application-specific approaches for their elimination. The discussion in Chapter 7 in [4] includes a complete list of the most important, relevant references.

The second difficulty, albeit related in some sense to the one above, was subtler and thus more difficult to identify, comprehend and overcome. It is associated with the inability of the node-based scalar bases to model properly the null space of the curl operator in the electromagnetic system. Clearly, the null space of the curl operator consists of all vector fields $\vec{F} = -\nabla\Psi$, where Ψ is an arbitrary scalar field; hence, its dimension is infinite. Considering the influence that the investigation of the issue of the occurrence of spurious modes in the finite element solution of vector field problems has had on the advancement of robust field solvers, it deserves some discussion. Our overview will be brief and predominantly qualitative. A more in-depth presentation, along with a comprehensive list of references, can be found in [4].

Some of the mathematical machinery of finite element approximation is needed to facilitate our discussion. We begin with the vector Helmholtz equation statement of the time-harmonic, electromagnetic boundary value problem, keeping the electric field intensity, $\vec{E}$, as the unknown vector quantity, and assuming a time-harmonic electric current source distribution, $\vec{J}_v$ of angular frequency ω in a linear, lossless, homogeneous domain, Ω, characterized by electric permittivity ϵ and magnetic permeability μ. The pertinent equation is obtained in a straightforward manner from the two Maxwell's curl equations and has the form

$$\nabla \times \nabla \times \vec{E} - \omega^2 \mu\epsilon\vec{E} = -j\omega\mu\vec{J}_v. \tag{2.1}$$

A complete statement of the boundary value problem requires the assignment of appropriate conditions at the boundary surface, Γ, of Ω to guarantee the uniqueness of the solution [6]. For the sake of simplicity and without loss of generality we will assume that the tangential components of the magnetic field intensity on Γ are zero. In the context of finite element methods this is recognized as the natural boundary condition for the electromagnetic boundary value problem [3].

This choice of homogeneous boundary conditions on Γ turns the domain Ω into a cavity. Hence, the solution of (2.1) may be constructed in terms of the eigenfunctions (or eigenmodes) of the cavity, which are obtained from the solution of the associated eigenvalue problem

$$\nabla \times \nabla \times \vec{E} - \beta^2 \vec{E} = 0, \tag{2.2}$$

subject to the same boundary conditions with (2.1), namely, zero tangential magnetic field components on Γ. Therefore, in our examination of the occurrence of parasitic, irrotational solutions to (2.1) it suffices to examine whether non-physical, irrotational eigenmodes are present in the spectrum of the eigenvalue problem of (2.2). The answer is obvious. There is an infinite number of such non-trivial, irrotational eigenmodes of (2.2) and their eigenvalue is zero. However, even though mathematically valid, these eigenmodes are not physically admissible as electromagnetic field eigenmodes. The reason for this is that in deriving (2.2) from Maxwell's curl equations, Gauss' law $\nabla \cdot \vec{E} = 0$ was not imposed. Consequently, they should not be utilized in the eigenfunction representation of the solution of (2.1).

A similar argument can be made for the so-called weak statement of (2.1), which is used to effect the weighted minimization of the error resulting from the approximation of the electric field intensity vector by its expansion in a given basis. The weak statement is obtained from (2.1) by taking its inner product with a weighting (or testing) function $\vec{w}$ and integrating over Ω. Through appropriate integration by parts and making use of the boundary conditions on Γ, the following weak statement of (2.1) is obtained

$$\int_\Omega (\nabla \times \vec{w}) \cdot (\nabla \times \vec{E}) dv - \left(\frac{\omega}{v_p}\right)^2 \int_\Omega \vec{w} \cdot \vec{E} dv = -j\omega\mu \int_\Omega \vec{w} \cdot \vec{J}_v dv, \tag{2.3}$$

where $v_p = 1/\sqrt{\mu\epsilon}$ is the wave phase velocity in Ω.

The weak statement in (2.3) has many interesting properties. Let us begin by pointing out that the left-most term in (2.3) contains the curl of the unknown electric field, which, for $\omega \neq 0$, is expected to be non-zero. More specifically, for the numerical solution of (2.1) we are interested in finite element approximations of $\vec{E}$ which satisfy the discrete version of (2.3) and do not belong in the null space of the curl operator. Hence, the important question is whether the discrete version of (2.3) admits any spurious irrotational solutions for $\omega \neq 0$ which may contaminate the approximation of the physically valid solution. Discretization error immediately comes to mind, suggesting that one of the key questions to be answered is how accurately the discrete form of the identity $\nabla \times (\nabla\Psi) \equiv 0$ is satisfied for a given choice of basis functions. In addition, considering that for (2.2) there exists an infinite number of irrotational eigenfunctions of zero eigenvalue, what needs to be examined further is whether a good choice of the basis for the finite element approximation can be made such that it provides for a sufficiently good representation of the null space of the curl operator. Clearly, the goodness of the representation can be quantified by the dimension of the $\nabla\Psi$ subspace in the finite-dimensional space spanned by the finite element basis functions.

In our search for answers to these questions it is helpful to consider the finite element matrix equation that results from the application of (2.3) with as many independent testing functions $\vec{w}$ as the number of degrees of freedom in the finite element approximation of $\vec{E}$.

$$Ax = \omega b \Rightarrow \left(A_1 - (\omega/v_{ph})^2 A_2\right) x = \omega b. \tag{2.4}$$

In (2.4) x denotes the unknown vector of dimension N, containing the N degrees of freedom in the approximation, b is the discrete form of the source term, while A_1 and A_2 are associated, respectively, with the discrete approximations of the first and second terms in (2.3). The associated matrix eigenvalue problem here is

$$A_1 e_k = \hat{\beta}_k^{\,2} A_2 e_k, \quad k = 1, 2, \ldots, N. \tag{2.5}$$

Let us assume, next, that the choice of the finite element basis is such that the subspace of irrotational fields $\nabla \Psi$ is approximated accurately. Then the eigenvectors, e_k^{irr}, that span this subspace satisfy $A_1 e_k^{\mathrm{irr}} = 0$; hence, despite the discretization, the correct zero value for their associated eigenvalue is recovered. In a manner similar to the continuous case, even though mathematically valid, these eigenvectors are not physically admissible since the constraint for solenoidal solution (Gauss' law) was not imposed in the development of the weak statement of the problem. Hence, they do represent non-physical eigenvectors and should not be involved in the representation of the numerical solution of the forced problem (2.1) with $\omega \neq 0$. However, contrary to the continuous case, excitation of such spurious, irrotational eigenvectors is possible in the discrete case when $\omega \neq 0$. One possibility is that some of the irrotational eigenmodes (recall that there is an infinite number of them) may not be represented accurately by the chosen basis (i.e., they fail to satisfy $A_1 e_k^{\mathrm{irr}} = 0$), and thus manifest themselves numerically as eigenvectors with non-zero eigenvalues. Consequently, they cannot be distinguished from the physically admissible eigenvectors.

Another possibility, more intimately related to the choice of the basis function, is that a potential source for these modes may result from the violation of the continuity of normal electric flux density at material interfaces caused by the discretization. Such a violation will manifest itself in the form of a spurious surface electric charge density, which will then act as the source for these irrotational modes. This seems to suggest that, in addition to their ability to model accurately the null space of the curl operator, their ability to prevent violation of the continuity of the normal component of the electric flux density at source-free, material interfaces is yet another desirable property of the finite element basis to be used for the expansion of vector fields.

As already mentioned, the occurrence of spurious modes soon became one of the most limiting difficulties in the use of node-based scalar elements for the finite element approximation of three-dimensional electromagnetic boundary value problems. This shortcoming motivated the emergence of a new class of finite elements, the so-called *edge elements*. Starting with the early ideas of Nédélec [7], [8] and Bossavit [9], [10], [11], the investigation of the advantages of edge elements as well as the research in systematic methodologies for their development and implementation in the finite element solution of vector field problems dominated the computational electromagnetic literature [12]-[15]. A collection of some of the first papers on the properties of edge elements and their application to the finite element solution of the vector Helmholtz equation can be found in [16]. The most comprehensive list of references is available in [4].

Edge elements are the basis functions in the lowest-order *tangentially continuous vector* (TV) space, which is the second important member in the family of finite elements. Its main advantage over the scalar space is that its elements enforce only the tangential field continuity at material interfaces while allowing for the normal component to be discontinuous. This is consistent with the discontinuity property of the normal components of the electric and magnetic field intensity at material interfaces. In addition, the space separates itself into the gradient subspace (null space of the curl operator) and the non-gradient subspace (range space of the curl operator). As mentioned above, this is a key attribute for rendering the finite element approximation free from spurious modes. Since the introduction of edge elements, several methodologies have been proposed for the construction of explicit forms of the interpolation functions in higher-order TV spaces (see, e.g., [17]-[30]). In electromagnetic field modeling applications, the TV space is most frequently used for the expansion of the field intensities in the finite element solution of the weak form of the vector Helmholtz equation. This finite element approximation results in an finite-element matrix equation with quadratic frequency dependence. It is discussed in detail in the next chapter and, for the purposes of this book, it will be referred to as the *field formulation* of the finite element approximation of the electromagnetic boundary value problem.

While the field formulation is the one most commonly used in computational electro-magnetics, our discussion will include a class of interesting applications of finite element methods, which are aimed toward the development of compact, network analysis-oriented, macromodels for passive electromagnetic devices [31]. For such applications an alternative formulation of the finite element approximation of the electromagnetic system is conve-nient. This formulation develops weak statements for the two Maxwell's curl equations. Thus, in addition to a finite element approximation of one of the filed intensities, an approx-imation of the flux density associated with the second field is needed. Since the electric flux density, $\vec{D}$, and the magnetic flux density, $\vec{B}$, are continuous in the direction normal to a material interfaces, their finite element representation should provide for such continuity. Thus their approximation is in terms of the third important family of finite element spaces, the so-called *normally continuous vector* (NV) space.

The attributes of the NV space are opposite to those of the TV spaces. Normal continuity of the flux is imposed, while tangential discontinuity is allowed. Furthermore, the NV space separates itself into a curl subspace (null space of the divergence operator) and a non-curl subspace (range space of the divergence operator). In addition to the aforementioned application of the NV space in the finite element approximation of the two Maxwell's curl equations for the purposes of electromagnetic device macro-modeling, its use is of importance to the development of time-domain finite element electromagnetic field solvers, an area of growing interest in electromagnetic field modeling in complex non-linear media [32]-[34].

Finally, the divergence of NV space results in another scalar space, which constitutes the fourth member in the family of finite element spaces of relevance to electromagnetic field modeling. It is used for the expansion of electric or magnetic charge distributions.

In summary, there are four members in the family of finite element spaces of relevance to electromagnetic modeling, namely, the scalar space for the expansion of scalar potential functions; the TV space for the expansion of electric and magnetic field intensities; the NV space for the expansion of electric and magnetic flux densities; and another scalar space for the expansion of charge distributions. These four spaces will also be referred to, respectively, as *potential*, *field*, *flux*, and *charge* space.

Bossavit proposed four Whitney forms [35], which are associated with the vertices, edges, facets and volume of a tetrahedron [9]-[11]. In essence, the proposed Whitney-0 from is the lowest order scalar potential space; the Whitney-1 form is the lowest order TV field space; the Whitney-2 form is the lowest order NV flux space; and the Whitney-3 form is the lowest order scalar charge space. There is a close relationship among these four spaces, which will be highlighted later in the chapter. Our objective is to present a systematic methodology for the construction of these four spaces for both triangles and tetrahedra, extending Bossavit's forms to higher order. In doing so, major emphasis will be placed on making the developed bases hierarchical, an attribute that is of key importance to the development of multigrid methods.

2.2 TWO-DIMENSIONAL FINITE ELEMENT SPACES

Consider a two-dimensional (2D) triangular element. Let us recall that a *simplex* element is defined as one for which the minimum number of nodes is one more than the dimension of space in which the element is defined. Thus for a planar triangle the minimum number of nodes in the element is three, equal to the number of its vertices. A commonly used set of coordinates on a simplex element is the set of *simplex* or *barycentric* coordinates [36]. The barycentric function for vertex m, where $m = 1, 2, 3$, is denoted as λ_m. It is a linear function, assuming the value of unity at node m and equal to zero at the remaining two nodes. The three barycentric functions in a planar triangle satisfy the relationship

$$\lambda_1 + \lambda_2 + \lambda_3 = 1. \tag{2.6}$$

2.2.1 Two-dimensional potential space

The pth-order scalar space W_s^p for the triangular element is

$$W_s^p = \left\{ \lambda_1^i \lambda_2^j \lambda_3^k \ \middle| \ \begin{array}{l} i = 0, 1, \cdots, p, \\ j = 0, 1, \cdots, p - i, \\ k = p - i - j \end{array} \right\} \tag{2.7}$$

$$\mathrm{Dim}(W_s^p) = \frac{(p+1)(p+2)}{2},$$

where $\mathrm{Dim}(\cdot)$ denotes the dimension of the space. The scalar space can be split into three subspaces defined as follows: The *node-type subspace*, $W_{s,n}^1$, contains the three basis functions λ_1, λ_2, λ_3, each one of which is non-zero at one of the three vertices and is referred to as node element. The *edge-type subspace*, $W_{s,e}^p$, contains those basis functions each of which is non-zero on one of the three edges of the triangle. These are

$$e_{m,n}^p = \left\{ \lambda_m \lambda_n^i \mid i = 1, 2, \cdots, p - 1 \right\},$$
$$W_{s,e}^p = \left\{ e_{1,2}^p, e_{1,3}^p, e_{2,3}^p \right\}, \tag{2.8}$$
$$\mathrm{Dim}(W_{s,e}^p) = 3(p - 1).$$

Finally, the *facet-type* subspace, $W_{s,f}^p$, contains the basis functions that are zero along the three edges of the triangle,

$$f_{1,2,3}^p = \left\{ \lambda_1 \lambda_2^i \lambda_3^j \ \middle| \ \begin{array}{l} i = 1, 2, \cdots, p - 2, \\ j = 1, 2, \cdots, p - 1 - i \end{array} \right\},$$
$$W_{s,f}^p = \left\{ f_{1,2,3}^p \right\}, \tag{2.9}$$
$$\mathrm{Dim}(W_{s,f}^p) = \frac{(p-1)(p-2)}{2}.$$

With this decomposition we have

$$W_s^p = W_{s,n}^1 \oplus W_{s,e}^p \oplus W_{s,f}^p. \tag{2.10}$$

As an example, consider the third-order scalar space W_s^3 ($p = 3$). From (2.7) we have

$$W_s^3 = \{\lambda_1^3, \lambda_2^3, \lambda_3^3, \lambda_1\lambda_2^2, \lambda_1^2\lambda_2, \lambda_1\lambda_3^2, \lambda_1^2\lambda_3, \lambda_2\lambda_3^2, \lambda_2^2\lambda_3, \lambda_1\lambda_2\lambda_3\}$$
$$\text{Dim}(W_s^3) = 10. \tag{2.11}$$

Obviously, the basis functions obtained from (2.7) are not hierarchical, that is, the basis functions of W_s^p will not appear explicitly in the set of basis functions for W_s^{p+1}. However, if W_s^p is constructed, instead, as the direct sum of $W_{s,n}^1$, (2.8), and (2.9) then the resulting basis is hierarchical . For $p = 3$ the resulting set of basis functions for W_s^3 is

$$W_s^3 = \{\underbrace{\lambda_1, \lambda_2, \lambda_3,}_{W_{s,n}^1} \underbrace{\lambda_1\lambda_2, \lambda_1\lambda_2^2, \lambda_1\lambda_3, \lambda_1\lambda_3^2, \lambda_2\lambda_3, \lambda_2\lambda_3^2,}_{W_{s,e}^3} \underbrace{\lambda_1\lambda_2\lambda_3}_{W_{s,f}^3}\}. \tag{2.12}$$

It can be shown that this set of basis functions can be obtained in terms of linear combinations of the basis functions of W_s^3 constructed using (2.11). For example, consider the basis functions λ_1 and $\lambda_1\lambda_2$ in (2.12). The fact that they can be written as linear combinations of the basis functions in (2.11) becomes obvious from the following two equations:

$$\begin{aligned}
\lambda_1 &= \lambda_1(\lambda_1 + \lambda_2 + \lambda_3)^2 \\
&= \lambda_1(\lambda_1^2 + \lambda_2^2 + \lambda_3^2 + 2\lambda_1\lambda_2 + 2\lambda_1\lambda_3 + 2\lambda_2\lambda_3) \\
\lambda_1\lambda_2 &= \lambda_1\lambda_2(\lambda_1 + \lambda_2 + \lambda_3) \\
&= \lambda_1^2\lambda_2 + \lambda_1\lambda_2^2 + \lambda_1\lambda_2\lambda_3.
\end{aligned} \tag{2.13}$$

A similar process can be followed to show that the rest of the basis functions in (2.11) can be written as linear combinations of the basis functions in (2.12). Hence, (2.10) holds for $p = 3$.

2.2.2 Two-dimensional field space

The two-dimensional field space of order p is the two-dimensional TV space W_{tv}^p. Its basis functions are vectors. The elements of W_{tv}^p can be obtained through the multiplication of every scalar basis in W_s^p with any two vectors selected from $\nabla\lambda_i$, where $i = 1, 2, 3$.

At this point it is useful to discuss the properties of the gradient of the hat function $\nabla\lambda_l$, where $l = 1, 2, 3$. It is straightforward to show that in a triangle (l, m, n) it is [39]

$$\nabla\lambda_l = \frac{\vec{t_l}}{2A}, \tag{2.14}$$

where the length of the vector $\vec{t_l}$ equals the length of edge (m, n), and its direction is perpendicular to edge (m, n) and pointing into the triangle. In the above equation A denotes the area of the triangle. The dot product of $\nabla\lambda_l$ with any of the two edge vectors $\vec{l}_{m,l}$ and $\vec{l}_{n,l}$, ending at vertex l , is one.

$$\nabla\lambda_l \cdot \vec{l}_{m,l} = \nabla\lambda_l \cdot \vec{l}_{n,l} = 1. \tag{2.15}$$

The vector bases induced from the scalar functions of (2.7-2.9) are

$$\lambda_1 \rightarrow \begin{pmatrix} (\lambda_1\nabla\lambda_2) \\ (\lambda_1\nabla\lambda_3) \end{pmatrix}, \quad \lambda_2 \rightarrow \begin{pmatrix} \lambda_2\nabla\lambda_1 \\ (\lambda_2\nabla\lambda_3) \end{pmatrix}, \quad \lambda_3 \rightarrow \begin{pmatrix} \lambda_3\nabla\lambda_1 \\ \lambda_3\nabla\lambda_2 \end{pmatrix},$$
$$e_{1,2}^p \rightarrow \begin{pmatrix} (e_{1,2}^p\nabla\lambda_2) \\ [e_{1,2}^p\nabla\lambda_3] \end{pmatrix}, \quad e_{1,3}^p \rightarrow \begin{pmatrix} e_{1,3}^p\nabla\lambda_2 \\ (e_{1,3}^p\nabla\lambda_3) \end{pmatrix}, \quad e_{2,3}^p \rightarrow \begin{pmatrix} e_{2,3}^p\nabla\lambda_1 \\ (e_{2,3}^p\nabla\lambda_3) \end{pmatrix}, \tag{2.16}$$
$$f_{1,2,3}^p \rightarrow \begin{pmatrix} f_{1,2,3}^p\nabla\lambda_1 \\ [f_{1,2,3}^p\nabla\lambda_3] \end{pmatrix}.$$

It is easy to confirm that the dimension of W_{tv}^p is $(p+1)(p+2)$. In the above equations some of the induced vector bases have been enclosed with parentheses and some with brackets. The reason for this distinction and will become clear in the following discussion.

In constructing W_{tv}^p we distinguish between two subspaces. The first one is called *edge-type* TV subspace, $W_{tv,e}^p$, and contains those vector basis functions with non-zero tangential components on one of the edges of the triangle. The second one is called the *facet-type* TV subspace, $W_{tv,f}^p$, and its elements have zero tangential components on all three edges. Thus we have

$$W_{tv}^p = W_{tv,e}^p \oplus W_{tv,f}^p. \tag{2.17}$$

Motivated by our preliminary comments on spurious, non-physical modes and their prevention through the proper representation of the null space of the curl operator, we would like to consider the decomposition of each one of the two subspaces in (2.17) into non-gradient subspaces and gradient subspaces. Since a gradient subspace of order p will be constructed from the gradients of scalar functions of order $p + 1$, the notation ∇W_s^{p+1} will be used for its representation. Thus we write

$$W_{tv}^p = \underbrace{W_{tv,e,ng}^p \oplus \nabla W_{s,e}^{p+1}}_{W_{tv,e}^p} \oplus \underbrace{W_{tv,f,ng}^p \oplus \nabla W_{s,f}^{p+1}}_{W_{tv,f}^p}, \tag{2.18}$$

where $W_{tv,e,ng}^p$ and $W_{tv,f,ng}^p$ denote the edge-type and facet-type non-gradient subspaces, respectively.

The desired decomposition is facilitated by the fact that the basis functions in the edge-type scalar subspace $W_{s,e}^{p+1}$ and the facet-type scalar subspace $W_{s,f}^{p+1}$ are available explicitly in (2.8) and (2.9). Thus by calculating their gradients and removing them from (2.16) the basis functions for the non-gradient subspaces, $W_{tv,e,ng}^p$ and $W_{tv,f,ng}^p$, are obtained.

The gradients of the basis functions in $W_{s,e}^{p+1}$ are

$$\nabla W_{s,e}^{p+1} = \left\{ \begin{array}{l} \nabla\left(\lambda_1\lambda_2^i\right) = \lambda_2^i\nabla\lambda_1 + i\lambda_1\lambda_2^{i-1}\nabla\lambda_2 \\ \nabla\left(\lambda_1\lambda_3^i\right) = \lambda_3^i\nabla\lambda_1 + i\lambda_1\lambda_3^{i-1}\nabla\lambda_3 \\ \nabla\left(\lambda_2\lambda_3^i\right) = \lambda_3^i\nabla\lambda_2 + i\lambda_2\lambda_3^{i-1}\nabla\lambda_3 \end{array} \right| \left. i = 1, 2, \cdots, p \right\}. \tag{2.19}$$

A comparison of (2.19) with (2.16) suggests that removal of the functions $\lambda_1\lambda_2^{i-1}\nabla\lambda_2$, $\lambda_1\lambda_3^{i-1}\nabla\lambda_3$, and $\lambda_2\lambda_3^{i-1}\nabla\lambda_3$ from (2.16), results in the removal of the gradient of the basis in $W_{s,e}^{p+1}$. The removed bases are the ones enclosed in parentheses in (2.16).

Similarly, the gradient of the basis functions in $W_{s,f}^{p+1}$ yields the following set,

$$\nabla W_{s,f}^{p+1} = \nabla\lambda_1\lambda_2^i\lambda_3^j = \lambda_2^i\lambda_3^j\nabla\lambda_1 + i\lambda_1\lambda_2^{i-1}\lambda_3^j\nabla\lambda_2 + j\lambda_1\lambda_2^i\lambda_3^{j-1}\nabla\lambda_3,$$
$$i = 1, 2, \cdots, p - 1, \tag{2.20}$$
$$j = 1, 2, \cdots, p - i.$$

The removal of the gradient of the basis functions in $W_{s,f}^{p+1}$ is achieved with the removal of the functions $\lambda_1\lambda_2^i\lambda_3^{j-1}\nabla\lambda_3$ from (2.16). The removed bases are the ones enclosed in brackets in (2.16).

The remaining functions in (2.16) constitute the basis for the non-gradient vector subspace. More specifically, it is

$$W_{tv,e,ng}^1 = \left\{ \lambda_2\nabla\lambda_1, \lambda_3\nabla\lambda_1, \lambda_3\nabla\lambda_2 \right\},$$
$$W_{tv,f,ng}^p = \left\{ e_{2,3}^p\nabla\lambda_1, e_{1,3}^p\nabla\lambda_2, f_{1,2,3}^p\nabla\lambda_1 \right\}. \tag{2.21}$$

It is noted that the edge-type, non-gradient subspace has only three basis functions.

Let us examine the lowest-order (first-order) TV space in some more detail. It only contains the edge-type basis functions, $W_{tv}^1 = W_{tv,e}^1$. Written in terms of its gradient/non-gradient subspace decomposition, it is

$$W_{tv,e}^1 = \underbrace{\{\lambda_2 \nabla\lambda_1, \ \lambda_3 \nabla\lambda_1, \ \lambda_3 \nabla\lambda_2\}}_{W_{tv,e,ng}^1} \oplus \underbrace{\{\nabla(\lambda_1\lambda_2), \ \nabla(\lambda_1\lambda_3), \ \nabla(\lambda_2\lambda_3)\}}_{\nabla W_{s,e}^2}. \qquad (2.22)$$

By construction, it contains explicitly the gradient of basis functions in the second-order scalar space. However, it also contains the gradients of the basis functions of the first-order scalar space, $\nabla W_{s,n}^1$, which are easily recovered through appropriate linear combinations of the basis functions in $W_{tv,e}^1$. Furthermore, we could make $\nabla W_{s,n}^1$ be solely associated with $W_{tv,e,ng}^1$ through the generation of a new set of basis functions for $W_{tv,e,ng}^1$. This is done through the linear combination of one basis function in $W_{tv,e,ng}^1$ with basis functions in $\nabla W_{s,e}^2$. In particular, this can be done in such a manner that the resulting set of functions maintains the non-gradient property of $W_{tv,e,ng}^1$. The resulting set of basis functions for $W_{tv,e,ng}^1$ is very-well known in the finite element literature. It is the most popular two-dimensional edge elements family [10], with explicit expressions for the basis functions as follows:

$$W_{tv,e,ng}^1 = \{\underbrace{\lambda_1 \nabla\lambda_2 - \lambda_2 \nabla\lambda_1}_{\vec{e}_{1,2}}, \underbrace{\lambda_1 \nabla\lambda_3 - \lambda_3 \nabla\lambda_1}_{\vec{e}_{1,3}}, \underbrace{\lambda_2 \nabla\lambda_3 - \lambda_3 \nabla\lambda_2}_{\vec{e}_{2,3}}\}. \qquad (2.23)$$

Clearly, $\vec{e}_{m,n}$ is obtained as $\nabla(\lambda_m\lambda_n) - 2\lambda_n\lambda_n$. It has a constant tangential component along edge (m, n). Its inner product with the edge-vector $\vec{l}_{m,n}$, directed from node m to n, yields

$$(\lambda_m \nabla\lambda_n - \lambda_n \nabla\lambda_m) \cdot \vec{l}_{m,n} = 1, \quad \text{on edge}(m, n). \qquad (2.24)$$

Thus the three TV-edge-elements are pictorially represented by arrows along the associated edges, as shown in the left drawing of Fig. 2.1.

The space $\nabla W_{s,n}^1$, spanned by the gradients of the basis functions of $W_{s,n}^1$, is solely contained in the new $W_{tv,e,ng}^1$. This becomes clearly apparent from the following expression:

$$[\nabla\lambda_1 \ \nabla\lambda_2 \ \nabla\lambda_3] = [\vec{e}_{1,2} \ \vec{e}_{1,3} \ \vec{e}_{2,3}] \begin{bmatrix} -1 & 1 & \\ -1 & & 1 \\ & -1 & 1 \end{bmatrix} \qquad (2.25)$$

$$\Rightarrow \nabla W_{s,n}^1 \subset W_{tv,e,ng}^1.$$

We conclude the discussion of the pth-order TV space, by providing a summary of its properties through the following formula:

$$W_{tv}^p = \underbrace{W_{tv,e,ng}^1 \oplus \nabla W_{s,e}^{p+1}}_{W_{tv,e}^p} \oplus \underbrace{W_{tv,f,ng}^p \oplus \nabla W_{s,f}^{p+1}}_{W_{tv,f}^p}, \qquad (2.26)$$

where

$$W_{tv,e,ng}^1 = \{\vec{e}_{1,2}, \vec{e}_{1,3}, \vec{e}_{2,3}\}, \qquad \mathrm{Dim}(W_{tv,e,ng}^1) = 3,$$

$$\nabla W_{s,e}^{p+1} = \left\{\nabla e_{1,2}^{p+1}, \nabla e_{1,3}^{p+1}, \nabla e_{2,3}^{p+1}\right\}, \qquad \mathrm{Dim}(\nabla W_{s,e}^{p+1}) = 3p,$$

$$W_{tv,f,ng}^p = \left\{e_{2,3}^p \nabla\lambda_1, e_{1,3}^p \nabla\lambda_2, f_{1,2,3}^p \nabla\lambda_1\right\}, \quad \mathrm{Dim}(W_{tv,f,ng}^p) = \frac{(p-1)(p+2)}{2},$$

$$\nabla W_{s,f}^{p+1} = \left\{\nabla f_{1,2,3}^{p+1}\right\}, \qquad \mathrm{Dim}(W_{tv,f}^{p+1}) = \frac{p(p-1)}{2}.$$

$$(2.27)$$

2.2.3 Two-dimensional flux space

The two-dimensional, normally continuous vector (NV) space, W_{nv}^p, will be referred to, for short, as the two-dimensional flux space. It can be constructed from the two-dimensional field space discussed in the previous chapter. Let $\hat{n}$ denote the unit vector perpendicular to the triangular element. The basis functions for the two dimensional NV space are obtained by taking the cross-product $\hat{n}$ with the basis functions of W_{tv}^e. Thus the two-dimensional flux space can be written as

$$W_{nv}^p = \hat{n} \times W_{tv}^p = \underbrace{\hat{n} \times W_{tv,e,ng}^1 \oplus \hat{n} \times \nabla W_{s,e}^{p+1}}_{\hat{n} \times W_{tv,e}^p} \oplus \underbrace{\hat{n} \times W_{tv,f,ng}^p \oplus \hat{n} \times \nabla W_{s,f}^{p+1}}_{\hat{n} \times W_{tv,f}^p}.$$

$$(2.28)$$

The space $\hat{n} \times W_{tv,e,ng}^1$ is a very well known space in computational electromagnetics because of its extensive use in the numerical approximation of surface integral equations. The cross-product of $\hat{n}$ with the three TV edge-elements in $W_{tv,e,ng}^1$ rotates their directions by ninety degrees, thus rendering them perpendicular to the corresponding edges. The resulting basis functions are referred to as normally continuous vector (NV) edge elements and are depicted by the three vectors perpendicular to the three edges of the triangle in the drawing on the right in Fig. 2.1. They are the well-known Rao-Wilton-Glisson (RWG) expansion functions, used extensively for the representation of surface electric current flow in electromagnetic field integral equations, due to their property of enforcing the continuity of the normal component of current flow across the common edge of adjacent triangles [37].

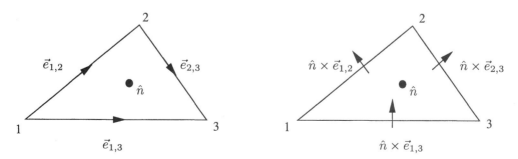

Figure 2.1   2D TV-edge-elements (*left*) and NV-edge-elements (*right*).

2.2.4 Two-dimensional charge space

The two-dimensional charge space is spanned by the functions obtained by taking the divergence of the basis functions of the two-dimensional flux space. This yields

$$W^p_{charge} = \nabla \cdot W^p_{nv} = \nabla \cdot (\hat{n} \times W^p_{tv}). \tag{2.29}$$

For the three NV edge-elements in $W^1_{nv,e,ng}$ it is

$$
\begin{aligned}
W^0_{charge} &= \nabla W^1_{nv,e,ng} = \nabla \cdot (\hat{n} \times W^1_{tv,e,ng}) \\
&= \{\nabla \cdot (\hat{n} \times \vec{e}_{1,2}),\ \nabla \cdot (\hat{n} \times \vec{e}_{1,3}),\ \nabla \cdot (\hat{n} \times \vec{e}_{2,3})\} \\
&= \{2\hat{n} \cdot \nabla\lambda_2 \times \nabla\lambda_1,\ 2\hat{n} \cdot \nabla\lambda_3 \times \nabla\lambda_1,\ 2\hat{n} \cdot \nabla\lambda_3 \times \nabla\lambda_2\} \\
&= \left\{ s\frac{1}{A},\ -s\frac{1}{A},\ s\frac{1}{A} \right\},
\end{aligned}
\tag{2.30}
$$

where $s = \mathrm{sgn}(\hat{n} \cdot \nabla\lambda_2 \times \nabla\lambda_1)$, and A denotes the area of the triangle. Thus, since the divergence of each one of the three NV edge-elements is constant over the triangle, we conclude that the only element in the space $\nabla \cdot W^1_{nv,e,ng}$ is the constant $1/A$. Hence, the pth-order charge space will be written as follows:

$$
\begin{aligned}
W^p_{charge} &= \left\{ \frac{1}{A} \right\} \oplus \nabla \cdot (\hat{n} \times W^{p+1}_{tv,f,ng}), \\
\mathrm{Dim}(W^p_{charge}) &= \frac{(p+1)(p+2)}{2}, \\
\mathrm{Dim}\left(\nabla \cdot (\hat{n} \times W^{p+1}_{tv,f,ng}) \right) &= \frac{p(p+3)}{2}.
\end{aligned}
\tag{2.31}
$$

2.3 RELATIONSHIP AMONG 2D FINITE ELEMENT SPACES

A relationship exists among the scalar, field, flux and charge spaces presented in the previous subsections. This relationship is derived and discussed next.

It is instructive to summarize the definition of the four two-dimensional spaces for finite element approximation on a triangle, casting them in a sequential order that reflects the construction of the basis functions for each space in terms of the basis functions of the previous space.

$$
\begin{aligned}
W^p_s &= W^1_{s,n} \oplus & W^p_{s,e} & \oplus & W^p_{s,f} \\
W^p_{tv} &= & \underbrace{W^1_{tv,e,ng} \oplus \nabla W^{p+1}_{s,e}}_{W^p_{tv,e}} & \oplus & \underbrace{W^p_{tv,f,ng} \oplus \nabla W^{p+1}_{s,f}}_{W^p_{tv,f}} \\
W^p_{nv} &= & \underbrace{\hat{n} \times W^1_{tv,e,ng} \oplus \hat{n} \times \nabla W^{p+1}_{s,e}}_{W^p_{nv,e}} \oplus & \underbrace{\hat{n} \times W^p_{tv,f,ng} \oplus \hat{n} \times \nabla W^{p+1}_{s,f}}_{W^p_{nv,f}} \\
W^p_{charge} &= & & & \{\tfrac{1}{A}\} \oplus \nabla \cdot (\hat{n} \times W^{p+1}_{tv,f,ng}).
\end{aligned}
$$

In the above expression we recognize the lowest-order spaces $W^1_{s,n}$, $W^1_{tv,e,ng}$, $\hat{n} \times W^1_{tv,e,ng}$ and W^0_{charge}, as the Whitney 0, 1, 2, and 3 forms, respectively, for the triangular element [35]. An interesting relationship among these Whitney forms, which is of relevance to the occurrence and prevention of spurious modes in finite element-based electromagnetic modeling and, in general, to the physically consistent representation of electromagnetic field quantities, is the following [9]:

$$\underbrace{W_{s,n}^1}_{\text{node-elements}} \xrightarrow[grad]{\nabla W_{s,n}^1 \subset} \underbrace{W_{tv,e,ng}^1}_{\text{TV-edge-elements}} \xrightarrow{\hat{n} \times W_{tv,e,ng}^1 =} \underbrace{\hat{n} \times W_{tv,e,ng}^1}_{\text{NV-edge-elements}} \xrightarrow[div]{\nabla \cdot (\hat{n} \times W_{tv,e,ng}^1)=} \underbrace{W_{charge}^0}_{\text{facet-elements}} .$$

(2.32)

The superscripts on the arrows indicate the operation on the basis functions of the previous space that is used to obtain basis functions in the space that follows. In particular, the superscript on the first arrow indicates the important result that the gradient of a 0-form basis functions forms a subset of the 1-form basis functions. In other words, the TV-edge-element basis functions are capable of representing the gradient of the 0-form basis functions.

In a similar manner, the 2-form basis functions (NV edge elements) are obtained from the 1-form basis functions through the cross-product operation with the unit normal vector on the surface of the triangle. Finally, the divergence of the 2-form basis functions produces the 3-form basis functions for the two-dimensional charge space. Finally, as indicated by the subscript under the arrow in the second equation, the basis functions for the charge space may be obtained directly from the 1-form basis functions as the component of the curl of the TV-edge-element basis functions in the direction normal to the surface of the triangle.

From the earlier development of the four spaces it is straightforward to see that the relationship of (2.32) among the Whitney 0, 1, 2 and 3 forms holds also for high-order FEM spaces ($p > 1$). Thus we have

$$W_s^p \xrightarrow[grad]{\nabla W_s^{p+1} \subset W_{tv}^p} W_{tv}^p \xrightarrow{\hat{n} \times W_{tv}^p = W_{nv}^p} W_{nv}^p \xrightarrow[div]{\nabla \cdot (\hat{n} \times W_{tv}^{p+1}) = W_{charge}^p} W_{charge}^p. \quad (2.33)$$

Let us consider a two-dimensional domain that has been discretized by means of triangular elements. Let N_n be the number of nodes, N_e the number of edges, and N_f the number of triangles (facets) in the domain. In the following we examine the relationship among the number of nodes, edges and facets in the resulting finite element grid to the dimensions of the Whitney forms.

Let λ_i denote the global nodal basis function associated with node i. For the node-element space it is

$$W_{s,n}^1 = \{\lambda_1, \lambda_2, \cdots, \lambda_{N_n}\},$$
$$\text{Dim}(W_{s,n}^1) = N_n, \quad (2.34)$$
$$\text{Dim}(\nabla W_{s,n}^1) = N_n - 1.$$

The fact that the dimension of $\nabla W_{s,n}^1$ is one less than that of $W_{s,n}^1$ follows immediately from (2.6).

Let $\vec{e}_i$ denote the TV-edge-element basis function associated with edge i. The TV-edge-element space is, then,

$$W_{tv,e,ng}^1 = \{\vec{e}_1, \vec{e}_2, \cdots, \vec{e}_{N_e}\},$$
$$\text{Dim}(W_{tv,e,ng}^1) = N_e. \quad (2.35)$$

The NV-edge-element space follows immediately from the TV-edge-element one,

$$\hat{n} \times W_{tv,e,ng}^1 = \{\hat{n} \times \vec{e}_1, \hat{n} \times \vec{e}_2, \cdots, \hat{n} \times \vec{e}_{N_e}\},$$
$$\text{Dim}(\hat{n} \times W_{tv,e,ng}^1) = N_e. \quad (2.36)$$

Finally, the facet-element space is

$$W^0_{charge} = \left\{ \frac{1}{A_1}, \frac{1}{A_2}, \cdots, \frac{1}{A_{N_f}} \right\},$$

$$\mathrm{Dim}(W^0_{chrage}) = N_f, \tag{2.37}$$

where A_i is the area of the ith triangle.

There is one more useful result that can be derived from the relationship that exists between the four Whitney forms. Recall that the charge basis functions (3-form) can be constructed from those of the field basis functions (1-form) through the operation $\nabla \cdot \hat{n} \times$ (i.e. $-\hat{n} \cdot \nabla \times$). However, since $N_n - 1$ of the elements of the 1-form space are the gradients of the basis functions of the potential (0-form) space (2.34), such an operation yields only $N_e - (N_n - 1)$ non-zero functions which equals N_f. This result, which may be cast in the form,

$$N_n - N_e + N_f = 1, \tag{2.38}$$

is called Euler's formula for the two-dimensional triangular discretization of a single-connected region. It provides for a useful means to verify the correct construction of a triangular mesh for a two-dimensional domain.

2.4   GRADIENT, CURL AND DIVERGENCE MATRICES FOR 2D FINITE ELEMENT SPACES

For the purposes of finite element matrix manipulations it is useful to obtain matrix representations for the operations involved in the relationship between the basis functions in the spaces associated with the four Whitney forms. More specifically, let $\{\nabla W^1_{s,n}\}$, $\{W^1_{tv,e,ng}\}$, $\{\nabla \cdot \hat{n} \times W^1_{tv,e,ng,}\}$, $\{\nabla \cdot W^1_{nv,f,nc}\}$, and $\{W^0_{charge}\}$ denote row vectors containing the basis functions for the corresponding spaces. We are interested in defining the matrices G, C and D in the following relations:

$$\nabla W^1_{s,n} \subset W^1_{tv,e,ng} \quad \Rightarrow \quad \{\nabla W^1_{s,n}\} = \{W^1_{tv,e,ng}\}G;$$
$$\nabla \cdot (\hat{n} \times W^1_{tv,e,ng}) = W^0_{charge} \quad \Rightarrow \quad \{\nabla \cdot (\hat{n} \times W^1_{tv,e,ng,})\} = \{W^0_{charge}\}C;$$
$$\nabla \cdot W^1_{nv,f,nc} = W^0_{charge} \quad \Rightarrow \quad \{\nabla \cdot W^1_{nv,f,nc}\} = \{W^0_{charge}\}D.$$

The *gradient matrix*, G, is an $N_e \times N_n$ matrix, each row of which corresponds to one edge and contains only two nonzero entries with values of $+1$ or -1. More specifically, the entry in the column corresponding to the start node of the edge is -1, while the entry in the column corresponding to the end node of the edge is $+1$. This is depicted in the leftmost drawing in Fig. 2.2. Hence, it is

$$
\begin{array}{ccc}
\text{node } m & \text{node } n & \\
\downarrow & \downarrow & \\
G_{N_e \times N_n}(i,:) = \begin{bmatrix} \quad -1 & \quad +1 \quad \end{bmatrix} & \leftarrow \text{edge } i.
\end{array}
\tag{2.39}
$$

Since the sum of the column vectors of G is the zero vector, the rank of G is $N_n - 1$.

The *divergence matrix*, D, is an $N_f \times N_e$ matrix, each row of which corresponds to a triangle and contains only three nonzero entries, of value $+1$ or -1. More specifically, the entry in the column corresponding to an NV-edge-element pointing outwards of the

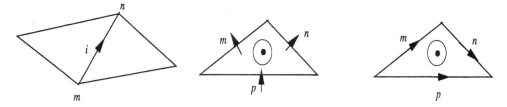

Figure 2.2   Visual interpretation of the gradient, curl, and divergence matrices.

triangular element has value $+1$, while the one in the column corresponding to an NV-edge-element pointing into the triangle has value -1 (see the middle drawing in Fig. 2.2). Hence, we write

$$\begin{array}{cccc} & \text{edge } m & \text{edge } n & \text{edge } p \\ & \downarrow & \downarrow & \downarrow \\ D_{N_f \times N_e}(i,:) = \; [& +1 & +1 & -1 \quad] \leftarrow \text{facet } i. \end{array}$$ (2.40)

Finally, the *curl matrix* or *circulation matrix*, C, is an $N_f \times N_e$ matrix, each row of which corresponds to a triangle and contains three non-zero entries, of value $+1$ or -1. The sign can be determined by the right hand rule. When the thumb points to the direction of $\hat{n}$, if the fingers rotate in the direction of the edge-element, the corresponding entry in the column is -1, otherwise it is $+1$. This is depicted in the rightmost drawing of Fig. 2.16. Hence, we write

$$\begin{array}{cccc} & \text{edge } m & \text{edge } n & \text{edge } p \\ & \downarrow & \downarrow & \downarrow \\ C_{N_f \times N_e}(i,:) = \; [& +1 & +1 & -1 \quad] \; \leftarrow \text{facet } i. \end{array}$$ (2.41)

For the two-dimensional case, the curl and divergence matrices are identical (i.e., $C = D$) because of the relationship between the TV and NV edge-elements. From (2.32), we have the following relationship between the matrices

$$D_{N_f \times N_e} \, G_{N_e \times N_n} = 0,$$
$$C_{N_f \times N_e} \, G_{N_e \times N_n} = 0.$$ (2.42)

2.5   THREE-DIMENSIONAL FINITE ELEMENT SPACES

With a systematic methodology in place for the construction of finite element spaces for the representation of potential, field intensity, flux density and charge in two dimensions, we now turn our attention to its extension to three dimensions. Intuitively, this is a much more challenging undertaking. The added complexity stems not only from the need to provide for interpolation in three dimensions but also by the desire to interpret the rotational and divergence properties of the electromagnetic fields in a manner that facilitates the meaningful assignment of degrees of freedom at the vertices, edges, facets and in the interior of an element. Our experience from the two-dimensional spaces suggests that hierarchy can be relied upon for tackling complexity. This will prove to be the case in our development. In particular, we will see that it pays to try to develop each space in terms of subspaces associated with the nodes, edges, facets and volume of the element.

The element of choice is the tetrahedron element. Like the triangle in two dimensions, the tetrahedron offers superior modeling versatility in three dimensions. The barycentric coordinates will be used here also [36]. The barycentric function (or hat function), λ_m, is associated with the vertex m of the tetrahedron. It is a linear function that assumes the value of 1 at node m and zero at the remaining three nodes. The four barycentric functions in a tetrahedron satisfy the relationship

$$\lambda_1 + \lambda_2 + \lambda_3 + \lambda_4 = 1. \tag{2.43}$$

The construction of the three-dimensional (3D) spaces will begin with the potential space. This will be followed by the development of 3D spaces for the expansion of field intensity, flux density and charge. Like in the 2D case, there are important relationships that exist between these spaces.

2.5.1 Three-dimensional potential space

The scalar potential distribution inside the tetrahedron can be expanded in terms of the scalar polynomial basis functions $\lambda_1^i \lambda_2^j \lambda_3^k \lambda_4^l$. Let W_s^p denote the pth-order scalar space in the tetrahedron, which contains polynomials in $\lambda_{1,2,3,4}$ of degree less than or equal to p. The explicit form of the scalar basis functions in W_s^p along with the dimension of the space are

$$W_s^p = \left\{ \lambda_1^i \lambda_2^j \lambda_3^k \lambda_4^l \; \middle| \; \begin{array}{l} i = 0, 1, \ldots, p, \\ j = 0, 1, \ldots, p - i, \\ k = 0, 1, \ldots, p - i - j, \\ l = p - i - j - k \end{array} \right\}, \tag{2.44}$$

$$\text{Dim}(W_s^p) = \frac{(p+1)(p+2)(p+3)}{6}.$$

In view of the properties of the barycentric coordinates, these basis functions could be assigned to nodes in the tetrahedron using the structure depicted in Fig. 2.44 for $p = 3$. The set of indices (i, j, k, l), is used to denote the position of each node in the tetrahedron, with i, j, k and l representing, respectively, the distance from the node to facet 1, 2, 3 and 4, where facet m is the one facing node m. Using this structure, a direct correspondence exists between each node and each basis function in W_s^p.

A moment's thought reveals that, in order to construct the set W_s^{p+1} for interpolation of order $p+1$, all the basis functions in the W_s^p must be modified and an extra $(p+2)(p+3)/2$ functions must be added to those used in W_s^p. Clearly, the need for modification of all the low-order basis functions is incompatible with the idea of a hierarchical development of higher-order finite element interpolations, which is essential to the development of hierarchical, multilevel preconditioners discussed in Chapter 6. Rather, the hierarchical development of finite element spaces should be such that the low order spaces are maintained as a subset of the next higher-order space. Such a hierarchical construction of scalar spaces is presented next.

In order to facilitate the construction of the hierarchical form of the scalar space, the basis functions in W_s^p are separated into four subspaces, namely, the node-type subspace $W_{s,n}^p$, the edge-type subspace $W_{s,e}^p$, the facet-type subspace $W_{s,f}^p$, and volume-type subspace $W_{s,v}^p$. Hence, we write

$$W_s^p = W_{s,n}^p \oplus W_{s,e}^p \oplus W_{s,f}^p \oplus W_{s,v}^p. \tag{2.45}$$

The basis functions included in each one of these four subspaces are presented in the next four subsections.

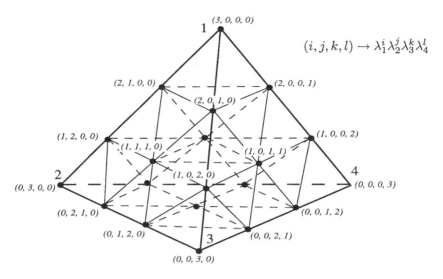

Figure 2.3   The node representation for the pth-order, non-hierarchical scalar basis, $(p = 3)$.

2.5.1.1 Node-type potential subspace

The node-type scalar subspace contains only the first-order basis functions. There are four basis functions in the node-type scalar subspace $W_{s,n}^1$,

$$W_{s,n}^1 = \{\, \lambda_1,\, \lambda_2,\, \lambda_3,\, \lambda_4 \,\},$$
$$\mathrm{Dim}(W_{s,n}^p) = 4. \tag{2.46}$$

Since λ_i, where $i = 1, 2, 3, 4$, has value of 1 at vertex i and is zero at the remaining three vertices, each one of the four basis functions is associated with the vertex of the tetrahedron at which it is non-zero and is referred to as a node element. This is depicted in Fig. 2.4.

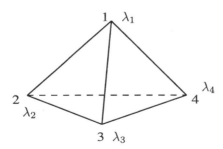

Figure 2.4   The four first-order, node-type scalar basis functions that form the subspace $W_{s,n}^1$, are assigned to the four vertices of a tetrahedron.

2.5.1.2 Edge-type potential subspace

Each one of the basis functions in the edge-type scalar subspace $W_{s,e}^p$ is nonzero on one of the edges of the tetrahedron and zero on all vertices and remaining edges. Let us adopt the notation (m, n), where $m < n$, to indicate the edge defined by vertices m and n. Then, on edge (m, n), the pth-order edge-type basis functions are

$$e_{m,n}^p = \{\, \lambda_m \lambda_n,\, \cdots,\, \lambda_m \lambda_n^{p-2},\, \lambda_m \lambda_n^{p-1} \,\}. \tag{2.47}$$

In view of the properties of the barycentric coordinates, these basis functions are assigned to nodes along the edge (m, n) as depicted in Fig. 2.5. The set of indices $(1, i)$ can be obtained from the position of the corresponding node on the edge, where i is the distance from facet n. Since there are six edges on a tetrahedron, the subspace $W_{s,e}^p$ is constructed as follows:

$$W_{s,e}^p = \{e_{1,2}^p, \ e_{1,3}^p, \ e_{1,4}^p, \ e_{2,3}^p, \ e_{2,4}^p, \ e_{3,4}^p\},$$

$$\text{Dim}(W_{s,e}^p) = 6(p - 1). \tag{2.48}$$

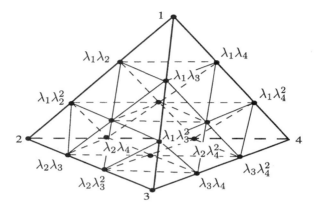

Figure 2.5   Assignment of edge-type scalar basis functions in $W_{s,e}^p$, $(p = 3)$, to nodes along the edges of the tetrahedron.

2.5.1.3 *Facet-type potential subspace*   Each of the basis functions in the facet-type scalar space $W_{s,f}^p$ is nonzero on one of the facets of the tetrahedron, while it assumes the zero value on all edges and the remaining facets. We will use the notation (l, m, n), where $l < m < n$, to denote the facet defined by the three vertices l, m, and n. On facet (l, m, n) the pth-order facet-type basis functions are

$$f_{l,m,n}^p = \left\{ \lambda_l \, \lambda_m^i \, \lambda_n^j \ \middle| \ \begin{array}{l} i = 1, 2, \cdots, p - 2, \\ j = 1, 2, \ldots, p - 1 - i \end{array} \right\} \tag{2.49}$$

The assignment of these basis functions using nodes on facet (l, m, n) is depicted in Fig. 2.6 for $p = 3$. The set of indices $(1, i, j)$ can be obtained from the position of the corresponding node on the facet, where i is the distance of the node to facet m and j is the distance of the node to facet n. Since there are four facets in a tetrahedron, the $W_{s,f}^p$ space is given by

$$W_{s,f}^p = \{ f_{1,2,3}^p, \ f_{1,2,4}^p, \ f_{1,3,4}^p, \ f_{2,3,4}^p \},$$

$$\text{Dim}(W_{s,f}^p) = 4\frac{(p - 2)(p - 1)}{2}. \tag{2.50}$$

2.5.1.4 *Volume-type potential subspace*   The basis functions in the volume-type potential subspace $W_{s,v}^p$ are zero on the four facets of the tetrahedron. We will use the

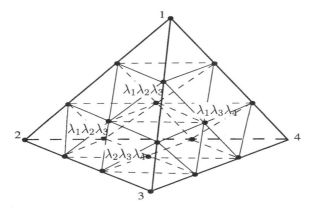

Figure 2.6   Assignment of facet-type scalar basis functions of $W_{s,f}^p$ for $(p = 3)$ to nodes on the facets of the tetrahedron.

notation $(1, 2, 3, 4)$ to denote the interior of the tetrahedron. The expressions for the volume-type basis functions are

$$
v_{1,2,3,4}^p = \left\{ \lambda_1 \lambda_2^i \lambda_3^j \lambda_4^k \; \middle| \; \begin{array}{l} i = 1, 2, \cdots, p - 3, \\ j = 1, 2, \cdots, p - 2 - i, \\ k = 1, 2, \cdots, p - 1 - i - j \end{array} \right\}. \tag{2.51}
$$

Clearly, the minimum order for volume-type potential basis functions is $p = 4$. Furthermore, the set of indices $(1, i, j, k)$ can be employed to assign each one of these basis functions to an appropriate node inside $(1, 2, 3, 4)$ as shown in Fig. 2.7 for the case $p = 4$, where i is the distance of the node to facet 2; j is the distance of the node to facet 3; k is the distance of the node to facet 4. The volume-type scalar space $W_{s,v}^p$ is

$$
\begin{aligned}
W_{s,v}^p &= \{ v_{1,2,3,4}^p \}, \\
\mathrm{Dim}(W_{s,v}^p) &= \frac{(p - 3)(p - 2)(p - 1)}{6}.
\end{aligned} \tag{2.52}
$$

With the node-type, edge-type, facet-type and volume-type potential subspaces defined, respectively, in (2.46), (2.48), (2.50), and (2.52), the hierarchical construction of high-order scalar potential spaces for the tetrahedron element can be effected in a very systematic fashion. The lowest order scalar space $(p = 1)$ contains only the four node-type basis functions (2.46). When the interpolation order increases from p to $p + 1$, all the basis functions in W_s^p are maintained, and the space W_s^{p+1} is completed through the addition of one extra edge function of (2.47) for every edge, $p - 1$ extra facet functions of (2.49) for each facet, and $(p - 1)(p - 1)/2$ extra volume functions of (2.51).

2.5.2   Three-dimensional field space

The three-dimensional field space is the tangentially continuous vector (TV) space. Like in the case of two-dimensional interpolation, the construction of the pth-order, three-dimensional field space will utilize the functions of the pth-order scalar space (2.44). More specifically, the pth-order TV space is obtained by multiplying each of the functions in

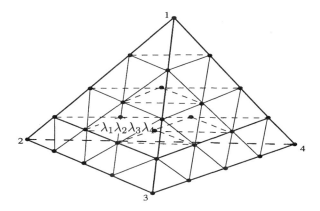

Figure 2.7   The node representations for volume-type scalar basis functions in $W_{s,v}^p$, $(p = 4)$.

(2.44) by any three of the four vectors $\nabla \lambda_i, i = 1, 2, 3, 4$. It follows immediately that the dimension of the pth-order vector space is $(p + 1)(p + 2)(p + 3)/2$.

At this point it is useful to discuss the properties of the gradient of the hat function $\nabla \lambda_k$, $k = 1, 2, 3, 4$. It is straightforward to show that [39]

$$\nabla \lambda_k = \frac{\vec{A}_k}{3V}, \tag{2.53}$$

where the length of the vector $\vec{A}_k$ equals the area of facet (l, m, n) and its direction is perpendicular to the facet and pointing into the tetrahedron. V denotes the volume of the tetrahedron. The dot product of $\nabla \lambda_k$ with any of the three edge vectors $\vec{l}_{l,k}$, $\vec{l}_{m,k}$, and $\vec{l}_{n,k}$ ending at vertex k, is equal to unity,

$$\nabla \lambda_k \cdot \vec{l}_{l,k} = \nabla \lambda_k \cdot \vec{l}_{m,k} = \nabla \lambda_k \cdot \vec{l}_{n,k} = 1. \tag{2.54}$$

The vector basis functions induced from the scalar ones through the aforementioned multiplication with the gradient of any three of the four hat functions can be separated into three subspaces, namely, the edge-type subspace $W_{tv,e}^p$, the facet-type subspace $W_{tv,f}^p$, and the volume-type subspace $W_{tv,v}^p$. Hence, we write

$$W_{tv}^p = W_{tv,e}^p \oplus W_{tv,f}^p \oplus W_{tv,v}^p. \tag{2.55}$$

The vector basis functions in the edge-type subspace $W_{tv,e}^p$ have tangential components along the edges of the tetrahedron. The pertinent mathematical expression of these functions is

$$W_{tv,e}^p = \left\{ \begin{array}{l} g_1^{p-1}(\lambda_m, \lambda_n) \lambda_m \nabla \lambda_n \\ g_2^{p-1}(\lambda_m, \lambda_n) \lambda_n \nabla \lambda_m \end{array} \middle| (m, n) = \left\{ \begin{array}{l} (1, 2), (1, 3), (1, 4) \\ (2, 3), (2, 4), (3, 4) \end{array} \right\} \right\}, \tag{2.56}$$

where g_1^{p-1} and g_2^{p-1} are two arbitrary $(p-1)$th-order polynomials of λ_m and λ_n. Clearly, these vector basis functions have nonzero tangential components only along edge (m, n). They are shown in Fig. 2.8.

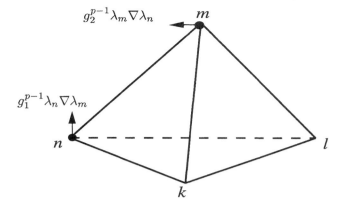

Figure 2.8   Vector basis functions in the edge-type TV subspace associated with edge (m, n).

The vector basis functions in the facet-type subspace $W_{tv,f}^p$ have tangential components on the facets of the tetrahedron. Their general forms are

$$W_{tv,f}^p = \left\{ \begin{array}{l} g_1^{p-2}(\lambda_m, \lambda_n, \lambda_k)\lambda_m\lambda_n\nabla\lambda_k \\ g_2^{p-2}(\lambda_m, \lambda_n, \lambda_k)\lambda_n\lambda_k\nabla\lambda_m \\ g_3^{p-2}(\lambda_m, \lambda_n, \lambda_k)\lambda_k\lambda_m\nabla\lambda_n \end{array} \middle| (m, n, k) = \left\{ \begin{array}{l} (1, 2, 3), (1, 2, 4) \\ (1, 3, 4), (2, 3, 4) \end{array} \right\} \right\},$$
(2.57)

where g_1^{p-2}, g_2^{p-2} and g_3^{p-2} are three arbitrary $(p-2)$th-order polynomials of λ_m, λ_n and λ_k. The vector basis functions only have nonzero tangential components on facet (m, n, k). They are depicted in Fig. 2.9.

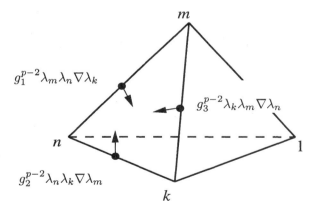

Figure 2.9   Vector basis functions in the facet-type TV subspace associated with facet (m, n, k).

The vector basis functions in the volume-type subspace $W_{tv,v}^p$ have zero tangential components on all the edges and facets of the tetrahedron. Their general forms are

$$W_{tv,v}^p = \left\{ \begin{array}{l} g_1^{p-3}(\lambda_1, \lambda_2, \lambda_3, \lambda_4)\lambda_2\lambda_3\lambda_4\nabla\lambda_1 \\ g_2^{p-3}(\lambda_1, \lambda_2, \lambda_3, \lambda_4)\lambda_1\lambda_3\lambda_4\nabla\lambda_2 \\ g_3^{p-3}(\lambda_1, \lambda_2, \lambda_3, \lambda_4)\lambda_1\lambda_2\lambda_4\nabla\lambda_3 \\ g_4^{p-3}(\lambda_1, \lambda_2, \lambda_3, \lambda_4)\lambda_1\lambda_2\lambda_3\nabla\lambda_4 \end{array} \right\},$$
(2.58)

where g_1^{p-3}, g_2^{p-3}, g_3^{p-3} and g_4^{p-3} are four arbitrary $(p-2)$th-order polynomials of λ_1, λ_2, λ_3 and λ_4. The vector basis functions in the volume-type subspace are depicted in Fig. 2.10.

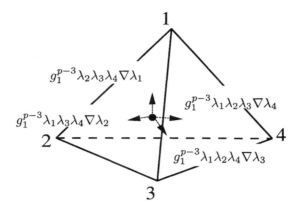

Figure 2.10   Vector basis functions in the volume-type TV subspace.

From our discussion of the development of TV field spaces for two-dimensional finite element interpolations we know that, besides the decomposition of (2.55), a decomposition of the TV space into a gradient subspace (null space of the curl operator) and a non-gradient subspace (range space of the curl operator) is possible,

$$W_{tv}^p = \nabla W_s^{p+1} \oplus W_{tv,ng}^p. \tag{2.59}$$

In this last expression, ∇W_s^{p+1} denotes the gradient of the $(p+1)$th-order scalar space, and $W_{tv,ng}^p$ is the pth-order non-gradient TV subspace. As in the two-dimensional case, the gradient vector basis functions are used to represent the contribution to the field intensities that is due to charges, while the non-gradient basis functions model the contribution to field intensities that is due to currents.

Combining the two decompositions (2.55) and (2.59) we may cast the complete decomposition of the TV space in the following fashion,

$$W_{tv}^p = \underbrace{W_{tv,e,ng}^p \oplus \nabla W_{s,e}^{p+1}}_{W_{tv,e}^p} \oplus \underbrace{W_{tv,f,ng}^p \oplus \nabla W_{s,f}^{p+1}}_{W_{tv,f}^p} \oplus \underbrace{W_{tv,v,ng}^p \oplus \nabla W_{s,v}^{p+1}}_{W_{tv,v}^p}. \tag{2.60}$$

In the subsequent sections, we undertake the systematic construction of the hierarchical TV space and its subspaces. The process we follow begins with the multiplication of each one of the scalar basis functions in W_s^p by any three of the four vectors $\nabla \lambda_i$, $i = 1, 2, 3, 4$. Subsequently, the decompositions (2.55) and (2.59) are properly employed to separate the vector basis functions into appropriate subspaces.

2.5.2.1   *Edge-type field subspace*

We consider first the scalar basis functions in the node-type and the edge-type scalar subspaces $W_{s,n}^1$ and $W_{s,e}^p$. Each one of the four scalar node basis functions λ_k in (2.46) is extended to three vector basis functions by multiplying with $\nabla \lambda_l$, $\nabla \lambda_m$, and $\nabla \lambda_n$, where $k \neq \{l, m, n\}$. The resulting vector basis functions are

$$\lambda_1 \to \begin{pmatrix} \lambda_1 \nabla \lambda_2 \\ \lambda_1 \nabla \lambda_3 \\ \lambda_1 \nabla \lambda_4 \end{pmatrix}, \; \lambda_2 \to \begin{pmatrix} \lambda_2 \nabla \lambda_1 \\ \lambda_2 \nabla \lambda_3 \\ \lambda_2 \nabla \lambda_4 \end{pmatrix}, \; \lambda_3 \to \begin{pmatrix} \lambda_3 \nabla \lambda_1 \\ \lambda_3 \nabla \lambda_2 \\ \lambda_3 \nabla \lambda_4 \end{pmatrix}, \; \lambda_4 \to \begin{pmatrix} \lambda_4 \nabla \lambda_1 \\ \lambda_4 \nabla \lambda_2 \\ \lambda_4 \nabla \lambda_3 \end{pmatrix}. \tag{2.61}$$

In a similar fashion, each one of the six sets of scalar edge basis functions $e^p_{k,l}$ in (2.48) is extended to vector basis functions by multiplying it with $\nabla\lambda_l$, $\nabla\lambda_m$ and $\nabla\lambda_n$. The vector basis functions induced from the scalar edge basis functions are

$$
e^p_{1,2} \rightarrow \begin{pmatrix} e^p_{1,2}\nabla\lambda_2 \\ (e^p_{1,2}\nabla\lambda_3) \\ (e^p_{1,2}\nabla\lambda_4) \end{pmatrix}, \; e^p_{1,3} \rightarrow \begin{pmatrix} e^p_{1,3}\nabla\lambda_3 \\ (e^p_{1,3}\nabla\lambda_2) \\ (e^p_{1,3}\nabla\lambda_4) \end{pmatrix}, \; e^p_{1,4} \rightarrow \begin{pmatrix} e^p_{1,4}\nabla\lambda_4 \\ (e^p_{1,4}\nabla\lambda_2) \\ e^p_{1,4}\nabla\lambda_3 \end{pmatrix},
$$
$$
e^p_{2,3} \rightarrow \begin{pmatrix} e^p_{2,3}\nabla\lambda_3 \\ (e^p_{2,3}\nabla\lambda_1) \\ (e^p_{2,3}\nabla\lambda_4) \end{pmatrix}, \; e^p_{2,4} \rightarrow \begin{pmatrix} e^p_{2,4}\nabla\lambda_4 \\ (e^p_{2,4}\nabla\lambda_1) \\ (e^p_{2,4}\nabla\lambda_3) \end{pmatrix}, \; e^p_{3,4} \rightarrow \begin{pmatrix} e^p_{3,4}\nabla\lambda_4 \\ (e^p_{3,4}\nabla\lambda_1) \\ (e^p_{3,4}\nabla\lambda_2) \end{pmatrix}.
$$

$$(2.62)$$

The functions in parentheses are those that do not fit the general form of edge-type basis functions (2.56). They are facet-type basis functions and will not be included in the assembly of the edge-type TV subspace $W^p_{tv,e}$. Combining the vector basis functions in (2.61) and the six legitimate edge-type functions in (2.62), we obtain the complete form of the edge-type TV subspace $W^p_{tv,e}$,

$$
W^p_{tv,e} = \left\{ \begin{array}{cccc} \{\lambda_1\nabla\lambda_2\} & \lambda_2\nabla\lambda_1 & \lambda_3\nabla\lambda_1 & \lambda_4\nabla\lambda_1 \\ \{\lambda_1\nabla\lambda_3\} & \{\lambda_2\nabla\lambda_3\} & \lambda_3\nabla\lambda_2 & \lambda_4\nabla\lambda_2 \\ \{\lambda_1\nabla\lambda_4\} & \{\lambda_2\nabla\lambda_4\} & \{\lambda_3\nabla\lambda_4\} & \lambda_4\nabla\lambda_3 \\ \{e^p_{1,2}\nabla\lambda_2\} & \{e^p_{1,3}\nabla\lambda_3\} & \{e^p_{1,4}\nabla\lambda_4\} & \\ \{e^p_{2,3}\nabla\lambda_3\} & \{e^p_{2,4}\nabla\lambda_4\} & \{e^p_{3,4}\nabla\lambda_4\} & \end{array} \right\},
$$
$$\text{Dim}(W^p_{tv,e}) = 12 + 6(p-1) = 6(p+1).$$

$$(2.63)$$

The reason for enclosing some of the functions in brackets has to do with our desire to decompose the space into two subspaces, namely, a non-gradient and a gradient subspace. This is done in a manner identical to the one used earlier for the development of spaces for two-dimensional finite element interpolations.

Since a gradient subspace of order p will be constructed from the gradients of scalar functions of order $p+1$, the notation $\nabla W^{p+1}_{s,e}$ will be used for its representation. Thus the desired decomposition is written as $W^p_{tv,e} = W^p_{tv,e,ng} \oplus \nabla W^{p+1}_{s,e}$, where it is

$$\nabla W^{p+1}_{s,e} = \{\nabla e^{p+1}_{1,2}, \; \nabla e^{p+1}_{1,3}, \; \nabla e^{p+1}_{1,4}, \; \nabla e^{p+1}_{2,3}, \; \nabla e^{p+1}_{2,4}, \; \nabla e^{p+1}_{3,4}\}. \qquad (2.64)$$

For brevity we consider the elements resulting from $\nabla e^{p+1}_{1,2}$ only,

$$\nabla e^{p+1}_{1,2} = \{\lambda_2\nabla\lambda_1 + \lambda_1\nabla\lambda_2, \; \cdots, \; \lambda^p_2\nabla\lambda_1 + p\lambda_1\lambda^{p-1}_2\nabla\lambda_2\}. \qquad (2.65)$$

To remove the subspace of $\nabla e^{p+1}_{1,2}$ from $W^p_{tv,e}$, we need to remove $\lambda_1\nabla\lambda_2$ and $e^p_{1,2}\nabla\lambda_2$ from (2.63). The subspaces induced by the remaining five gradients in (2.63) are removed in a similar manner. The removed vector basis functions are shown enclosed in brackets in (2.63). After the removal, the non-gradient edge-type TV subspace results,

$$W^1_{tv,e,ng} = \{\lambda_2\nabla\lambda_1, \; \lambda_3\nabla\lambda_1, \; \lambda_4\nabla\lambda_1, \; \lambda_3\nabla\lambda_2, \; \lambda_4\nabla\lambda_2, \; \lambda_4\nabla\lambda_3\}. \qquad (2.66)$$

As can be seen, the non-gradient, edge-type TV subspace has only six first-order polynomial vector bases.

Next, we would like to relate the first-order, edge-type TV subspace with the popular 3D edge elements that have become the basis functions of choice in the finite element approximation of vector field problems [4]. We begin by writing the first-order, edge-type TV subspace in terms of its decomposition in gradient and non-gradient subspaces,

$$
\begin{aligned}
W_{tv,e}^1 &= \nabla W_{s,e}^2 \oplus W_{tv,e,ng}^1 \\
&= \left\{ \begin{array}{llllll} \nabla(\lambda_1\lambda_2), & \nabla(\lambda_1\lambda_3), & \nabla(\lambda_1\lambda_4), & \nabla(\lambda_2\lambda_3), & \nabla(\lambda_2\lambda_4), & \nabla(\lambda_3\lambda_4), \\ \lambda_2\nabla\lambda_1, & \lambda_3\nabla\lambda_1, & \lambda_4\nabla\lambda_1, & \lambda_3\nabla\lambda_2, & \lambda_4\nabla\lambda_2, & \lambda_4\nabla\lambda_3, \end{array} \right\}.
\end{aligned} \tag{2.67}
$$

Next, a linear combination of elements from $\nabla W_{s,e}^2$ with elements from $W_{tv,e,ng}^1$ is used to generate a new form of $W_{tv,e,ng}^1$. Multiplication of each function in $W_{tv,e,ng}^1$ by two and subtraction of the result from the corresponding basis function in $\nabla W_{s,e}^2$ yields the following new form of $W_{tv,e,ng}^1$

$$
W_{tv,e,ng}^1 = \left\{ \begin{array}{lll} \underbrace{\lambda_1\nabla\lambda_2 - \lambda_2\nabla\lambda_1}_{\vec{e}_{1,2}} & \underbrace{\lambda_1\nabla\lambda_3 - \lambda_3\nabla\lambda_1}_{\vec{e}_{1,3}} & \underbrace{\lambda_1\nabla\lambda_4 - \lambda_4\nabla\lambda_1}_{\vec{e}_{1,4}} \\[2ex] \underbrace{\lambda_2\nabla\lambda_3 - \lambda_3\nabla\lambda_2}_{\vec{e}_{2,3}} & \underbrace{\lambda_2\nabla\lambda_4 - \lambda_4\nabla\lambda_2}_{\vec{e}_{2,4}} & \underbrace{\lambda_3\nabla\lambda_4 - \lambda_4\nabla\lambda_3}_{\vec{e}_{3,4}} \end{array} \right\}. \tag{2.68}
$$

The resulting set of vector basis functions is recognized as the popular set of 3D edge-elements. The indicated notation $\vec{e}_{m,n}$ is the one used in the literature. The function $\vec{e}_{m,n}$ has a constant tangential component along the edge (m, n) of the tetrahedron. More specifically, its dot product with the edge-vector $\vec{l}_{m,n}$, directed from node m to n, yields

$$
\vec{e}_{m,n} \cdot \vec{l}_{m,n} = (\lambda_m\nabla\lambda_n - \lambda_n\nabla\lambda_m) \cdot \vec{l}_{m,n} = 1, \quad \text{on edge}(m, n). \tag{2.69}
$$

Hence, the six edge-elements are indicated by arrows along the associated edges of the tetrahedron as depicted in Fig. 2.11.

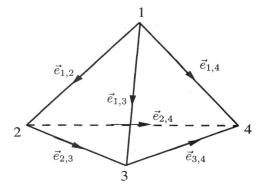

Figure 2.11   The six edge-elements.

There is one more important comment that needs to be made at this point. In our construction of the edge-type TV subspace we made use of the gradient of the functions in the edge-type scalar subspace. No mention was made of the gradient of the elements of the node-type scalar space $\nabla W_{s,n}^1$. To explore this point, consider the form of the elements of $\nabla W_{s,n}^1$ (2.46). It is immediately apparent that their gradient can be recovered through linear combinations of the elements of $W_{tv,e}^1$ (2.67). However, with the 3D edge-element space of (2.68) in place, it is straightforward to show that the space $\nabla W_{s,n}^1$ is solely contained in

$W_{tv,e,ng}^1$. More specifically, $\nabla\lambda_m$, $m = 1, 2, 3, 4$, is obtained from the linear combination of the three edge-elements pointing away from the node m. In matrix form,

$$[\nabla\lambda_1, \nabla\lambda_2, \nabla\lambda_3, \nabla\lambda_4]$$

$$= [\vec{e}_{1,2}, \vec{e}_{1,3}, \vec{e}_{1,4}, \vec{e}_{2,3}, \vec{e}_{2,4}, \vec{e}_{3,4}]\begin{bmatrix} -1 & 1 & & & \\ -1 & & 1 & & \\ -1 & & & 1 & \\ & -1 & 1 & & \\ & -1 & & 1 & \\ & & -1 & 1 \end{bmatrix}. \tag{2.70}$$

Hence, we write

$$\nabla W_{s,n}^1 \subset W_{tv,e,ng}^1. \tag{2.71}$$

In conclusion, contrary to the old form of (2.66), the new form of $W_{tv,e,ng}^1$, namely, the 3D edge-element space, contains not only the first-order, non-gradient basis functions, but also the zeroth-order gradient basis functions. This is an important advantage in the quest for the removal of spurious solutions in the finite element approximation of three-dimensional vector fields.

2.5.2.2 *Facet-type field subspace*

The construction of the facet-type TV subspace begins with the multiplication of the functions $f_{l,m,n}^p$ in the facet-type scalar subspace $W_{s,f}^p$ with $\nabla\lambda_l$, $\nabla\lambda_n$ and $\nabla\lambda_k$,

$$f_{1,2,3}^p \to \begin{pmatrix} f_{1,2,3}^p\nabla\lambda_1 \\ f_{1,2,3}^p\nabla\lambda_3 \\ (f_{1,2,3}^p\nabla\lambda_4) \end{pmatrix}, \quad f_{1,2,4}^p \to \begin{pmatrix} f_{1,2,4}^p\nabla\lambda_1 \\ f_{1,2,4}^p\nabla\lambda_4 \\ (f_{1,2,4}^p\nabla\lambda_3) \end{pmatrix},$$

$$f_{1,3,4}^p \to \begin{pmatrix} f_{1,3,4}^p\nabla\lambda_1 \\ f_{1,3,4}^p\nabla\lambda_4 \\ (f_{1,3,4}^p\nabla\lambda_2) \end{pmatrix}, \quad f_{2,3,4}^p \to \begin{pmatrix} f_{2,3,4}^p\nabla\lambda_2 \\ f_{2,3,4}^p\nabla\lambda_3 \\ (f_{2,3,4}^p\nabla\lambda_1) \end{pmatrix}. \tag{2.72}$$

The functions in parentheses in (2.72) do not fit the general form of the facet-type basis functions in (2.57). Rather, they are volume-type functions. The combination of the functions in parentheses in (2.62) with the eight (the two top functions from each column) facet-type functions in (2.72), yields a complete set of the functions that constitute the facet-type TV subspace $W_{tv,f}^p$,

$$W_{tv,f}^p = \begin{Bmatrix} \{e_{1,2}^p\nabla\lambda_3\} & e_{1,3}^p\nabla\lambda_2 & e_{1,4}^p\nabla\lambda_2 & f_{1,2,3}^p\nabla\lambda_1 & f_{1,2,4}^p\nabla\lambda_1 \\ \{e_{1,2}^p\nabla\lambda_4\} & \{e_{1,3}^p\nabla\lambda_4\} & e_{1,4}^p\nabla\lambda_1 & \{f_{1,2,3}^p\nabla\lambda_3\} & \{f_{1,2,4}^p\nabla\lambda_4\} \\ e_{2,3}^p\nabla\lambda_1 & e_{2,4}^p\nabla\lambda_1 & e_{3,4}^p\nabla\lambda_1 & f_{1,3,4}^p\nabla\lambda_1 & f_{2,3,4}^p\nabla\lambda_2 \\ \{e_{2,3}^p\nabla\lambda_4\} & e_{2,4}^p\nabla\lambda_3 & e_{3,4}^p\nabla\lambda_2 & \{f_{1,3,4}^p\nabla\lambda_4\} & \{f_{2,3,4}^p\nabla\lambda_4\} \end{Bmatrix},$$

$$\text{Dim}(W_{tv,f}^p) = 12(p-1) + 8\frac{(p-1)(p-2)}{4}$$

$$= 4(p-1)(p+1). \tag{2.73}$$

The reason for enclosing some of the functions in the above expression in brackets is our desire to decompose the space $W_{tv,f}^p$ into a gradient and a non-gradient subspace, $W_{tv,f}^p = \nabla W_{s,f}^{p+1} \oplus W_{tv,f,ng}^p$. The procedure follows the one used earlier. First, we consider the subspace $\nabla W_{s,f}^{p+1}$, the elements of which are the gradients of the basis functions in $W_{s,f}^{p+1}$,

$$\nabla W_{s,f}^{p+1} = \left\{ \nabla f_{1,2,3}^{p+1}, \ \nabla f_{1,2,4}^{p+1}, \ \nabla f_{1,3,4}^{p+1}, \ \nabla f_{2,3,4}^{p+1} \right\}. \tag{2.74}$$

Consider the function $\nabla f_{1,2,3}^{p+1}$. In view of (2.49) it is

$$\nabla f_{1,2,3}^{p+1} = \left\{ \lambda_2^i \lambda_3^j \nabla \lambda_1 + i \lambda_1 \lambda_2^{i-1} \lambda_3^j \nabla \lambda_2 + j \lambda_1 \lambda_2^i \lambda_3^{j-1} \nabla \lambda_3 \ \middle| \ \begin{array}{l} i = 1, 2, \cdots, p-1 \\ j = 1, 2, \cdots, p-i \end{array} \right\}. \tag{2.75}$$

In view of the expressions (2.47) and (2.49) for $e_{m,n}^p$ and $f_{l,m,n}^p$, it is rather straightforward to see that, in order to remove the functions in the subspace $\nabla f_{1,2,3}^{p+1}$ from $W_{tv,f}^p$ it suffices to remove only the following functions:

$$\lambda_1 \lambda_2^i \lambda_3^{j-1} \nabla \lambda_3, \quad (i = 1, 2, \cdots, p-1, j = 1, 2, \cdots, p-i). \tag{2.76}$$

These are nothing else but $f_{1,2,3}^p \nabla \lambda_3$ and $e_{1,2}^p \nabla \lambda_3$, which are two of the bracketed terms in (2.73). The remaining six bracketed terms correspond to the removal of the remaining three sets of gradient functions in (2.75), following a similar reasoning as the one above for the removal of $\nabla f_{1,2,3}^{p+1}$. Once removed, the remaining basis functions form the non-gradient, facet-type TV subspace $\vec{W}_{tv,f,ng}^p$

$$W_{tv,f,ng}^p = \left\{ e_{l,n}^p \nabla \lambda_m, e_{n,m}^p \nabla \lambda_l, f_{l,m,n}^p \nabla \lambda_l \ \middle| \ (l, m, n) = \left\{ \begin{array}{l} (1,2,3), (2,3,4) \\ (1,2,4), (2,3,4) \end{array} \right\} \right\},$$

$$\mathrm{Dim}(W_{tv,f,ng}^p) = 8(p-1) + 4\frac{(p-2)(p-1)}{2}$$

$$= 4\frac{(p-1)(p+2)}{2}. \tag{2.77}$$

This completes the decomposition of the facet-type TV subspace in its gradient and its non-gradient subspaces. In summary, we write

$$W_{tv,f}^p = \nabla W_{s,f}^{p+1} \oplus W_{tv,f,ng}^p,$$

$$\mathrm{Dim}(\nabla W_{s,f}^{p+1}) = 4\frac{(p-1)p}{2}, \tag{2.78}$$

$$\mathrm{Dim}(W_{tv,f,ng}^p) = 4\frac{(p-1)(p+2)}{2}.$$

2.5.2.3 *Volume-type field subspace*

The construction of the volume-type TV subspace, $W_{tv,v}^p$, we consider the scalar basis functions $v_{1,2,3,4}^p$ of $W_{s,v}^p$. They are extended to vector basis functions by multiplying them with $\nabla \lambda_1$, $\nabla \lambda_2$ and $\nabla \lambda_4$,

$$v_{1,2,3,4}^p \quad \rightarrow \quad \begin{pmatrix} v_{1,2,3,4}^p \nabla \lambda_1 \\ v_{1,2,3,4}^p \nabla \lambda_2 \\ v_{1,2,3,4}^p \nabla \lambda_4 \end{pmatrix}. \tag{2.79}$$

The resulting vector basis functions are combined with the terms in parentheses in (2.72) to yield the complete form for the volume-type TV subspace $W_{tv,v}^p$,

$$W_{tv,v}^p = \left\{ \begin{array}{ccc} \{f_{1,2,3}^p \nabla \lambda_4\} & f_{1,2,4}^p \nabla \lambda_3 & v_{1,2,3,4}^p \nabla \lambda_1 \\ f_{1,3,4}^p \nabla \lambda_2 & f_{2,3,4}^p \nabla \lambda_1 & v_{1,2,3,4}^p \nabla \lambda_2 \\ & & \{v_{1,2,3,4}^p \nabla \lambda_4\} \end{array} \right\},$$

$$\text{Dim}(W_{tv,v}^p) = 4\frac{(p-1)(p-2)}{2} + 3\frac{(p-3)(p-2)(p-1)}{6} \qquad (2.80)$$

$$= \frac{(p-1)(p-2)(p+1)}{2}.$$

Again, the terms in brackets are the ones that must be removed in order to obtain the rotational (non-gradient) subspace from $W_{tv,v,}^p$, . More specifically, let us denote the decomposition of $W_{tv,v}^p$ into its gradient and non-gradient subspaces as follows, $W_{tv,v}^p = \nabla W_{s,v}^{p+1} \oplus W_{tv,v,ng}^p$. The elements of $\nabla W_{s,v}^{p+1}$ are easily obtained as the gradients of the functions $v_{1,2,3,4}^{p+1}$,

$$\nabla W_{s,v}^{p+1} =$$

$$\left\{ \begin{array}{l} \lambda_2^i \lambda_3^j \lambda_4^k \nabla \lambda_1 + i\lambda_1 \lambda_2^{i-1} \lambda_3^j \lambda_4^k \nabla \lambda_2 \\ + j\lambda_1 \lambda_2^i \lambda_3^{j-1} \lambda_4^k \nabla \lambda_3 + k\lambda_1 \lambda_2^i \lambda_3^j \lambda_4^{k-1} \nabla \lambda_4 \end{array} \left| \begin{array}{l} i = 1, 2, \cdots, p-2, \\ j = 1, 2, \cdots, p-1-i, \\ k = 1, 2, \cdots, p-i-j \end{array} \right. \right\}. \qquad (2.81)$$

Hence, for the construction of the subspace $W_{tv,v,ng}^p$ we must remove from $W_{tv,v}^p$ the functions

$$\lambda_1 \lambda_2^i \lambda_3^j \lambda_4^{k-1} \nabla \lambda_4. \qquad (2.82)$$

These are $f_{1,2,3}^p \nabla \lambda_4$ and $v_{1,2,3,4}^p \nabla \lambda_4$, which are the bracketed terms in (2.80). Once removed, the subspace $W_{tv,v,ng}^p$ is obtained

$$W_{tv,v,ng}^p = \left\{ \begin{array}{cc} f_{2,3,4}^p \nabla \lambda_1 & v_{1,2,3,4}^p \nabla \lambda_1 \\ f_{1,3,4}^p \nabla \lambda_2 & v_{1,2,3,4}^p \nabla \lambda_2 \\ f_{1,2,4}^p \nabla \lambda_3 & \end{array} \right\},$$

$$\text{Dim}(W_{tv,v,ng}^p) = 3\frac{(p-1)(p-2)}{2} + 2\frac{(p-3)(p-2)(p-1)}{6} \qquad (2.83)$$

$$= \frac{(p-1)(p-2)(2p+3)}{6}.$$

This completes the decomposition of the volume-type TV subspace into its gradient and non-gradient subspaces. In summary, we write

$$W_{tv,v}^p = \nabla W_{s,v}^{p+1} \oplus W_{tv,v,ng}^p,$$

$$\text{Dim}(\nabla W_{s,v}^{p+1}) = \frac{(p-1)(p-1)p}{6}, \qquad (2.84)$$

$$\text{Dim}(W_{tv,f,ng}^p) = \frac{(p-2)(p-1)(2p+3)}{6}.$$

2.5.3 Three-dimensional flux space

The three-dimensional flux space is the normally continuous vector (NV) space. The dimension of the pth-order NV space is three times the dimension of the pth-order scalar space, $(p+1)(p+2)(p+3)/2$. The vector basis functions in NV space can be induced from the scalar ones through the multiplication of the latter by three of the six vectors, $\nabla\lambda_m \times \nabla\lambda_n$, $(m, n) = \{(1,2), (1,3), (1,4), (2,3), (2,4), (3,4)\}$. A useful result is the following:

$$\nabla\lambda_m \times \nabla\lambda_n = \frac{1}{6V}s(-1)^{k+l}\vec{l}_{k,l}, \quad (m < n \text{ and } k < l)$$

$$s = \mathrm{sgn}(\vec{l}_{2,3} \times \vec{l}_{2,4} \cdot \vec{l}_{2,1}),$$

(2.85)

where $\vec{l}_{k,l}$ is the edge-vector from vertex k to vertex q and V is the volume of the tetrahedron. The sign s is determined by the orientation of the tetrahedron. For example, for the tetrahedron depicted on the left of Fig. 2.12, it is $s = +1$; hence, the six vectors of (2.85) become

$$\nabla\lambda_1 \times \nabla\lambda_2 = -\frac{1}{6V}\vec{l}_{3,4}$$

$$\nabla\lambda_1 \times \nabla\lambda_3 = \frac{1}{6V}\vec{l}_{2,4}$$

$$\nabla\lambda_1 \times \nabla\lambda_4 = -\frac{1}{6V}\vec{l}_{2,3}$$

$$\nabla\lambda_2 \times \nabla\lambda_3 = -\frac{1}{6V}\vec{l}_{1,4}$$

(2.86)

$$\nabla\lambda_2 \times \nabla\lambda_4 = \frac{1}{6V}\vec{l}_{1,3}$$

$$\nabla\lambda_3 \times \nabla\lambda_4 = -\frac{1}{6V}\vec{l}_{1,2}.$$

On the other hand, for the tetrahedron depicted on the right of Fig. 2.12, it is $s = -1$; hence, the six vectors of (2.85) are given by

$$\nabla\lambda_1 \times \nabla\lambda_2 = \frac{1}{6V}\vec{l}_{3,4}$$

$$\nabla\lambda_1 \times \nabla\lambda_3 = -\frac{1}{6V}\vec{l}_{2,4}$$

$$\nabla\lambda_1 \times \nabla\lambda_4 = \frac{1}{6V}\vec{l}_{2,3}$$

$$\nabla\lambda_2 \times \nabla\lambda_3 = \frac{1}{6V}\vec{l}_{1,4}$$

(2.87)

$$\nabla\lambda_2 \times \nabla\lambda_4 = -\frac{1}{6V}\vec{l}_{1,3}$$

$$\nabla\lambda_3 \times \nabla\lambda_4 = \frac{1}{6V}\vec{l}_{1,2}.$$

Since the vectors of (2.85) are directed along the edges of the tetrahedron, a convenient visual representation is possible by assigning them to the corresponding edges as depicted in Fig. 2.12. The direction of the cross-products in (2.85) can be determined using the right hand rule. For example, for $\nabla\lambda_m \times \nabla\lambda_n$, if the fingers rotate in the direction from node m to n, the thumb points in the direction of the cross-product.

The three vectors from (2.85) chosen for the construction of the flux space must be linearly independent. This constraint is satisfied by making sure that the three vectors do not lie on the same facet of the tetrahedron. This is easily seen by recognizing that the three

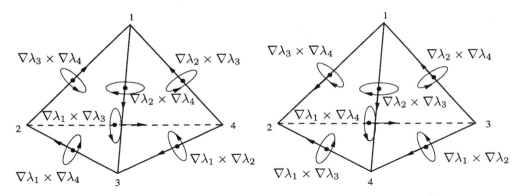

Figure 2.12   The six cross-products of (2.85) depicted for two different orientations of the tetrahedron.

vectors ($\nabla\lambda_m \times \nabla\lambda_n$, $\nabla\lambda_m \times \nabla\lambda_k$, and $\nabla\lambda_m \times \nabla\lambda_l$) lie on the facet opposite to vertex m and it is

$$\nabla\lambda_m \times \nabla\lambda_n + \nabla\lambda_m \times \nabla\lambda_k + \nabla\lambda_m \times \nabla\lambda_l = -\nabla\lambda_m \times \nabla\lambda_m = 0. \qquad (2.88)$$

The vector basis functions induced from the scalar ones can be decomposed into two subspaces, namely, the facet-type subspace, $W_{nv,f}^p$, and the volume-type subspace, $W_{nv,v}^p$; hence, we write

$$W_{nv}^p = W_{nv,f}^p \oplus W_{nv,v}^p. \qquad (2.89)$$

The vector functions in the facet-type subspace $W_{nv,f}^p$ have normal components to a facet of the tetrahedron. More specifically, their general form is

$$W_{nv,f}^p = \left\{ \begin{matrix} g_1^{p-1}(\lambda_m,\lambda_n,\lambda_k)\lambda_m\nabla\lambda_n \times \nabla\lambda_k \\ g_2^{p-1}(\lambda_m,\lambda_n,\lambda_k)\lambda_n\nabla\lambda_k \times \nabla\lambda_m \\ g_3^{p-1}(\lambda_m,\lambda_n,\lambda_k)\lambda_k\nabla\lambda_m \times \nabla\lambda_n \end{matrix} \middle| (m,n,k) = \left\{ \begin{matrix} (1,2,3),(1,2,4) \\ (1,3,4),(2,3,4) \end{matrix} \right\} \right\}, \qquad (2.90)$$

where g_1, g_2 and g_3 are three arbitrary $(p-1)$th-order polynomials of λ_m, λ_n, and λ_k. These three sets of vector basis functions have normal components to facet (m,n,k), and are depicted in the left drawing of Fig. 2.13.

The vector bases in $W_{nv,v}^p$ have zero normal components on all facets. Their general forms are

$$W_{nv,v}^p = \left\{ g^{p-2}(\lambda_1,\lambda_2,\lambda_3,\lambda_4)\lambda_m\lambda_n\nabla\lambda_k \times \nabla\lambda_l \right\}, \qquad (2.91)$$

where g is a $(p-2)$th-order polynomial of λ_1, λ_2, λ_3, and λ_4. A graphical illustration of the volume-type, NV vector basis functions is given in the right drawing in Fig. 2.13.

Since these functions will be used for the representation of the electric and magnetic flux densities in Maxwell's equations, it is desired to decompose the NV space into a curl subspace (null space of the divergence operator) and a non-curl subspace (range space of divergence operator). For example, in view of the physics of the electromagnetic fields, the curl subspace is suitable for the representation of the magnetic flux density, while the non-curl subspace can be used to represent electrostatic flux densities generated by electrical charge distributions in space. Such a decomposition is written as follows:

$$W_{nv}^p = \nabla \times W_{tv}^{p+1} \oplus W_{nv,nc}^p = \nabla \times W_{tv,ng}^{p+1} \oplus W_{nv,nc}^p, \qquad (2.92)$$

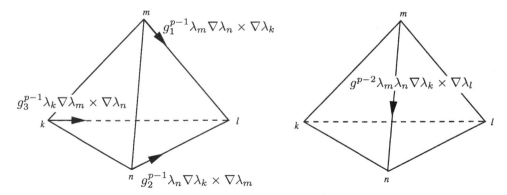

Figure 2.13 Facet-type (*left*) and volume-type (*right*) three-dimensional normally continuous vector (NV) basis functions.

where $\nabla \times W_{tv,ng}^{p+1}$ denotes the space containing the curl of the elements of the $(p+1)$th-order TV space, and $W_{nv,nc}^{p}$ stands for the pth-order, non-curl NV subspace. The two decompositions of (2.89 and 2.92) can be combined to yield a complete decomposition of the NV space as follows:

$$W_{nv}^{p} = \underbrace{W_{nv,f,nc}^{p} \oplus \nabla \times W_{tv,f,ng}^{p+1}}_{W_{nv,f}^{p}} \oplus \underbrace{W_{nv,v,nc}^{p} \oplus \nabla \times W_{tv,v,ng}^{p+1}}_{W_{nv,v}^{p}}. \tag{2.93}$$

In what follows, the systematic procedure used for the development of the hierarchical TV space is utilized for the construction of a hierarchical NV space. The two key steps involved may be summarized as follows. First, each of the scalar basis functions in W_s^p is multiplied by three independent vectors in the set (2.85). Then, use is made of the two decompositions (2.89 and 2.92) to separate the induced vector functions into appropriate subspaces.

2.5.3.1 Facet-type flux subspace

First, let us consider the vector basis functions induced by the node-type scalar space $W_{s,n}^1$, the edge-type scalar space $W_{s,e}^p$, and the facet-type scalar space $W_{s,f}^p$ through the multiplication of their basis functions by the vector functions of (2.85).

The four scalar basis functions of (2.46) induce the following set of vector functions:

$$\lambda_1 \rightarrow \begin{pmatrix} \lambda_1 \nabla\lambda_2 \times \nabla\lambda_3 \\ \lambda_1 \nabla\lambda_3 \times \nabla\lambda_4 \\ \lambda_1 \nabla\lambda_4 \times \nabla\lambda_2 \end{pmatrix}, \qquad \lambda_2 \rightarrow \begin{pmatrix} \lambda_2 \nabla\lambda_1 \times \nabla\lambda_3 \\ \lambda_2 \nabla\lambda_3 \times \nabla\lambda_4 \\ \lambda_2 \nabla\lambda_4 \times \nabla\lambda_1 \end{pmatrix},$$
$$\lambda_3 \rightarrow \begin{pmatrix} \lambda_3 \nabla\lambda_1 \times \nabla\lambda_2 \\ \lambda_3 \nabla\lambda_2 \times \nabla\lambda_4 \\ \lambda_3 \nabla\lambda_4 \times \nabla\lambda_1 \end{pmatrix}, \qquad \lambda_4 \rightarrow \begin{pmatrix} \lambda_4 \nabla\lambda_1 \times \nabla\lambda_2 \\ \lambda_4 \nabla\lambda_2 \times \nabla\lambda_3 \\ \lambda_4 \nabla\lambda_3 \times \nabla\lambda_1 \end{pmatrix}. \tag{2.94}$$

The edge-type scalar basis functions of (2.48) induce the following set of vector functions:

$$
e_{1,2}^p \rightarrow \begin{pmatrix} e_{1,2}^p \nabla \lambda_3 \times \nabla \lambda_1 \\ e_{1,2}^p \nabla \lambda_4 \times \nabla \lambda_1 \\ (e_{1,2}^p \nabla \lambda_4 \times \nabla \lambda_3) \end{pmatrix}, \quad
e_{1,3}^p \rightarrow \begin{pmatrix} e_{1,3}^p \nabla \lambda_3 \times \nabla \lambda_2 \\ e_{1,3}^p \nabla \lambda_3 \times \nabla \lambda_1 \\ (e_{1,3}^p \nabla \lambda_4 \times \nabla \lambda_2) \end{pmatrix},
$$

$$
e_{1,4}^p \rightarrow \begin{pmatrix} e_{1,4}^p \nabla \lambda_4 \times \nabla \lambda_3 \\ e_{1,4}^p \nabla \lambda_4 \times \nabla \lambda_2 \\ (e_{1,4}^p \nabla \lambda_3 \times \nabla \lambda_2) \end{pmatrix}, \quad
e_{2,3}^p \rightarrow \begin{pmatrix} e_{2,3}^p \nabla \lambda_4 \times \nabla \lambda_2 \\ e_{2,3}^p \nabla \lambda_3 \times \nabla \lambda_1 \\ (e_{2,3}^p \nabla \lambda_4 \times \nabla \lambda_1) \end{pmatrix}, \quad (2.95)
$$

$$
e_{2,4}^p \rightarrow \begin{pmatrix} e_{2,4}^p \nabla \lambda_4 \times \nabla \lambda_1 \\ e_{2,4}^p \nabla \lambda_4 \times \nabla \lambda_3 \\ (e_{2,4}^p \nabla \lambda_3 \times \nabla \lambda_1) \end{pmatrix}, \quad
e_{3,4}^p \rightarrow \begin{pmatrix} e_{3,4}^p \nabla \lambda_4 \times \nabla \lambda_1 \\ e_{3,4}^p \nabla \lambda_4 \times \nabla \lambda_2 \\ (e_{3,4}^p \nabla \lambda_2 \times \nabla \lambda_1) \end{pmatrix}.
$$

The terms enclosed in parentheses in the above equation do not fit the general form of the facet-type NV basis functions. Rather, they belong to the volume-type NV subspace.

The scalar facet-type functions of (2.50) induce the following vector functions:

$$
f_{1,2,3}^p \rightarrow \begin{pmatrix} f_{1,2,3}^p \nabla \lambda_3 \times \nabla \lambda_1 \\ f_{1,2,3}^p \nabla \lambda_4 \times \nabla \lambda_1 \\ (f_{1,2,3}^p \nabla \lambda_4 \times \nabla \lambda_2) \end{pmatrix}, \quad
f_{1,2,4}^p \rightarrow \begin{pmatrix} f_{1,2,4}^p \nabla \lambda_4 \times \nabla \lambda_1 \\ f_{1,2,4}^p \nabla \lambda_4 \times \nabla \lambda_3 \\ (f_{1,2,4}^p \nabla \lambda_3 \times \nabla \lambda_1) \end{pmatrix},
$$

$$
f_{1,3,4}^p \rightarrow \begin{pmatrix} f_{1,3,4}^p \nabla \lambda_4 \times \nabla \lambda_1 \\ f_{1,3,4}^p \nabla \lambda_4 \times \nabla \lambda_2 \\ (f_{1,3,4}^p \nabla \lambda_2 \times \nabla \lambda_1) \end{pmatrix}, \quad
f_{2,3,4}^p \rightarrow \begin{pmatrix} f_{2,3,4}^p \nabla \lambda_4 \times \nabla \lambda_2 \\ f_{2,3,4}^p \nabla \lambda_4 \times \nabla \lambda_1 \\ (f_{2,3,4}^p \nabla \lambda_2 \times \nabla \lambda_1) \end{pmatrix}. \quad (2.96)
$$

The terms enclosed in parentheses in the above equation fit the general form of the volume-type NV subspace. Combination of the vector functions in (2.94) with the first two terms in every column in (2.95) and (2.96) yields the facet-type NV subspace. Thus we write

$$
W_{nv,f}^p =
$$

$$
\left\{ \begin{matrix}
\{\lambda_1 \nabla \lambda_2 \times \nabla \lambda_3\} \; \{\lambda_2 \nabla \lambda_1 \times \nabla \lambda_3\} \; \lambda_3 \nabla \lambda_1 \times \nabla \lambda_2 \; \lambda_4 \nabla \lambda_1 \times \nabla \lambda_2 \\
\{\lambda_1 \nabla \lambda_3 \times \nabla \lambda_4\} \; \{\lambda_2 \nabla \lambda_3 \times \nabla \lambda_4\} \; \{\lambda_3 \nabla \lambda_2 \times \nabla \lambda_4\} \; \lambda_4 \nabla \lambda_2 \times \nabla \lambda_3 \\
\{\lambda_1 \nabla \lambda_4 \times \nabla \lambda_2\} \; \{\lambda_2 \nabla \lambda_4 \times \nabla \lambda_1\} \; \{\lambda_3 \nabla \lambda_4 \times \nabla \lambda_1\} \; \lambda_4 \nabla \lambda_3 \times \nabla \lambda_1 \\
[e_{1,2}^p \nabla \lambda_3 \times \nabla \lambda_1] \quad \{e_{1,3}^p \nabla \lambda_3 \times \nabla \lambda_2\} \quad \{e_{1,4}^p \nabla \lambda_4 \times \nabla \lambda_2\} \\
[e_{1,2}^p \nabla \lambda_4 \times \nabla \lambda_1] \quad [e_{1,3}^p \nabla \lambda_4 \times \nabla \lambda_1] \quad \{e_{1,4}^p \nabla \lambda_4 \times \nabla \lambda_3\} \\
[e_{2,3}^p \nabla \lambda_4 \times \nabla \lambda_2] \quad \{e_{2,4}^p \nabla \lambda_4 \times \nabla \lambda_3\} \quad \{e_{3,4}^p \nabla \lambda_4 \times \nabla \lambda_1\} \\
\{e_{2,3}^p \nabla \lambda_3 \times \nabla \lambda_1\} \quad \{e_{2,4}^p \nabla \lambda_4 \times \nabla \lambda_1\} \quad \{e_{3,4}^p \nabla \lambda_4 \times \nabla \lambda_2\} \\
[f_{1,2,3}^p \nabla \lambda_3 \times \nabla \lambda_1] \quad [f_{1,2,4}^p \nabla \lambda_4 \times \nabla \lambda_1] \\
[f_{1,3,4}^p \nabla \lambda_4 \times \nabla \lambda_1] \quad [f_{2,3,4}^p \nabla \lambda_4 \times \nabla \lambda_2]
\end{matrix} \right\} \quad (2.97)
$$

$$
\mathrm{Dim}(W_{nv,v}^p) = 12 + 12(p-1) + 4\frac{(p-2)(p-1)}{2}
$$

$$
= 4\frac{(p+2)(p+1)}{2}.
$$

The reason for enclosing some of the terms in the equation above in parentheses and some in square brackets is related to the assembly of the non-curl and the curl, facet-type NV subspaces. This assembly is discussed next.

In order to build the non-curl, facet-type NV subspace $W_{nv,f,nc}^p$, we must remove the curl of the basis functions of $W_{tv,f,ng}^{p+1}$ from $W_{nv,f}^p$. The space formed by taking the curl of the basis functions of $W_{tv,f,ng}^{p+1}$ is

$$\nabla \times W_{tv,f,ng}^{p+1} = \left\{ \begin{array}{l} \nabla \times e_{m,k}^{p+1} \nabla \lambda_n, \nabla \times e_{n,k}^{p+1} \nabla \lambda_m, \nabla \times f_{m,n,k}^{p+1} \nabla \lambda_m \\ (m,n,k) = \{(1,2,3),(1,3,4),(1,2,4),(2,3,4)\} \end{array} \right\}. \qquad (2.98)$$

The curl of $e_{m,k}^{p+1} \nabla \lambda_n$ and $e_{n,k}^{p+1} \nabla \lambda_m$ yields

$$\begin{aligned}
\nabla \times (e_{m,k}^{p+1} \nabla \lambda_n) &= \lambda_k^i \nabla \lambda_m \times \nabla \lambda_n + i \lambda_m \lambda_k^{i-1} \nabla \lambda_k \times \lambda_n, \\
\nabla \times (e_{n,k}^{p+1} \nabla \lambda_m) &= \lambda_k^i \nabla \lambda_n \times \nabla \lambda_m + i \lambda_n \lambda_k^{i-1} \nabla \lambda_k \times \lambda_m.
\end{aligned} \qquad (2.99)$$

To remove the vector functions resulting from the above two curl operations from $W_{nv,f}^p$, we need to remove the following sets of functions from (2.97):

$$\left\{ \begin{array}{l} \lambda_m \lambda_k^{i-1} \nabla \lambda_k \times \lambda_n \\ \lambda_n \lambda_k^{i-1} \nabla \lambda_k \times \lambda_m \end{array} \middle| \begin{array}{l} i = 1,2,\cdots,p, \\ (m,n,k) = \left\{ \begin{array}{l} (1,2,3),(1,2,4) \\ (1,3,4),(2,3,4) \end{array} \right\} \end{array} \right\}. \qquad (2.100)$$

The removed functions are the ones enclosed in brackets in (2.97). The curl of $f_{m,n,k}^{p+1} \nabla \lambda_m$ in (2.98) is

$$\nabla \times \lambda_m \lambda_n^i \lambda_k^j \nabla \lambda_m = i \lambda_m \lambda_n^{i-1} \lambda_k^j \nabla \lambda_n \times \nabla \lambda_m + j \lambda_m \lambda_n^i \lambda_k^{j-1} \nabla \lambda_k \times \nabla \lambda_m. \qquad (2.101)$$

To remove the vector functions resulting from the above operation from $W_{nv,f}^p$, we need to remove the following set of functions from (2.97)

$$\left\{ \lambda_m \lambda_n^i \lambda_k^{j-1} \nabla \lambda_k \times \nabla \lambda_m, \middle| \begin{array}{l} i = 1,2,\cdots,p-1 \\ j = 1,2,\cdots,p-i \\ (m,n,k) = \left\{ \begin{array}{l} (1,2,3),(1,2,4) \\ (1,3,4),(2,3,4) \end{array} \right\} \end{array} \right\}. \qquad (2.102)$$

The removed functions are the ones enclosed in square brackets in (2.97). After their removal, there are only four first-order vector functions remaining in $W_{nv,f}^p$. They define the first-order, non-curl, facet-type NV subspace

$$W_{nv,f,nc}^1 = \left\{ \begin{array}{l} \lambda_4 \nabla \lambda_3 \times \nabla \lambda_2, \lambda_4 \nabla \lambda_3 \times \nabla \lambda_1, \\ \lambda_4 \nabla \lambda_2 \times \nabla \lambda_1, \lambda_3 \nabla \lambda_2 \times \nabla \lambda_1 \end{array} \right\}, \qquad (2.103)$$

$$\text{Dim}(W_{nv,f,nc}^1) = 4.$$

In view of (2.98), the complete first-order, facet-type, NV subspace is

$$\begin{aligned}
W_{nv,f}^1 &= \nabla \times W_{tv,f,ng}^2 \oplus W_{nv,f,nc}^1 \\
&= \left\{ \begin{array}{l} \nabla \times \lambda_1 \lambda_3 \nabla \lambda_2, \nabla \times \lambda_1 \lambda_4 \nabla \lambda_2, \nabla \times \lambda_1 \lambda_4 \nabla \lambda_3, \nabla \times \lambda_2 \lambda_4 \nabla \lambda_3 \\ \nabla \times \lambda_2 \lambda_3 \nabla \lambda_1, \nabla \times \lambda_2 \lambda_4 \nabla \lambda_1, \nabla \times \lambda_3 \lambda_4 \nabla \lambda_1, \nabla \times \lambda_3 \lambda_4 \nabla \lambda_2 \\ \lambda_3 \nabla \lambda_2 \times \nabla \lambda_1, \lambda_4 \nabla \lambda_2 \times \nabla \lambda_1, \lambda_4 \nabla \lambda_3 \times \nabla \lambda_1, \lambda_4 \nabla \lambda_3 \times \nabla \lambda_2 \end{array} \right\}.
\end{aligned} \qquad (2.104)$$

In the process used above for the construction of the facet-type, non-curl, NV subspace $W_{nv,f,nc}^p$, only the functions obtained by taking the curl of the basis functions of the non-gradient, facet-type, TV subspace $\nabla \times W_{tv,f,ng}^p$ were removed. Hence, we expect that the functions associated with the curl of the basis functions of the non-gradient, edge-type, TV subspace $\nabla \times W^1_{tv,e,ng}$ are contained in $W^1_{nv,f}$ but not in $W^1_{nv,f,nc}$. However, it is possible to introduce an alternative form for the subspace $W^1_{nv,f,nc}$ such that the subspace $\nabla \times W^1_{tv,e,ng}$ is contained solely in it. To do so, let us calculate the elements of $\nabla \times W^1_{tv,e,ng}$. From (2.68) we obtain

$$\nabla \times W^1_{tv,e,ng} = \left\{ \begin{array}{l} \nabla \times \vec{e}_{1,2} = 2\nabla\lambda_1 \times \nabla\lambda_2, \\ \nabla \times \vec{e}_{1,3} = 2\nabla\lambda_1 \times \nabla\lambda_3 \\ \nabla \times \vec{e}_{1,4} = 2\nabla\lambda_1 \times \nabla\lambda_4, \\ \nabla \times \vec{e}_{2,3} = 2\nabla\lambda_2 \times \nabla\lambda_3 \\ \nabla \times \vec{e}_{2,4} = 2\nabla\lambda_2 \times \nabla\lambda_4, \\ \nabla \times \vec{e}_{3,4} = 2\nabla\lambda_3 \times \nabla\lambda_4 \end{array} \right\}. \tag{2.105}$$

Next, a linear combination of the bases in $W^1_{nv,f}$ is used to obtain the following new form of $W^1_{nv,f,nc}$:

$$W^1_{nv,f,nc} = \left\{ \begin{array}{l} \vec{f}_{1,2,3} = 2\left(\lambda_1\nabla\lambda_2 \times \nabla\lambda_3 + \lambda_2\nabla\lambda_3 \times \nabla\lambda_1 + \lambda_3\nabla\lambda_1 \times \nabla\lambda_2\right) \\ \vec{f}_{1,2,4} = 2\left(\lambda_1\nabla\lambda_2 \times \nabla\lambda_4 + \lambda_2\nabla\lambda_4 \times \nabla\lambda_1 + \lambda_4\nabla\lambda_1 \times \nabla\lambda_2\right) \\ \vec{f}_{1,3,4} = 2\left(\lambda_1\nabla\lambda_3 \times \nabla\lambda_4 + \lambda_3\nabla\lambda_4 \times \nabla\lambda_1 + \lambda_4\nabla\lambda_1 \times \nabla\lambda_3\right) \\ \vec{f}_{2,3,4} = 2\left(\lambda_2\nabla\lambda_3 \times \nabla\lambda_4 + \lambda_3\nabla\lambda_4 \times \nabla\lambda_2 + \lambda_4\nabla\lambda_2 \times \nabla\lambda_3\right) \end{array} \right\}. \tag{2.106}$$

Each one of the basis functions $\vec{f}_{n,k,l}$ defined above has constant normal component on facet m. Moreover, its dot-product with the normal area vector $\vec{A}_m$ yields

$$\vec{f}_{n,k,l} \cdot \vec{A}_m = (-1)^{m+1}s, \tag{2.107}$$

where the sign s is determined by the orientation of the tetrahedron as defined earlier in (2.85). We refer to the basis functions in $W^1_{nv,f,nc}$ as *facet elements*. They are represented pictorially by the normal vectors on each facet, as depicted in Fig. 2.14.

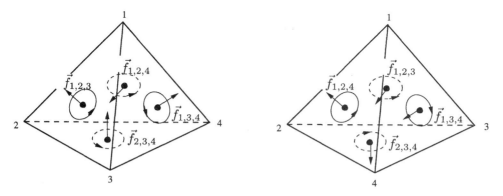

Figure 2.14   The four facet elements for the two different orientations of the tetrahedron.

The direction of the facet elements may also be determined using the right-hand rule. With the fingers in the direction $n \to k \to l$, the thumb points in the direction of the facet element $\vec{f}_{n,k,l}$ as shown in Fig. 2.14.

It is straightforward to show that the functions obtained by taking the curl of the basis functions of $W^1_{tv,f,nc}$ are contained solely in the new form of $W^1_{nv,f,nc}$ in (2.106). Hence, we write

$$\nabla \times W^1_{tv,e,ng} \subset W^1_{nv,f,nc}. \tag{2.108}$$

More specifically, the curl of $\vec{e}_{m,n}$ is equal to the sum of the facet elements associated with the two neighboring facets (m, n, k) and (m, n, l). This is depicted pictorially in Fig. 2.15. Mathematically, this relationship may be cast in the following form

$$[\nabla \times \vec{e}_{1,2}, \nabla \times \vec{e}_{1,3}, \nabla \times \vec{e}_{1,4}, \nabla \times \vec{e}_{2,3}, \nabla \times \vec{e}_{2,4}, \nabla \times \vec{e}_{3,4}]$$

$$= \left[\vec{f}_{1,2,3}, \vec{f}_{1,2,4}, \vec{f}_{1,3,4}, \vec{f}_{2,3,4}\right] \begin{bmatrix} 1 & -1 & & 1 & & \\ 1 & & -1 & & 1 & \\ & & 1 & -1 & & 1 \\ & & & 1 & -1 & 1 \end{bmatrix}. \tag{2.109}$$

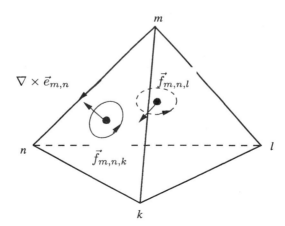

Figure 2.15 Construction of the curl of an edge element from the linear combination of the two facet elements associated with the two facets that share the specific edge.

2.5.3.2 Volume-type flux subspace
The development of the volume-type flux subspace begins with the construction of a new set of vector functions induced from the volume-type scalar space $W^p_{s,f}$ of (2.50). They are

$$v^p_{1,2,3,4} \rightarrow \begin{pmatrix} v^p_{1,2,3,4} \nabla \lambda_4 \times \nabla \lambda_1 \\ v^p_{1,2,3,4} \nabla \lambda_4 \times \nabla \lambda_2 \\ v^p_{1,2,3,4} \nabla \lambda_2 \times \nabla \lambda_1 \end{pmatrix}. \tag{2.110}$$

These functions are combined with the functions in parentheses in (2.95) and (2.96) to yield the complete form of the volume-type NV subspace

$$
W_{nv,v}^p = \left\{
\begin{array}{l}
\{e_{1,2}\nabla\lambda_4 \times \nabla\lambda_3\} \quad \{e_{1,3}\nabla\lambda_4 \times \nabla\lambda_2\} \quad e_{1,4}\nabla\lambda_3 \times \nabla\lambda_2 \\
\{e_{2,3}\nabla\lambda_4 \times \nabla\lambda_1\} \quad e_{2,4}\nabla\lambda_3 \times \nabla\lambda_1 \quad e_{3,4}\nabla\lambda_2 \times \nabla\lambda_1 \\
[f_{1,2,3}\nabla\lambda_4 \times \nabla\lambda_1] \quad \{f_{1,2,4}\nabla\lambda_4 \times \nabla\lambda_3\} \\
[f_{1,2,3}\nabla\lambda_4 \times \nabla\lambda_2] \quad f_{1,2,4}\nabla\lambda_3 \times \nabla\lambda_1 \\
\{f_{1,3,4}\nabla\lambda_4 \times \nabla\lambda_2\} \quad \{f_{2,3,4}\nabla\lambda_4 \times \nabla\lambda_1\} \\
f_{1,3,4}\nabla\lambda_2 \times \nabla\lambda_1 \quad f_{2,3,4}\nabla\lambda_2 \times \nabla\lambda_1 \\
[v_{1,2,3,4}^p\nabla\lambda_4 \times \nabla\lambda_1] \quad [v_{1,2,3,4}^p\nabla\lambda_4 \times \nabla\lambda_2] \quad v_{1,2,3,4}^p\nabla\lambda_2 \times \nabla\lambda_1
\end{array}
\right\},
$$

$$
\mathrm{Dim}(W_{tv,v}^p) = 6(p-1) + 8\frac{(p-2)(p-1)}{2} + 3\frac{(p-3)(p-2)(p-1)}{6}
$$
$$
= \frac{(p-1)(3p^2 + 9p + 6)}{6}.
$$

$$(2.111)$$

Following our earlier pattern, we have enclosed in brackets and square brackets, respectively, those functions that will need to be removed later in the process of the decomposition of $W_{nv,v}^p$ into a non-curl and a curl subspace, $W_{nv,v,nc}^p \oplus \nabla \times W_{tv,v,ng}^{p+1}$. Toward this objective let us calculate the curl of the elements of $W_{tv,v,ng}^{p+1}$. In view of (2.83) the resulting space is

$$
\nabla \times W_{tv,v,ng}^{p+1} = \left\{
\begin{array}{l}
\nabla \times f_{2,3,4}^{p+1}\nabla\lambda_1, \\
\nabla \times f_{1,3,4}^{p+1}\nabla\lambda_2, \\
\nabla \times f_{1,2,4}^{p+1}\nabla\lambda_3, \\
\nabla \times v_{1,2,3,4}^{p+1}\nabla\lambda_1, \\
\nabla \times v_{1,2,3,4}^{p+1}\nabla\lambda_2
\end{array}
\right\}.
$$

$$(2.112)$$

The calculation of the curl of $f_{2,3,4}^{p+1}\nabla\lambda_1$, $f_{1,3,4}^{p+1}\nabla\lambda_2$, $f_{1,2,4}^{p+1}\nabla\lambda_3$ yields

$$
\nabla \times f_{2,3,4}^{p+1}\nabla\lambda_1 = \nabla \times (\lambda_2\lambda_3^i\lambda_4^j\nabla\lambda_1)
$$
$$
= \lambda_3^i\lambda_4^j\nabla\lambda_2 \times \nabla\lambda_1 + i\lambda_2\lambda_3^{i-1}\lambda_4^j\nabla\lambda_3 \times \nabla\lambda_1 + j\lambda_2\lambda_3^i\lambda_4^{j-1}\nabla\lambda_4 \times \nabla\lambda_1,
$$
$$
\nabla \times f_{1,3,4}^{p+1}\nabla\lambda_2 = \nabla \times (\lambda_1\lambda_3^i\lambda_4^j\nabla\lambda_2)
$$
$$
= \lambda_3^i\lambda_4^j\nabla\lambda_1 \times \nabla\lambda_2 + i\lambda_1\lambda_3^{i-1}\lambda_4^j\nabla\lambda_3 \times \nabla\lambda_2 + j\lambda_1\lambda_3^i\lambda_4^{j-1}\nabla\lambda_4 \times \nabla\lambda_2,
$$
$$
\nabla \times f_{1,2,4}^{p+1}\nabla\lambda_3 = \nabla \times (\lambda_1\lambda_2^i\lambda_4^j\nabla\lambda_3)
$$
$$
= \lambda_2^i\lambda_4^j\nabla\lambda_1 \times \nabla\lambda_3 + i\lambda_1\lambda_2^{i-1}\lambda_4^j\nabla\lambda_3 \times \nabla\lambda_2 + j\lambda_1\lambda_2^i\lambda_4^{j-1}\nabla\lambda_4 \times \nabla\lambda_3.
$$

This result suggests that in order to remove these functions from (2.111) we must remove the following terms:

$$
\left\{
\begin{array}{l}
\lambda_2\lambda_3^i\lambda_4^{j-1}\nabla\lambda_4 \times \nabla\lambda_1 \\
\lambda_1\lambda_3^i\lambda_4^{j-1}\nabla\lambda_4 \times \nabla\lambda_2 \\
\lambda_1\lambda_2^i\lambda_4^{j-1}\nabla\lambda_4 \times \nabla\lambda_3
\end{array}
\;\middle|\;
\begin{array}{l}
i = 1, 2, \cdots, p-1, \\
j = 1, 2, \cdots, p-i
\end{array}
\right\}.
$$

$$(2.113)$$

The removed vector functions are the ones enclosed in brackets in (2.111).

In a similar fashion, let us calculate the curl of $v_{1,2,3,4}^{p+1}\nabla\lambda_1$ and $v_{1,2,3,4}^{p+1}\nabla\lambda_2$ in (2.112). The result is

$$\nabla \times v_{1,2,3,4}^{p+1}\nabla\lambda_1 = \nabla \times (\lambda_1\lambda_2^i\lambda_3^j\lambda_4^k\nabla\lambda_1)$$
$$= i\lambda_1\lambda_2^{i-1}\lambda_3^j\lambda_4^k\nabla\lambda_2 \times \nabla\lambda_1 + j\lambda_1\lambda_2^i\lambda_3^{j-1}\lambda_4^k\nabla\lambda_3 \times \nabla\lambda_1 + k\lambda_1\lambda_2^i\lambda_3^j\lambda_4^{k-1}\nabla\lambda_4 \times \nabla\lambda_1,$$
$$\nabla \times v_{1,2,3,4}^{p+1}\nabla\lambda_2 = \nabla \times (\lambda_1\lambda_2^i\lambda_3^j\lambda_4^k\nabla\lambda_2)$$
$$= i\lambda_1\lambda_2^{i-1}\lambda_3^j\lambda_4^k\nabla\lambda_1 \times \nabla\lambda_2 + j\lambda_1\lambda_2^i\lambda_3^{j-1}\lambda_4^k\nabla\lambda_3 \times \nabla\lambda_2 + k\lambda_1\lambda_2^i\lambda_3^j\lambda_4^{k-1}\nabla\lambda_4 \times \nabla\lambda_2.$$

In view of this result, the removal of the remaining elements of $\nabla \times W_{tv,v,ng}^{p+1}$ from $W_{nv,v}^p$ is effected through the removal of the following vector functions from (2.111)

$$\left\{\begin{matrix}\lambda_1\lambda_2^i\lambda_3^j\lambda_4^{k-1}\nabla\lambda_4 \times \nabla\lambda_1 \\ \lambda_1\lambda_2^i\lambda_3^j\lambda_4^{k-1}\nabla\lambda_4 \times \nabla\lambda_2\end{matrix}\left|\begin{matrix}i = 1, 2, \cdots, p-2 \\ j = 1, 2, \cdots, p-1-i, \\ k = 1, 2, \cdots, p-i-j\end{matrix}\right.\right\}. \tag{2.114}$$

The removed functions are the ones enclosed in square brackets in (2.111). The remaining vector functions constitute the pth-order non-curl, volume-type NV space

$$W_{nv,v,nc}^p = \left\{\begin{matrix}e_{1,4}^p\nabla\lambda_3 \times \nabla\lambda_2 \\ e_{2,4}^p\nabla\lambda_3 \times \nabla\lambda_1 \\ e_{3,4}^p\nabla\lambda_2 \times \nabla\lambda_1 \\ f_{2,3,4}^p\nabla\lambda_2 \times \nabla\lambda_1 \\ f_{1,3,4}^p\nabla\lambda_2 \times \nabla\lambda_1 \\ f_{1,2,4}^p\nabla\lambda_3 \times \nabla\lambda_1 \\ v_{1,2,3,4}^p\nabla\lambda_2 \times \nabla\lambda_1\end{matrix}\right\},$$

$$\text{Dim}(W_{nv,v,nc}^p) = 3(p-1) + 3\frac{(p-2)(p-1)}{4} + \frac{(p-3)(p-2)(p-1)}{6}$$
$$= \frac{(p-1)(p^2 + 4p + 6)}{6}. \tag{2.115}$$

2.5.4   Three-dimensional charge space

The divergence of the elements of the three-dimensional NV space generates a set of functions that define another scalar space in the finite element family of interpolation functions. This space is used for the expansion of the charge density in the tetrahedron. In view of the decomposition of the NV space discussed in the previous section, the p-th order charge space is written as follows:

$$W_{charge}^p = \nabla \cdot W_{nv}^p = \nabla \cdot W_{nv,f,nc}^1 \oplus \nabla \cdot W_{nv,v,nc}^{p+1}. \tag{2.116}$$

The space $\nabla \cdot W_{nv,f,nc}^1$ is

$$\nabla \cdot W_{nv,f,nc}^1 = \left\{\nabla \cdot \vec{f}_{1,2,3}, \nabla \cdot \vec{f}_{1,2,4}, \nabla \cdot \vec{f}_{1,3,4}, \nabla \cdot \vec{f}_{2,3,4}\right\}. \tag{2.117}$$

The explicit forms for the basis functions induced by the divergence operation are easy to compute from (2.49). They are

$$
\begin{aligned}
\nabla \cdot \vec{f}_{1,2,3} &= 2\left(\nabla\lambda_1 \cdot \nabla\lambda_2 \times \nabla\lambda_3 + \nabla\lambda_2 \cdot \nabla\lambda_3 \times \nabla\lambda_1 + \nabla\lambda_3 \cdot \nabla\lambda_1 \times \nabla\lambda_2\right) \\
&= \frac{s}{V}, \\
\nabla \cdot \vec{f}_{1,2,4} &= 2\left(\nabla\lambda_1 \cdot \nabla\lambda_2 \times \nabla\lambda_4 + \nabla\lambda_2 \cdot \nabla\lambda_4 \times \nabla\lambda_1 + \nabla\lambda_4 \cdot \nabla\lambda_1 \times \nabla\lambda_2\right) \\
&= \frac{-s}{V}, \\
\nabla \cdot \vec{f}_{1,3,4} &= 2\left(\nabla\lambda_1 \cdot \nabla\lambda_3 \times \nabla\lambda_4 + \nabla\lambda_3 \cdot \nabla\lambda_4 \times \nabla\lambda_1 + \nabla\lambda_4 \cdot \nabla\lambda_1 \times \nabla\lambda_3\right) \\
&= \frac{s}{V}, \\
\nabla \cdot \vec{f}_{2,3,4} &= 2\left(\nabla\lambda_2 \cdot \nabla\lambda_3 \times \nabla\lambda_4 + \nabla\lambda_3 \cdot \nabla\lambda_4 \times \nabla\lambda_2 + \nabla\lambda_4 \cdot \nabla\lambda_2 \times \nabla\lambda_3\right) \\
&= \frac{-s}{V},
\end{aligned}
\tag{2.118}
$$

where V is the volume of the tetrahedron and the sign function, s, is defined in (2.85). Another method of determining the sign of the divergence of each facet element $\vec{f}_{k,l,m}$, is by using the right hand rule. If the direction of the thumb is pointing outwards, away from the tetrahedron, as the fingers rotate in the direction of the vertices $k \to l \to m$, then $\nabla \cdot \vec{f}_{k,l,m}$ is positive; otherwise, it is negative.

To summarize, the three-dimensional charge space for the tetrahedron is

$$
\begin{aligned}
W^p_{charge} &= W^0_{charge} \oplus \nabla \cdot W^{p+1}_{nv,v,nc}, \quad W^0_{charge} = \left\{\frac{1}{V}\right\}, \\
\mathrm{Dim}(W^p_{charge}) &= \frac{(p+1)(p+2)(p+3)}{6}, \\
\mathrm{Dim}(\nabla \cdot W^{p+1}_{nv,v,nc}) &= \frac{p^3 + 6p^2 + 11p}{6}.
\end{aligned}
\tag{2.119}
$$

2.6 RELATIONSHIP AMONG 3D FINITE ELEMENT SPACES

In the previous sections the construction of the four classes of finite element interpolation functions for the tetrahedron was presented. The construction was carried out in a manner that allows for the systematic, hierarchical transition from a space of order p to the next higher order $p + 1$. In a manner similar to the case of finite element interpolation in two dimensions, it was found that the four spaces, namely, the potential, field, flux, and charge space, are related. This relationship is discussed in a more focused manner in this section.

First, let us summarize our result for the four spaces in a manner that not only shows explicitly their decomposition but also indicates their relationship.

$$
\begin{aligned}
W^p_s &= W^1_{s,n} \oplus && W^p_{s,e} && \oplus && W^p_{s,f} && \oplus && W^p_{s,v} \\
W^p_{tv} &= && \underbrace{W^1_{tv,e,ng} \oplus \nabla W^{p+1}_{s,e}}_{W^p_{tv,e}} \oplus && \underbrace{W^p_{tv,f,ng} \oplus \nabla W^{p+1}_{s,f}}_{W^p_{tv,f}} \oplus && \underbrace{W^p_{tv,v,ng} \oplus \nabla W^{p+1}_{s,v}}_{W^p_{tv,v}} \\
W^p_{nv} &= && && \underbrace{W^1_{nv,f,nc} \oplus \nabla \times W^{p+1}_{tv,f,ng}}_{W^p_{nv,f}} \oplus && \underbrace{W^p_{nv,v,nc} \oplus \nabla \times W^{p+1}_{tv,v,ng}}_{W^p_{nv,v}} \\
W^p_c &= && && && \underbrace{W^0_{charge} \oplus \nabla \cdot W^{p+1}_{nv,v,nc}}.
\end{aligned}
\tag{2.120}
$$

The four Whitney forms, $W_{s,n}^1$, $W_{tv,e,ng}^1$, $W_{nv,f,nc}^1$ and W_{charge}^0, proposed by Bossavit [9] are the lowest order subspaces of the four spaces. They are, respectively, the 0-form, 1-form, 2-form and 3-form. The structure above suggests how these four spaces can be extended to any arbitrary order p in a systematic, hierarchical fashion following the methodology elaborated in the previous sections.

At this point we turn our attention to the relationship that exists among Bossavit's four Whitney forms. We will use the following diagram to represent this relationship in compact form,

$$\underbrace{W_{s,n}^1}_{\text{node-elements}} \xrightarrow[grad]{\nabla W_{s,n}^1 \;\subset} \underbrace{W_{tv,e,ng}^1}_{\text{edge-elements}} \xrightarrow[curl]{\nabla \times W_{tv,e,ng}^1 \;\subset} \underbrace{W_{nv,f,nc}^1}_{\text{facet-elements}} \xrightarrow[div]{\nabla \cdot W_{nv,f,nc}^1 \;=} \underbrace{W_{charge}^0}_{\text{tetra-elements}},$$

$$(2.121)$$

where the vector operation in parentheses above the arrow (which is also stated explicitly under the arrow) indicates how the basis functions of the space on the left are operated upon to obtain the basis functions of the space on the right. Also indicated with the symbol $\subset$ or $=$ is whether the space resulting from operating with the specific operator on the basis functions of the space on the left of the arrow is either a subset or equal to the space on the right. As already mentioned earlier, the basis functions in $W_{s,n}^1$, $W_{tv,e,ng}^1$, $W_{nv,f,nc}^1$ and W^0, are commonly referred to as node elements, edge elements, facet elements, and tetra elements, respectively. Explicit forms for these basis functions are provided in (2.46), (2.68), (2.106) and (2.118).

The relationship between the four Whitney forms also holds for higher-order spaces. More specifically, for $p > 1$, it may be cast in the following compact form

$$W_s^p \xrightarrow[grad]{\nabla W_s^{p+1} \subset W_{tv}^p} W_{tv}^p \xrightarrow[curl]{\nabla \times W_{tv}^{p+1} \subset W_{nv}^p} W_{nv}^p \xrightarrow[div]{\nabla \cdot W_{nv}^{p+1} = W_{charge}^p} W_{charge}^p. \qquad (2.122)$$

In a manner similar to the two-dimensional case of finite element interpolation on triangles, Euler's formula can be derived for the three-dimensional finite-element interpolation using tetrahedra and the four Whitney forms. Let N_n denote the total number of nodes, N_e the total number of edges, N_f the total number of facets, and N_t the total number of tetrahedra in the computational domain. The dimension of the node-element space is N_n, while the dimension of the space formed by taking the gradient of its basis functions is $N_n - 1$.

The fact that the dimension of the gradient space $\nabla W_{s,n}^1$ is one less than that of the node-elements, follows from the fact that the node elements contain the zeroth-order scalar space whose dimension is one, as shown in (2.43). The dimension of the edge-element space $W_{tv,e,ng}^1$ is N_e, while the dimensions of the facet-element space $W_{nv,f,nc}^1$ and the tetra-element space W^0 are, respectively, N_f and N_t.

In view of (2.121) and the decompositions indicated in (2.120), we notice that the dimension of W^0 equals the dimension of the $W_{nv,f,nc}^1$ minus the dimension of the subspace formed by the curl of the elements of $W_{tv,e,ng}^1$. Recognizing that the latter is equal to the dimension of $W_{tv,e,ng}^1$ minus the dimension of the subspace formed by the gradient of the elements of $W_{s,n}^1$, we have

$$N_t = N_f - (N_e - (N_n - 1))$$
$$\Rightarrow N_n - N_e + N_f - N_t = 1. \qquad (2.123)$$

Euler's formula can be used to check the correctness of the three-dimensional finite element mesh.

2.7 GRADIENT, CURL AND DIVERGENCE MATRICES FOR 3D FINITE ELEMENT SPACES

For the purposes of finite element matrix manipulations it is convenient to cast the relationship between the Whitney forms in matrix form. This is achieved using the same process that was applied in the two-dimensional case (see Section 2.1.6). Thus, once again, with $\{\nabla W^1_{s,n}\}$, $\{W^1_{tv,e,ng}\}$, $\{\nabla \times W^1_{tv,e,ng},\}$, $\{W^1_{nv,f,nc}\}$, $\{\nabla \cdot W^1_{nv,f,nc}\}$, and $\{W^0\}$ denoting the row vectors containing the basis functions for the corresponding spaces, we are interested in defining the gradient matrix G, the curl matrix C and the divergence matrix D in the following relations for the case of three-dimensional finite element interpolation using tetrahedra and the four Whitney forms.

$$
\begin{aligned}
\nabla W^1_{s,n} \subset W^1_{tv,e,ng} & \quad\Rightarrow\quad \{\nabla W^1_{s,n}\} = \{W^1_{tv,e,ng}\}G, \\
\nabla \times W^1_{tv,e,ng} \subset W^1_{nv,f,nc} & \quad\Rightarrow\quad \{\nabla \times W^1_{tv,e,ng},\} = \{W^1_{nv,f,nc}\}C, \\
\nabla \cdot W^1_{nv,f,nc} = W^0_{charge} & \quad\Rightarrow\quad \{\nabla \cdot W^1_{nv,f,nc}\} = \{W^0_{charge}\}D.
\end{aligned}
\tag{2.124}
$$

The gradient matrix G is of dimension $N_e \times N_n$. Each row of G corresponds to an edge in the mesh and contains two nonzero entries with values $+1$ and -1. The entry in the column that corresponds to the start node of the edge is -1, while the entry in the column that corresponds to the end node of the edge is $+1$. This is depicted in the left-most plot in Fig. 2.16. Hence, for the ith row of G we have

$$
\begin{array}{cc}
\text{node } m & \text{node } n \\
\downarrow & \downarrow
\end{array}
$$
$$
G_{N_e \times N_n}(i,:) = \quad [\quad\quad -1 \quad\quad\quad +1 \quad\quad] \quad \leftarrow \text{edge } i.
\tag{2.125}
$$

Since the summation of column vectors in G is the zero vector, the rank of G is N_{node-1}.

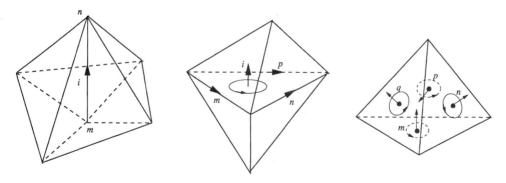

Figure 2.16 Visual aid for the interpretation of the way the gradient, curl, and divergence matrices are constructed.

The curl matrix C is of dimension $N_f \times N_e$. Each row of C corresponds to a facet and has only three nonzero entries of value ± 1. The columns at which these nonzero values occur are associated with the edges that form the specific facet. Let i be the facet and m,

n, p the three edges associated with it, with orientation as depicted in the center plot of Fig. 2.16. Using the right hand rule, with the unit normal on facet i pointing in the direction of the thumb of the right hand, the value of the entry associated with each edge is decided as follows. If the orientation of the edge is in the direction of the right hand rule, then the corresponding entry has value $+1$; otherwise, its value is -1. Thus the ith row of C is

$$C_{N_f \times N_e}(i,:) = \begin{array}{ccc} \text{edge } m & \text{edge } n & \text{edge } p \\ \downarrow & \downarrow & \downarrow \\ [\quad +1 & +1 & -1 \quad] \end{array} \leftarrow \text{facet } i. \tag{2.126}$$

The divergence matrix D is of dimension $N_t \times N_f$. Each row of D corresponds to a tetrahedron and has only four nonzero entries of value ± 1. The nonzero values are associated with the four facets of the tetrahedron. Referring to the right-most plot in Fig. 2.16, which depicts the ith tetrahedron with facets m, n, p, q, the sign of each nonzero entry is decided as follows. Using the right hand rule, with the fingers pointing in the direction of the assigned sense of rotation on each facet, if the thumb is pointing outwards, away from the tetrahedron, then the value is positive; otherwise it is negative. Thus we write for the ith row of D

$$D_{N_t \times N_f}(i,:) = \begin{array}{cccc} \text{face } m & \text{face } n & \text{face } p & \text{face } q \\ \downarrow & \downarrow & \downarrow & \downarrow \\ [\quad -1 & +1 & -1 & +1 \quad] \end{array} \leftarrow \text{tetra } i. \tag{2.127}$$

From (2.121), we have the following relationship between the matrices:

$$\begin{aligned} D_{N_t \times N_f} C_{N_f \times N_e} &= 0, \\ C_{N_f \times N_e} G_{N_e \times N_n} &= 0. \end{aligned} \tag{2.128}$$

2.8 THE SPACES $\mathcal{H}^P(CURL)$ AND $\mathcal{H}^P(DIV)$

At this point we have completed the presentation of the methodology for the systematic development of hierarchical finite element basis functions for two- and three-dimensional field approximations in domains discretized using triangles and tetrahedron, respectively. In this section, we would like to turn our attention to the relationships that exist among the constructed spaces.

The most popular finite element formulation of the electromagnetic boundary value problem involves the vector Helmholtz equation. The unknown quantity is either the electric field intensity or the magnetic field intensity. Hence, the TV space is the one used for its finite element approximation. The popular notation for the space of the basis function used for this purpose is $\mathcal{H}^p(curl)$, where the superscript p indicates that the functions involved along with their rotation are vector polynomials complete to the pth order. Let us take a closer look at these spaces.

The zeroth-order space, $\mathcal{H}^0(curl)$, is the space of the edge elements. It is the same as the $W_{tv,e,ng}^1$ in our notation. Its basis functions and their rotation are polynomials complete to zeroth order.

$$\begin{aligned} \mathcal{H}^0(curl) &= \{\lambda_m \nabla \lambda_n - \lambda_n \nabla \lambda_m \; ; \; m < n\}, \\ \text{Dim} \mathcal{H}^0(curl) &= 6. \end{aligned} \tag{2.129}$$

In order to form the next higher-order TV space, $\mathcal{H}^1(curl)$, we add the basis functions from $\nabla W_{s,e}^2$ to make the space complete to the first order, along with the basis functions from

$W_{tv,f,ng}^2$ to make the curl of the elements in the space complete to first order, also. Thus the complete $\mathcal{H}^1(curl)$ TV space is

$$
\begin{aligned}
&\mathcal{H}^1(curl) = \mathcal{H}^0(curl) \oplus \nabla W_{s,e}^2 \oplus W_{tv,f,ng}^2, \\
&\text{Dim}\mathcal{H}^1(curl) = 20, \\
&\nabla W_{s,e}^2 = \{\, \nabla(\lambda_m \lambda_n)\,;\, m < n \,\}, \\
&W_{tv,f,ng}^2 = \{\, \lambda_m \lambda_k \nabla \lambda_n,\ \lambda_n \lambda_k \nabla \lambda_m\,;\, m < n < k \,\}.
\end{aligned} \tag{2.130}
$$

Another form of $W_{tv,f,ng}^2$ that is commonly used is [40]

$$
W_{tv,f,ng}^2 = \left\{ \begin{array}{l} 4\lambda_m(\lambda_n \nabla \lambda_k - \lambda_k \nabla \lambda_n) \\ 4\lambda_n(\lambda_k \nabla \lambda_m - \lambda_m \nabla \lambda_k) \end{array} \middle| \; m < n < k \right\}. \tag{2.131}
$$

It is only slightly different from the one in (2.130), and its elements can be written as linear combination of the basis functions of $W_{tv,f,ng}^2$ and $\nabla W_{s,f}^3$.

The construction of $\mathcal{H}^2(curl)$ is carried out in a similar manner. First, the second-order basis functions of $\nabla W_{s,e}^3$ and $\nabla W_{s,f}^3$ are added to $\mathcal{H}^1(curl)$ to make the space complete to second order. Then, the third-order basis functions from the non-gradient spaces $W_{tv,f,ng}^3$ and $W_{tv,v,ng}^3$ are added to make the curl of the elements of the space complete to second order. The complete $\mathcal{H}^2(curl)$ space is

$$
\begin{aligned}
&\mathcal{H}^2(curl) = \mathcal{H}^0(curl) \oplus \nabla W_{s,e}^3 \oplus \nabla W_{s,f}^3 \oplus W_{tv,f,ng}^3 \oplus W_{tv,v,ng}^3, \\
&\text{Dim}\mathcal{H}^2(curl) = 45, \\
&\nabla W_{s,e}^3 = \{\, \nabla(\lambda_m \lambda_n),\ \nabla(\lambda_m \lambda_n^2)\,;\, m < n \,\}, \\
&\nabla W_{s,f}^3 = \{\, \nabla(\lambda_m \lambda_n \lambda_k)\,;\, m < n < k \,\}, \\
&W_{tv,f,ng}^3 = \left\{ \begin{array}{l} \lambda_m \lambda_k \nabla \lambda_n,\ \lambda_n \lambda_k \nabla \lambda_m, \\ \lambda_m \lambda_k^2 \nabla \lambda_n,\ \lambda_n \lambda_k^2 \nabla \lambda_m, \\ \lambda_m \lambda_n \lambda_k \nabla \lambda_m \end{array} \middle| \; m < n < k \right\}, \\
&W_{tv,v,ng}^3 = \{\, \lambda_1 \lambda_2 \lambda_3 \nabla \lambda_0,\ \lambda_0 \lambda_2 \lambda_3 \nabla \lambda_1,\ \lambda_0 \lambda_1 \lambda_3 \nabla \lambda_2 \,\}.
\end{aligned} \tag{2.132}
$$

The above procedure can be repeated for the construction of the basis functions for the TV space $\mathcal{H}^p(curl)$ for any order p. In compact form, the recipe for the construction of $\mathcal{H}^p(curl)$ is

$$
\mathcal{H}^p(curl) = \mathcal{H}^0(curl) \oplus \underbrace{\nabla\left(W_{s,e}^{p+1} \oplus W_{s,f}^{p+1} \oplus W_{s,v}^{p+1}\right)}_{p\text{th-order gradient bases}} \oplus \underbrace{\left(W_{tv,f,ng}^{p+1} \oplus W_{tv,v,ng}^{p+1}\right)}_{(p+1)\text{th-order non-gradient bases}},
$$

$$
\text{Dim}(\mathcal{H}^0(curl)) = 6,
$$

$$
\text{Dim}(\mathcal{H}^p(curl)) = \frac{p^3 + 7p^2 + 22p + 10}{2}, \quad (p \geq 1).
$$

Next, we turn our attention to the construction of spaces for the expansion of the electric or magnetic flux densities. Clearly, it is the NV space that will be used in this case. For the applications considered in this book, the NV space will be used to model the magnetic flux density $\vec{B}$. Since its divergence is zero, the expansion space for $\vec{B}$ is $\mathcal{H}^p(div)$, which is a subspace of the NV space with its divergence complete to pth-order polynomial functions.

First, we construct the space $\mathcal{H}^0(div)$. This is the space of facet elements,

$$\mathcal{H}^0(div) = \{2\,(\lambda_m \nabla \lambda_n \times \nabla \lambda_k + \lambda_n \nabla \lambda_k \times \nabla \lambda_m + \lambda_k \nabla \lambda_m \times \nabla \lambda_n)\,;\, m < n < k\}\,,$$
$$\mathrm{Dim}(\mathcal{H}^0(div)) = 4.$$

The construction of the next order space, $\mathcal{H}^1(div)$, requires the addition of new basis functions from $\nabla \times W^2_{tv,f,ng}$. The result is

$$\mathcal{H}^1(div) = \mathcal{H}^0(div) \oplus \nabla \times W^2_{tv,f,ng},$$
$$\mathrm{Dim}(\mathcal{H}^1(div)) = 12, \tag{2.133}$$
$$\nabla \times W^2_{tv,f,ng} = \{\,\nabla \times (\lambda_m \lambda_n \nabla \lambda_k),\ \nabla \times (\lambda_n \lambda_k \nabla \lambda_m)\,;\, m < n < k\,\}.$$

The construction of $\mathcal{H}^2(div)$ requires the addition of basis functions from $\nabla \times W^3_{tv,f,ng}$ and $\nabla \times W^3_{tv,v,ng}$. The complete $\mathcal{H}^2(div)$ space is

$$\mathcal{H}^2(div) = \mathcal{H}^0(div) \oplus \nabla \times W^3_{tv,f,ng} \oplus \nabla \times W^3_{tv,v,ng},$$
$$\mathrm{Dim}(\mathcal{H}^2(div)) = 27,$$
$$\nabla \times W^3_{tv,f,ng} = \left\{ \begin{array}{l} \nabla \times (\lambda_m \lambda_n \nabla \lambda_n),\ \nabla \times (\lambda_n \lambda_k \nabla \lambda_m) \\ \nabla \times (\lambda_m \lambda_n \lambda_k \nabla \lambda_m) \\ \nabla \times (\lambda_m \lambda_n^2 \nabla \lambda_n),\ \nabla \times (\lambda_n \lambda_k^2 \nabla \lambda_m) \end{array} \right\vert\ m < n < k \right\},\tag{2.134}$$
$$\nabla \times W^3_{tv,v,ng} = \left\{ \begin{array}{l} \nabla \times (\lambda_1 \lambda_2 \lambda_3 \nabla \lambda_0) \\ \nabla \times (\lambda_0 \lambda_2 \lambda_3 \nabla \lambda_1) \\ \nabla \times (\lambda_0 \lambda_1 \lambda_3 \nabla \lambda_2) \end{array} \right\}.$$

The procedure can be applied repeatedly toward the construction of the basis functions for the flux space $\mathcal{H}^p(div)$ of any arbitrary order p. The general form of $\mathcal{H}^p(div)$ is

$$\mathcal{H}^p(div) = \mathcal{H}^0(div) \quad \oplus \quad \underbrace{\nabla \times \left(W^{p+1}_{tv,f,ng} \oplus W^{p+1}_{tv,v,ng} \right)}_{p\text{th-order curl basis functions}},\tag{2.135}$$
$$\mathrm{Dim}(\mathcal{H}^p(div)) = \frac{2p^3 + 15p^2 + 31p + 24}{6}, \quad (p \geq 0).$$

2.9 THE ISSUE OF ORTHOGONALITY IN HIERARCHICAL BASES

The convenience of the systematic development of hierarchical bases of any desired order for the finite element approximation of electromagnetic quantities comes at a cost that, at first sight, may be perceived as a numerical nuisance. It has to do with the condition number of the mass matrix. More specifically, as the order of the approximation increases, the condition of the mass matrix worsens. The reason for this worsening is the increasing similarity between the added higher-order basis functions to the lower-order ones, which becomes especially problematic as the order increases. This suggests that orthogonalization of the basis functions becomes necessary prior to their use in increasing the order of the finite element approximation.

At this point it is important to point out that interpolatory basis functions [21] do not exhibit this apparent weakness of the proposed hierarchical basis functions. However, even though the mass matrix resulting from the use of interpolatory basis functions is much better conditioned, the price that one has to pay for their use is that they are not hierarchical. Hence, they are not suitable for use in conjunction with multi-level preconditioners and p-adaptive methods. Since the development of multi-level preconditioners and p-adaptive methods is the ultimate objective and the main theme of this book, the way orthogonalization should be applied to the hierarchical basis functions is discussed in the following paragraphs.

The orthogonalization procedure we adopt follows the one presented in [24]. Unfortunately, complete orthogonality in arbitrarily shaped tetrahedra is not possible. Thus, relying on the fact that continuing advances in mesh generation emphasize the construction of meshes with high-quality tetrahedra, we will settle for partial orthogonalization in an equilateral tetrahedron.

We begin with the scalar space. The node-type scalar space of (2.46) remains unchanged. Scalar basis functions are made orthogonal to the ones in the same-type scalar space; hence, the edge-type scalar basis functions of (2.48) are orthogonalized only with respect to other edge-type scalar basis functions on the same edge; the facet-type scalar basis functions of (2.50) are orthogonalized only with respect to other facet-type scalar basis functions on the same facet; and the volume-type scalar basis functions of (2.52) are orthogonalized only with respect to other volume-type scalar basis functions. The orthogonalization uses is the popular Gram-Schmidt orthogonalization.

Let $\tilde{\mathbf{b}} = [\tilde{b}_1, \tilde{b}_2, \ldots, \tilde{b}_n]$ denote the vector of the new set of orthonormal basis functions, while $\mathbf{b} = [b_1, b_2, \ldots, b_n]$ is the vector of the old set of basis functions. In matrix form, the orthogonalization procedure is written as

$$\tilde{\mathbf{b}} = \mathbf{b}\,T, \qquad (2.136)$$

where T is upper triangular transformation matrix. With the notation $\langle \mathbf{a_1}^T, \mathbf{a_2} \rangle$ understood as a matrix with elements the inner products of the elements of the column vectors $\mathbf{a_1}$ and $\mathbf{a_2}$, the orthonormality of the new basis yields

$$\langle \tilde{\mathbf{b}}^T, \tilde{\mathbf{b}} \rangle = T^T \langle \mathbf{b}^T, \mathbf{b} \rangle\, T = I \Rightarrow \left(TT^T \right)^{-1} = \langle \mathbf{b}^T, \mathbf{b} \rangle. \qquad (2.137)$$

Hence, the transformation matrix T can be obtained from the Cholesky factorization of the mass matrix constructed using the old basis functions.

For the TV space, gradient basis functions are orthogonalized with respect to gradient basis functions of same or lower order, while non-gradient basis functions are orthogonalized with respect to both gradient and non-gradient basis functions of same or lower order. Let $\tilde{\mathbf{b}}_g$ and $\tilde{\mathbf{b}}_{ng}$ denote the column vectors containing the new gradient and non-gradient basis functions, while $\mathbf{b}_g$ and $\mathbf{b}_{ng}$ are column vectors containing the old gradient and non-gradient basis functions. In matrix form, the orthogonalization process is written as

$$\begin{aligned} \tilde{\mathbf{b}}_g &= \mathbf{b}_g\,T_1, \\ \tilde{\mathbf{b}}_{ng} &= \mathbf{b}_{ng}\,T_2 + \mathbf{b}_g\,T_3, \end{aligned} \qquad (2.138)$$

where T_1, T_2 and T_3 are upper triangular transformation matrices. Since the new gradient basis functions are orthonormal, it is

$$\langle \tilde{\mathbf{b}}_g^T, \tilde{\mathbf{b}}_g \rangle = T_1^T \langle \mathbf{b}_g^T, \mathbf{b}_g \rangle T_1 = I \Rightarrow \left(T_1 T_1^T \right)^{-1} = \langle \mathbf{b}_g^T, \mathbf{b}_g \rangle. \qquad (2.139)$$

Hence, the transformation matrix T_1 can be obtained from the Cholesky factorization of the mass matrix $\langle \mathbf{b}_g, \mathbf{b}_g^T \rangle$.

The new non-gradient basis functions are orthonormal. Furthermore, they are orthogonal to the new gradient basis functions. Hence, it is

$$
\begin{aligned}
\langle \tilde{\mathbf{b}}_{ng}^T, \tilde{\mathbf{b}}_g \rangle &= T_2^T \langle \mathbf{b}_{ng}^T, \mathbf{b}_g \rangle T_1 + T_3^T \langle \mathbf{b}_g^T, \mathbf{b}_g \rangle T_1 = 0 \\
\langle \tilde{\mathbf{b}}_{ng}^T, \tilde{\mathbf{b}}_{ng} \rangle &= T_2^T \langle \mathbf{b}_{ng}^T, \mathbf{b}_{ng} \rangle T_2 + T_3^T \langle \mathbf{b}_g^T, \mathbf{b}_{ng} \rangle T_2 = I.
\end{aligned}
\tag{2.140}
$$

The solution of the above system yields the relationships needed for the construction of the transformation matrices T_2 and T_3.

$$
\begin{aligned}
\left(T_2 T_2^T \right)^{-1} &= \langle \mathbf{b}_{ng}^T, \mathbf{b}_{ng} \rangle - \langle \mathbf{b}_{ng}^T, \mathbf{b}_g \rangle \langle \mathbf{b}_g^T, \mathbf{b}_g \rangle^{-1} \langle \mathbf{b}_g^T, \mathbf{b}_{ng} \rangle, \\
T_3^T &= -T_2^T \langle \mathbf{b}_{ng}^T, \mathbf{b}_g \rangle \langle \mathbf{b}_g^T, \mathbf{b}_g \rangle^{-1}.
\end{aligned}
\tag{2.141}
$$

The same procedure is used to construct orthonormal basis for the NV space. The only difference in the expressions above will be the change of the subscripts g and ng, indicating gradient and non-gradient type basis functions, respectively, to c and nc, which will now indicate, respectively, curl and non-curl basis functions.

The inner product used in the above orthogonalization is the integral over the volume of the tetrahedron for the three-dimensional case, or the integral over the surface of the triangle for the two-dimensional case. Two important results, which are useful in performing the orthogonalization integrals, are the following:

$$
\begin{aligned}
\iint_{face} \lambda_m^i \lambda_n^j \lambda_p^k ds &= \frac{i!\, j!\, k!}{(2+i+j+k)!} 2S, \\
\iiint_{tetra} \lambda_m^i \lambda_n^j \lambda_p^k \lambda_q^l dv &= \frac{i!\, j!\, k!\, l!}{(3+i+j+k+l)!} 6V,
\end{aligned}
\tag{2.142}
$$

where S denotes the area of the triangle and V denotes the volume of the tetrahedron.

The implementation of the aforementioned orthogonalization of the hierarchical basis functions results in a better conditioned mass matrix, thus improving the convergence of the iterative solution of the approximate problem. The price we pay for this orthogonalization is the increased complexity of orthogonalization coefficients for the new basis functions. Those who appreciate and enjoy the clarity of the proposed hierarchical basis functions would rather not have to pay such a price. Clearly, their only alternative is the availability of a very good preconditioner, capable of handling the poor conditioning of the mass matrix that results from the use of the hierarchical basis functions. It turns out that such preconditioners can be constructed using the multigrid/multilevel techniques that constitute the main theme of this book. It will be shown in later chapters that such preconditioners are much more effective than orthogonalization in improving the condition of the finite element matrix and, thus, can dramatically improve the convergence of the iterative solver. Hence, the orthogonalization of the hierarchical basis functions becomes unnecessary if such a preconditioner is available.

In this chapter, we presented a systematic methodology for the construction of a hierarchical finite element space family for three-dimensional tetrahedron and two-dimensional triangular elements. This space family includes the scalar, TV, NV, and the charge space. These spaces can be used for the modeling of electromagnetic scalar potentials, electric and magnetic field intensities, electric and magnetic flux densities, and electric charge, respectively. Explicit mathematical expressions for basis functions of arbitrary order have been derived for each space. In addition, the properties of each space as well as useful relationships between the spaces have been discussed in detail. These spaces will serve as the foundation for the development of the multigrid and multi-level methods presented in later chapters in the book.

REFERENCES

1. J. T. Ogden and G. F. Carey, *Finite Elements: Mathematical Aspects*, Englewood Cliffs, NJ: Prentice-Hall, Inc., 1983.

2. B. Szabó and I. Babuška, *Finite Element Analysis*, New York: John Wiley & Sons, Inc., 1991.

3. J. M. Jin, *The Finite Element Method in Electromagnetics,* 2nd ed., New York: John Wiley & Sons, Inc., 2002.

4. M. S. Palma, T. K. Sarkar, L.E. García-Castillo, T. Roy, and A. Djordjević, *Iterative and Self-Adaptive Finite-Elements in Electromagnetic Modeling*, Norwood, MA: Artech House, 1998.

5. R. Dyczij-Edlinger, G. Peng, and J. F. Lee, "A fast vector-potential method using tangentially continuous vector finite elements," *IEEE Trans. Microwave Theory Tech.* vol. 46, pp. 863-868, Jun. 1998.

6. R. F. Harrington, *Time-Harmonic Electromagnetic Fields*, Piscataway, NJ: IEEE Press Series on Electromagnetic Wave Theory, 2001.

7. J. C. Nédélec, "Mixed finite elements in R^3," *Numer. Math.*, vol. 35, pp. 315-34, 1980.

8. J. C. Nédélec, "A new family of mixed finite elements in R^3," *Numer. Math.*, vol. 50, pp. 57-81, 1986.

9. A. Bossavit, "Whitney forms: A class of finite elements for three-dimensional computation in electromagnetics," *IEE Proceedings*, vol. 135, pt. A, pp. 493-500, 1988.

10. A. Bossavit, "A rationale for edge elements in 3-D fields computations," *IEEE Trans. Magn.*, vol. 24, pp. 74-79, Jan. 1988.

11. A. Bossavit and I. Mayergoyz, "Edge-elements for scattering problems," *IEEE Trans. Magnetics*, vol. 25, pp. 2816-2821, Jul. 1989.

12. Z. J. Cendes, "Vector finite elements for electromagnetic field computations," *IEEE Trans. Magnetics*, vol. 27, pp. 3958-3966, Sep. 1991.

13. J. F. Lee, "Tangential vector finite elements and their application to solving electromagnetic scattering problems," *Applied Computational Electromagnetic Society Newsletter.* vol. 10, pp. 52-75, Mar. 1995.

14. J. F. Lee, D. K. Sun and Z. J. Cendes, "Tangential vector finite elements for electromagnetic field computation," *IEEE Trans. Magnetics*, vol. 27, pp. 4032-4035, Sep. 1991.

15. J. P. Webb, "Edge elements and what they can do for you," *IEEE Trans. Magnetics*, vol. 29, pp. 1460-1465, Mar. 1993.

16. P. P. Silvester and G. Pelosi (editors), *Finite Elements for Wave Electromagnetics: Methods and Techniques.* Piscataway, NJ: IEEE Press, 1994.

17. J. P. Webb and B.Forghani, "Hierarchical scalar and vector tetrahedra," *IEEE Trans. Magn.*, vol. 29, pp. 1495-1498, Mar. 1993.

18. T. V. Yioultsis and T. D. Tsiboukis, "Multiparametric vector finite elements: a systematic approach to the construction of three-dimensional, higher order, tangential vector shape functions," *IEEE Trans. Magnetics.* vol. 32, pp. 1389-1494, May 1996.

19. T. V. Yioultsis and T. D. Tsiboukis, "Development and implementation of second and third order vector finite elements in various 3-D electromagnetic field problems," *IEEE Trans. Magnetics*, vol. 33, pp. 1812-1815, Mar. 1997.

20. T. V. Yioultsis and T. D. Tsiboukis, C. S. Antonopoulos, and T. D. Tsiboukis, "A fully explicit Whitney element - time domain scheme with higher order vector finite elements for three-diemensional high frequency proble3ms" *IEEE Trans. Magnetics*, vol. 34, pp. 3288-3291, Sep. 1998.

21. R. D. Graglia, D. R. Wilton, and A. F. Peterson, "Higher order interpolatory vector bases for computational electromagnetics," *IEEE Trans. Antenna and Propagation.* vol. 45, pp. 329-342, Mar. 1997.

22. A. F. Peterson and D. R. Wilton, "Curl-conforming mixed-order edge elements for discretizing the 2D and 3D vector Helmholtz equation," *Finite Element Software for Microwave Engineering*, T. Itoh, G. Pelosi and P. P. Silvester, Eds., New York: Wiley, 1996, ch. 5, pp. 101-124.

23. J. S. Savage and A. F. Peterson, "Higher order vector finite elements for tetrahedral cells," *IEEE Trans. Microwave Theory Tech.*, vol. 44, pp. 874-879, June 1996.

24. J. P. Webb, "Hierarchal vector basis functions of arbitrary order for triangular and tetrahedral finite elements," *IEEE Trans. Antennas and Propagation*, vol. 47, pp. 1244-1253, Aug. 1999.

25. Y. Zhu and A. C. Cangellaris, "Hierarchical finite element basis function spaces for tetrahedral elements," *Proc. of Applied Computational Electromagnetics Society Meeting*, Monterey, CA, Mar. 2001.

26. L. S. Andersen and J. L. Volakis, "Hierarchical tangential vector finite elements for tetrahedra," *IEEE, Microwave Guided Wave Lett.*, vol. 8, pp. 127-129, Mar. 1998.

27. L. S. Andersen and J. L. Volakis, "Development and application of a novel class of hierarchical tangential vector finite elements for electromagnetics," *IEEE Trans. Antenna Propagat.*, vol. 47, pp. 112-120, Jan. 1999.

28. L. S. Andersen and J. L. Volakis, "Accurate and efficient simulation of antennas using hierarchical mixed-order tangential vector finite elements for tetrahedra," *IEEE Trans. Antenna Propagat.*, vol. 47, pp. 1240-1243, Aug. 1999.

29. J. Wang and J. P. Webb, "Hierarchical vector boundary elements and p-adaption for 3-D electromagnetic scattering," *IEEE Trans. Antenna Propagat.*, vol. 45, pp. 1869-1879, Dec. 1997.

30. D. A. White, "Orghogonal vector basis functions for time-domain finite element solution of the vector wave equation," *IEEE Trans. Magnetics*, vol. 35, pp. 1458-1461, May 1999.

31. Y. Zhu and A. C. Cangellaris, "A New FEM Formulation for Reduced Order Electromagnetic Modeling," *IEEE Microwave and Guided Wave Lett.*, May 2001.

32. K. Mahadevan, R. Mittra, and P. M. Vaidya, "Use of Whitney's edge and face elements for efficient finite element time domain solution of Maxwell's equations, " *J. Electromagnetic Waves and Appl.*, vol. 8, no. 9/10, pp. 1178-1191, 1994.

33. J. F. Lee and Z. Sacks, "Whitney elements time domain (WETD) methods," *IEEE. Trans. Magnetics*, vol. 31, pp. 1325-1329, May 1995.

34. J. F. Lee, R. Lee, and A. C. Cangellaris, "Time-domain finite element methods," *IEEE Trans. Antennas Propagat.*, vol. AP-45, pp. 430-442, Mar. 1997.

35. H. Whitney, *Geometric Integration Theory*, Princeton NJ: Princeton University Press, 1957.

36. P.P. Silvester and R.L. Ferrari, *Finite Elements for Electrical Engineering*, Cambridge: Cambridge University Press, 1983.

37. S. M. Rao, D. R. Wilton, and A. W. Glisson, "Electromagnetic scattering by surfaces of arbitrary shape," *IEEE Trans. Antennas Propagat.*, vol. 30, pp. 409-418, May 1982.

38. J. M. Gil and J. P. Webb, "A new edge element for the modeling of field singularities in transmission lines and waveguides," *IEEE. Trans. Microwave Theory Tech.*, vol. 45, pp. 2125-2130, Dec. 1997.

39. J. F. Lee and R. Mittra, "A note on the application of edge-elements for modeling three-dimensional inhomogeneously-filled cavities," *IEEE Trans. Microwave Theory Tech.*, vol. 40, pp. 1767-1773, Sep. 1992.

40. G. Peng, R. Dyczij-Edlinger, and J. F. Lee, "Hierarchical methods for solving matrix equations from TVFEMs for microwave components," *IEEE Trans. Magnetics*, vol. 35, pp. 1474-1477, May 1999.

CHAPTER 3

FINITE ELEMENT FORMULATIONS OF ELECTROMAGNETIC BVPS

This chapter is devoted to the development of finite element formulations of various types of electromagnetic boundary value problems (BVPs). Formulations pertinent to both static and dynamic problems, and their subsequent reduction into matrix equation statements of their discrete (finite element) approximation using the Galerkin process, are presented. The development of several possible formulations, used commonly for the finite element-based solution of real-world problems, is guided by the properties of the hierarchical function spaces discussed in the previous chapter. Our development includes formulations in terms of both the electric or magnetic field vectors and the electromagnetic scalar and vector potentials. Special emphasis is placed on the numerical attributes of the generated approximate statements of the BVPs. Of particular concern is the susceptibility of the approximate finite element statement to the occurrence of spurious solutions that may render the numerical results useless. As one might expect, this issue is intimately connected to the degree of accuracy with which the requirements for the uniqueness of the solution are satisfied in the finite element approximation of the continuous BVP.

Electrostatic and magnetostatic BVPs are considered first, followed by quasi-magnetostatic problems pertinent to the numerical modeling of magnetic field diffusion phenomena in conductors. This is followed by the development of finite element formulations for the numerical solution of full-wave, electrodynamic BVPs. In all cases, the media involved are assumed to be linear and time-invariant.

The discussion concludes with a slight divergence from the central theme of finite element-based formulations of electromagnetic BVPs to discuss a topic of significant importance in computational electromagnetics, namely, the relevance of the scalar and vector function spaces of Chapter 2 to the development of numerically stable, integral equation formulations of electromagnetic BVPs. This topic has received special attention recently in the context of application of integral equation-based fast solvers for the broadband (from dc to tens of GHz) electromagnetic analysis of interconnect structures in high-density integrated electronic systems. Our discussion will be in the context of the Partial Element Equivalent Circuit (PEEC) formulation, one of the most popular integral equation formulations used for the electromagnetic modeling of such interconnect structures.

One issue that is not addressed in our development is the error resulting from the approximation of the computational domain. This error is caused by the replacement of the physical domain of interest with the computational one, resulting from the union of the finite elements over which the unknown fields are approximated. Clearly, the geometric attributes of the elements govern the fidelity with which the actual boundaries of the physical domain are represented and, hence, the accuracy with which the physical boundary conditions are approximated in the finite element statement of the BVP.

To offer a specific example of such an error in the context of the modeling of electromagnetic devices, let us consider the application of finite elements for the eigenvalue analysis of a circular cylindrical cavity. If a finite element mesh of tetrahedra is utilized for the discretization of the geometry, the actual circular boundary of the cavity will be replaced by a polygon, giving rise to a new cavity geometry the eigenvalues of which will be different from those of the original one. Since, in general, we will not be able to calculate the eigenvalues of the actual cavity, a measure of the error in their approximation due to the inexact modeling of the boundaries is most desirable.

The quantification of this boundary approximation error has been studied extensively since the early days of the application of finite element methods for the numerical modeling of physical phenomena. Of particular value are results from these studies that quantify the dependence of the approximation error on the type of element used and the order of the polynomial interpolation. Such results, available in the form of appropriate norms, can be found in [1]-[3]. A nice discussion of these results in the context of finite element modeling of electromagnetic BVPs is given in [4]. Even though limited to specific classes of elements and interpolation functions, the derived error norms capture a trend that is intuitively expected, namely, that the error decreases as the finite element mesh is refined.

The problem is, of course, that mesh refinement increases the degrees of freedom in the approximation. To counteract such an increase, curved elements may be implemented to provide for piecewise, higher-order polynomial approximation of curved boundaries. Reference [4] offers a detailed account of the development of interpolation functions on curved elements, while more recent work addresses their use in conjunction with vector interpolation functions (e.g., [5]). Application of curved elements comes at the cost of added numerical overhead in the computation of the elements of the finite element matrix. Furthermore, since higher-order interpolation functions are used in conjunction with such elements, a meaningful assessment of the benefits from the utilization of curved elements is best done within the framework of the so-called $h - p$ finite element methodologies (e.g., [6], [7]).

Our discussion refrains from the explicit consideration of the boundary approximation error. Rather, we will call upon the aforementioned trend of this error to decrease through mesh refinement, to make the argument that the multigrid finite element solution machinery developed in this book offers a means toward a systematic and computationally efficient way for its reduction and control. In addition, we will be given the chance to assess its impact through the application of our finite element solution machinery to the analysis of several canonical electromagnetic structures for which analytic solutions exist.

3.1 ELECTROSTATIC BOUNDARY VALUE PROBLEMS

3.1.1 Governing equations and boundary conditions

The governing equations for an electrostatic boundary value problem are

$$\begin{cases} \nabla \times \vec{E}(\vec{r}) = 0 \\ \nabla \cdot \vec{D}(\vec{r}) = \rho_v(\vec{r}), \end{cases} \tag{3.1}$$

where $\vec{E}(\vec{r})$, $\vec{D}(\vec{r})$ denote, respectively, the electric field and the electric flux density vectors, while $\rho_v(\vec{r})$ denotes the electric charge density. Under the assumption of linear, isotropic media, $\vec{D}$ and $\vec{E}$ are related through the constitutive relation

$$\vec{D}(\vec{r}) = \epsilon_r(\vec{r})\epsilon_0 \vec{E}(\vec{r}), \tag{3.2}$$

where ϵ_0 is the electric permittivity of free space, while the relative electric permittivity, $\epsilon_r(\vec{r})$, is, in general, position dependent. In the following, the dependence of the fields and the electric permittivity on the position vector $\vec{r}$ will be suppressed for simplicity, unless its use is needed for clarity.

The irrotational nature of the electric field intensity under static conditions prompts the introduction of the electrostatic scalar potential, Φ, for its alternative computation through the relation,

$$\vec{E} = -\nabla\Phi. \tag{3.3}$$

Substitution of (3.3) in the second of the equations in (3.1), results in Poisson's equation statement for the electrostatic potential,

$$\nabla \cdot (\epsilon_r \nabla\Phi) = -\frac{\rho_v}{\epsilon_0}. \tag{3.4}$$

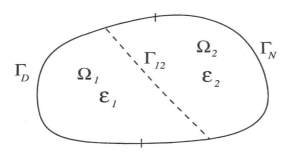

Figure 3.1   Domain for the statement of the electrostatic BVP.

A complete statement of the electrostatic BVP requires the definition of the domain within which the solution is sought. Depicted in Fig. 3.1 is a simple, representative domain consisting of two volumes, Ω_1 and Ω_2, that share the interior boundary Γ_{12}. The outer boundary is denoted as $\Gamma_D \bigcup \Gamma_N$. To ensure the uniqueness of the solution, appropriate boundary conditions must be imposed on all boundaries in the domain. More specifically, on the enclosing boundary of the solution domain the tangential component of the $\vec{E}$ must be specified on one portion of the boundary, while the normal component of $\vec{D}$ must be specified on the remaining portion. Thus, for example, for the simple geometry of Fig. 3.1, without loss of generality, we specify boundary conditions as follows:

$$\begin{cases} \hat{n} \times \vec{E} = \vec{E}_s & \text{on } \Gamma_D \\ \hat{n} \cdot \vec{D} = \rho_s & \text{on } \Gamma_N, \end{cases} \tag{3.5}$$

where $\hat{n}$ denotes the outward-pointing unit normal vector on the boundary, while ρ_s serves as the representation of the outward normal component of the electric flux density on the boundary. In terms of the electrostatic potential, Φ, these boundary conditions assume the form,

$$\begin{cases} \Phi = \Phi_s & \text{on } \Gamma_D \\ \hat{n} \cdot (\epsilon_r \nabla \Phi) = -\dfrac{\rho_s}{\epsilon_0} & \text{on } \Gamma_N. \end{cases} \tag{3.6}$$

The scalar function Φ_s defines the scalar potential value at any point on the boundary, Γ_D; hence, it constitutes a Dirichlet-type boundary condition and we refer to this portion of the boundary as the Dirichlet boundary. The second of the equations in (3.6) specifies the normal derivative of the scalar potential to assume a value consistent with the normal component of the electric flux density distribution on the remaining portion of the boundary, Γ_N; hence, it constitutes a Neumann-type boundary condition, and we refer to Γ_N as the Neumann boundary.

Boundary conditions are required also at the interior boundary Γ_{12}. The pertinent equations state the continuity of the tangential electric field intensity and the normal component of the electric flux density across the interface. Their mathematical statement, in the absence of any surface electric charge density on Γ_{12}, is

$$\begin{cases} \hat{n}_{21} \times (\vec{E}_2 - \vec{E}_1) = 0 \\ \hat{n}_{21} \cdot (\vec{D}_2 - \vec{D}_1) = 0, \end{cases} \tag{3.7}$$

where the subscripts 1 and 2 are used to denote the field quantities inside regions Ω_1 and Ω_2. Written in terms of the scalar potential these boundary conditions assume the form,

$$\begin{cases} \Phi_2 = \Phi_1 \\ \hat{n}_{21} \cdot (\epsilon_{r,2} \nabla \Phi_2 - \epsilon_{r,1} \nabla \Phi_1) = 0. \end{cases} \tag{3.8}$$

It is instructive to review the mathematical process used for proving the uniqueness of the electrostatic BVP defined by (3.4), (3.6) and (3.8) for the specific case of the domain depicted in Fig. 3.1. Assume that there exist two solutions, Φ_A and Φ_B, satisfying the given source and boundary conditions. Their difference, $\Phi = \Phi_A - \Phi_B$, satisfies $\nabla \cdot \epsilon_r \nabla \Phi = 0$ inside the domain $\Omega_1 \cup \Omega_2$, $\Phi = 0$ on Γ_D, $\partial \Phi / \partial n = 0$ on Γ_N, and (3.8) on Γ_{12}. Application of Green's first identity (A.17) for the functions Φ and $\epsilon_r \Phi$ in each one of the two domains, Ω_1 and Ω_2, followed by the summation of the resulting equations and the application of the continuity conditions (3.8) at Γ_{12}, yields the following equation:

$$\iiint_{\Omega_1 \cup \Omega_2} \epsilon_r |\nabla \Phi|^2 dv = \oiint_{\Gamma_D \cup \Gamma_N} \Phi \left(\epsilon_r \frac{\partial \Phi}{\partial n} \right) ds. \tag{3.9}$$

In view of the boundary conditions on Γ_D and Γ_N, the right-hand side of (3.9) is zero. Consequently, assuming that the relative permittivity is positive, it follows that $\nabla \Phi = 0$ everywhere inside $\Omega_1 \cup \Omega_2$. Thus Φ is constant inside $\Omega_1 \cup \Omega_2$. However, in view of the

Dirichlet boundary condition $\Phi = 0$ on Γ_D, we conclude that $\Phi = 0$ and, hence, $\Phi_A = \Phi_B$ inside $\Omega_1 \cup \Omega_2$. Thus the solution of the BVP defined by (3.4), (3.6) and (3.8) is unique.

The special case where a Neumann boundary condition is imposed over the entire enclosing boundary of the domain requires special care for the following reason. Since in this case $\partial\Phi/\partial n = 0$ everywhere on the boundary, uniqueness of the solution is guaranteed apart from an additive constant. For the continuous case this constant is of no importance and is chosen arbitrarily as the reference electrostatic potential value. However, in the discrete approximation of the electrostatic BVP, this arbitrariness of the reference potential value implies that the numerical solution of the matrix statement of the associated discrete BVP will not be unique; hence, the matrix is singular. This situation is easily rectified by explicitly "pinning" the value of one of the degrees of freedom in the approximation of the electrostatic potential. In other words, a reference potential value is explicitly assigned. The equation associated with this degree of freedom is subsequently eliminated from the matrix statement of the BVP, resulting in an invertible matrix.

3.1.2 Weak statement of the electrostatic BVP

The development of the weak statement of the electrostatic BVP will be carried out using Galerkin's method [8]. Multiplication of (3.4) by a scalar weighting function w, followed by the integration of the resulting statement over the computational domains of Ω_1 and Ω_2, yields, respectively,

$$
\begin{aligned}
\iiint_{\Omega_1} \nabla w \cdot \epsilon_{r1} \nabla\Phi\, dv - \oiint_{\partial\Omega_1} w\, \hat{n} \cdot \epsilon_{r1} \nabla\phi\, ds &= \frac{1}{\epsilon_0} \iiint_{\Omega_1} w\, \rho_v\, dv, \\
\iiint_{\Omega_2} \nabla w \cdot \epsilon_{r2} \nabla\Phi\, dv - \oiint_{\partial\Omega_2} w\, \hat{n} \cdot \epsilon_{r2} \nabla\phi\, ds &= \frac{1}{\epsilon_0} \iiint_{\Omega_2} w\rho_v\, dv,
\end{aligned}
\tag{3.10}
$$

where use was made of Green's first identity (A.17). The sum of the above two equations and use of the boundary conditions on the boundaries yields

$$
\iiint_{\Omega} \nabla w \cdot \epsilon_r \nabla\Phi\, dv = \frac{1}{\epsilon_0} \left(\iiint_{\Omega} w\, \rho_v\, dv - \iint_{\Gamma_N} w\, \rho_s\, ds \right)
\tag{3.11}
$$

where $\Omega = \Omega_1 \cup \Omega_2$. The above equation is used to guide us in the proper choice of the class of testing (or weighting) functions w. More specifically, since in the Galerkin's process the testing functions are the same with the functions used for the approximation of Φ over the computational domain, for the integrals in (3.11) to have meaning, w and its first derivatives must be square-integrable over Ω and w must vanish on Γ_D. The mathematical statement of this choice is

$$
w \in \left\{ w \,|\, w \in L^2(\Omega), \nabla w \in L^2(\Omega), w = 0 \text{ on } \Gamma_D \right\},
\tag{3.12}
$$

where $L^2(\Omega)$ denotes the space of square-integrable scalar functions in Ω.

The weak statement (3.11) contains explicitly the source terms in the BVP, namely, the electric charge density ρ_v inside Ω and the normal derivative of the electric flux density on Γ_N defined by the term ρ_s. While initially disconcerting, due to the unknown value of $\partial\Phi/\partial n$ on Γ_D, the additional term on the right-hand side of (3.11) associated with the Dirichlet boundary does not pose any particular difficulties in the development of the matrix statement. Since the value of the potential is specified on Γ_D, the subset of testing functions used for the development of the matrix statement is chosen such that its elements are zero on

Γ_D. The information about the value of the potential on Γ_D enters the approximate statement of the problem through the expansion of the potential. More specifically, appropriate values are assigned a-priori to the degrees of freedom in the expansion associated with the Dirichlet boundary. This point is elaborated further in the following paragraphs.

The development of the finite element approximation of the electrostatic BVP continues with the choice of the space of expansion functions for the unknown potential. From the discussion in Chapter 2 we know that the natural candidate for this purpose is the p-th order scalar space W_s^p. More specifically, and without loss of generality, the first-order scalar space, W_s^1, of the well-known node elements is chosen for the expansion of the potential over the computational domain discretized using tetrahedra. Let N_n denote the number of nodes in the discretization, from which N_{Dn} nodes are on Γ_D. For clarity in the discussion, the numbering of the nodes is taken to be such that the indices of the nodes on Γ_D are the last in the sequence of N_n. Thus, in accordance with the discussion in the previous paragraph, the approximation of the electrostatic potential assumes the form

$$\Phi = \sum_{i=1}^{N_n - N_{Dn}} w_{n,i}\, v_i + \sum_{i=N_n - N_{Dn}+1}^{N_n} w_{n,i}\, \phi_{s,i}, \tag{3.13}$$

where the coefficients $v_1, \ldots, v_{N_n - N_{Dn}}$ constitute the unknown degrees of freedom in the approximation, while the coefficients $\phi_{s,N_n - N_{Dn}+1}, \ldots, \phi_{s,N_n}$ are known, with values consistent with the assigned potential Φ_s on Γ_D. The subspace $W_{n,\Omega-\Gamma_D}$ of the testing functions consists of the following node elements:

$$W_{n,\Omega-\Gamma_D} = \{w_{n,1}, w_{n,2}, \cdots, w_{n,N_n - N_{Dn}}\}. \tag{3.14}$$

Substitution of (3.13) in (3.11), followed by testing with each one of the elements of $W_{\Omega-\Gamma_D}$, results in the matrix statement of the finite element approximation of the BVP

$$S_v x_v = f_v, \tag{3.15}$$

where $x_v = [v_1, \cdots, v_{N_n - N_{Dn}}]^T$ is the vector of the unknown degrees of freedom, S_v is the finite element matrix of dimension $(N_n - N_{Dn}) \times (N_n - N_{Dn})$, and f_v is the forcing vector, also of length $(N_n - N_{Dn})$. Their elements are calculated through the following integrals

$$(S_v)_{i,j} = \iiint_\Omega \nabla w_{n,i} \cdot \epsilon_r \nabla w_{n,j} dv \tag{3.16}$$

$$f_{v,i} = \frac{1}{\epsilon_0} \left(\iiint_\Omega w_{n,i}\rho_v dv - \iint_{\Gamma_N} w_{n,i}\rho_s dv \right)$$
$$- \sum_{j=N_n - N_{Dn}+1}^{N_n} \phi_j \iiint_\Omega \nabla w_{n,i} \cdot \epsilon_r \nabla w_{n,j} dv. \tag{3.17}$$

The solution of (3.15) yields the coefficients in the finite element approximation (3.13) for the electrostatic potential and, thus, a polynomial approximation of the solution over the computational domain. As already mentioned, the system is solvable only with a reference potential value explicitly defined in the finite element approximation. In other words, a Dirichlet-type boundary condition must always be present in the finite element statement of the electrostatic BVP. For example, in most cases of practical interest the potential value on the surface of one or more conductors in the computational domain is assigned.

It is immediately evident from the expression for the elements of S_v that the finite element matrix is real and symmetric. Whether the matrix is also positive definite can be examined by considering the scalar $v^T S_v v$, $v \neq 0$. First, let us consider the case where no Dirichlet condition is assigned; hence, $N_{Dn} = 0$ and the entire enclosing boundary is a Neumann boundary. It follows immediately from the above equation that the choice $v = [1, 1, \ldots, 1]^T$ results in

$$v^T S_v v = \iiint_{\Omega} \sum_{i=1}^{N_n} \nabla w_{n,i} \cdot \epsilon_r \sum_{j=1}^{N_n} \nabla w_{n,j} \, dv = 0, \tag{3.18}$$

since $\sum_{i=1}^{N_n} w_{n,i} = 1$ in Ω. Clearly, this result indicates that the vector $v = [1, 1, \ldots, 1]^T$ is an eigenvector of S_v with eigenvalue zero.

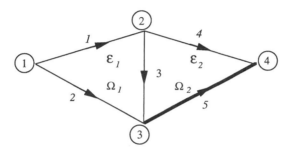

Figure 3.2 A two-dimensional domain consisting of two elements.

Consider a simple two-dimensional example with only two elements as depicted in 3.2. The four nodes are ordered according to the number assigned to them. Without loss of generality, let us consider the product of the second row of S_v with $v = [1, 1, \ldots, 1]^T$ which is

$$
\begin{aligned}
\sum_{i=1}^{4} S_v(2, i) \quad &= \langle w_2, \epsilon_1 w_1 \rangle_{\Omega_1} + [\langle w_2, \epsilon_1 w_2 \rangle_{\Omega_1} + \langle w_2, \epsilon_2 w_2 \rangle_{\Omega_2}] \\
&\quad + [\langle w_2, \epsilon_1 w_3 \rangle_{\Omega_1} + \langle w_2, \epsilon_2 w_3 \rangle_{\Omega_1}] + \langle w_2, \epsilon_2 w_4 \rangle_{\Omega_2} \\
&= \langle w_2, \epsilon_1 (w_1 + w_2 + w_3) \rangle_{\Omega_1} + \langle w_2, \epsilon_2 (w_2 + w_3 + w_4) \rangle_{\Omega_2} \\
&= \langle w_2, \epsilon_1 1 \rangle_{\Omega_1} + \langle w_2, \epsilon_2 1 \rangle_{\Omega_2} = 0,
\end{aligned}
$$

where $\langle w_i, \epsilon_s w_j \rangle_{\Omega_s}$ denotes $\int_{\Omega_s} \nabla w_i \cdot \epsilon_s \nabla w_j \, ds$ and use was made of the fact that $w_1 + w_2 + w_3 = 1$ in Ω_1 and $w_2 + w_3 + w_4 = 1$ in Ω_2. Hence, the matrix S_v is singular and non-invertible. This result is consistent with our comments earlier in this section about the difficulties encountered in the numerical solution of an electrostatic BVP for which a reference potential has not been assigned explicitly. However, this situation is easily rectified by assigning non-zero values to one or more of the degrees of freedom in the finite element approximation of the potential; hence, $N_{Dn} \neq 0$, and the right-hand side of (3.18) is always non-zero for any non-zero vector v, even in the limiting case where a Neumann boundary condition is imposed over the entire enclosing boundary. Thus we conclude that, the presence of a Dirichlet boundary condition in the statement of the electrostatic BVP suffices to guarantee that the finite element matrix of (3.15) is positive definite.

3.1.3 The case of unbounded domains

The finite element solution of an electrostatic BVP for which one of the boundaries is at infinity necessitates the use of special practices for the development of a numerical approximation involving a finite number of degrees of freedom, while reproducing with acceptable engineering accuracy the impact of the unbounded domain on the field behavior.

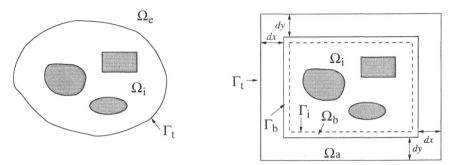

Figure 3.3   a: Pictorial description of an unbounded domain. b: Definition of buffer and absorption layers for the application of coordinate-stretching based domain truncation.

To fix ideas, consider the geometry depicted on the left side of Fig. 3.3. The mathematical boundary Γ_t is introduced to separate the unbounded domain into two subdomains. Subdomain Ω_i contains all sources and the finite conducting and dielectric volumes associated with the electrostatic BVP. Ω_e denotes the exterior domain which is assumed to be simple enough for its Green's function to be readily available. For example, two cases of significant practical interest are the case of a homogeneous unbounded medium and the case of an unbounded planar stratified medium.

The natural way to approach the numerical approximation of such a BVP is through the combination of a finite element approximation of the problem in Ω_i with a boundary integral statement of the exterior problem in Ω_e, expressed in terms of the unknown potential and its normal derivative on Γ_t. Such hybrid, finite element/boundary integral methods offer the most effective way for handling the unbounded domain in a rigorous manner [8]. The price one pays is the added complexity of the matrix statement of the approximate problem, caused by the need to augment the sparse finite element matrix statement of the interior BVP with the dense matrix statement resulting from the numerical approximation of the boundary integral statement of the exterior BVP. Reference [8] presents several approaches used in practice to facilitate the solution of the system of equations resulting from the approximation of such hybrid finite element/integral equation statements of unbounded BVPs.

Alternatives to the above hybrid formulations, aimed at the development of the approximate statement relying solely on finite elements, exploit the fact that, in most practical applications, it is the *near field* that is of interest. For example, for the purposes of calculating the capacitance matrix for a set of conductors, the quantity of interest is the normal component of the electrostatic potential on the surfaces of the conductors. Thus, with the boundary Γ_t placed at a distance sufficiently far away from the conducting and/or dielectric finite volumes and sources, a way is sought for approximating, in the context of the finite element formalism, the behavior of the solution in Ω_e.

Toward this objective, a popular approach has been the application of semi-infinite elements [9]. One way of defining such elements is as maps of elements of finite size to semi-infinite ones in the direction along which the domain of interest extends to infinity.

With reference to Fig. 3.3, these semi-infinite elements will be the outermost elements in the finite element discretization of Ω_i. In this manner, the boundary condition for the potential at infinity (which, in most cases of practical interest, is of the Dirichlet type with the potential due to a finite distribution of charge assuming the value of zero at infinity) is enforced directly at the truncation boundary Γ_t. In addition to the discussion in [9], [4] offers extensive information on the construction of semi-infinite elements and their application to the finite element solution of unbounded electrostatic BVPs.

An alternative approach for grid truncation has been proposed in [10] in the context of the application of the method of finite differences for the solution of unbounded electrostatic problems. However, the methodology is quite general, thus making its utilization in the context of other types of differential equation-based numerical methods fairly straightforward. In the following, the development of the method in the context of finite elements is presented.

A simple analytical problem will be used to illustrate the concept of coordinate stretching and its suitability for truncation of unbounded domains. The pertinent two-dimensional BVP concerns the solution of Poisson's equation in a homogeneous strip in the xy-plane,

$$
\begin{cases}
\nabla^2 \Phi(x, y) = -\dfrac{1}{\epsilon_0} \delta(x)\delta(y - b), & (0 \le y \le a, -\infty \le x \le \infty); \\
\Phi(x, 0) = \Phi(x, a) = 0.
\end{cases}
\tag{3.19}
$$

The solution to this problem is easily found to be

$$
\Phi(x, y) = \sum_{n=1}^{\infty} \frac{1}{n\pi} \sin \frac{n\pi b}{a} \sin \frac{n\pi y}{a} \exp\left(-\frac{n\pi |x|}{a}\right).
\tag{3.20}
$$

Consider, next, the following boundary value problem,

$$
\begin{cases}
\nabla^2 \Phi_i(x, y) = -\dfrac{1}{\epsilon_0} \delta(x)\delta(y - b), & (0 \le y \le a, |x| \le c); \\
\Phi_i(x, 0) = \Phi_i(x, a) = 0, \\
\nabla_s \cdot (\epsilon_0 \nabla_s \Phi_e(x, y)) = 0, & (0 \le y \le a, |x| \ge c); \\
\Phi_e(x, 0) = \Phi_e(x, a) = 0, \\
\Phi_i(\pm c, y) = \Phi_e(\pm c, y); \quad \hat{x} \cdot (\epsilon_0 \nabla \Phi_i - \epsilon_0 \nabla_s \Phi_e) \text{, at } x = \pm c.
\end{cases}
\tag{3.21}
$$

In the above statement, the operator ∇_s is defined as follows:

$$
\nabla_s = \hat{x} \frac{1}{s_x(x)} \frac{\partial}{\partial x} + \hat{y} \frac{1}{s_y(y)} \frac{\partial}{\partial y} + \hat{z} \frac{1}{s_z(z)} \frac{\partial}{\partial z}.
\tag{3.22}
$$

We will refer to this operator as the *stretched* gradient operator, following suit with the work in [11] where the idea of coordinate stretching was introduced in conjunction with unbounded grid truncation for the numerical solution of electromagnetic radiation problems using finite methods.

Returning to the two-dimensional BVP of (3.21), let $s_y(y) = 1$ and $s_x(x) = s$, where s is a real constant. It is easily checked that the solution of (3.21) is as follows:

$$\Phi_i(x, y) = \sum_{n=1}^{\infty} \frac{1}{n\pi} \sin \frac{n\pi b}{a} \sin \frac{n\pi y}{a} \exp \left(-\frac{n\pi |x|}{a} \right),$$

$$(0 \le y \le a, |x| \le c);$$

$$\Phi_e(x, y) = \sum_{n=1}^{\infty} \frac{1}{n\pi} \sin \frac{n\pi b}{a} \sin \frac{n\pi y}{a} \exp \left(-\frac{n\pi c}{a} \right) \exp \left(-s\frac{n\pi |x - c|}{a} \right),$$

$$(0 \le y \le a, |x| \le c).$$

(3.23)

It is immediately apparent from (3.23) and (3.20) that the solution in the interior domain, bounded by the planes $x = \pm c$ and enclosing the source, of the modified BVP (3.21) is identical to the solution of the original BVP (3.19). However, in the exterior domain, $|x| \ge c$, the exponential decay of the solution of the modified problem can be enhanced by selecting $s > 1$. Thus, with a proper choice of the stretching variable s, a rapid decay of the potential can be achieved over a reasonably short distance, d, away from the interior domain, without any impact on the solution for the potential in the interior domain. With d and s chosen in a manner such that the exponential decay is strong enough to render the potential value essentially zero at the planes $x = \pm(c+d)$, it is reasonable to expect that use of the Dirichlet boundary condition $\Phi_e(x = \pm(c+d), y) = 0$ to truncate the infinite extent of the exterior domain in the BVP (3.21) will have negligible impact on the solution for the potential in the interior domain.

The results from this simple canonical problem suggest that coordinate stretching may be used to replace, for the purposes of the numerical solution, an electrostatic BVP in an unbounded domain with one defined over a bounded domain. More specifically, referring to the unbounded geometry depicted on the left-hand side of Fig. 3.3, a bounded-domain version of the geometry is defined as depicted on the right-hand side of the figure. While, for simplicity, the pictorial description is for a two-dimensional geometry, the mathematical development of the approximate bounded-domain statement of the electrostatic BVP is given for the general three-dimensional case.

Let Γ_i denote the boundary enclosing the interior domain Ω_i that contains all finite dielectric and conducting volumes (i.e., all material inhomogeneities) along with all the sources defining the electrostatic BVP. A *buffer* region, Ω_b, is introduced, with boundaries Γ_i and Γ_b, where Γ_b is taken to be the surface of a rectangular box enclosing Ω_i. The material properties of Ω_b are the same with those in the same domain in the original (unbounded) domain. Subsequently, an *attenuation layer* Ω_a is defined, bounded by Γ_b and an exterior boundary Γ_t, taken to be the surface of a rectangular box that constitutes the truncation boundary for the modified bounded problem. The distances d_x, d_y, d_z of Γ_t from Γ_b along $\pm x, \pm y$ and $\pm z$, respectively, are used to define the thickness of the attenuation layer.

The bounded-domain approximation of the original unbounded problem relies upon the implementation of the stretched ∇ operator, ∇_s, of (3.22) in place of the physical ∇ operator, with the stretching parameters s_x, s_y and s_z taken to be equal to 1 inside Ω_i and Ω_b, while assuming values greater than 1 within the attenuation domain Ω_a, in a position-dependent sense to be described in the following paragraph. In addition to all original source and boundary conditions at all finite boundaries in the original problem, the Dirichlet boundary condition $\Phi(\vec{r}) = 0, \vec{r} \in \Gamma_t$, is also imposed. This completes the statement of the bounded-domain approximation of the original unbounded problem. The weak statement and the finite-element matrix approximation of the resulting BVP can be deduced directly from the corresponding expressions (3.11) and (3.17) through the replacement of ∇ with ∇_s.

The parameters that govern the implementation of this grid truncation methodology are the thickness of the buffer region, the thickness of the attenuation layer, and the functional dependence of the stretching parameters s_x, s_y and s_z on position. With respect to the latter, and with reference to Fig. 3.3, we choose $s_y = s_z = 1$ for the left and right attenuation layers, while s_x is taken to be a function of x only, assuming values greater than 1. Similarly, for the top and bottom attenuation layers the choice $s_x = s_z = 1$, $s_y(y) > 1$ is made, while the choice $s_x = s_y = 1$, $s_z(z) > 1$ is employed for the attenuation layers perpendicular to the z axis. With regards to the functional dependence of the stretching coefficients, the results in [10] suggest that a variation of the form

$$s(d) = 1 + s_{max} \left(\frac{d}{D} \right)^n, \quad s_{max} > 0, \quad 0 \le d \le D \tag{3.24}$$

with $n = 1$ or 2, should provide for numerical solution of satisfactory accuracy. In the above equation, D denotes the thickness of the attenuation layer. While the specific choice of D and s_{max} is, in general, dependent on the geometric and material attributes of the BVP under consideration, the studies in [10] suggest that very good performance should be expected by ten-cell thick attenuation layers with average values, s_{ave}, of 10 or greater for the stretching coefficient. From (3.24), s_{ave} can be expressed in terms of s_{max}, and n as follows, $s_{ave} = 1 + (s_{max}/(n+1))$.

With regards to the thickness of the buffer region, numerical studies of [10] concerning the calculation of the per-unit-length capacitance of stripline structures suggest that a buffer layer of minimum thickness of five cells should be sufficient for good numerical solution accuracy in the interior domain.

3.2 MAGNETOSTATIC BOUNDARY VALUE PROBLEMS

3.2.1 Governing equations and boundary conditions

Under static conditions magnetic phenomena in linear media are governed by the following system of equations:

$$\begin{cases} \nabla \times \vec{H} = \vec{J}_v \\ \nabla \cdot \vec{B} = 0 \end{cases} \quad \text{and} \quad \vec{B} = \mu_r \mu_0 \vec{H}, \tag{3.25}$$

where $\vec{J}_v$ is the impressed volume electric current density, μ_r is the relative magnetic permeability of the medium, and μ_0 denotes the magnetic permeability of free space. For simplicity, the dependence of $\vec{J}_v$, μ_r, the magnetic field intensity $\vec{H}$, and the magnetic flux density $\vec{B}$, on position has been suppressed.

At first glance, the system (3.25) appears to be over-determined since it consists of four scalar equations for the three components of the magnetic field. One approach to resolve this concern is to recall Helmholtz's theorem which, loosely speaking, states that any vector field, $\vec{F}$, in a domain Ω is uniquely expressed as the sum of the gradient of a scalar field, Ψ, and the curl of a solenoidal vector field $\vec{G}$ [12],

$$\vec{F} = \nabla \Psi + \nabla \times \vec{G}, \quad \nabla \cdot \vec{G} = 0. \tag{3.26}$$

Furthermore, from the divergence of $\vec{F}$ and through the application of simple vector calculus identities, it follows immediately that Ψ can be specified uniquely once the divergence of $\vec{F}$

inside the domain along with its normal component on the enclosing boundary are specified. In a similar fashion, from the curl of $\vec{F}$ it can be deduced that $\vec{G}$ can be specified uniquely once the curl of $\vec{F}$ along with its tangential components on the enclosing boundary are specified.

From the above we conclude that (3.25) is not an over-determined system, since both the curl and the divergence of the vector field must be known for its determination. Furthermore, the conditions that must be imposed on boundaries to ensure uniqueness of the solution of the BVP will involve the normal and tangential components of the field.

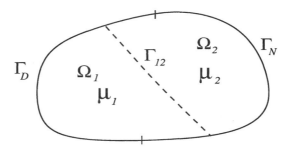

Figure 3.4 Domain for the statement of the magnetostatic BVP.

The complete statement of a well-posed magnetostatic BVP for which a solution can be uniquely derived requires the assignment of appropriate boundary conditions on material interface boundaries and the enclosing boundary of the domain of interest. Their discussion is presented in the context of the geometry of Fig. 3.4, where μ_{r1} and μ_{r2} denote the relative magnetic permeabilities for the two subdomains.

We consider first the material interface boundary conditions at Γ_{12}. These are well known, derived directly from the integral form of the equations in (3.25), and given by

$$\begin{cases} \hat{n}_{21} \times \left(\vec{H}_2 - \vec{H}_1 \right) = 0 \\ \hat{n}_{21} \cdot \left(\vec{B}_2 - \vec{B}_1 \right) = 0, \end{cases} \tag{3.27}$$

where it is assumed that no surface electric current density is present on Γ_{12}.

Of particular interest to our purposes are two types of boundary conditions. The first involves the assignment of the tangential components of the magnetic field on a portion of the boundary. The second boundary condition involves the assignment of the normal component of the magnetic flux density on a portion of the boundary. In particular, of interest is the case where the normal component is set to zero on the boundary. From the second equation in (3.27), it is straightforward to conclude that this choice is associated with the region exterior to the portion of the boundary on which this condition is imposed having infinite permeability.

In summary, and with reference to the geometry depicted in Fig. 3.4, we will be concerned with magnetostatic BVPs with the following homogeneous boundary conditions imposed on the enclosing boundary:

$$\begin{cases} \hat{n} \times \vec{H} = 0, & \text{on } \Gamma_N, \\ \hat{n} \cdot \vec{B} = 0, & \text{on } \Gamma_D. \end{cases} \tag{3.28}$$

3.2.1.1 Solvability Conditions At this point, and prior to discussing the definition of appropriate boundary conditions on the enclosing boundary, it is useful to examine the so-called solvability conditions for the system (3.25). The first condition requires that the impressed current source, $\vec{J}_v$, is solenoidal,

$$\nabla \cdot \vec{J}_v = 0, \tag{3.29}$$

a result that follows immediately from the first of the equations in (3.25) by taking the divergence of both sides. Furthermore, and in view of this constraint, additional constraints can be deduced for the normal component of $\vec{J}_v$ on the enclosing boundary.

Use of (3.29) yields the result that the net flux of $\vec{J}_v$ through $\Gamma_D \cup \Gamma_N$ is zero,

$$\oiint_{\Gamma_D \cup \Gamma_N} \hat{n} \cdot \vec{J}_v ds = 0. \tag{3.30}$$

To derive the second constraint on the normal component of $\vec{J}_v$ on the enclosing boundary, we will make use of the following result for a vector field $\vec{H}$ on Γ_N:

$$\text{If } \hat{n} \times \vec{H} = 0 \text{ on } \Gamma_N, \quad \text{then } \hat{n} \cdot \nabla \times \vec{H} = 0 \text{ on } \Gamma_N. \tag{3.31}$$

The proof of the above result follows immediately by taking a contour integral over the small neighborhood of area $\Delta\Gamma$ surrounding a point $\vec{r}_s$ on Γ_N and applying Stokes theorem (A.21)

$$\oint_{\Delta c} \vec{H} \cdot \delta\vec{\ell} = \iint_{\Delta\Gamma} \hat{n} \cdot (\nabla \times \vec{H}) ds = 0, \tag{3.32}$$

where Δc is the contour of $\Delta\Gamma$. Consequently, we obtain the additional solvability constraint for $\vec{J}_v$,

$$\hat{n} \cdot \vec{J}_v = 0 \quad \text{on } \Gamma_N. \tag{3.33}$$

In view of this last result we conclude that the surface integration in (3.30) involves only the surface of Γ_D.

Equations (3.29), (3.30) and (3.33) constitute the constraints that must be imposed on the electric current source density for the magnetostatic BVP to be solvable. Use of these constraints will be made later in this section, in the context of the proper definition and subsequent finite element approximation of the electric current source density used as excitation of the magnetostatic BVP.

3.2.1.2 Uniqueness of Solution The uniqueness of the solution of the system (3.25) subject to the boundary conditions (3.28) and (3.27) can be deduced through the following argument. Let $(\vec{H}_a, \vec{B}_a)$ and $(\vec{H}_b, \vec{B}_b)$ be two different solutions of the BVP. Since the two solutions satisfy the same boundary and source conditions, their difference, $\vec{H}_d = \vec{H}_a - \vec{H}_b$, $\vec{B}_d = \vec{B}_a - \vec{B}_b$, in addition to satisfying the boundary conditions, satisfies the following equations inside Ω,

$$\begin{cases} \nabla \times \vec{H}_d = 0, \\ \nabla \cdot \vec{B}_d = 0. \end{cases} \tag{3.34}$$

Then, it follows immediately from Helmholtz's theorem that $\vec{H}_d = 0$ inside Ω, thus proving the uniqueness of the solution of (3.25) subject to the boundary conditions.

3.2.1.3 The Magnetic Vector Potential An alternative statement of the magneto-static BVP can be derived in terms of the magnetic vector potential, $\vec{A}$. Taking advantage of the solenoidal character of the magnetic flux density $\vec{B}$, the magnetic vector potential is introduced in the following manner:

$$\vec{B} = \nabla \times \vec{A}. \tag{3.35}$$

Use of (3.35) in Ampere's law in (3.25) yields the partial differential equation for $\vec{A}$,

$$\nabla \times \frac{1}{\mu_r} \nabla \times \vec{A} = \mu_0 \vec{J}_v. \tag{3.36}$$

Furthermore, in view of (3.35) and (3.31) the boundary conditions (3.28) on the enclosing boundary, Γ, assume the form

$$\begin{cases} \hat{n} \times \dfrac{1}{\mu_r} \nabla \times \vec{A} = 0, & \text{on } \Gamma_N, \\ \hat{n} \times \vec{A} = 0, & \text{on } \Gamma_D. \end{cases} \tag{3.37}$$

Similarly, the boundary conditions on the material interface boundary Γ_{12}, are cast in the following form:

$$\begin{cases} \hat{n}_{21} \times \left(\dfrac{1}{\mu_{r,2}} \nabla \times \vec{A}_2 - \dfrac{1}{\mu_{r,1}} \nabla \times \vec{A}_1 \right) = 0, \\ \hat{n}_{21} \times \left(\vec{A}_2 - \vec{A}_1 \right) = 0. \end{cases} \tag{3.38}$$

In view of the Helmholtz theorem, unless the divergence of $\vec{A}$ is specified, the uniqueness of the solution $\vec{A}$ to a given magnetostatic BVP cannot be guaranteed. More specifically, if $\vec{A}$ is a solution to the magnetostatic BVP of (3.36), (3.37) and (3.38), then $\vec{A} + \nabla\Psi$, where Ψ is a scalar field subject to the boundary conditions consistent with (3.37) and (3.38) is also a solution to the magnetostatic BVP resulting in the same magnetic field. This non-uniqueness of the solution of (3.36), which is intimately connected with the fact that the curl operator has a non-trivial null space that includes the gradient of any scalar field, is of major importance in the development of the finite element approximation of the magnetostatic BVP and will be discussed at length later in this section.

3.2.2 Weak statement of the magnetostatic BVP

The magnetic vector potential statement of the magnetostatic BVP will be employed for the development of a weak statement to be used for the subsequent construction of a finite element approximation [13]-[17]. For this purpose, the inner product of a vector weighting function, $\vec{w}$, with (3.36) is integrated over the union of the domains Ω_1 and Ω_2 of Fig. 3.4. This, followed by the application of (A.19) and the enforcement of the continuity of the tangential component of the magnetic field at Γ_{12} yields

$$\iiint_\Omega \nabla \times \vec{w} \cdot \frac{1}{\mu_r} \nabla \times \vec{A} \, dv = -\oiint_{\Gamma_D \cup \Gamma_N} \vec{w} \cdot \left(\hat{n} \times \frac{1}{\mu_r} \nabla \times \vec{A} \right) ds + \mu_0 \iiint_\Omega \vec{w} \cdot \vec{J}_v \, dv. \tag{3.39}$$

The surface integral term in the above weak statement of the magnetostatic BVP may be split into two parts as follows:

$$-\iint_{\Gamma_D} (\hat{n} \times \vec{w}) \cdot \left(\frac{1}{\mu_r} \nabla \times \vec{A} \right) ds + \iint_{\Gamma_N} \vec{w} \cdot \left(\hat{n} \times \frac{1}{\mu_r} \nabla \times \vec{A} \right) ds. \tag{3.40}$$

In view of (3.37), the integral over the Neumann boundary Γ_N vanishes. The Dirichlet condition $\hat{n} \times \vec{A} = 0$ on Γ_D is imposed by choosing the testing functions $\vec{w}$ such that $\hat{n} \times \vec{w} = 0$ on Γ_D. Thus the weak statement becomes

$$\iiint_\Omega \nabla \times \vec{w} \cdot \frac{1}{\mu_r} \nabla \times \vec{A} \, dv = \mu_0 \iiint_\Omega \vec{w} \cdot \vec{J}_v \, dv. \tag{3.41}$$

The appropriate choice of expansion and testing functions for the development of the finite element approximation of the BVP is dictated by the physical attributes of the magnetic vector potential and the boundary conditions it satisfies. Clearly, the TV space discussed in Chapter 2 is the appropriate space for its expansion. In particular, since the sources for $\vec{A}$ are solenoidal current densities, the appropriate space for its expansion is the non-gradient subspace of the TV space. In the following, without loss of generality, $\vec{A}$ is approximated in the first-order TV space $W_{tv,e,ng}^1$, i.e., the edge-element space.

Let N_e denote the total number of edges, and N_{De} the number of edges on Γ_D on which the homogeneous Dirichlet condition in (3.37) for the tangential component of $\vec{A}$ is imposed. The magnetic vector potential $\vec{A}$ is expanded as

$$\vec{A} = \sum_{i=1}^{N_e - N_{De}} \vec{w}_{e,i} a_i = \vec{W}_{e,\Omega - \Gamma_D} x_a, \tag{3.42}$$

where $\vec{w}_{e,i}$ denotes the edge-elements excluding those associated with Γ_D, x_a is the vector of expansion coefficients a_i, and $\vec{W}_{e,\Omega - \Gamma_D}$ is a row vector with elements the functions of the edge-element space excluding those on Γ_D

$$\vec{W}_{e,\Omega - \Gamma_D} = [\vec{w}_{e,1}, \vec{w}_{e,2}, \cdots, \vec{w}_{e,N_e - N_{De}}]. \tag{3.43}$$

Substitution of (3.42) into (3.41), followed by testing with all elements of $\vec{W}_{e,\Omega - \Gamma_D}$, results in the finite-element matrix statement of the magnetostatic BVP,

$$S_a x_a = f_a. \tag{3.44}$$

The elements for the system matrix, S_a, and the forcing vector, f_a, are calculated in terms of the following integrals:

$$(S_a)_{i,j} = \iiint_\Omega \nabla \times \vec{w}_{e,i} \cdot \frac{1}{\mu_r} \nabla \times \vec{w}_{e,j} \, dv,$$
$$f_{a,i} = \mu_0 \iiint_\Omega \vec{w}_{e,i} \cdot \vec{J}_v \, dv. \tag{3.45}$$

The dimension of the system matrix S_a is equal to the total number of edges in the finite element mesh excluding those on the Dirichlet boundary, Γ_D.

Prior to attempting the numerical inversion of (3.44), it is important to consider the properties of S_a. In particular, we will consider two cases pertinent to several practical applications. The first concerns the case where no Dirichlet condition is imposed on the enclosing boundary. The second is the case where a Dirichlet condition is imposed.

When no Dirichlet boundary condition is imposed on the enclosing boundary of the domain the dimension of S_a is N_e, equal to the total number of edges in the finite element mesh. Let N_n denote the total number of nodes in the mesh. From Chapter 2 we recall the following relationship between the basis functions of node elements and edge elements,

$$[\nabla w_{n,1}, \nabla w_{n,2}, \cdots, \nabla w_{n,N_n}] = [\vec{w}_{e,1}, \vec{w}_{e,2}, \cdots, \vec{w}_{e,N_e}] G_{N_e \times N_n}, \tag{3.46}$$

where $G_{N_e \times N_n}$ is the *gradient matrix*, introduced in Chapter 2, of (column) rank N_n-1. Taking the curl of the above equation yields

$$\left[\nabla \times \vec{w}_{e,1}, \nabla \times \vec{w}_{e,2}, \cdots, \nabla \times \vec{w}_{e,N_e}\right] G_{N_e \times N_n} = 0. \tag{3.47}$$

In view of this result and the expression for the elements of S_a in (3.45), we arrive at the following matrix statement:

$$S_a G = 0. \tag{3.48}$$

Furthermore, since the column rank of G is N_n-1, we conclude that S_a has N_n-1 zero eigenvalues with corresponding independent eigenvectors provided by any set of N_n-1 of the columns of G.

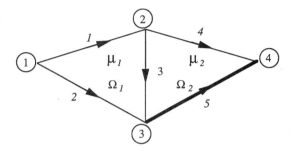

Figure 3.5   A two-dimensional domain consisting of only two elements.

This is an important result of relevance to all finite element approximations of electromagnetic BVPs with unknowns vector fields that exhibit tangential continuity at material boundaries, and, thus are most suitably expanded using edge elements.

Let us confirm this result for the case of the simple two-dimensional finite element mesh depicted in Fig. 3.5. It consists of two elements, Ω_1, and Ω_2, with magnetic permeabilities μ_1 and μ_2, respectively. In the absence of any Dirichlet boundary conditions on the four edges that constitute the enclosing boundary, the finite element matrix, S_a, is of dimension 5, while the matrix G is a 5×4 matrix of the following form:

$$G = \begin{bmatrix} -1 & 1 & 0 & 0 \\ -1 & 0 & 1 & 0 \\ 0 & -1 & 1 & 0 \\ 0 & -1 & 0 & 1 \\ 0 & 0 & -1 & 1 \end{bmatrix}. \tag{3.49}$$

Note that the sum of the columns of G is the zero column vector; hence, the column rank of G is 3. Without loss of generality, let us consider the product of the third row of S_a, with the second column of G,

$$\sum_{i=1}^{5} S_a(3,i)G(i,2)$$
$$= \langle \vec{w}_3, \mu_1 \vec{w}_1 \rangle_{\Omega_1} - \left(\langle \vec{w}_e, \mu_1 \vec{w}_3 \rangle_{\Omega_1} + \langle \vec{w}_3, \mu_2 \vec{w}_3 \rangle_{\Omega_2} \right) - \langle \vec{w}_3, \mu_2 \vec{w}_4 \rangle_{\Omega_2} \tag{3.50}$$
$$= \langle \vec{w}_3, \mu_1 (\vec{w}_1 - \vec{w}_3) \rangle_{\Omega_1} - \langle \vec{w}_3, \mu_2 (\vec{w}_3 + \vec{w}_4) \rangle_{\Omega_2}$$
$$= \langle \vec{w}_3, \mu_1 \nabla w_2 \rangle_{\Omega_1} + \langle \vec{w}_3, \mu_2 \nabla w_2 \rangle_{\Omega_2} = 0.$$

In the above, $\langle \vec{a}, \mu_s \vec{b} \rangle_{\Omega_s}$ denotes $\int_{\Omega_s} \nabla \times \vec{a} \cdot \mu_s^{-1} \nabla \times \vec{b} ds$, and use was made of the property that $\vec{w}_1 - \vec{w}_3 = \nabla w_2$ in Ω_1 and $\vec{w}_3 + \vec{w}_4 = -\nabla w_2$ in Ω_2.

A similar manipulation shows that the product of each one of the rows of S_a with each one of the columns of G equals zero. Hence, $S_a G = 0$. This, together with the fact that the column rank of G is 3, implies that S_a has three zero eigenvalues, with corresponding independent eigenvectors any three of the four columns of G.

The matrix S_a is also singular in the case where a Dirichlet boundary condition is imposed on a portion of the enclosing boundary. We begin the proof of this result by noting that the dimension of S_a is $N_e - N_{De}$, where N_{De} is the number of edges on the Dirichlet boundary Γ_D. Furthermore, let N_{Dn} denote the number of nodes on Γ_D. Then, it is

$$\left[\nabla w_{n,1}, \nabla w_{n,2}, \cdots, \nabla w_{n,N_n - N_{Dn}} \right] = \left[\vec{w}_{e,1}, \vec{w}_{e,2}, \cdots, \vec{w}_{e,N_e - N_{De}} \right] G, \qquad (3.51)$$

where the gradient matrix, G, has dimensions $(N_e - N_{De}) \times (N_n - N_{Dn})$. Note that, because of the existence of the Dirichlet boundary, the column rank of G is $N_n - N_{Dn}$.

A reasoning similar to the one described in the paragraph following (3.46) yields the result $S_a G = 0$. Hence, we conclude that when a Dirichlet condition is imposed over a portion of the enclosing boundary of the domain of a magnetostatic BVP, the system matrix S_a has $N_n - N_{Dn}$ zero eigenvalues with corresponding eigenvectors the column vectors of G.

The explicit verification of this result is possible using, again, the simple two-element domain of Fig. 3.5. Let the highlighted edge 5 be the one where the Dirichlet boundary condition is imposed. The gradient matrix, G, in this case is obtained from (3.49) by removing the row associated with edge 5 and the two columns associated with the two nodes, 3 and 4, which are the endpoints of edge 5. Hence, the following (4×2) matrix, of column rank 2, results:

$$G = \begin{bmatrix} -1 & 1 \\ -1 & \\ & -1 \\ & -1 \end{bmatrix}. \qquad (3.52)$$

Since the number of unconstrained edges is 4, the dimension of S_a is 4. It is now a straightforward algebraic manipulation to verify that the product of any one of the rows of S_a with any column of (3.52) is zero, thus proving that $S_a G = 0$.

Since S_a is singular, the solvability of (3.44) and the uniqueness of the solution obtained must be considered. In a manner analogous to the solvability and uniqueness of the continuous magnetostatic BVP, discussed in the previous subsection, the solvability of (3.44) is intimately connected with the properties and constraints imposed on the source terms, while the uniqueness of the solution is tied with the unique definition of the magnetic vector potential through the specification of both its curl and its divergence. In the following subsections we elaborate on ways in which these issues must be addressed within the finite element approximation framework to ensure the robust, accurate and unique solution of (3.44).

3.2.3 Existence of solution (solvability)

We recall that the constraints for solution existence (or solvability) of the magnetostatic BVP, included the solenoidal (divergence-free) nature of the source current densities and the closely-related requirement that the net current flux through the enclosing boundary is zero. From (3.45) it is immediately apparent that these constraints enter the finite element approximation through the forcing vector f_a. Therefore, prior to addressing the way these constraints are imposed in the finite element approximation, we need to examine the solvability conditions for the matrix equation (3.44).

Toward this objective, let us review some basic definitions and results pertinent to linear matrix equations of the form $Sx = f$. First, recognizing the matrix-vector product Sx as a linear combination of the columns of S with weights the corresponding elements of x, it follows that the system will have a solution if and only if the forcing vector f is a linear combination of the columns of S. We then say that the system of linear equations $Sx = f$ is *consistent*, and that f lies in the *range* of S, which will be denoted as $\mathcal{R}(S)$; hence, $f \in \mathcal{R}(S)$.

The *null* space, $\mathcal{N}(S)$, of a matrix, S, is the space of all nontrivial solutions of the equation $Sx = 0$. Hence, $\mathcal{N}(S)$ will be nonempty if and only if S is singular.

Returning to the specific case of the magnetostatic matrix S_a, we note that since it is a positive-indefinite matrix, it is singular and has a nonempty null space, $\mathcal{N}(S_a)$. Furthermore, its null space is orthogonal to its range space,

$$\mathcal{N}(S_a) \perp \mathcal{R}(S_a). \tag{3.53}$$

To prove this useful result, let $y \in \mathcal{R}(S_a)$. Then $S_a z = y$, where z is a non-trivial vector. Then, the orthogonality of v, ($v \in \mathcal{N}(S_a)$), and y is deduced through the following matrix manipulations: $v^T y = v^T S_a z = (S_a^T v)^T z = (S_a v)^T z = 0$, where use was made of the fact that S_a is symmetric.

Furthermore, since the independent eigenvectors of S_a associated with its zero eigenvalues are the linearly independent columns of G, we conclude that $\mathcal{N}(S_a) = \mathcal{R}(G)$. It follows, then, that

$$\mathcal{R}(S_a) \perp \mathcal{R}(G), \tag{3.54}$$

from which we deduce that a (nontrivial) vector f_a in the range of S_a is orthogonal to the range space of G,

$$f_a \in \mathcal{R}(S_a) \quad \Rightarrow \quad f_a \perp \mathcal{R}(G). \tag{3.55}$$

In view of this result we conclude that if $G^T f_a = 0$, then $S_a x_a = f_a$ is solvable.

3.2.3.1 *Numerical approximation of divergence-free condition*   At this point we are ready to embark on the development of a numerical approximation of a solenoidal source current density, and show, subsequently, that success in this endeavor will help ensure the solvability of (3.44).

We begin with the observation that since the volume current density, $\vec{J}_v$, is solenoidal, it satisfies the property of normal continuity across a material interface; hence, its finite element approximation should be in terms of the elements of the NV space discussed in Chapter 2. Without loss of generality, we will use the first-order NV space for this expansion, that is, the facet element space of (2.121). Let N_f denote the total number of facets in the finite element mesh, while N_{Nf} denotes the facets on the Neumann boundary Γ_N. To facilitate the mathematical notation in the expressions that follow, we order the indices of facets in a manner such that the facets on Γ_N are listed last. Thus $\vec{J}_v$ is expanded as follows:

$$\vec{J}_v = \sum_{i=1}^{N_f - N_{Nf}} \vec{w}_{f,i} j_{v,i} = \vec{W}_{f,\Omega - \Gamma_N} j_v, \tag{3.56}$$

where $\vec{W}_{f,\Omega-\Gamma_N}$ is a row vector containing the facet elements excluding those on Γ_N,

$$\vec{W}_{f,\Omega-\Gamma_N} = [\vec{w}_{f,1}, \vec{w}_{f,2}, \cdots, \vec{w}_{f,N_f-N_{Nf}}] \tag{3.57}$$

and j_v is the vector of the expansion coefficients, $j_v = [j_{v,1}, \cdots, j_{v,N_f-N_{Nf}}]^T$. The reason we exclude the facet elements on Γ_N is because the normal component of $\vec{J}_v$ is zero on Γ_N as proven in (3.33). The divergence-free property of $\vec{J}_v$ leads to

$$\nabla \cdot \vec{J}_v = \nabla \cdot \vec{W}_{f,\Omega-\Gamma_N} \, j_v = 0, \tag{3.58}$$

where the notation $\nabla \cdot \vec{W}_{f,\Omega-\Gamma_N}$ is used to denote the row vector with elements the divergence of the elements of $\vec{W}_{f,\Omega-\Gamma_N}$. Use of (2.124) from Chapter 2 allows us to write

$$\nabla \cdot \vec{J}_v = \nabla \cdot \vec{W}_{f,\Omega-\Gamma_N} \, j_v = W^0_{charge} D \, j_v. \tag{3.59}$$

Recall that W^0_{charge} is a row vector containing the elements of the zeroth-order Whitney-3 form defined in (2.119), while D is the *divergence matrix* defined in (2.127). This last results leads us to the conclusion that the finite-element approximation of the solenoidal character of the imposed electric current density assumes the form

$$D \, j_v = 0. \tag{3.60}$$

It is instructive to point out that this relationship describes the simple fact that the net current flux through the surface of each tetrahedron in the mesh is zero. This is depicted pictorially in Fig. 3.6, and can be deduced immediately by recalling the following properties of D. Its number of rows is equal to the number of tetrahedra, N_{tetra}, in the mesh, while its number of columns is equal to the number of facets, $N_f - N_{Nf}$. Each row of D has only four nonzero entries of value ± 1, associated with the four facets of the tetrahedron corresponding to the specific row. Hence, with reference to Fig. 3.6, the i-th row of (3.60) has the form

$$(-j_{v,1} + j_{v,2} - j_{v,3} + j_{v,4})_i = 0, \tag{3.61}$$

where $j_{v,i}, i = 1, \cdots, 4$, denote the expansion coefficients associated with the four facets of tetrahedron i. It is clear from this expression that one can interpret (3.60) as a Kirchoff's current law statement for the finite element approximation of the impressed volume current density on the "circuit" formed by the finite element mesh.

3.2.3.2 *Discrete divergence-free source and solvability* It is now time to explore whether the finite-element representations (3.60) of the solenoidal volume electric current densities yield forcing vectors, f_a, in the range space of S_a, thus guaranteeing the solvability of (3.44).

Toward this, let us consider f_a in (3.45), which can be written in the following compact form:

$$f_a = \mu_0 \langle \vec{W}^T_{e,\Omega-\Gamma_D}, \vec{J}_v \rangle = \mu_0 \langle \vec{W}^T_{e,\Omega-\Gamma_D}, \vec{W}_{f,\Omega-\Gamma_N} \rangle j_v. \tag{3.62}$$

In the above equation it is $\langle a, b \rangle \equiv \int a \cdot b \, dv$. Also, use was made of the discrete form of $\vec{J}_v$ given in (3.56). Multiplication of the above equation on the left by G^T yields

$$\begin{aligned} G^T f_a &= \mu_0 G^T \langle \vec{W}^T_{e,\Omega-\Gamma_D}, \vec{W}_{f,\Omega-\Gamma_N} \rangle j_v \\ &= \mu_0 \langle \nabla W^T_{n,\Omega-\Gamma_D}, \vec{W}_{f,\Omega-\Gamma_N} \rangle j_v, \end{aligned} \tag{3.63}$$

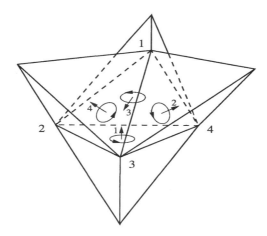

Figure 3.6   Geometry for the visualization of the numerical approximation of a solenoidal volume current density.

where use was also made of (3.46). The elements of the matrix have the form

$$\langle \nabla W_{n,\Omega-\Gamma_D}^T, \vec{W}_{f,\Omega-\Gamma_N} \rangle_{i,j} = \iiint_\Omega \nabla w_{n,i} \cdot \vec{w}_{f,j} dv. \tag{3.64}$$

Making use of the divergence theorem in (A.5), we have

$$\iiint_\Omega \nabla w_{n,i} \cdot \vec{w}_{f,j} dv = - \iiint_\Omega w_{n,i} \nabla \cdot \vec{w}_{f,j} dv + \oiint_{\Gamma_D \cup \Gamma_N} \hat{n} \cdot w_{n,i} \vec{w}_{f,j} ds. \tag{3.65}$$

The surface integral over Γ_D vanishes since no nodal elements associated with Γ_D are included in the expansion. Also, the surface integral over Γ_N vanishes since, in accordance to (3.33), the facet elements used for the expansion of $\vec{J}_v$ exclude elements associated with Γ_N (see (3.56)). Thus, making use of (3.60), (3.63) becomes

$$G^T f_a = \mu_0 G^T \langle \vec{W}_{e,\Omega-\Gamma_D}^T, \vec{W}_{f,\Omega-\Gamma_N} \rangle j_v$$
$$= -\mu_0 \langle W_{n,\Omega-\Gamma_D}^T, \nabla \cdot \vec{W}_{f,\Omega-\Gamma_N} \rangle j_v \tag{3.66}$$
$$= 0.$$

In view of (3.58), which expresses the finite-element statement of the solenoidal attribute of the impressed volume current density, we conclude that the forcing vector, f_a, is exactly in the range space of S_a. Thus the solution to $S_a x_a = f_a$ exists when the numerical divergence-free condition (3.60) is satisfied.

The next issue to be dealt with concerns the computation of the expansion coefficients, j_v, in a manner such that (3.60) is satisfied. To fix ideas, consider the expansion for the volume current in the tetrahedron of Fig. 3.6,

$$\vec{J}_v = \vec{w}_{f,1} j_{v,1} + \vec{w}_{f,2} j_{v,2} + \vec{w}_{f,3} j_{v,3} + \vec{w}_{f,4} j_{v,4}. \tag{3.67}$$

Since each facet element has only the normal component on its associated facet, multiplication of the above equation with the unit normal vector $\hat{n}_i$ on facet i, followed by the integration of the result over the facet and making use of (2.107) yields

$$j_{v,i} = \iint_{facet-i} \hat{n}_i \cdot \vec{J}_v ds. \tag{3.68}$$

In the above equation $\hat{n}_i$ points along the direction of the facet element. If in the calculation of the expansion coefficient $j_{v,i}$ the above integral is computed exactly, the numerical divergence-free condition (3.60) will be satisfied exactly. However, the error due to the finite precision of the numerical integration results in (3.60) being satisfied only approximately. Thus f_a may not be exactly in the range space of S_a, a result that manifests itself through the stalling of the conjugate gradient method after an initial rapid convergence.

Intuitively it is easy to argue that the violation of (3.60) due to roundoff amounts to the presence of a spurious electric charge, which is "numerically" introduced in the computational domain. One approach toward canceling the effect of this spurious charge is through the intentional introduction of a fictitious charge density during the numerical integration. More specifically, consider the modified current density $\hat{j}_v = j_v + D^T q$, where the term $D^T q$ is such that

$$D\hat{j}_v = D(j_v + D^T q) = 0. \tag{3.69}$$

Then, the (discrete) spurious charge, q, due to roundoff is obtained from the solution of the following equation:

$$(DD^T)q = -Dj_v. \tag{3.70}$$

The matrix DD^T is an $N_t \times N_t$ sparse matrix, where N_t is the number of tetrahedra in the mesh. Once q is obtained from the above equation, the vector, j_v, of expansion coefficients for a source current that satisfies (3.60) is modified as follows:

$$\hat{j}_v \leftarrow j_v + D^T q. \tag{3.71}$$

An alternative approach exploits the fact that the divergence-free nature of $\vec{J}_v$ allows for the introduction of a current vector potential, $\vec{T}_v$ as follows [20]:

$$\vec{J}_v = \nabla \times \vec{T}_v. \tag{3.72}$$

The inverse of this relation is readily obtained from Biot-Savart's law through the following integral:

$$\vec{T}_v = \frac{1}{4\pi} \iiint_\Omega \frac{\vec{J}_v(\vec{r}') \times (\vec{r} - \vec{r}')}{|\vec{r} - \vec{r}'|^3} dv. \tag{3.73}$$

The above expression allows for the calculation of $\vec{T}_v$ for a given current distribution $\vec{J}_v$. Subsequently, the calculated current vector potential is expanded in terms of the elements of the TV space in the form

$$\vec{T}_v = \sum_{i=1}^{N_e} \vec{w}_{e,i} t_{v,i} = \vec{W}_{e,\Omega} t_v. \tag{3.74}$$

In the above expression $t_{v,i}$ are the expansion coefficients and t_v is the vector with elements these coefficients. The finite element approximation of the source current density is then obtained as the curl of $\vec{T}_v$,

$$\vec{J}_v = \nabla \times \vec{T}_v = \nabla \times \vec{W}_{e,\Omega} t_v = \vec{W}_{f,\Omega} C t_v, \tag{3.75}$$

where C is the *curl matrix* of size $N_f \times N_e$ introduced in (2.126). In view of this last result it follows immediately that the vector of the expansion coefficients in the finite element approximation of the source current density is given by

$$j_v = C t_v. \tag{3.76}$$

The advantage of this form is readily recognized by recalling the relationship $DC = 0$ satisfied by the *curl matrix* and the *divergence matrix*. Clearly, in view of this relationship, the finite element approximation (3.76) of the given source current density satisfies the numerical divergence-free condition of (3.60) by construction.

3.2.4 Uniqueness of solution

When f_a is in the range space of S_a, a solution to the system $S_a x_a = f_a$ exists but may not be unique, since the null space of S_a is not empty. We recognize here the well-known analytic result that for the magnetic vector potential to be uniquely defined both its curl and its divergence must be specified. Thus, unless $\nabla \cdot \vec{A}$ is specified, $\vec{A} + \nabla\Phi$ for any scalar function Φ is also an acceptable vector potential function for the unique definition of the magnetic flux density. To render the solution for $\vec{A}$ unique, its divergence must be specified through the introduction of an appropriate *gauge*.

In a similar manner, if x_a is a non-zero solution to $S_a x_a = f_a$, then $x_a + G\phi$ is also a solution, where ϕ is a non-zero vector, since G is the null space of S_a. In this section, we discuss several techniques followed in practice for the numerical introduction of a gauge to render the numerical solution x_a unique.

3.2.4.1 Ungauged Solution The uniqueness of the magnetic vector potential $\vec{A}$ is not important for the uniqueness of the magnetic flux density, $\vec{B}$. Thus, similarly with the continuous case, the numerical curl operation on the finite element solution x_a for $\vec{A}$ will remove any gradient component present and will result in a unique vector for the finite element approximation of the magnetic flux density.

In view of the above we conclude that all that is required for an iterative solution to the equation $S_a x_a = f_a$ to converge is for f_a to be in the range of S_a, even in the case where S_a is singular. Thus the most commonly used gauging technique is the so-called *ungauged* method. The *ungauged* method relies, simply, on the use of an iterative solver such as the conjugate gradient method for the solution, with a right-hand side vector, f_a constructed such that the divergence-free condition of (3.60) is satisfied.

3.2.4.2 Zero-tree gauge The tree-cotree gauge technique was proposed in [19]. In the following, the mathematical development of the equations used for its numerical implementation is presented in the context of the finite element spaces of Chapter 2.

We begin with a review of some definitions and properties from graph theory. Since a *graph* is understood as a set of nodes and a set of branches such that each branch terminates at each end into a node, the finite element grid constitutes a graph. A graph is called *connected* if there is at least one path along the branches of the graph between any two nodes in the graph.

Let $\mathcal{G}$ be a graph. Then $\mathcal{G}_1$ is called a *subgraph* of $\mathcal{G}$, if a)$\mathcal{G}_1$ is itself a graph; b) every node of $\mathcal{G}_1$ is a node of $\mathcal{G}$; c) every branch of $\mathcal{G}_1$ is a branch of $\mathcal{G}$.

A subgraph of a graph is a *loop* if the subgraph is connected and precisely two branches of the subgraph are incident with each node.

A subgraph of a connected graph, $\mathcal{G}$, is called a *tree* if a) it is a connected subgraph; b) it contains all the nodes of $\mathcal{G}$; c) it contains no loops. In the following, the letter $\mathcal{T}$ will be used to denote a tree.

Given a connected graph, $\mathcal{G}$, and a tree, $\mathcal{T}$, we will call the branches of $\mathcal{T}$ the *tree branches*. The remaining branches of $\mathcal{G}$ constitute the *cotree* of $\mathcal{T}$.

In the context of a finite element mesh, branches correspond to edges of the elements. Thus, in the following, the word *edge* will be used in place of *branch* in reference to trees and cotrees of the finite element mesh.

Next, let us consider the case of a computational domain without a Dirichlet boundary. Let N_e be the number of edges in the finite element mesh of tetrahedra used for the discretization of the domain, while N_n is the number of nodes. As shown in Chapter 2, the curl of the edge elements in the domain satisfies the condition

$$[\nabla \times \vec{w}_{e,1}, \cdots, \nabla \times \vec{w}_{e,N_e}]G = 0. \tag{3.77}$$

The size of G is $N_e \times (N_n - 1)$, since a column vector is removed from it.

Consider, next, a tree T of the graph defined by the finite element mesh. Clearly, the number of edges in the tree is $N_{e,t} = N_n - 1$. Also, let $N_{e,c}$ denote the remaining number of edges in the graph, that is, the edges associated with the cotree of T. Obviously, $N_{e,c} + N_{e,t} = N_e$. With the numbering of the elements adjusted such that the edges in the cotree are listed first, (3.77) is cast in the following form:

$$[\underbrace{\nabla \times \vec{w}_{e,1}, \cdots, \nabla \times \vec{w}_{e,N_{e,c}}}_{\nabla \times \vec{W}_{e,c}}, \underbrace{\nabla \times \vec{w}_{e,N_{e,c}+1}, \cdots, \vec{w}_{e,N_e}}_{\nabla \times \vec{W}_{e,t}}] \begin{bmatrix} G_c \\ G_t \end{bmatrix} = 0, \tag{3.78}$$

where $\vec{W}_{e,c}$ is a row vector containing the edge elements in the cotree and $\vec{W}_{e,t}$ is a row vector containing the edge elements in T. The G matrix is split into two sub-matrices, with the sub-matrix G_t containing the rows of G associated with the edges of T, and G_c containing the rows associated with the edges of the cotree.

As an example, consider the simple mesh of Fig. 3.5. Using the edge numbers to define the tree and the cotree, let the tree be chosen as $1 - 3 - 5$. Its cotree is, then, $2 - 4$. The matrices G_c and G_t are

$$G_c = \begin{bmatrix} -1 & 0 & 1 \\ 0 & -1 & 0 \end{bmatrix}, \quad G_t = \begin{bmatrix} -1 & 1 & 0 \\ 0 & -1 & 1 \\ 0 & 0 & -1 \end{bmatrix}. \tag{3.79}$$

The matrix G_t is an $(N_n-1) \times (N_n-1)$ full-rank matrix; thus its inverse exists and we obtain from (3.78),

$$\nabla \times \vec{W}_{e,t} = -\nabla \times \vec{W}_{e,c} G_c G_t^{-1}. \tag{3.80}$$

The last result is a general result that can be used to write S_a and f_a of (3.44) in the following form:

$$\begin{aligned} S_a &= \begin{bmatrix} S_{cc} & S_{ct} \\ S_{tc} & S_{tt} \end{bmatrix} = \begin{bmatrix} I \\ -(G_c G_t^{-1})^T \end{bmatrix} [S_{cc}] \begin{bmatrix} I & -G_c G_t^{-1} \end{bmatrix} \\ f_a &= \begin{bmatrix} f_{a,c} \\ f_{a,t} \end{bmatrix} = \begin{bmatrix} I \\ -(G_c G_t^{-1})^T \end{bmatrix} f_{a,c}. \end{aligned} \tag{3.81}$$

Hence, the matrix equation becomes

$$\begin{bmatrix} S_{cc} & S_{ct} \\ S_{tc} & S_{tt} \end{bmatrix} \begin{bmatrix} x_{a,c} \\ x_{a,t} \end{bmatrix} = \begin{bmatrix} f_{a,c} \\ f_{a,t} \end{bmatrix}. \tag{3.82}$$

Clearly, the sub-matrix S_{cc} is associated with the degrees of freedom in the cotree only, while S_{tt} is associated with the degrees of freedom in the tree.

The above result suggests a gauging technique that sets all the degrees of freedom associated with the tree edges to zero; hence, $x_{a,t} = 0$. Thus the solution of the problem is obtained in terms of the solution of a reduced matrix equation involving as unknowns only the degrees of freedom in the cotree

$$S_{cc}x_{a,c} = f_{a,c}. \tag{3.83}$$

In view of the reduced size of S_{cc} better convergence and thus improved numerical speed of the iterative solution is expected. Unfortunately, the reduced matrix equation is ill-conditioned if the random-tree choice reported in [20] and [21] is utilized; hence, convergence is not improved, despite the smaller dimension of the reduced matrix. In general, the success of this gauging technique is dependent on the choice of the "ideal" tree that will result in a well-conditioned matrix S_{cc}. As such, and considering the fact that complex meshes result in graphs containing a very large number of trees, this approach lacks the robustness needed for the development of an effective and fast iterative solver.

3.2.4.3 Coulomb gauge
The Coulomb gauge amounts to imposing the condition that the divergence of the magnetic vector potential is zero, $\nabla \cdot \vec{A} = 0$. A way in which this constraint is imposed within the finite element approximation of the unknown vector potential is described next.

We begin the discussion by noting that if the vector expansion functions used for the approximation of $\vec{A}$ were "divergence-free" then the finite element approximation of the vector potential would satisfy the Coulomb gauge by construction.

From vector calculus we know that if a vector field can be written as the curl of another vector field, then the field is divergence-free (solenoidal). This result suggests that expansion functions free from any gradient components, which can be written entirely in terms of the curl of a vector field, would be the ideal choice for the expansion of a Coulomb gauge-constrained magnetic vector potential.

Considering that edge elements are one of the most popular set of expansion functions for the finite element approximation of vector fields exhibiting tangential continuity, it is worth examining whether their divergence is zero. From Chapter 2 we recall that the elements of the zeroth-order space of edge elements, $\mathcal{H}^0(curl)$, are polynomials complete to order zero. The same is true for the curl of the elements. Hence, it is straightforward to see that, since $\nabla \cdot \nabla \lambda_m = 0$, the divergence of the zeroth-order edge element is zero. For example, for the edge element $\lambda_m \nabla \lambda_n - \lambda_n \nabla \lambda_m$, associated with edge (m,n) it is

$$\nabla \cdot (\lambda_m \nabla \lambda_n - \lambda_n \nabla \lambda_m) = \nabla \lambda_m \cdot \nabla \lambda_n - \nabla \lambda_n \cdot \nabla \lambda_m = 0. \tag{3.84}$$

However, this result should not be interpreted to mean that edge elements are solenoidal. Actually, it is apparent from their mathematical form that, in addition to a curl component they contain the zeroth-order gradient component. Consequently, the notion that edge elements are divergence free is wrong. Furthermore, since a constant gradient can be expanded in terms of edge elements, non-uniqueness difficulties may be encountered when edge elements are used for the finite element approximation of the magnetic vector potential. Clearly, a remedy to this problem is necessary.

One approach, suggested in [19], makes use of the tree-cotree splitting of the unknowns discussed in the previous section. The method is motivated by the observation that uniqueness of the solution to the matrix equation $S_a x_a = f_a$ when S_a is singular is guaranteed by constraining x_a to be orthogonal to the null space of S_a. It was shown in Section 3.2.2 that the null space of S_a consists of the columns of the gradient matrix G. Hence, uniqueness is guaranteed by enforcing the constraint $G^T x_a = 0$.

More specifically, making use of the split form of x_a in terms of the cotree and tree degrees of freedom, and the corresponding split form of G defined in the previous section, the above constraint is cast in the form

$$\begin{bmatrix} x_{a,c} \\ x_{a,t} \end{bmatrix} = \begin{bmatrix} I \\ -G_t^{-T} G_c^T \end{bmatrix} x_{a,c}. \tag{3.85}$$

Substitution of this result into (3.44) and making use of the tree-cotree split form of S_a yields a reduced matrix equation with unknowns the degrees of freedom associated with the cotree,

$$\begin{bmatrix} I & -G_c G_t^{-1} \end{bmatrix} \begin{bmatrix} S_{cc} & S_{ct} \\ S_{tc} & S_{tt} \end{bmatrix} \begin{bmatrix} I \\ -G_t^{-T} G_c^T \end{bmatrix} x_{a,c} = \begin{bmatrix} I & -G_c G_t^{-1} \end{bmatrix} \begin{bmatrix} f_{a,c} \\ f_{a,t} \end{bmatrix}. \tag{3.86}$$

Once $x_{a,c}$ has been obtained, $x_{a,t}$ is recovered from (3.85).

An alternative approach introduces a pure gradient field, $\nabla\Phi$, which is added to $\vec{A}$ in order to cancel its gradient component according to the equation

$$\nabla \cdot \left(\vec{A} + \nabla\Phi \right) = 0. \tag{3.87}$$

The weak statement of the above equation is obtained by multiplying with a nodal element, $w_{n,i}$, followed by an integration over the computational domain,

$$\iiint_\Omega \nabla w_{n,i} \cdot \nabla\Phi \, dv = - \iiint_\Omega \nabla w_{n,i} \cdot \vec{A} \, dv. \tag{3.88}$$

The finite element matrix statement follows readily from this weak statement through a standard Galerkin's process. With $\vec{A}$ expanded in terms of edge elements and Φ in terms of nodal elements, the matrix equation is of the form

$$T x_\phi = -P x_a, \tag{3.89}$$

where the elements of the matrices T and P are computed through the integrals

$$\begin{aligned} T_{i,j} &= \iiint_\Omega \nabla w_{n,i} \cdot \nabla w_{n,j} \, dv, \\ P_{i,j} &= \iiint_\Omega \nabla w_{n,i} \cdot \vec{w}_{e,j} \, dv. \end{aligned} \tag{3.90}$$

The vector x_ϕ contains the expansion coefficients for Φ. Once obtained, the vector x_a is updated as

$$x_a \leftarrow x_a + G x_\phi. \tag{3.91}$$

To conclude, let us point out that, rather than using x_ϕ in such a postprocessing sense to render x_a divergence free, the process described by (3.89) and (3.91) is most suitable for use in conjunction with the iterative methods described in the following chapter.

3.3 MAGNETO-QUASI-STATIC (EDDY-CURRENT) PROBLEMS

Magneto-quasi-static (MQS) or eddy-current problems constitute an important class of electromagnetic BVPs with numerous practical applications. The underlying physical phenomenon of interest is the behavior of a time-varying electromagnetic field inside a conducting medium when the displacement current density is negligible compared to the conduction current density. The conditions under which the MQS approximation to Maxwell's equations is valid are represented in a convenient manner through the so-called *loss tangent*, $\tan \delta$, of the medium, which is defined as the ratio of the magnitude of the conduction current density to the displacement current density. Under time-harmonic conditions, this ratio is given by $\tan \delta = \sigma/(\omega\epsilon)$, where $\omega = 2\pi f$ is the angular frequency of the time-harmonic variation of the fields and σ is the conductivity of the medium.

Of particular interest in practice is the case of field penetration in good conductors. For a good conductor the loss tangent is much greater than unity, indicating the dominance of the conduction current over the displacement current. This dominance is observed for most frequencies of practical interest. Under such conditions, the field behavior inside the good conductor exhibits exponential attenuation with depth; thus the term *magnetic diffusion* is often used to describe electromagnetic field behavior under MQS conditions.

The exponential decay of the field magnitude inside the conductor is most conveniently quantified in terms of the *penetration depth*, d_p, in the medium, which is defined as the distance over which the magnitude of a uniform plane wave propagating inside the conductor decreases by $1/e$. The penetration depth is given in terms of the properties of the media and the frequency as follows:

$$d_p = (\pi f \mu \sigma)^{-1/2}. \tag{3.92}$$

Clearly, the higher the frequency the smaller the penetration depth. For example, for aluminum with conductivity 4×10^7 S/m and permeability that of free space, the penetration depth at 1 GHz is approximately 2.5 μm. Thus field penetration is restricted to a very thin layer beneath the conductor surface, a phenomenon known as *skin effect*.

The finite element method has been applied extensively to the solution of MQS problems of practical interest, ranging from electric motor design, to underground electromagnetic field probing, to the quantification of conductor loss-induced signal distortion and attenuation in high-speed/high-frequency integrated circuits [20]-[25]. In the following we discuss the two most popular classes of finite element formulations for the MQS BVP, namely, the electric field intensity formulation and the potential formulation.

3.3.1 Governing equations and boundary conditions

Consider the solution of the MQS BVP inside the domain depicted in Fig. 3.7. The electromagnetic properties of the linear media are assumed to be position dependent. Furthermore, the presence of material interfaces is allowed, across which abrupt changes occur in the electromagnetic properties of the media. Such an interface is depicted in Fig. 3.7 as the boundary Γ_I between region Ω_c and the remaining of the domain of interest Ω.

Under time-harmonic conditions and with the displacement current density assumed negligible, Maxwell's equations in phasor form are approximated as follows:

$$\begin{cases} \nabla \times \vec{E} = -j\omega\vec{B}, \\ \nabla \times \vec{H} = \vec{J_v} + \vec{J_e}, \\ \nabla \cdot \vec{B} = 0. \end{cases} \tag{3.93}$$

In the above equations $\vec{J_v}$ is the impressed volume current density and $\vec{J_e}$ is the induced current density in the conducting media. It should be noted that the divergence of the second equation yields the conservation of charge equation

$$\nabla \cdot \left(\vec{J_v} + \vec{J_e} \right) = 0. \tag{3.94}$$

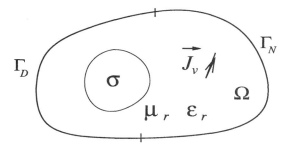

Figure 3.7   Domain for the statement of the magneto-quasi-static BVP.

Under the assumption of linear, isotropic media, the following constitutive relations hold:

$$\vec{B} = \mu_r \mu_0 \vec{H}, \quad \vec{J}_e = \sigma \vec{E}. \tag{3.95}$$

The complete statement of the MQS BVP requires the assignment of appropriate boundary conditions on the enclosing boundary of the domain. For our purposes, and with reference to Fig. 3.7, the following homogeneous boundary conditions will be considered,

$$\begin{cases} \hat{n} \times \vec{E} = 0, & \text{on } \Gamma_D, \\ \hat{n} \times \vec{H} = 0, & \text{on } \Gamma_N. \end{cases} \tag{3.96}$$

Boundary conditions are required also at any media interfaces across which the electromagnetic properties undergo abrupt changes. The pertinent boundary conditions are well known and require the continuity of the tangential components of the electric and magnetic field intensities.

3.3.2 Electric field formulation

Elimination of $\vec{H}$ in (3.93) and use of the constitutive relations (3.95) results in the following equation for the electric field

$$\nabla \times \frac{1}{\mu_r} \nabla \times \vec{E} = -j\omega\mu_0 \vec{J}_v - j\omega\sigma\mu_0 \vec{E}. \tag{3.97}$$

Furthermore, use of the first of (3.93) allows us to cast the boundary conditions of (3.96) solely in terms of $\vec{E}$ as follows:

$$\begin{cases} \hat{n} \times \vec{E} = 0, & \text{on } \Gamma_D, \\ \hat{n} \times \dfrac{1}{\mu_r} \nabla \times \vec{E} = 0, & \text{on } \Gamma_N. \end{cases} \tag{3.98}$$

The development of the weak statement of (3.97) calls for the multiplication of (3.97) by an edge element $\vec{w}_e$, followed by integration over the volume of the domain Ω. Use of the vector identity (A.19) and the homogeneous boundary condition (3.96) for the tangential component of the magnetic field on Γ_N, results in the following equation:

$$\iiint_\Omega \nabla \times \vec{w}_e \cdot \frac{1}{\mu_r} \nabla \times \vec{E} \, dv + j\omega\mu_0 \iiint_\Omega \sigma \vec{w}_e \cdot \vec{E} \, dv$$
$$= -j\omega\mu_0 \iiint_\Omega \vec{J}_v \cdot \vec{w}_e \, dv + j\omega\mu_0 \iint_{\Gamma_D} \vec{w}_e \cdot \hat{n} \times \vec{H} \, ds. \tag{3.99}$$

The electric field $\vec{E}$ is expanded in terms of edge elements subject to the constraint that the degrees of freedom associated with the edges on Γ_D are set explicitly to zero, in accordance with the first of the boundary conditions in (3.96). Hence, the boundary integral term on the right-hand side of the above weak statement vanishes. Use of edge elements ensures the boundary condition of tangential electric field continuity at material interface boundary. Testing of the reduced form of the weak statement with every edge element in the set used for the expansion of $\vec{E}$ yields the following finite element system,

$$(S + j\omega Z)x_e = -j\omega f_e, \tag{3.100}$$

where it is

$$S_{i,j} = \iiint_\Omega \nabla \times \vec{w}_{e,i} \cdot \frac{1}{\mu_r} \nabla \times \vec{w}_{e,j} dv$$

$$Z_{i,j} = \mu_0 \iiint_\Omega \sigma \vec{w}_{e,i} \cdot \vec{w}_{e,j} dv \tag{3.101}$$

$$f_{e,i} = \mu_0 \iiint_\Omega \vec{J}_v \cdot \vec{w}_{e,i} dv$$

and x_e is the vector containing the expansion coefficients for the electric field.

If there is no conducting region, the matrix equation is the same with the one encountered in the case of the magnetostatic BVPs. If only a portion of the computational domain is lossy, the resultant matrix is also singular. Consequently, the issues associated with solution existence and uniqueness, discussed in the context of finite element approximations of magnetostatic BVPs, apply also to the MQS BVP case.

3.3.3 Potential formulation

An alternative formulation for the finite element solution of the MQS BVP, which is of significant importance to the multigrid techniques discussed in this book, is the potential formulation [26]. Because of the solenoidal character of $\vec{B}$, the introduction of the magnetic vector potential $\vec{A}$ as

$$\vec{B} = (-j\omega)^{-1} \nabla \times \vec{A} \tag{3.102}$$

is combined with Faraday's law of induction to cast the electric field vector in terms of $\vec{A}$ and the electric scalar potential, V, as follows:

$$\vec{E} = \vec{A} + \nabla V. \tag{3.103}$$

Substitution of (3.102) and (3.103) in Ampere's law yields

$$\nabla \times \frac{1}{\mu_r} \nabla \times \vec{A} + j\omega\mu_0\sigma \left(\vec{A} + \nabla V \right) = -j\omega\mu_0 \vec{J}_v. \tag{3.104}$$

Furthermore, taking the divergence of the above equation we arrive at the following result:

$$\nabla \cdot (\sigma \vec{A} + \sigma \nabla V) = -\nabla \cdot \vec{J}_v. \tag{3.105}$$

The homogeneous boundary conditions on Γ_D and Γ_N may be cast, in terms of the potentials, in the following form:

$$\begin{cases} \hat{n} \times (\vec{A} + \nabla V) = 0, & \text{on } \Gamma_D, \\ \hat{n} \times \frac{1}{\mu_r} \nabla \times \vec{A} = 0, & \text{on } \Gamma_N. \end{cases} \tag{3.106}$$

The finite element approximation of the system of (3.104) and (3.105) begins with the expansion of $\vec{A}$ and V in terms of edge and nodal elements, respectively. Recognizing that (3.104) is a vector equation while (3.105) is a scalar equation, their weak statements are obtained through multiplication by an edge element w_e and a nodal element w_n, respectively, followed by integration over the computational domain, Ω. Thus the resulting weak statements assume the form

$$\iiint_\Omega (\nabla \times \vec{w}_e) \cdot \left(\frac{1}{\mu_r} \nabla \times \vec{A} \right) dv + j\omega\mu_0 \iiint_\Omega \vec{w}_e \cdot \sigma \left(\vec{A} + \nabla V \right) dv$$

$$= - \oiint_{\Gamma_D \cup \Gamma_N} \vec{w}_e \cdot \left(\hat{n} \times \frac{1}{\mu_r} \nabla \times \vec{A} \right) ds - j\omega\mu_0 \iiint_\Omega \vec{w}_e \cdot \vec{J}_v dv,$$

$$\iiint_\Omega \nabla w_n \cdot \sigma \left(\vec{A} + \nabla V \right) dv = \oiint_{\Gamma_D \cup \Gamma_N} w_n \sigma \hat{n} \cdot \left(\vec{A} + \nabla V \right) ds + \iiint_\Omega w_n \nabla \cdot \vec{J}_v dv.$$

$$\text{(3.107)}$$

The final step concerns the application of the boundary conditions on Γ_D and Γ_N. The pertinent boundary equation constraints are given in terms of $\vec{A}$ and ∇V in (3.106). The boundary condition of zero tangential electric field on Γ_D is imposed explicitly by excluding the edge elements and nodal elements on Γ_D from the set of expansion functions used in the approximation of $\vec{A}$ and V, respectively.

A term that requires some discussion is the flux density term $\hat{n} \cdot (\vec{A} + \nabla V)$, which may also be cast in terms of the magnetic field vector as follows, $\hat{n} \cdot \left(\nabla \times \vec{H} \right)$. In view of (3.31), this term vanishes on Γ_N. Furthermore, the solenoidal nature of the total (impressed and induced) current flow in the MQS problem requires that the current density flux through Γ_D be zero unless an impressed source is imposed on Γ_D. In the absence of such an impressed source, the surface integral term over $\Gamma_N \cup \Gamma_D$ involving $\hat{n} \cdot \left(\vec{A} + \nabla V \right)$ is zero.

Finally, the weak statements for the development of the potential-based finite element approximation of the MQS problem are given by

$$\iiint_\Omega \nabla \times \vec{w}_e \cdot \frac{1}{\mu_r} \nabla \times \vec{A} dv + j\omega\mu_0 \iiint_\Omega \vec{w}_e \cdot \sigma \left(\vec{A} + \nabla V \right) dv = -j\omega\mu_0 \iiint_\Omega \vec{w}_e \cdot \vec{J}_v dv,$$

$$j\omega\mu_0 \iiint_\Omega \nabla w_n \cdot \sigma \left(\vec{A} + \nabla V \right) dv = j\omega\mu_0 \iiint_\Omega w_n \nabla \cdot \vec{J}_v dv.$$

Substitution of the expansions for $\vec{A}$ and V in the above equation, followed by Galerkin's testing yields the finite element matrix statement of the MQS BVP

$$\begin{bmatrix} M_{aa} & M_{av} \\ M_{va} & M_{vv} \end{bmatrix} \begin{bmatrix} x_a \\ x_v \end{bmatrix} = -j\omega \begin{bmatrix} f_a \\ f_v \end{bmatrix}, \quad \text{(3.108)}$$

where x_a, x_v are, respectively, the vectors of the expansion coefficients for $\vec{A}$ and V, while the elements in all remaining matrices and vectors are computed in terms of the following integrals:

$$
\begin{aligned}
(M_{aa})_{i,j} &= \iiint_\Omega \nabla \times \vec{w}_{e,i} \cdot \frac{1}{\mu_r} \nabla \times \vec{w}_{e,i} dv + j\omega\mu_0 \iiint_\Omega \vec{w}_{e,i} \cdot \sigma \vec{w}_{e,j} dv, \\
(M_{vv})_{i,j} &= j\omega\mu_0 \iiint_\Omega \nabla w_{n,i} \cdot \sigma \nabla w_{n,j} dv, \\
(M_{av})_{i,j} &= j\omega\mu_0 \iiint_\Omega \vec{w}_{e,i} \cdot \sigma \nabla w_{n,j} dv, \quad (M_{va})_{i,j} = (M_{av})_{j,i}, \\
f_a &= f_e, \quad (f_v)_i = -\mu_0 \iiint_\Omega w_{n,i} \nabla \cdot \vec{J}_v dv.
\end{aligned}
\tag{3.109}
$$

It is to be expected that a relationship exists between the potential-based finite element matrix statement of the MQS problem and the one obtained from the electric field formulation. This relationship can be deduced by making use of the relationship $\vec{E} = \vec{A} + \nabla V$, and the result in (3.51). It follows immediately that the coefficient vectors in the expansions for $\vec{E}$, $\vec{A}$ and V satisfy the relationship

$$
x_e = \begin{bmatrix} I & G \end{bmatrix} \begin{bmatrix} x_a \\ x_v \end{bmatrix}.
\tag{3.110}
$$

Furthermore, a direct comparison of (3.109) with (3.101) reveals that it is

$$
\begin{aligned}
M_{aa} &= S + j\omega Z, \quad M_{av} = j\omega Z G, \\
M_{va} &= j\omega G^T Z, \quad M_{vv} = j\omega G^T Z G.
\end{aligned}
\tag{3.111}
$$

Making use of the fact that $SG = 0$, these relations may be cast in matrix form as follows:

$$
\begin{bmatrix} M_{aa} & M_{av} \\ M_{va} & M_{vv} \end{bmatrix} = \begin{bmatrix} I \\ G^T \end{bmatrix} [S + j\omega Z] \begin{bmatrix} I & G \end{bmatrix}.
\tag{3.112}
$$

Furthermore, the use of the vector identity (A.5) allows us to recast the expression for the elements of f_v in the following form:

$$
(f_v)_i = -\mu_0 \iiint_\Omega w_{n,i} \nabla \cdot \vec{J}_v dv = \mu_0 \iiint_\Omega \nabla w_{n,i} \cdot \vec{J}_v dv,
\tag{3.113}
$$

where use was made of the fact that $\hat{n} \cdot w_{n,i} \vec{J}_v = 0$ on the boundary of $\Gamma_D \cup \Gamma_N$. In view of the last of the equations in (3.101), this result may be cast in matrix form as follows, $f_v = G^T f_e$. Thus we conclude that the forcing vector for the finite element approximation of the potential formulation can be expressed in terms of that for the electric field formulation in the following form:

$$
\begin{bmatrix} f_a \\ f_v \end{bmatrix} = \begin{bmatrix} I \\ G^T \end{bmatrix} f_e.
\tag{3.114}
$$

Finally, substitution of (3.110) into (3.100) and multiplication on the left by $[I\ G]^T$ yields

$$
\begin{bmatrix} I \\ G^T \end{bmatrix} [S + j\omega Z] \begin{bmatrix} I & G \end{bmatrix} \begin{bmatrix} x_a \\ x_v \end{bmatrix} = \begin{bmatrix} I \\ G^T \end{bmatrix} f_e.
\tag{3.115}
$$

The equivalence between the two finite element formulations will be exploited for improving the computational efficiency of the iterative solution techniques discussed in later chapters.

3.4 FULL-WAVE BOUNDARY VALUE PROBLEMS

As the name "full-wave" suggests, this is the case where the electromagnetic BVP is considered in its completeness with both time-dependent terms in the two curl equations, namely, the time derivative of the magnetic flux density in Faraday's law and the time derivative of the electric flux density (i.e., the displacement current density term) in Ampére's law, present in the statement of the BVP.

3.4.1 Governing equations and boundary conditions

Consider the domain depicted in Fig. 3.7. We are interested in the development of finite element approximations to the full-wave system of Maxwell's equations under the assumption of time-harmonic excitation and linear, isotropic media. The pertinent governing equations are

$$
\begin{aligned}
\nabla \times \vec{E} &= -j\omega\vec{B}, \\
\nabla \times \vec{H} &= \vec{J_v} + \vec{J_c} + j\omega\vec{D}, \\
\nabla \cdot \vec{D} &= \rho_v, \\
\nabla \cdot \vec{B} &= 0,
\end{aligned}
\tag{3.116}
$$

where the source terms, $\vec{J_v}$ and ρ_v, are assumed known. The electric and magnetic flux densities, $\vec{D}$, $\vec{B}$, respectively, and the conduction current density, $\vec{J_e}$, are related to the electric and magnetic field intensities through the constitutive relations

$$
\begin{aligned}
\vec{B} &= \mu_r\mu_0\vec{H}, \\
\vec{D} &= \epsilon_r\epsilon_0\vec{E}, \\
\vec{J_e} &= \sigma\vec{E}.
\end{aligned}
\tag{3.117}
$$

It is noted that the electromagnetic properties of the media are assumed to be position dependent, including abrupt changes at material interfaces. Furthermore, the conductivity, σ, is understood to account for both conductor and dielectric media loss.

For the BVP to be well-posed boundary conditions must be imposed on all boundaries in the domain Ω. On the material interface and in the absence of any surface source current density the pertinent boundary conditions are the continuity of the tangential components of the electric and magnetic fields. On the enclosing boundary the following homogeneous boundary conditions are imposed

$$
\begin{cases}
\hat{n} \times \vec{H} = 0 & \text{on } \Gamma_N, \\
\hat{n} \times \vec{E} = 0 & \text{on } \Gamma_D.
\end{cases}
\tag{3.118}
$$

Full-wave electromagnetic BVPs of practical interest include radiation and scattering problems involving unbounded domains. The finite element formulation of such problems must properly account for the correct physical behavior of the fields in the unbounded region. Since finite element methods require the definition of a bounded domain for their application, numerous methodologies have been proposed and implemented for making their application possible to the solution of radiation and scattering problems.

One class of methods utilizes a rigorous statement of the electromagnetic BVP in the exterior of the finite element domain. The two statements, one for the exterior and one for the interior, are linked through the enforcement of field continuity conditions at the mathematical boundary used for the definition of the interior finite element domain. Simultaneous solution of the numerical approximations of the two problems yields the solution to the electromagnetic BVP. For an overview of numerous such methodologies proposed over the years, including a fairly thorough literature review, the reader is referred to the texts [4], [8] and [28].

A second class of methods relies on the application of a local boundary condition on the truncation boundary, which attempts to represent in a physically consistent manner the outgoing (radiating) attributes of the fields that are propagating away from the domain, either due to scattering of an incident field by the media inside the domain or as direct radiation from any antenna structures present inside the computational domain. The references in the previous paragraph offer a rich overview of the history and latest developments in this class of methods.

The latest addition to methodologies for unbounded domain truncation calls for the utilization of a properly constructed layer of absorbing material, which is of finite thickness and is wrapped around the computational domain. First proposed by Berenger [29], the absorbing layers are constructed to be *matched* to the media in the computational domain, in the sense that the waves impinging on them penetrate with negligible reflection. Hence, they are most commonly referred to as *perfectly matched layers* (PML). Following a spur of early research effort toward improving the understanding and performance of PMLs (e.g., [11], [30]-[32]), more recent work has focused on the development of conformal PMLs suitable for the truncation of cylindrical, spherical and, in general, arbitrarily curved (convex) truncation boundaries. A comprehensive list of relevant publications can be found in [8].

Use of absorbing boundary condition and related grid truncation schemes will be made in later chapters, in the context of specific applications of iterative solvers for the solution of electromagnetic scattering and radiation problems. However, for the purposes of the development of the various types of finite element formulations for the solution of (3.116), only the simple pair of homogeneous boundary conditions of (3.118) on the enclosing boundary will be considered.

3.4.2   Electric field formulation

Cancelation of the magnetic field intensity from the pair of the two curl equations in (3.116) results in the vector Helmholtz equation for the electric field vector

$$\nabla \times \frac{1}{\mu_r} \nabla \times \vec{E} + j\omega\mu_0\sigma\vec{E} - \omega^2\mu_0\epsilon_0\epsilon_r\vec{E} = -j\omega\mu_0\vec{J_v}. \tag{3.119}$$

The weak statement of the BVP is obtained through multiplication of the above equation by the edge element $\vec{w}_e$, followed by integration over the volume of the domain.

$$\iiint_\Omega \nabla \times \vec{w}_e \cdot \frac{1}{\mu_r} \nabla \times \vec{E}\, dv + j\omega\mu_0 \iiint_\Omega \vec{w}_e \cdot \sigma\vec{E}\, dv - \oiint_{\Gamma_D \cup \Gamma_N} \hat{n} \cdot (\vec{w}_e \times$$
$$\frac{1}{\mu_r}\nabla \times \vec{E})ds - \omega^2\mu_0\epsilon_0 \iiint_\Omega \vec{w}_e \cdot \epsilon_r\vec{E}dv = -j\omega\mu_0 \iiint_\Omega \vec{w}_e \cdot \vec{J_v}\, dv. \tag{3.120}$$

The development of the finite element approximation continues with the expansion of the electric field intensity in terms of edge elements. In accordance with the Dirichlet boundary condition of zero tangential electric field on Γ_D, all edge elements in the finite element grid over Ω except for those associated with the edges on Γ_D are used in the expansion. Thus use of a Galerkin's testing procedure leads to the following FEM system:

$$M_{ee}x_e = -j\omega f_e, \tag{3.121}$$

where x_e denotes the vector of the unknown expansion coefficients of $\vec{E}$, f_e is the approximation of the source term associated with the impressed current $\vec{J}_v$,

$$(f_e)_i = \mu_0 \iiint_\Omega \vec{w}_{e,i} \cdot \vec{J}_v \, dv \tag{3.122}$$

and M_{ee} is the FEM matrix with elements,

$$(M_{ee})_{i,j} = \iiint_\Omega \nabla \times \vec{w}_{e,i} \cdot \frac{1}{\mu_r} \nabla \times \vec{w}_{e,j} \, dv + j\omega\mu_0 \iiint_\Omega \vec{w}_{e,i} \cdot \sigma\vec{w}_{e,j} \, dv \\ - \omega^2\epsilon_0\mu_0 \iiint_\Omega \vec{w}_{e,i} \cdot \epsilon_r\vec{w}_{e,j} \, dv. \tag{3.123}$$

The matrix M_{ee} may also be written in terms of three matrices as follows:

$$M_{ee} = (S + j\omega Z - \omega^2 T), \tag{3.124}$$

where the specific forms of the entries in the matrices S, Z, (both of which are identical to those derived in the previous section for the MQS case), and T become immediately evident through a direct, term by term comparison of (3.124) with (3.123).

A similar process may be followed for the development of a finite element approximation to a vector Helmholtz equation for the magnetic field. The electric field formulation is the one most commonly used in practice. However, the application of iterative matrix solvers to (3.121) is plagued by slow convergence or even convergence failure. One of the primary causes of this unpredictable convergence behavior is the fact that the DC modes associated with the null space of S appear in the matrix M_{ee} as eigenmodes with negative eigenvalues. This is clearly evident from (3.124). This numerical difficulty can be tackled by eliminating these modes from the space of admissible solutions. This, in turn, requires the explicit enforcement of Gauss' law for the electric field, $\nabla \cdot \vec{D} = \rho_v$. The way this is done in the context of the potential formulation is discussed next.

3.4.3 Potential formulation

The development of the potential formulation for the full-wave BVP follows closely the one used in the case of the MQS BVP. Thus, adopting the notation in [33], the electric field is written in terms of the magnetic vector potential $\vec{A}$ and the electric scalar potential V in the form

$$\vec{E} = \vec{A} + \nabla V. \tag{3.125}$$

Thus, from Maxwell's equations, the following vector wave equation is obtained:

$$\nabla \times \frac{1}{\mu_r} \nabla \times \vec{A} + j\omega\mu_0\sigma \left(\vec{A} + \nabla V\right) - \omega^2\mu_0\epsilon_0\epsilon_r \left(\vec{A} + \nabla V\right) = -j\omega\mu_0\vec{J}_v. \tag{3.126}$$

This equation is supplemented by Gauss' law for the electric field which, making use of the conservation of charge equation, may be cast in the following form:

$$-\nabla \cdot (j\omega\epsilon_r\epsilon_0 + \sigma)\left(\vec{A} + \nabla V\right) = \nabla \cdot \vec{J}_v. \tag{3.127}$$

The development of the weak statement for (3.126) and (3.127) follows closely the one for the MQS BVP. Thus all intermediate steps are suppressed and only the final results are given.

$$\iiint_\Omega \nabla \times \vec{w}_e \cdot \frac{1}{\mu_r} \nabla \times \vec{A} \, dv + j\omega\mu_0 \iiint_\Omega \vec{w}_e \cdot \sigma \left(\vec{A} + \nabla V \right) dv$$

$$- \omega^2 \mu_0 \epsilon_0 \iiint_\Omega \vec{w}_e \cdot \epsilon_r \left(\vec{A} + \nabla V \right) dv = -j\omega\mu_0 \iiint_\Omega \vec{w}_e \cdot \vec{J}_v \, dv,$$

$$j\omega\mu_0 \iiint_\Omega \nabla w_n \cdot \sigma \left(\vec{A} + \nabla V \right) dv - \omega^2 \epsilon_0 \mu_0 \iiint_\Omega \nabla w_n \cdot \epsilon_r \left(\vec{A} + \nabla V \right) dv$$

$$= -j\omega\mu_0 \iiint_\Omega \nabla w_n \cdot \vec{J}_v \, dv. \tag{3.128}$$

From the above weak statement of the BVP, a finite element approximation in the Galerkin's sense yields the following finite element matrix:

$$\begin{bmatrix} M_{aa} & M_{av} \\ M_{va} & M_{vv} \end{bmatrix} \begin{bmatrix} x_a \\ x_v \end{bmatrix} = -j\omega \begin{bmatrix} f_a \\ f_v \end{bmatrix}, \tag{3.129}$$

where x_a and x_v are the expansion coefficients in the finite element approximations of vector and scalar potentials, respectively, and the elements of the matrices and forcing vectors are given by

$$M_{aa} = M_{ee}, \quad M_{va} = M_{av}^T,$$

$$(M_{av})_{i,j} = j\omega\mu_0 \iiint_\Omega \vec{w}_{e,i} \cdot \sigma \nabla w_{n,j} \, dv - \omega^2 \mu_0 \epsilon_0 \iiint_\Omega \vec{w}_{e,i} \cdot \epsilon_r \nabla w_{n,j} \, dv,$$

$$(M_{vv})_{i,j} = j\omega\mu_0 \iiint_\Omega \nabla w_{n,i} \cdot \sigma \nabla w_{n,j} \, dv - \omega^2 \epsilon_0 \mu_0 \iiint_\Omega \nabla w_{n,i} \cdot \epsilon_r \nabla w_{n,j} \, dv,$$

$$f_a = f_e, \quad f_{v,i} = \mu_0 \iiint_\Omega \nabla w_{n,i} \cdot \vec{J}_v \, dv.$$

In view of (3.110) and the relationship between the finite element matrices and forcing vectors for the electric field and potential formulations, the following matrix relationships can be deduced:

$$\begin{bmatrix} M_{aa} & M_{av} \\ M_{va} & M_{vv} \end{bmatrix} = \begin{bmatrix} I \\ G^T \end{bmatrix} M_{ee} \begin{bmatrix} I & G \end{bmatrix}, \quad \begin{bmatrix} f_a \\ f_v \end{bmatrix} = \begin{bmatrix} I \\ G^T \end{bmatrix} f_e. \tag{3.130}$$

We conclude that the field formulation (3.121) and potential formulation (3.129) are mathematically equivalent. However, their iterative solutions exhibit different convergence behavior. This issue is explored in more detail in later chapters.

3.4.4 Field-flux formulation

The field-flux formulation develops a finite element approximation for the system of Maxwell's curl equations [34]. Thus, for the case of linear, isotropic and frequency-independent media, the derived discrete model will exhibit linear dependence on the angular frequency ω. Because of this linear dependence, this formulation is most suitable for use in a variety of electromagentic modeling applications for which discrete state-space forms of Maxwell's hyperbolic system are advantageous. A special class of such applications concerns the utilization of Krylov subspace-based techniques for the systematic reduction of state-space approximations of electromagnetic BVPs, and the development of compact macromodels for complex electromagnetic devices and circuits in terms of multi-port frequency-dependent macromodels [35]. This class of applications is discussed in detail in Chapter 11.

In the field-flux formulation the electric field intensity, $\vec{E}$, and magnetic flux density, $\vec{B}$, are expanded, respectively, in the tangentially continuous vector space, W_{tv}, and the normally continuous vector space, W_{nv} (see Chapter 2 for details). The development begins by casting the two curl Maxwell's equations in the form

$$\begin{cases} \nabla \times \vec{E} = -j\omega\vec{B}, \\ \nabla \times \left(\vec{B}/\mu_r\right) = j\omega\mu_0\epsilon_0\epsilon_r\vec{E} + \mu_0\sigma\vec{E} + \mu_0\vec{J}_v. \end{cases} \tag{3.131}$$

Let $\vec{w}_e$ denote a basis function of W_{tv} and $\vec{w}_f$ a basis function of W_{nv}. Application of Galerkin's method, where the curl equations (3.131) are multiplied by $\vec{w}_f$ and $\vec{w}_e$, respectively, and integration over the domain of interest, yields

$$\iiint_\Omega \nabla \times \vec{E} \cdot \frac{1}{\mu_r}\vec{w}_f \, dv = -j\omega \iiint_\Omega \vec{w}_f \cdot \frac{1}{\mu_r}\vec{B} \, dv,$$

$$\iiint_\Omega \nabla \times \vec{w}_e \cdot \frac{1}{\mu_r}\vec{B} \, dv = j\omega\epsilon_0\mu_0 \iiint_\Omega \vec{w}_e \cdot \epsilon_r\vec{E} \, dv \tag{3.132}$$

$$+ \mu_0 \iiint_\Omega \vec{w}_e \cdot \sigma\vec{E} \, dv + \mu_0 \iiint_\Omega \vec{w}_e \cdot \vec{J}_v \, dv.$$

The above weak statements of Maxwell's curl equations yield, upon substitution of $\vec{E}$ and $\vec{B}$ with their corresponding expansions, and subsequent testing with each one of the functions used in the expansions, the following finite element matrix statement:

$$\begin{bmatrix} 0 & D \\ -D^T & Z \end{bmatrix} \begin{bmatrix} x_b \\ x_e \end{bmatrix} = -j\omega \begin{bmatrix} P & 0 \\ 0 & T \end{bmatrix} \begin{bmatrix} x_b \\ x_e \end{bmatrix} - \begin{bmatrix} 0 \\ f_e \end{bmatrix}, \tag{3.133}$$

where the vectors x_e and x_b contain the expansion coefficients for $\vec{E}$ and $\vec{B}$, respectively, and

$$D_{i,j} = \iiint_\Omega \frac{1}{\mu_r}\vec{w}_{f,i} \cdot \nabla \times \vec{w}_{e,j} \, dv,$$

$$P_{i,j} = \iiint_\Omega \vec{w}_{f,i} \cdot \frac{1}{\mu_r}\vec{w}_{f,j} \, dv. \tag{3.134}$$

The elements of the matrices Z and T are the same with those derived in the electric field formulation of the full-wave problem (see (3.124)).

One expects an equivalence between the finite element statement of (3.133) with the one in (3.121), derived from the electric field formulation. To explore this equivalence, in a manner similar to the one used for deriving the vector Helmholtz equation for the electric field from the system of Maxwell's curl equations, we proceed to eliminate x_b from the system in (3.133). A simple matrix manipulation leads to the following equation:

$$\left(D^T P^{-1} D + j\omega Z - \omega^2 T\right) x_e = -j\omega f_e. \tag{3.135}$$

Hence, to prove the equivalence of (3.121) and (3.133) we need to prove that $S = D^T P^{-1} D$. This is shown by first recalling the form of the elements of S given in (3.128), and then recognizing that, in the spirit of finite element approximation, the position-dependent relative magnetic permeability, μ_r, will be discretized in a manner such that its value is constant over each element. These, combined with the fact that, as shown in Chapter 2, it is $\nabla \times W_{tv} \subset W_{nv}$, lead to the desirable result.

3.5 PARTIAL ELEMENT EQUIVALENT CIRCUIT MODEL

It should be apparent from the discussion in the previous sections that the finite element approximation of electromagnetic vector functions (i.e., fields, fluxes, current and charge sources) in terms of expansion functions that capture correctly the vector properties of these functions and associated boundary conditions at material interfaces, provide for a physically consistent framework for the development of discrete approximations of differential equation-based electromagnetic BVPs. Furthermore, the spaces developed in Chapter 2 for the approximation of electromagnetic fields, fluxes and sources in a discretized domain are suitable not only for use in conjunction with the finite element framework, but also for use in the development of discrete models using any one of the popular differential equation-based and integral equation-based statements of the electromagnetic BVPs. As an example, we apply in this section the results from Chapter 2 to the development of a physically consistent, discrete approximation of the electric field integral equation (EFIE) statement of the electromagnetic BVP over a volume discretized in terms of tetrahedra. In particular, the development will be carried out in the spirit of the Partial Element Equivalent Circuit (PEEC) approximation, which is the most commonly used method for the EFIE-based numerical modeling of the signal distribution network in high-speed/high-frequency integrated electronic systems [36]-[40].

The objective of this development is two-fold. First, it provides a systematic extension of the PEEC methodology to arbitrarily shaped volumes discretized in terms of tetrahedra. Second, it helps interpret the ill-conditioning of the EFIE matrix observed at low frequencies in terms of the properties of the expansion functions and the associates spaces, and leads us to effective remedies for mitigating numerical instability and ensuring numerical robustness of the matrix solution from almost DC to very high frequencies.

3.5.1 Electric field integral equation

For the purposes of this discussion we limit ourselves to the development of the electric field integral equation statement for a set of conductors in free space. Figure 3.8 depicts the geometry of interest. The conductors are characterized by conductivity σ_i and permittivity ϵ_i, while their magnetic permeability is assumed to be that of free space. Without loss of generality, the excitation is provided by point time-harmonic current sources attached to the surface of the conductors. Other types of excitation, such as impressed voltage sources and incident electromagnetic fields can be taken into account through techniques described in detail in the PEEC literature. The electric current density and the electric charge density induced in the conductors constitute the unknowns in the electromagnetic BVP.

Under the assumption of time-harmonic excitation with angular frequency ω, the governing system of Maxwell's equations is

$$\nabla \times \vec{E} = -j\omega\mu_0\vec{H} \tag{3.136}$$

$$\nabla \times \vec{H} = j\omega\epsilon\vec{E} + \sigma\vec{E} \tag{3.137}$$

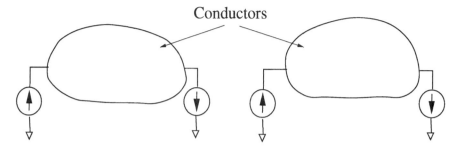

Figure 3.8   A set of conductors in free space excited by current sources.

$$\nabla \cdot \left(\epsilon \vec{E}\right) = 0 \tag{3.138}$$

$$\nabla \cdot \left(\mu_0 \vec{H}\right) = 0, \tag{3.139}$$

where it is understood that ϵ and σ are position-dependent parameters and no distributed impressed sources are present. It will be shown later that the assumed current sources, through which the structure is driven, are brought into the integral equation statement of the BVP and its discrete form through the enforcement of the conservation of charge equation. The induced electric current and charge densities in the conducting regions can be explicitly shown in the above set of equations by recasting them in the following form:

$$\nabla \times \vec{E} = -j\omega\mu_0\vec{H} \tag{3.140}$$

$$\nabla \times \vec{H} = j\omega\epsilon_0\vec{E} + (\sigma + j\omega(\epsilon - \epsilon_0))\,\vec{E} \tag{3.141}$$

$$\nabla \cdot \left(\epsilon_0\vec{E}\right) = -\nabla \cdot \left((\epsilon - \epsilon_0)\vec{E}\right) \tag{3.142}$$

$$\nabla \cdot \left(\mu_0\vec{H}\right) = 0. \tag{3.143}$$

The induced electric current density, $\vec{J}$, and electric charge density, ρ, appearing in the above equations are readily identified as

$$\vec{J} = (\sigma + j\omega(\epsilon - \epsilon_0))\,\vec{E}, \quad \rho = -\rho_v - \nabla \cdot \left((\epsilon - \epsilon_0)\vec{E}\right). \tag{3.144}$$

We will accompany this system with the conservation of charge statement for the induced current and charge densities at all points inside the conducting regions,

$$\nabla \cdot \vec{J} = -j\omega\rho. \tag{3.145}$$

This equation is readily derived by taking the divergence of (3.137) and making use of (3.138). It is noted that the conservation of charge statement must be modified in the vicinity of the point current sources attached to the surface of the conductors, to account for the current flux provided by the sources. This point will be discussed in more detail in the next section.

It is evident from the system of (3.140)-(3.143) that the induced electric current and charge densities act as sources for a secondary electromagnetic field radiating in free space. Hence, the standard potential formulation for time-harmonic electromagnetic field radiation in free space can be used to obtain the electromagnetic fields at any point in space [41]. More specifically, the resulting magnetic and electric fields are obtained in terms of the magnetic vector potential, $\vec{A}$, and the electric scalar potential, Φ, through the equations

$$\vec{H} = \frac{1}{\mu_0}\nabla \times \vec{A}, \tag{3.146}$$

$$\vec{E} = -j\omega\vec{A} - \nabla\Phi. \tag{3.147}$$

The two potentials satisfy the following Helmholtz equations

$$\nabla^2\vec{A} + \omega^2\epsilon_0\mu_0\vec{A} = -\mu_0\vec{J}, \tag{3.148}$$

$$\nabla^2\Phi + \omega^2\epsilon_0\mu_0\Phi = -\frac{1}{\epsilon_0}\rho, \tag{3.149}$$

and are taken to be related through the Lorenz gauge relationship

$$\nabla \cdot \vec{A} = -j\omega\epsilon_0\mu_0\Phi. \tag{3.150}$$

The solutions to (3.148) and (3.149) are

$$\begin{aligned}\vec{A}(\vec{r}) &= \mu_0 \iiint_\Omega \vec{J}(\vec{r}')G(\vec{r},\vec{r}')dv' \\ \Phi(\vec{r}) &= \frac{1}{\epsilon_0}\iiint_\Omega \rho(\vec{r}')G(\vec{r},\vec{r}')dv',\end{aligned} \tag{3.151}$$

where Ω denotes the volume of all conductors and the free-space Green's function, $G(\vec{r},\vec{r}')$, is given by

$$G(\vec{r},\vec{r}') = \frac{e^{-jk_0|\vec{r}-\vec{r}'|}}{4\pi|\vec{r}-\vec{r}'|}, \tag{3.152}$$

where $k = \omega\sqrt{\epsilon_0\mu_0}$. When the maximum dimension of the computational domain is sufficiently smaller than the wavelength such that $k_0|\vec{r}-\vec{r}'| \ll 1$, the approximation $e^{-jk_0|\vec{r}-\vec{r}'|} \approx 1$ is used to reduce the full-wave kernel of (3.152) to its quasi-static form

$$G(\vec{r},\vec{r}') \approx \frac{1}{4\pi|\vec{r}-\vec{r}'|}. \tag{3.153}$$

Substitution of (3.151) in (3.147) yields the expression for the electric field at any point in space in terms of the induced electric current and charge densities,

$$\vec{E}(\vec{r}) = -j\omega\mu_0 \iiint_\Omega \vec{J}(\vec{r}')G(\vec{r},\vec{r}')dv' - \frac{1}{\epsilon_0}\nabla\iiint_\Omega \rho(\vec{r}')G(\vec{r}.\vec{r}')dv'. \tag{3.154}$$

The integral equation statement for the induce current density in the conductors is derived by enforcing the above equations at points inside the conductors, while making use of the first one of the equations in (3.144) The resulting equation is

$$\frac{\vec{J}(\vec{r})}{\sigma + j\omega(\epsilon - \epsilon_0)} + j\omega\mu_0 \iiint_\Omega \vec{J}(\vec{r}')G(\vec{r},\vec{r}')dv' + \frac{1}{\epsilon_0}\nabla\iiint_\Omega \rho(\vec{r}')G(\vec{r},\vec{r}')dv' = 0. \tag{3.155}$$

3.5.2 Development of the PEEC model

The discrete approximation of (3.155) begins with the discretization of the conducting volumes by means of tetrahedra. Since the unknown current density, $\vec{J}$, is a flux quantity exhibiting normal continuity at (surface charge-free) interfaces between adjacent elements, its approximation is in terms of elements of the normally continuous vector space W_{nv}. Without loss of generality, let us consider the lowest-order NV basis functions, i.e, the facet elements of Chapter 2, as the elements used for the expansion of $\vec{J}_v$. Thus we write

$$\vec{J} = \sum_{i=1}^{N_f} \vec{w}_{f,i} I_i, \tag{3.156}$$

where $\vec{w}_{f,i}$ denotes the facet element on the i-th facet, I_i is the corresponding expansion coefficient, and N_f is the number of facets inside the conductors excluding the facets on the surfaces of the conductors (since the normal component of the current density is zero on the conductor surface).

At this point it is useful to recall the following property of facet elements

$$\iint_{facet-i} \hat{n} \cdot \vec{w}_{f,i} ds = 1, \tag{3.157}$$

where $\hat{n}$ is the unit normal on the i-th facet, pointing in the direction of the facet element $\vec{w}_{f,i}$. In view of this result it is immediately evident that the expansion coefficient I_i is the net current flux through the i-th facet. Hence, the expansion (3.156) is over the facets interior to the conducting volumes only.

Prior to discussing the expansion of the electric charge density it is appropriate to mention that for the case of homogeneous good conductors any electric charge present will accumulate on the surface of the conductors. To elaborate, the time-dependent form of the conservation of charge equation inside the conductor, written in terms of the electric charge density only, describes the time evolution of the charge density

$$\frac{\partial \rho}{\partial t} + \frac{\sigma}{\epsilon} \rho = 0. \tag{3.158}$$

Assuming a non-zero initial charge distribution, ρ_0, inside the conductor at $t = 0$, the solution of the above equation is readily found to be

$$\rho(t) = \rho_0 \exp\left(-t/T_r\right), \tag{3.159}$$

where $T_r = \epsilon/\sigma$ is the *relaxation time* for the conducting material. For good conductors the relaxation time is significantly smaller than the characteristic times of relevance to engineering applications. For example, for aluminum with conductivity 4×10^7 S/m the relaxation time is $\approx 2.2 \times 10^{-19}$ s. Hence, the relaxation of electric charge from the interior of the conductor to its surface can be assumed instantaneous for most practical applications.

Returning to the expansion of the unknown charge density, let us consider, first, the case where the volume charge density is assumed nonzero. The special case of good conductors for which electric charge accumulation is on the conductor surface only will be addressed in the next section. In accordance with the development in Section 2.5.4, the elements of the three-dimensional charge space W_{charge}^0 are used for the expansion of the volume charge density

$$\rho = \sum_{i=1}^{N_t} w_{t,i} Q_i, \tag{3.160}$$

where N_t is the number of tetrahedra inside the conductors and each expansion function $w_{t,i}$ is given by

$$w_{t,i} = \frac{1}{V_i}, \tag{3.161}$$

where V_i is the volume of the ith tetrahedron. Thus the expansion coefficient Q_i is readily recognized as the charge stored in the ith tetrahedron.

Multiplication of (3.155) by $\vec{w}_{f,i}$, followed by integration over the conducting volumes yields

$$\iiint_\Omega \vec{w}_{f,i} \cdot \frac{1}{\sigma + j\omega(\epsilon - \epsilon_0)} \vec{J} dv + j\omega\mu_0 \iiint_\Omega \vec{w}_{f,i} \cdot \left(\iiint_\Omega G(\vec{r},\vec{r}') \vec{J}(\vec{r}') dv' \right) dv$$
$$- \frac{1}{\epsilon_0} \iiint_\Omega \nabla \cdot \vec{w}_{f,i} \left(\iiint_\Omega G(\vec{r},\vec{r}') \rho(\vec{r}') dv' \right) dv = 0. \tag{3.162}$$

In deriving the above equation use was made of the vector identity $\vec{w}_{f,i} \cdot \nabla\Psi = \nabla \cdot (\vec{w}_{f,i}\Psi) - \Psi\nabla \cdot \vec{w}_{f,i}$ along with the divergence theorem (A.14) and the fact that $\hat{n} \cdot \vec{w}_{f,i} = 0$ on the boundary of the conductors to obtain the following result:

$$\frac{1}{\epsilon_0} \iiint_\Omega \vec{w}_{f,i} \nabla \left(\iiint_\Omega G(\vec{r},\vec{r}') \rho(\vec{r}') dv' \right) dv$$
$$= -\frac{1}{\epsilon_0} \iiint_\Omega \nabla \cdot \vec{w}_{f,i} \left(\iiint_\Omega G(\vec{r},\vec{r}') \rho(\vec{r}') dv' \right) dv. \tag{3.163}$$

We have shown in Chapter 2 that the divergence of facet elements can be written as the linear combination of the elements of W_{charge}^0 as follows:

$$\begin{bmatrix} \nabla \cdot \vec{w}_{f,1}, & \nabla \cdot \vec{w}_{f,2}, & \cdots, \nabla \cdot \vec{w}_{f,N_f} \end{bmatrix} = \begin{bmatrix} w_{t,1}, & w_{t,2}, & \cdots, w_{t,N_t} \end{bmatrix} D_{N_t \times N_f}, \tag{3.164}$$

where $D_{N_f \times N_t}$ is the divergence matrix discussed in (2.127). Substitution of (3.156), (3.160) and (3.164) into (3.162), followed by testing with all expansion functions used in the approximation of $\vec{J}$, yields the following matrix equation:

$$(R + j\omega L)I_p + D^T PQ = 0, \tag{3.165}$$

where I_p is the vector containing the coefficients, I_i, in the expansion of the induced current density, and Q is the vector containing the coefficients, Q_i, in the expansion of the induced charge density. The matrices R, L and P have elements given by the following integrals:

$$R_{i,j} = \iiint_\Omega \vec{w}_{f,i} \cdot \frac{1}{\sigma + j\omega(\epsilon - \epsilon_0)} \vec{w}_{f,j} dv,$$

$$L_{i,j} = \mu_0 \iiint_\Omega \vec{w}_{f,i}(\vec{r}) \cdot \left(\iiint_\Omega G(\vec{r},\vec{r}') \vec{w}_{f,j}(\vec{r}') dv' \right) dv, \tag{3.166}$$

$$P_{i,j} = -\frac{1}{\epsilon_0} \iiint_\Omega w_{t,i}(\vec{r}) \left(\iiint_\Omega G(\vec{r},\vec{r}') w_{t,j}(\vec{r}') dv' \right) dv.$$

The resulting system consists of N_f equations for the $N_f + N_t$ unknowns. The additional N_t equations needed to close the system of (3.165) are provided from the discrete form of the conservation of charge statement (3.145). This is obtained by substituting in (3.145) the approximations for $\vec{J}$ and ρ given by (3.156) and (3.160), respectively, and enforcing the equation at each one of the tetrahedra in the conducting regions. In doing so, we must also take into account the presence of the point current sources attached to the surface of the conductors. The resulting set of equations assumes the form

$$DI_p + j\omega Q = I_s. \tag{3.167}$$

where I_s, the source current vector, contains the strengths of the impressed point current sources.

Combining (3.165) and (3.167) yields the matrix statement of the PEEC approximation of the problem

$$\begin{bmatrix} R + j\omega L & D^T P \\ D & j\omega I_{N_t \times N_t} \end{bmatrix} \begin{bmatrix} I_p \\ Q \end{bmatrix} = \begin{bmatrix} 0 \\ I_s \end{bmatrix}, \tag{3.168}$$

where $I_{N_t \times N_t}$ is the identity matrix of dimension N_t.

3.5.3   The case of surface current flow

It is often the case in practice that electric current flow can be assumed to occur only on a surface of a volume. The most commonly encountered case is the case of good conductors at frequencies high enough for the skin effect to be well developed, such that the skin depth is much smaller than the conductor dimensions. Thus current flow can be assumed to occur right at the conductor surface, within a layer of thickness equal to one skin depth. For such cases, it is common to utilize the surface impedance, η_s, for good conductors [41]

$$\eta_s = (1 + j)\sqrt{\frac{\pi f \mu}{\sigma}} \tag{3.169}$$

to impose the following impedance condition between the tangential electric field, $\vec{E}_t$, and the induced current density, $\vec{J}_s = \hat{n} \times \vec{H}$ on the conductor surface,

$$\vec{E}_t = \eta_s \vec{J}_s. \tag{3.170}$$

For such cases it is clear that only the surface of the conducting volumes needs to be discretized. For this purpose, a mesh of triangular elements is used and the induced current density is expanded in terms of the elements in the two-dimensional NV space discussed in Section 2.2.3. Without loss of generality, the lowest-order 2D NV space is used. These are the well-known Rao-Wilton-Glisson basis functions [42]:

$$\vec{J}_s = \sum_i^{N_e} \vec{w}_{ne,i} I_i, \tag{3.171}$$

where $\vec{w}_{ne,i}$ denotes the RWG basis associated with ith-edge (m, n). As discussed in Section 2.2.3, it is $\vec{w}_{ne,i} = \hat{n} \times \vec{e}_{m,n}$. N_e is the number of edges on the surface of the conductors.

The surface charge density, ρ_s, is expanded in terms of the elements of the lowest-order two-dimensional charge space, W^0_{charge}, as follows:

$$\rho_s = \sum_i^{N_f} w_{f,i} Q_i, \tag{3.172}$$

where $w_{f,i}$ is the basis function on the facet i, given by

$$w_{f,i} = \frac{1}{A_i}, \tag{3.173}$$

where A_i is the area of facet i. N_f is the number of facets (triangles) on the surface of the conducting volumes.

As shown in Chapter 2, the two sets of expansion functions are related as follows:

$$[\nabla \cdot \vec{w}_{ne,1}, \nabla \cdot \vec{w}_{ne,2}, \cdots, \nabla \cdot \vec{w}_{ne,N_e}] = [w_{f,1}, w_{f,2}, \cdots, w_{f,N_f}] D_{N_f \times N_e}, \tag{3.174}$$

where the divergence matrix $D_{N_f \times N_e}$ in two dimensions was discussed in (2.40).

The integral equation statement for the induced surface current density is similar to the one in (3.162) where the volume integrals are replaced by surface integrals over the surfaces of the conducting volumes. The only term that needs some discussion is the one associated with the contribution of the scalar potential Φ. Upon testing with $\vec{w}_{ne,i}$, this term assumes the form

$$\frac{1}{\epsilon_0} \iint_{\Gamma_c} \vec{w}_{ne,i} \nabla_s \left(\iint_{\Gamma_c} G(\vec{r}, \vec{r}') \rho_s(\vec{r}') ds' \right) ds, \tag{3.175}$$

where Γ_c denotes the boundary of the conducting regions, and ∇_s indicates that the gradient operation is performed on the surface of the conductor. This integral may be recast in a form similar to the one in (3.163) by making use of the identity $\nabla_s \cdot (\Psi \vec{w}_{ne,i}) = \Psi \nabla_s \cdot \vec{w}_{ne,i} + \vec{w}_{ne,i} \nabla_s \Psi$ along with the following result: [43],

$$\iint_{\Gamma_c} \nabla_s \cdot (\Psi \vec{w}_{ne,i}) ds = \int_C \Psi \vec{w}_{ne,i} \cdot \hat{p} d\ell - \iint_{\Gamma_c} K_1 (\Psi \vec{w}_{ne,i} \cdot \hat{n}) ds, \tag{3.176}$$

where C is the curve enclosing the surface Γ_c, K_1 is the first curvature on the surface Γ_c. The two unit vectors $\hat{n}$ and $\hat{p}$ are perpendicular to each other and to the curve C and, together with a unit vector $\hat{t}$ tangent to C, define a right-handed local coordinate system as follows, $\hat{t} = \hat{n} \times \hat{p}$ [43].

For the case of a closed surface, since $\vec{w}_{ne,i}$ is tangential to the surface, the integral vanishes. Furthermore, for the case of interest, since the normal component of the surface current density is zero at the edges of the conductor, the integral vanishes also for the case of open conducting surfaces (e.g., conducting strips and plates). Thus we conclude that (3.175) may be cast in the form

$$\frac{1}{\epsilon_0} \iint_{\Gamma_c} \vec{w}_{ne,i} \nabla_s \left(\iint_{\Gamma_c} G(\vec{r}, \vec{r}') \rho_s(\vec{r}') ds' \right) ds$$
$$= -\frac{1}{\epsilon_0} \iint_{\Gamma_c} \nabla_s \cdot \vec{w}_{ne,i} \left(\iint_{\Gamma_c} G(\vec{r}, \vec{r}') \rho_s(\vec{r}') ds' \right) ds. \tag{3.177}$$

Finally, the integral equation corresponding to (3.162) for the case of surface electric current and charge distribution becomes

$$\iint_{\Gamma_c} \vec{w}_{ne,i} \cdot \eta_s \vec{J}_s ds + j\omega\mu_0 \iint_{\Gamma_c} \vec{w}_{ne,i} \cdot \left(\iint_{\Gamma_c} G(\vec{r}, \vec{r}') \vec{J}_s(\vec{r}') ds' \right) ds$$
$$- \frac{1}{\epsilon_0} \iint_{\Gamma_c} \nabla_s \cdot \vec{w}_{ne,i} \left(\iint_{\Gamma_c} G(\vec{r}, \vec{r}') \rho_s(\vec{r}') ds' \right) ds = 0, \tag{3.178}$$

where use was made of (3.174).

The PEEC approximation is obtained by substituting (3.171) and (3.172) in (3.178) and testing with all expansion functions in the approximation of $\vec{J}_s$. The resulting PEEC matrix statement is identical in form to (3.165) with matrix elements given in terms of the following integrals:

$$(R_s)i,j = \iint_{\Gamma_c} \vec{w}_{ne,i} \cdot \eta_s \vec{w}_{ne,j} ds,$$

$$(L_s)i,j = \mu_0 \iint_{\Gamma_c} \vec{w}_{ne,i}(\vec{r}) \cdot \left(\iint_{\Omega} G(\vec{r},\vec{r}') \vec{w}_{ne,j}(\vec{r}') ds' \right) ds, \qquad (3.179)$$

$$(P_s)i,j = -\frac{1}{\epsilon_0} \iint_{\Gamma_c} w_{f,i}(\vec{r}) \left(\iint_{\Gamma_c} G(\vec{r},\vec{r}') w_{f,j}(\vec{r}') ds' \right) ds.$$

The resulting system of N_e equations involves N_e+N_f unknowns. The remaining N_f equations, required to close the system, are obtained by enforcing the charge conservation equation at each one of the N_f triangles. The resulting system of equations is identical in form with the one in (3.167). This completes the PEEC model for the case of surface current density flow. The pertinent matrix equation is

$$\begin{bmatrix} R_s + j\omega L_s & D^T P_s \\ D & j\omega I_{N_f \times N_f} \end{bmatrix} \begin{bmatrix} I_p \\ Q \end{bmatrix} = \begin{bmatrix} 0 \\ I_s \end{bmatrix}, \qquad (3.180)$$

where $I_{N_f \times N_f}$ is the identity matrix of dimension N_f.

3.5.4 Low-frequency numerical instability

It is common practice in the numerical solution of electromagnetic integral equations to eliminate the unknown vector, Q, in the system of (3.168) or (3.180) and solve the resulting matrix equation

$$\left(R + j\omega L - \frac{1}{j\omega} D^T P D \right) I_p = -\frac{1}{j\omega} D^T P I_s \qquad (3.181)$$

for the unknown vector current I_p. It is well known that at low frequencies the above system is ill-conditioned. In [44] and [45] a comprehensive review is provided of earlier efforts on the investigation of this low-frequency numerical instability and efforts toward its mitigation. In the same papers the authors discuss effective methodology for overcoming this problem. In the context of the PEEC formulation of the EFIE, [46] and [47] discuss the low-frequency numerical instability, and discuss methodologies in which such instability can be rectified.

In this section, we examine the low-frequency instability of (3.181) in the light of the properties of the spaces used for the expansion of the electric current density and the electric charge density. In particular, we show that the utilization of the now popular loop/star and loop/tree decomposition of the unknown current density as a remedy for dealing with the low-frequency instability is intimately connected with the properties of the divergence matrix D.

We begin our discussion by examining the properties of the divergence matrix D, derived in Chapter 2. For a three-dimensional mesh of tetrahedra, let us assign a node at the center of each tetrahedron. These nodes, together with the branches that connect them, define a three-dimensional circuit. The number of rows of the matrix D, which in circuit theory is known as the *incidence matrix*, equals the number of nodes (i.e., the number of tetrahedra) in the circuit. The number of columns of D equals the number of branches (i.e., the number of facets, except for the facets on the boundary of the mesh). The only nonzero elements in each row are associated with the branches that connect to the node corresponding to the row, and their value is +1 if the branch leaves the node and −1 if the branch enters the node. From the discussion in Section 2.7, we have $D_{N_t \times N_f} C_{N_f \times N_e} = 0$ (see (2.42)). This implies that each column of C is an element of the null space of D. At this point, it is useful to recall from Section 2.7 that each column of C is associated with an edge in the mesh, and its non-zero elements are associated with those branches (in the corresponding "circuit" for the mesh) which form a loop around the edge. The ith row of D and ith column of C are illustrated in Fig. 3.9.

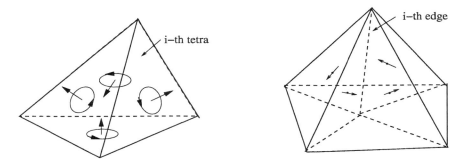

Figure 3.9   a: A row of incident matrix D. b: A column of loop matrix C.

The relationship between D and C can be used to facilitate the investigation of the properties of the matrix in (3.181). At low frequencies the third term on the left-hand side of (3.181) is dominant. Hence, since the null space of D is not empty, the matrix is ill conditioned. More specifically, the problematic modes are the ones associated with the loops described by the columns of C.

In order to remedy this situation, it makes sense to single out the loop currents in the expansion of $\vec{J}$. Availability of C allows us to identify these loops and, hence, construct the associated expansion functions in a straightforward fashion. What remains is the augmentation of these *loop expansion functions* with another set of expansion functions which are orthogonal to the loop expansion functions. This is easily accomplished by noting that the range space of D^T is orthogonal to the null space of D (see Section 3.2.3 for the proof of this result). Thus, with M_1 denoting the null space of D, $M_1 = \mathcal{N}(D)$, and M_2 denoting the range space of D^T, $M_2 = \mathcal{R}(D^T)$, the decomposition of I_p in terms of the elements of the two spaces assumes the form

$$I_p = M_1 I_1 + M_2 I_2. \tag{3.182}$$

Substitution of (3.182) into (3.181) yields

$$(R + j\omega L)(M_1 I_1 + M_2 I_2) - \frac{1}{j\omega}(D^T P D)M_2 I_2 = -\frac{1}{j\omega}D^T P I_s. \tag{3.183}$$

By multiplying on the left the above equation by M_1^T and M_2^T, respectively, the following system of equations is obtained:

$$\begin{bmatrix} M_1^T(R + j\omega L)M_1 & M_1^T(R + j\omega L)M_2 \\ M_2^T(R + j\omega L)M_1 & M_2^T\left(R + j\omega L - \frac{1}{j\omega}D^T P D\right)M_2 \end{bmatrix} \begin{bmatrix} I_1 \\ I_2 \end{bmatrix} = \begin{bmatrix} 0 \\ -\frac{M_2^T D^T P}{j\omega}I_s \end{bmatrix}. \tag{3.184}$$

The resulting system is much better conditioned than the original system of (3.181).

The same process can be applied for the case of a surface current flow, for which a two-dimensional mesh will be utilized. In this case, the null space of the divergence matrix is the gradient matrix, as is easily concluded from the results in Section 2.4 (see (2.128). Thus for 2D meshes, the gradient matrix G is the loop matrix. Each column of G corresponds to an internal node, and its non-zero elements are associated with those branches (in the corresponding "circuit" for the mesh) which form a loop around the node. Thus the construction of the loop expansion functions for the current in 2D is obtained in a straightforward manner in terms of the columns of the gradient matrix G.

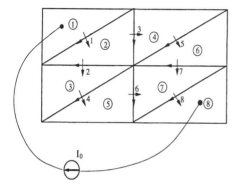
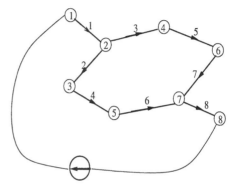

Figure 3.10 Triangular mesh for a metal strip conductor.　　**Figure 3.11** Its corresponding circuit.

To illustrate these results with an example let us consider the infinitesimally thin metal strip shown in Fig. 3.10. The unknown currents are associated with the edges of a two-dimensional mesh of triangles. The edges on the boundary are excluded because the normal component of current on the boundary is zero. The corresponding circuit, shown on the right of Fig. 3.11, is developed with nodes assigned to each triangle and branches connecting the nodes assigned for each edge shared by two adjacent elements. The driving current source, I_0, is connected to patches 1 and 8. Thus the source vector I_s is given by

$$I_s = \begin{bmatrix} I_0 & 0 & 0 & 0 & 0 & 0 & 0 & 0 & -I_0 \end{bmatrix}^T. \tag{3.185}$$

The matrix D is easily found to be

$$D_{8\times8} = \begin{bmatrix} 1 & & & & & & & \\ -1 & 1 & 1 & & & & & \\ & -1 & & 1 & & & & \\ & & -1 & & 1 & & & \\ & & & -1 & & 1 & & \\ & & & & -1 & & 1 & \\ & & & & & -1 & -1 & 1 \\ & & & & & & & -1 \end{bmatrix}. \tag{3.186}$$

Since there is only one internal node, the G matrix is given by

$$G_{8\times1} = \begin{bmatrix} 0 & -1 & 1 & -1 & 1 & -1 & 1 & 0 \end{bmatrix}^T. \tag{3.187}$$

It is easy to verify that $D_{8\times8}G_{8\times1} = 0$. Therefore, for 2D and 3D meshes the loop matrix can be constructed, from the gradient or the curl matrix, respectively. For 2D meshes there is a loop associated with each internal node, while for 3D meshes there is a loop associated with each internal edge. The loops found by the gradient or curl matrix are referred to as *local loops*, since the associated branches in the corresponding circuit are in the immediate vicinity of the corresponding node or edge.

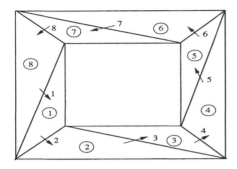 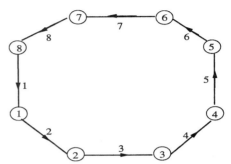

Figure 3.12   A conducting strip loop.   **Figure 3.13**   Its corresponding circuit.

In addition to local loops, *global loops* (or super-loops) may be present in the corresponding circuit model for a mesh, as easily demonstrated by the conducting strip loop structure depicted in Fig. 3.12. The corresponding circuit is depicted in Fig. 3.13. The divergence matrix is easily found to be

$$
D_{8\times8} = \begin{bmatrix}
1 & & & & & & & -1 \\
-1 & 1 & & & & & & \\
& -1 & 1 & & & & & \\
& & -1 & 1 & & & & \\
& & & -1 & 1 & & & \\
& & & & -1 & 1 & & \\
& & & & & -1 & 1 & \\
& & & & & & -1 & 1
\end{bmatrix}.
\tag{3.188}
$$

Its gradient matrix is zero since the mesh has no internal nodes. Nevertheless, the structure has a super-loop, as evident by its corresponding circuit, which is defined by

$$
\begin{bmatrix} 1 & 1 & 1 & 1 & 1 & 1 & 1 & 1 \end{bmatrix}.
\tag{3.189}
$$

This super-loop must be found and used in the decomposition of the unknown current vector for the development of the numerically stable matrix approximation of (3.184).

Local loops are easily found from the curl and gradient matrices of the 3D and the 2D meshes, respectively. However the identification of super-loops, which are most often encountered in multiply-connected domains, requires special algorithms for their identification. A methodology based on a tree-cotree decomposition of the graph defined by the corresponding circuit can be found in [40]

REFERENCES

1. G. Strang and G. J. Fix, *An Analysis of the Finite Element Method*, Englewood Cliffs, NJ: Prentice-Hall, Inc., 1973.

2. J.T. Oden and G.F. Carey, *Finite Elements: Mathematical Aspects, Vol. IV*, Englewood Cliffs, NJ: Prentice-Hall, Inc., 1983.

3. O.C. Zienkiewicz and A. Graig, "Adaptive refinement, error estimates, multigrid solution, and hierarchic finite element method concept," in *Accuracy Estimates and Adaptive Refinements in Finite Element Computations*, I. Babuska, O.C. Zienkiewich, J. Gago and E.R. de A. Oliveira, (Eds.), London: John Wiley & Sons, 1986.

4. M. Salazar-Palma, T.K. Sarkar, L.-E. Garcia-Castillo, T. Roy, and A. Djordjevic, *Iterative and Self-Adaptive Finite Elements in Electromagnetic Modeling*, Norwood, MA: Artech House, Inc., 1989.

5. R. D. Graglia and G. Lombardi, "New higher order two-dimensional singular elements for FEM and MOM applications," *Proc. of the ICEAA 03*, pp. 87-89, Torino, Italy, Sep. 2003.

6. I. Babuska, "The *p- and* h-p versions of the finite element method: The state of the art," in *Finite Elements Theory and Applications*, D. L. Dwoyer, M.Y. Hussain and R.G. Voigt (Eds.), New York: Springer-Verlag, 1988.

7. I. Babuska and B.Q. Guo, "The *h-p* version of the finite element method for domains with curved boundaries," *SIAM J. Numer. Anal.*, vol. 25, pp. 837-861, 1988.

8. J. Jin, *The Finite Element Method in Electromagnetics*, 2nd. ed., New York: John Wiley & Sons, Inc., 2002.

9. P. Bettess and J. A. Bettess, "Infinite elements for static problems," *Eng. Comp.*, vol. 1, pp. 4-15, 1984.

10. W. Pinello, M. Gribbons, and A. C. Cangellaris, "A new numerical grid truncation scheme for the Finite Difference/Finite Element solution of Laplace's equation," *IEEE Trans. Magnetics*, vol. 32, no. 3, pp. 1397-1400, May 1996.

11. W. C. Chew and W. H. Weedon, "A 3D perfectly matched medium from modified Maxwell's equations with stretched coordinates," *Microwave and Optical Tech. Lett.*, vol. 7, pp. 599-604, Sep. 1994.

12. P. M. Morse and H. Feshbach, *Methods of Theoretical Physics: Part I*, New York: McGraw-Hill Book Company, Inc., 1978.

13. J-L. Coulomb, "Finite element three diemnsional magnetic field computation," *IEEE Trans. Magn.*, vol. 17, pp. 3241-3246, Nov. 1981.

14. M. L. Barton and Z. J. Cendes, "New vector finite elements for three-dimensional magnetic field computation," *J. Appl. Phys.*, vol. 61, pp. 3919-3921, 1987.

15. K. Preis, I. Bardi, O. Biro, C. Magele, W. Renhart, K. R. Richter, and G. Vrisk, 'Numerical analysis of 3D magnetostatic fields," *IEEE Trans. Magn.*, vol. 27, pp. 3798-3803, Sep. 1991.

16. K. Preis, I. Bardi, O. Biro, C. Magele, G. Vrisk, and K. R. Richter, "Different finite element formulations of 3D magnetostatic fields," *IEEE Trans. Magn.*, vol. 28, pp. 1056-1059, Mar. 1992.

17. O. Biro, K. Pries, and K. R. Richter, "On the use of the magnetic vector potential in the nodal and edge finite element analysis of 3D magnetostatic problems," *IEEE Trans. Magn.*, vol. 32, no. 3, pp. 651-654, May 1996.

18. J. R. Brauer, S. M. Schaefer, J. F. Lee, and R. Mittra, "Asympotic boundary condition for three dimensional magnetostatic finite elements," *IEEE. Trans. Magn.*, vol. 27, no. 6, pp. 5013-5017, Nov. 1991.

19. J. B. Manges and Z. Cendes, "A generalized tree-cotree gauge for magnetic field computation," *IEEE Trans. Magn.*, vol. 31, no. 3, pp. 1342-1347, May 1995.

20. O. Biro and K. Preis, "On the use of magnetic vector potential in the finit element analysis of 3-D eddy currents", *IEEE Trans. Magn.*, vol. 25, pp. 3145-359, Jul. 1989.

21. A. Kameari, "Study on 3D eddy current analysis using FEM," *4th International IGTE Symposium and European TEAM Workshop*, Graz, Austria, 10-12, Oct. 1990, pp. 91-99.

22. R. Albanese and G. Rubinacci, "Solution of three dimensional eddy current problems by integral and differential methods", *IEEE Trans. Magn.*, vol. 26, pp. 702-705, Mar. 1990.

23. O. Biro, "Edge edge formulations of eddy current problems," *Computer Methods in Applied Machanics and Engineering*, vol. 169, no. 3-4, pp. 391-405, 1999.

24. O. Biro, P. Bohm, K. Preis, and G. Wachutka, "Edge finite element analysis of transient skin effect problems," *IEEE Trans. Magn.*, vol. 36, no. 4, pp. 835-839, July. 2000.

25. I. D. Mayergoyz, "A new approach to the calculation of three-dimensional skin effect problems," *IEEE Trans. Magn.*, vol. 19, pp. 2198-2200, Sep. 1993.

26. K. Fujiwara, T. Nakata, and H. Ohashi, "Improvement of convergence characteristics of ICCG method for the $A - \phi$ method using edge elements," *IEEE Trans. Magn.*, vol. 32, pp. 804-807, 1996.

27. R. Dyczij-Edlinger and O. Biro, "A joint vector and scalar potential formulation for driven highfrequency problems using hybrid edge and nodal finite elements," *IEEE Trans. Microwave Theory Tech.*, vol. 44, pp. 15-23, Jan. 1995.

28. P. P. Silvester and G. Pelosi, (Editors), *Finite Elements for Electromagnetics: Methods and Techniques*, Piscataway, NJ: IEEE Press, 1994.

29. J.-P. Berenger, "A perfectly matched layer for the absorption of electromagnetic waves," *J. Comput. Phys.*, vol. 114, no. 2, pp. 185-200, 1994.

30. Z. S. Sacks, D. M. Kingsland, R. Lee, and J.-F. Lee, "A perfectly matched anisotropic absorber for use as an absorbing boundary condition," *IEEE Trans. Antennas Propagat.*, vol. AP-43, pp. 1460-1463, Dec. 1995.

31. L. Zhao and A. C. Cangellaris, "GT-PML: generalized theory of perfectly matched layers and its application to the reflectionless truncation of finite-difference time-domain grids," *IEEE Trans. Microwave Theory Tech.*, vol. 44, no. 12, pp. 2555- 2563, Dec. 1996.

32. S. D. Gedney, "An anisotropic perfectly matched layer absorbing medium for the truncation of FDTD lattices," *IEEE Trans. Antennas Propagat.*, vol. AP-44, pp. 1630-1639, Dec. 1996.

33. R. Dyczij-Edlinger, G. Peng, and J. F. Lee "A fast vector-potential method using tangentially continuous vector finite elements," *IEEE Trans. Microwave Theory Tech.* vol. 46, pp. 863-868, Jun. 1998.

34. Y. Zhu and A. C. Cangellaris, "A new FEM formulation for reduced order electromagnetic modeling," *IEEE Microw. Guid. Wave Lett.*, May 2001.

35. A. Cangellaris, M. Celik, S. Pasha, and L. Zhao, "Electromagnetic model order reduction for system-level modeling," *IEEE Trans. Microw. Theory Tech.*, vol. 47, no. 6, pp. 840-849, June 1999.

36. A. Ruehli, "Equivalent circuit models for three-dimensional multiconductor system," *IEEE Trans. Microwave Theory Tech.*, vol. MTT-22, pp. 216-221, Mar. 1974

37. H. Heeb and A. E. Ruehli, "Three-dimensional interconnect analysis using partial element equivalent circuits," *IEEE Trans. Circuits and Systems*, vol. 39, no. 11, pp. 974-982, Nov. 1992.

38. W. Pinello, A. C. Cangellaris, and A. Ruehli, "Hybrid electromagnetic modeling of noise interactions in packaged electronics based on the Partial-Element Equivalent-Circuit formulation," *IEEE Trans. Microwave Theory Tech.*, vol. 45, no. 10, pp. 1889-1896, Oct. 1997.

39. A. E. Ruehli and A. C. Cangellaris, "Progress in the methodologies for the electrical modeling of interconnects and electronic packages," *Proc. of the IEEE*, vol. 89, no. 5, pp. 740-771, May 2001.

40. A. Rong, A. C. Cangellaris, and L. Dong, "Comprehensive broadband electromagnetic modeling of on-chip interconnects with a surface discretization-based generalized PEEC model," *IEEE Trans. Advanced Packaging*, vol. 28, no. 3, pp. 434-444, Aug. 2005.

41. R. F. Harrington, *Time-Harmonic Electromagnetic Fields*, Piscataway, NJ: IEEE Press Series on Electromagnetic Wave Theory, 2001.

42. S. M. Rao, D. R. Wilton, and A. W. Glisson, "Electromagnetic scattering by surfaces of arbitrary shape," *IEEE Trans. Antennas Propagat.*, vol. 30, pp. 409-418, May 1982.

43. J. Van Bladel, *Electromagnetic Fields*, New York: Hemisphere Publishing Corporation, 1985.

44. W. Wu, A. W. Glisson, and D. Kajfez, "A study of two numerical solution procedures for the electric field integral equation at low frequency," *Appl. Computat. Electromag. Soc. J.*, vol. 10, no. 3, pp. 69-80, Nov. 1995.

45. J.-S. Zhao and W. C. Chew, "Integral equation solution of Maxwell's equations from zero frequency to microwave frequencies," *IEEE Trans. Antennas Propagat.*, vol. 48, no. 10, pp. 1635-1645, Oct. 2000.

46. M. Kammon, N. A. Marques, L. M. Silveira, and J. White, "Automatic generation of accurate circuit models of 3-D interconnect," *IEEE Trans. Comp. Pack. & Manufacturing Tech. Part B*, vol. 21, no. 3, pp. 225-234, Aug. 1998.

47. A. Rong and A. C. Cangellaris, "Electromagnetic modeling of interconnects for mixed-signal integrated circuits from DC to multi-GHz frequencies," *Proc. IEEE Intl. Microwave Symposium*, vol.3, pp. 1893-1896, 2002.

CHAPTER 4

ITERATIVE METHODS, PRECONDITIONERS, AND MULTIGRID

Like most numerical methods, the final step in the finite element solution of a boundary value problem is the solution of a system of linear (or nonlinear, if the original system is nonlinear) equations with unknowns the degrees of freedom (DOF) in the finite element approximation. For problems of relevance to practical engineering applications, the number of DOF, and, hence, the dimension of the resulting system is very large for direct matrix inversion techniques to be of practical use. Therefore, iterative methods must be employed for its solution.

It is the purpose of this chapter to review the tools that the computational mathematics community has put in place to assist us in this objective. The literature on iterative methods for the numerical solution of the types of systems of equations resulting from approximations to BVPs using finite methods is very rich. Classic texts such as [1] and [2] provide a comprehensive presentation of the state-of-the-art in iterative methods for sparse linear systems, accompanied by a thorough list of references.

The presentation in this chapter will serve as a brief overview of some of the methods that we have found useful and employed for the classes of electromagnetic BVPs considered in this book. We begin the discussion with a review of key definitions and results from linear matrix theory. This is followed by the review of some very popular iterative solution processes, which have been found to be very effective in practical applications. Our presentation emphasizes the importance of preconditioning the matrix for improving the effectiveness and speed of convergence of the iterative process. This discussion leads naturally to the concept of multigrid as a preconditioning process. Even though the specifics of the implementation of multigrid in the context of iterative solution of electromagnetic boundary value problems is presented in later chapters, this chapter concludes with a discussion of the basic ideas behind multigrid and its algorithmic implementation.

4.1 DEFINITIONS

4.1.1 Vector space, inner product, and norm

With $\mathbb{R}$ denoting the set of real numbers, he set of all the N-tuples of real numbers is a vector space designated as $\mathbb{R}^N$; the set of all the $M \times N$ real matrices is a vector space denoted by $\mathbb{R}^{M \times N}$. Similarly, with $\mathbb{C}$ denoting the set of complex numbers, the set of all the N-tuples of complex numbers is a vector space designated as $\mathbb{C}^N$; the set of all the $M \times N$ complex matrices is a vector space denoted $\mathbb{C}^{M \times N}$.

The *inner product* is an operation on two vectors which produces a scalar. Consider the vector space $\mathbb{R}^N$, its inner product is denoted as $\mathbb{R}^N \times \mathbb{R}^N \to \mathbb{R}$. The standard *Euclidean inner product* is defined as

$$\langle x, y \rangle = x^T y. \tag{4.1}$$

Consider the vector space $\mathbb{C}^N$, its inner product is denoted as $\mathbb{C}^N \times \mathbb{C}^N \to \mathbb{C}$. Motivated by the desire to have the real associated norm on a complex vector, usually the inner product for complex vectors is the standard *Hermitian inner product* defined as

$$\langle x, y \rangle = x^H y. \tag{4.2}$$

The aforementioned inner products are used to define the following *vector norm* known as *2-norm*:

$$\|x\|_2 = \sqrt{\langle x, x \rangle}. \tag{4.3}$$

For a complex vector, its norm induced by the Hermitian inner-product is a real number, while the one induced by Euclidean inner-product is a complex number.

Another norm which is often used in practice is the *infinity norm* is defined as

$$\|x\|_\infty = \max_{1 \le i \le N} |x_i|. \tag{4.4}$$

Finally, we will also make use of the *A-norm* denoted as $\|x\|_A$, for a vector $x \in \mathbb{C}^N$, defined as

$$\|x\|_A = \sqrt{\langle x, Ax \rangle}, \tag{4.5}$$

where $A \in \mathbb{C}^{N \times N}$.

Vector norms are used extensively for estimating the error in the numerical solution of matrix equations. For example, let $\hat{x}$ be an approximation to the vector x. Then, the *absolute error* in $\hat{x}$ is calculated as

$$\epsilon_{abs} = \|\hat{x} - x\|, \tag{4.6}$$

where either the 2-norm or the infinity norm is used. Similarly, the *relative error* in x is calculated as follows:

$$\epsilon_{rel} = \frac{\|\hat{x} - x\|}{\|x\|}. \tag{4.7}$$

Matrix norms are also used extensively in iterative algorithms for the numerical manipulation of matrices. Of particular relevance to the topics discussed in this book are the *p-norms*, which are defined through the following expression:

$$\|A\|_p = \max_{x \neq 0} \frac{\|Ax\|_p}{\|x\|_p}. \tag{4.8}$$

A useful property of the *p*-norms is the following:

$$\|Ax\|_p \leq \|A\|_p \|x\|_p. \tag{4.9}$$

Examples of matrix *p*-norms are the 2-norm and the infinity norm mentioned above.

In our subsequent discussion we will make use of the aforementioned vector and matrix norms without necessarily making any specific reference to which norm is being used. Unless indicated otherwise, the 2-norm will be assumed to be the norm of choice in the development of all algorithms presented in this and later chapters.

4.1.2 Matrix eigenvalues and eigenvectors

The eigenvectors and eigenvalues of a matrix hold center stage in the development and assessment of iterative solution processes [3]. They are obtained from the solution of the following matrix equation:

$$Av_i = \lambda_i v_i, \quad (i = 1, 2, \cdots, N), \tag{4.10}$$

where $A \in \mathbb{C}^{N \times N}$, $\lambda_i \in \mathbb{C}$ is the eigenvalue of A, and the vector $v_i \in \mathbb{C}^N$ is called the *right eigenvector* of A associated with λ_i. The set of all eigenvalues of A is referred to as the *spectrum* of A. An alternative statement of (4.10) is obtained by defining the matrix V with columns the right eigenvectors of A as $V = [v_1, v_2, \cdots, v_N]$:

$$AV = V\Lambda, \tag{4.11}$$

where $\Lambda = \text{diag}(\lambda_1, \lambda_2, \cdots, \lambda_N)$. The above constitutes the *eigen decomposition* statement of A.

The eigenvalues of A are the roots of its *characteristic polynomial*, $p(s) \equiv \det(sI - A)$, where I denotes the identity matrix and $s \in \mathbb{C}$. Hence, it is

$$p(s) = \prod_{i=1}^{N}(s - \lambda_i) = s^N + a_1 s^{N-1} + a_2 s^{N-2} + \cdots + a_{N-1}s + a_N. \tag{4.12}$$

It follows immediately from the above equations that

$$\det(A) = \prod_{i=1}^{N} \lambda_i. \tag{4.13}$$

Furthermore, the following result may be deduced from (4.12) for the *trace* of A, defined as, $\text{tr}(A) = \sum_{i=1}^{N} A_{ii}$:

$$\text{tr}(A) = \sum_{i=1}^{N} \lambda_i. \tag{4.14}$$

A matrix A is *symmetric* if it remains unaltered under transposition; hence, $A^T = A$. If $A \in \mathbb{C}^{M \times N}$, the complex conjugate of its transpose is often used. In the following, we will use the term *adjoint* to refer to the Hermitian conjugate of a matrix, and we will use the short-hand notation A^H for its symbolic representation; hence, $A^H = (A^*)^T$. A complex matrix is called *Hermitian* if it is equal to its adjoint, $A^H = A$. It follows immediately that the diagonal elements of a Hermitian matrix are real numbers.

The *left eigenvector w_i* is defined as follows:

$$w_i^H A = w_i^H \hat{\lambda}_i, \quad (i = 1, 2, \cdots, N). \tag{4.15}$$

When written in matrix eigen decomposition form, the above equation becomes

$$W^H A = \hat{\Lambda} W^H \quad \text{or} \quad A^H W = W \hat{\Lambda}^* \tag{4.16}$$

where $W = [w_1, w_2, \cdots, w_N]$. Throughout the book, unless otherwise stated, "eigenvector" will imply "right eigenvector."

Next, we make use of the inner product to define a class of complex scalars that are most useful in practice are the *Rayleigh quotients*, defined for any nonzero $x \in \mathbb{C}^N$, as follows:

$$\mu(x) = \frac{\langle x, Ax \rangle}{\langle x, x \rangle}. \tag{4.17}$$

It follows immediately from (4.10) that an eigenvalue, λ_i, of a matrix A may be written as the following Rayleigh quotient:

$$\lambda_i = \frac{\langle v_i, Av_i \rangle}{\langle v_i, v_i \rangle}, \tag{4.18}$$

where v_i is the corresponding eigenvector.

4.1.3   Properties of Hermitian matrices

If A is Hermitian, it is straightforward to show that its eigenvalues are real and its eigenvectors form an orthonormal basis. The fact that the eigenvalues are real follows immediately from (4.10) through the following sequence of operations:

$$\lambda_i = v_i^H A v_i = v_i^H A^H v_i = (v_i^H A v_i)^* = \lambda_i^*. \tag{4.19}$$

With all eigenvectors normalized such that, $\|v_i\|_2 = 1$, $i = 1, 2, \cdots, N$, let v_i and v_j be two eigenvectors corresponding to distinct eigenvalues λ_i and λ_j, $\lambda_i \neq \lambda_j$. Then, making use of the fact that A is Hermitian and its eigenvalues real, we have from (4.10),

$$\lambda_i \langle v_j, v_i \rangle = \langle v_j, Av_i \rangle = v_j^H A v_i = (Av_j)^H v_i = \lambda_j^* \langle v_j, v_i \rangle$$
$$\Rightarrow (\lambda_i - \lambda_j)\langle v_j, v_i \rangle = 0 \tag{4.20}$$
$$\Rightarrow \langle v_j, v_i \rangle = 0.$$

The above result may be written in matrix form as

$$V^H V = I. \tag{4.21}$$

A matrix satisfying (4.21) is called *unitary*. From (4.21) it follows immediately that the inverse of a unitary matrix is equal to its complex transpose.

Returning to the case of left eigenvectors, it follows immediately from (4.11) and (4.16) that, if A is Hermitian, then $W = V$ and $\Lambda = \hat{\Lambda}$.

4.1.4 Positive definite matrices

A matrix $A \in \mathbb{C}^{N \times N}$ that satisfies $\langle x, Ax \rangle > 0 \ \forall x \in \mathbb{C}^N$ and $x \neq 0$, is called *positive definite*. From the discussion in the previous subsection it follows immediately that the eigenvalues of a Hermitian positive definite matrix are positive.

A matrix that satisfies $\langle x, Ax \rangle \geq 0 \ \forall x \in \mathbb{C}^N$, is called *positive semi-definite*. If A is also Hermitian it follows immediately that its eigenvalues are non-negative.

An example of a positive semi-definite matrix is the Hermitian matrix $A^H A$. Given a matrix $A \in \mathbb{C}^{N \times N}$, the square roots of the eigenvalues of $A^H A$ are called the *singular values* of A and are denoted by σ_i.

4.1.5 Independence, invariant subspaces, and similarity transformations

A set of vectors $\{y_1, y_2, \ldots, y_m\}$ in $\mathbb{C}^N$ is said to be *linearly independent* if the statement $\sum_{i=1}^m \alpha_i y_i = 0$ and $\alpha_i \in \mathbb{C}$ implies $\alpha_i = 0$, $i = 1, 2, \ldots, m$. It follows immediately from (4.21) that the eigenvectors of a Hermitian matrix are linearly independent.

A subspace of $\mathbb{C}^N$ is a subset that is also a vector space. Given a set of vectors $\{y_1, y_2, \cdots, y_m\}$ in $\mathbb{C}^N$, the set of all linear combinations of these vectors forms a subspace S that will be called the *span* of $\{y_1, y_2, \cdots, y_m\}$. We write

$$\text{span}\{y_1, y_2, \cdots, y_m\} = \left\{ \sum_{i=1}^m \alpha_i y_i, \quad \alpha_i \in \mathbb{C} \right\}. \tag{4.22}$$

Of particular interest is the case where the vectors $\{y_1, y_2, \cdots, y_m\}$ are linearly independent. In this case, they constitute a set of *bases* for S, and their number, m, is the *dimension* of S.

A subspace $S \subseteq \mathbb{C}^N$ such that

$$y \in S \Rightarrow Ay \in S \tag{4.23}$$

is said to be *invariant* with respect to multiplication by A. For example, a set of eigenvectors of A defines an invariant subspace.

Next, let us consider the matrices $A \in \mathbb{C}^{N \times N}$, $P \in \mathbb{C}^{N \times n}$, and $B \in \mathbb{C}^{n \times n}$, satisfying the relation $AP = PB$. Let y be an eigenvector of B with corresponding eigenvalue λ, that is, $By = \lambda y$. In view of the above relationship, it is $APy = PBy = \lambda Py$. This implies that the eigenvalues of B are also eigenvalues of A. For the special case where P is a square, non-singular matrix (hence, A, B, and P are all of the same dimension), the relation $AP = PB$ implies that A and B have the same eigenvalues. The matrices A and B are called *similar* and P is referred to as the *similarity transformation* between A and B according to the relationship,

$$B = P^{-1}AP. \tag{4.24}$$

Similarity transformations are particularly useful for the reduction of a given matrix A to any one of several *canonical forms*. One of the most important canonical form of matrices is the *Jordan form*. The pertinent result for the Jordan decomposition of a matrix A is the following. Given a matrix $A \in \mathbb{C}^{N \times N}$, then there exists a nonsingular matrix $X \in \mathbb{C}^{N \times N}$

such that $X^{-1}AX = \text{diag}(J_1, J_2, \cdots, J_t)$, where the *Jordan block*

$$J_i = \begin{bmatrix} \lambda_i & 1 & & \\ & \lambda_i & \ddots & \\ & & \ddots & 1 \\ & & & \lambda_i \end{bmatrix} \tag{4.25}$$

is an $m_i \times m_i$ matrix and $\sum_{i=1}^{i=t} m_i = N$.

Another important canonical form, which will be found useful in dealing with the solution of matrix eigenvalue problems, is the *Schur canonical form* [3]. If $A \in \mathbb{C}^{N \times N}$, then there exists a unitary matrix $Q \in \mathbb{C}^{N \times N}$ such that

$$Q^H A Q = R, \tag{4.26}$$

where R is upper triangular. The column vectors forming the matrix Q are referred to as *Schur* vectors. If A is Hermitian, the upper triangular matrix R is actually the diagonal eigenvalue matrix Λ, and $W = V = Q$. Hence, for Hermitian matrices, Schur decomposition and eigen-decomposition are the same.

4.2   ITERATIVE METHODS FOR THE SOLUTION OF LARGE MATRICES

The finite element approximation of a linear boundary value problem results in the following, complex matrix equation

$$Ax = f, \tag{4.27}$$

where $A \in \mathbb{C}^{N \times N}$, $x \in \mathbb{C}^N$, is the unknown vector with elements the DOF in the approximation, and $b \in \mathbb{C}^N$ represents the excitation vector. When N is sufficiently small, direct matrix decomposition methods, such as Gaussian elimination and Cholesky factorization can be employed [3]. However, what is of concern here is the case where N is large enough for such direct methods to be computationally prohibitive.

Because of the compact support of the expansion functions used in the development of finite element approximations of boundary value problems, the resulting matrices are very sparse. More specifically, the number of nonzero entries in each row of a finite element matrix is in the order of tens. Consequently, significant savings in memory can be achieved by storing only the nonzero entries of the A along with information about the row and column indices for each nonzero entry. Ways in which such compact storage is employed are discussed in [1].

Another important finite element matrix operation that is required quite often prior to solution is *re-ordering*. Re-ordering of a matrix is prompted by the desire to contain the number of fill-ins during its Gaussian elimination (or its incomplete LU decomposition, which is often required in the iterative solution process). Numerous heuristic methods, such as the minimum degree algorithm, are available for identifying and implementing effective re-ordering of a given matrix. A good overview is given in [1].

When the number of unknowns is large enough for the direct solution of the matrix to be impractical, an iterative process is utilized for obtaining an approximation to the solution. There are two general classes of iterative matrix solution methods, *stationary* methods and *non-stationary* methods. Stationary methods include the Jacobi, Gauss-Seidel, and successive over relaxation (SOR) processes. The convergence of these methods is not guaranteed for all types of matrices. However, they are known to be very effective when applied to matrices resulting from the finite element approximation of elliptic partial differential equations. Thus they are suitable for the solution of electrostatic and magnetostatic boundary value problems.

4.2.1   Stationary methods

To present the basic principles of stationary methods we will examine, briefly, some of the members of this class of methods. We begin by splitting the matrix A as follows:

$$A = D - L - U, \tag{4.28}$$

where D is a diagonal matrix with elements the diagonal entries of A; $-L$ is its strict lower part; and $-U$ is its strict upper part. The Jacobi iteration determines the $(k+1)$th iterate of the solution, x_{k+1}, from the previous iterate, x_k, through the diagonal sweep,

$$x_{k+1} = D^{-1} \left[f + (L + U) x_k \right]. \tag{4.29}$$

Its implementation requires two separate arrays for storing x_{k+1} and x_k.

A more compact form of the Jacobi iteration results by defining the Jacobi iteration matrix, R_J, as follows:

$$R_J = D^{-1} (L + U). \tag{4.30}$$

A simple matrix manipulation of the above equation, making use of (4.28), yields the following alternative form for R_J:

$$R_J = I - D^{-1}A, \tag{4.31}$$

where I is the identity matrix. Using these definitions, (4.29) is cast in the following form:

$$x_{k+1} = R_J x_k + D^{-1} f. \tag{4.32}$$

The *weighted Jacobi* method is obtained from (4.32) through a slight modification of the iteration process as follows:

$$x_{k+1} = \left[(1 - \omega)I + \omega R_J \right] x_k + \omega D^{-1} f, \tag{4.33}$$

where the weighting factor ω is introduced with the objective of improving the convergence rate of the iteration. The weighted Jacobi iteration matrix

$$R_{J\omega} = \left[(1 - \omega)I + \omega R_J \right] \tag{4.34}$$

is introduced to cast (4.33) in a more compact form as follows:

$$x_{k+1} = R_{J\omega} x_k + \omega D^{-1} f. \tag{4.35}$$

It is useful to note that, in view of (4.31), the weighted Jacobi iteration matrix may be cast in the following form:

$$R_{J\omega} = I - \omega D^{-1}A. \tag{4.36}$$

The Gauss-Seidel iteration determines x_{k+1} from x_k through either forward or backward substitutions. For the forward Gauss-Seidel we have

$$(D - L)x_{k+1} = U x_k + f \ \Rightarrow \ x_{k+1} = (D - L)^{-1} \left[f + U x_k \right]. \tag{4.37}$$

On the other hand, the backward Gauss-Seidel performs the backward substitution,

$$(D - U)x_{k+1} = Lx_k + f \implies x_{k+1} = (D - U)^{-1}[f + Lx_k]. \quad (4.38)$$

Clearly, a single array suffices for storing the updated iterate x_{k+1} and the previous one x_k. In the forward substitution the entries in the solution array are updated in ascending order; in the backward substitution they are updated in descending order.

4.2.2 Convergence of iterative methods

From the above discussion it is clear that the development of an iterative method begins with casting the original matrix equation in the form

$$x = Bx + g. \quad (4.39)$$

Subsequently, an approximation, $\hat{x}$, to x is sought through the iteration

$$\hat{x}_{k+1} = B\hat{x}_k + g, \quad k = 0, 1, 2, \cdots, \quad (4.40)$$

where the subscript k denotes the kth iterate of the approximation and $\hat{x}_0$ is an initial guess for x.

Let $e_k = x - \hat{x}_k$ denote the error in the approximation of x at iteration k. Then, it follows from (4.39) and (4.40) that the error after k iterations satisfies the following equation:

$$e_k = B^k e_0. \quad (4.41)$$

A bound for the error after k iterations can be obtained making use of (4.9),

$$\|e_k\| \leq \|B\|^k \|e_0\|. \quad (4.42)$$

Clearly, convergence in the iteration process, manifested through reduction in the norm of the error, is achieved provided that $\|B\| < 1$. This result allows the definition of an asymptotic convergence factor for an iterative method in terms of the *spectral radius*, $\rho(B)$, of B, which is defined in terns of the eigenvalues of B by

$$\rho(B) = \max |\lambda(B)|. \quad (4.43)$$

It can be shown in [1] that, in any norm, it is

$$\rho(B) = \lim_{k \to \infty} \|B^k\|^{1/k}. \quad (4.44)$$

Thus, in view of (4.42), we conclude that the iterative method with iteration matrix B will converge for any (non-trivial) initial guess if and only if $\rho(B) < 1$.

The spectral radius is also useful as an estimate of the number of iterations, k, needed to reduce the error by a factor of 10^q. From (4.42), in view of (4.44), it is

$$\frac{\|e^{(k)}\|}{\|e^{(0)}\|} \leq 10^{-q} \implies (\rho(B))^k \leq 10^{-q} \implies k \geq -\frac{q}{\log_{10}(\rho(B))}. \quad (4.45)$$

For the above results to be useful in practice one must be able to either calculate or estimate the spectral radius of the matrix. For a very large matrix calculation of its eigenvalues is computationally expensive. Thus an estimate of the largest eigenvalue is sought instead.

For this purpose, the following useful result, known as Gershgorin's theorem, can be used. Gershgorin's theorem states that any eigenvalue of a matrix B, $B \in \mathbb{C}^{N \times N}$, is located within one of the discs on the complex plane with centers B_{ii} and radii, r_i, given by

$$r_i = \sum_{j=1, j \neq i}^{j=N} |B_{ij}|. \tag{4.46}$$

4.2.3 Non-stationary methods

Non-stationary methods are based on the so-called *projection techniques*. Let $V = [v_1, \cdots, v_m]$ denote an $N \times m$ matrix whose column vectors form a set of bases for subspace $\mathcal{K}$, and $W = [w_1, \cdots, w_m]$ denote an $N \times m$ matrix whose column vectors form a set of bases for subspace $\mathcal{L}$. The objective of a projection method is, given an initial guess x_0, to find an approximate solution, $\hat{x}$, such that $\hat{x} \in x_0 + \mathcal{K}$ and $(f - A\hat{x}) \perp \mathcal{L}$. Therefore, $\mathcal{K}$ is referred to as the *expansion* space, and $\mathcal{L}$ as the *projection* space. The approximate solution is written as

$$\hat{x} = x_0 + Vy. \tag{4.47}$$

Thus the orthogonality condition $(f - A\hat{x}) \perp \mathcal{L}$ leads to the following equation for y:

$$W^T(f - Ax_0 - AVy) = 0 \quad \Rightarrow \quad W^T AVy = W^T r_0, \tag{4.48}$$

where $r_0 = f - Ax_0$ is the initial residual. In view of the above, the approximate solution is written in the form

$$\hat{x} = x_0 + V(W^T AV)^{-1} W^T r_0. \tag{4.49}$$

It must be noted that the matrix $W^T AV$, the inversion of which leads to the calculation of the approximate solution, is a square matrix of dimension m. The key objective of a projection method is to construct subspaces $\mathcal{K}$ and $\mathcal{L}$ with the smallest dimension possible in order to expedite the computation of the approximate solution.

One important class of projection methods is the class of Krylov subspace methods. In this class of methods the expansion space $\mathcal{K}$ is constructed as follows:

$$K_m(A, r_0) = \text{span}\left\{r_0, Ar_0, A^2 r_0, \cdots, A^{m-1} r_0\right\}. \tag{4.50}$$

This subspace will be referred to as the Krylov subspace. Two popular Krylov subspace-based iterative methods, namely, the generalized minimum residual method and the conjugate gradient method, are presented next.

4.3 GENERALIZED MINIMUM RESIDUAL METHOD

The generalized minimum residual method (GMRES) [1] attempts to find a vector $\hat{x}$, such that the norm of the residual $r = f - A\hat{x}$ is minimized. More specifically, using the notation and definitions in the previous section, let x_0 be an initial guess and $\hat{x} \in x_0 + \mathcal{K}$. Then, it is

$$\hat{x} = x_0 + Vy \tag{4.51}$$

and the norm of the residual is written as

$$\|r\| = \|f - A\hat{x}\| = \|f - Ax_0 - AVy\| = \|r_0 - AVy\|, \tag{4.52}$$

where $r_0 = f - Ax_0$. For the definition of the norm, it is straightforward to deduce that the minimization of the residual $\|r\|$ is equivalent to solving for y such that the residual vector r is perpendicular to AV,

$$\min \|r\| \equiv (r_0 - AVy) \perp AV. \tag{4.53}$$

A comparison of (4.53) with (4.48), suggests that GMRES can be considered as a projection method where the projection space W is related to the expansion space V as $W = AV$, and V is taken to be the m-th Krylov subspace of (4.50).

From this introductory discussion it becomes apparent that the GMRES method may be broken down into two steps. The first step involves the construction of the Krylov subspace. The second step deals with the least-squares minimization process.

The construction of the Krylov subspace K_m entails the generation of a set of orthonormal vectors $v_1, v_2, \cdots, v_m$ for K_m. These vectors are often referred to as *Arnoldi vectors*. The generation of the Arnoldi vectors from (4.50) is facilitated through the Gram-Schmidt formula

$$\tilde{v}_{i+1} = Av_i - h_{1,i}v_1 - h_{2,i}v_2 \cdots - h_{i,i}v_i, \tag{4.54}$$

where the coefficients $h_{p,q}$ are chosen such that $\tilde{v}_{i+1}$ is orthogonal to the earlier Arnoldi vectors and its norm is unity. It is straightforward to show that this is the case if

$$\begin{aligned} h_{j,i} &= \langle v_j, Av_i \rangle, \quad j = 1, 2, \cdots i, \\ h_{i+1,i} &= \|\tilde{v}_{i+1}\|, \\ v_{i+1} &= \tilde{v}_{i+1}/h_{i+1,i}. \end{aligned} \tag{4.55}$$

In matrix form, the above process after m steps yields the following result:

$$\begin{aligned} A &\underbrace{[v_1\ v_2\ \cdots\ v_m]}_{V_m} \\ &= \underbrace{[v_1\ v_2\ \cdots\ v_m]}_{V_m} \underbrace{\begin{bmatrix} h_{1,1} & h_{1,2} & \cdots & h_{1,m-1} & h_{1,m} \\ h_{2,1} & h_{2,2} & \cdots & h_{2,m-1} & h_{2,m} \\ & \ddots & & & \vdots \\ & & & h_{m,m-1} & h_{m,m} \end{bmatrix}}_{H_{m,m}} + h_{m+1,m}v_{m+1}e_m^T \\ &= \underbrace{[v_1\ v_2\ \cdots\ v_m\ v_{m+1}]}_{V_{m+1}} \underbrace{\begin{bmatrix} H_{m,m} \\ h_{m+1,m}e_m^T \end{bmatrix}}_{H_{m+1,m}}, \end{aligned} \tag{4.56}$$

where $H_{m,m}$ is an $m \times m$ upper Hessenberg matrix. (A matrix H is upper Hessenberg if $H(i,j) = 0$ for $i > j + 1$.) The matrix $H_{m+1,m}$ is simply $H_{m,m}$ augmented by one extra row. The vector e_m is a vector of zeros except for its m-th row entry which has a value of one.

The following is a pseudo-code description of the algorithm that can be used for the generation of the Arnoldi vectors v_m.

Algorithm (4.1): Generation of Arnoldi Vectors.
```
1 v₁ ← r₀/‖r₀‖;
2 i ← 1;
3 for i = 1, 2, ⋯, m
```

```
3.a  ṽ_{i+1} ← Av_i;
3.b  for  j = 1, 2, ··· i,   h_{j,i} = ⟨v_j, ṽ_{i+1}⟩;  ṽ_{i+1} ← ṽ_{i+1} − h_{j,i}v_i;
3.c  h_{i+1,i} = ‖ṽ_{i+1}‖;
3.d  v_{i+1} = ṽ_{i+1}/h_{i+1,i}.
```

Once the Arnoldi vectors have been generated, use of the relationship in (4.56) in the expression (4.52) for the norm of the residual yields the following expression for the functional to be minimized:

$$\|r\| = \|r_0 - AV_m y\| = \|r_0 - V_{m+1}H_{m+1,m}y\| = \|\|r_0\|e_1 - H_{m+1,m}y\|, \quad (4.57)$$

where use was made of the fact that $V_{m+1}^T r_0 = \|r_0\|$ due to Step 1 in Algorithm (4.1).

The second step of GMRES concerns the determination of the vector y that minimizes $\|r\|$. The process for the solution of this least-squares minimization problem begins with the transformation of the upper Hessenberg matrix $H_{m+1,m}$ into an upper triangular matrix. This can be achieved through a series of left-multiplications of the matrix with a properly chosen series of rotation matrices Ω_i $(i = 1, 2, \cdots, m)$. To reveal the structure of these rotation matrices, consider the form of the transformed matrix after multiplication by the $(i-1)$th rotation matrix,

$$\begin{bmatrix} \tilde{h}_{1,1} & & \cdot & \cdots & \tilde{h}_{1,m-1} & \tilde{h}_{1,m} \\ & \ddots & & & & \vdots \\ & & \tilde{h}_{i,i} & \cdots & \tilde{h}_{i,m-1} & \tilde{h}_{i,m} \\ & & h_{i+1,i} & \cdots & h_{i+1,m-1} & h_{i+1,m} \\ & & & \ddots & & \vdots \\ & & & & h_{m,m-1} & h_{m,m} \\ & & & & & h_{m+1,m} \end{bmatrix}, \quad (4.58)$$

where it is shown that the sub-diagonal elements $h_{j,j-1}$ $(j = 2, 3, \cdots, i)$ have been eliminated. The $\sim$ on top of the elements in the resulting matrix up to and including the ith row, indicates that these elements have been transformed from their original values. The next step involves the elimination of $h_{i+1,i}$. For this, we left-multiply the matrix by the rotation matrix Ω_i,

$$\Omega_i = \begin{bmatrix} 1 & & & & & & & \\ & \ddots & & & & & & \\ & & 1 & & & & & \\ & & & c_i & s_i & & & \\ & & & -s_i & c_i & & & \\ & & & & & 1 & & \\ & & & & & & \ddots & \\ & & & & & & & 1 \end{bmatrix}, \quad (4.59)$$

where it is

$$s_i = \frac{h_{i+1,i}}{\sqrt{\tilde{h}_{i,i}^2 + h_{i+1,i}^2}},$$

$$c_i = \frac{\tilde{h}_{i,i}}{\sqrt{\tilde{h}_{i,i}^2 + h_{i+1,i}^2}}. \quad (4.60)$$

The rotation matrices Ω_i, which are often referred to as *Givens* rotation matrices, are unitary matrices. Left multiplication of Ω_i with $\tilde{H}_{m+1,m}$ eliminates the $h_{i+1,i}$ and changes the entries on the i and $(i+1)$-th rows. Continuous multiplication with rotation matrices transforms $H_{m+1,m}$ to a triangular form

$$\underbrace{\Omega_m \Omega_{m-1} \cdots \Omega_1}_{Q_m} H_{m+1,m} = \underbrace{\begin{bmatrix} \tilde{h}_{1,1} & \cdots & \tilde{h}_{1,m} \\ & \ddots & \vdots \\ & & \tilde{h}_{m+1,m} \\ & & 0 \end{bmatrix}}_{R_{m+1,m}}. \tag{4.61}$$

Thus the functional to be minimized assumes the simpler form (note that the $\|r_0\|e_1$ is also multiplied on the left by the rotation matrices),

$$\min \|r\| = \min \| \|r_0\|e_1 - H_{m+1,m}y\| = \min \| \|r_0\|Q_m e_1 - R_{m+1,m}y\|, \tag{4.62}$$

where use was made of the fact that Q_m is unitary.

The least-squares solution to this minimization problem is straightforward to obtain by noticing that, if we set $\|r_0\|Q_m e_1 = g = [\tilde{g}_m^T, \delta]^T$, where it should be clear that the vector $\tilde{g}_m$ of length m has the same m first elements of $\|r_0\|Q_m e_1$, then it is

$$\| \|r_0\|Q_m e_1 - R_{m+1,m}y\| = |\delta| + \|\tilde{g}_m - R_{m,m}y\|, \tag{4.63}$$

where $R_{m,m}$ is obtained from $R_{m+1,m}$ by deleting the last row. From this last equation it is now evident that the minimum is achieved when the second term on the right-hand side is zero. Hence, it is

$$y = R_{m,m}^{-1}\tilde{g}_m. \tag{4.64}$$

Furthermore, the minimum residual of $\|r_0\|e_1 - H_{m+1,m}y\|$ is $|\delta|$.

The GMRES algorithm results from the merging of the Arnoldi process for the construction of an orthonormal basis for the Krylov subspace with the aforementioned least-squares minimization process. A pseudo-code description of the algorithm is given below.

Algorithm (4.2): GMRES
1 $r_0 = b - Ax_0$, $v_1 \leftarrow r_0/\|r_0\|$, $g = \|r_0\|e_1$
2 $i \leftarrow 1$;
3 do
    3.a $\tilde{v}_{i+1} \leftarrow Av_i$;
    3.b for $j = 1 \cdots i$, 3.b.1 $h_{j,i} = \langle v_j, \tilde{v}_{i+1}\rangle$; $\tilde{v}_{i+1} \leftarrow \tilde{v}_{i+1} - h_{j,i}v_i$;
    3.c $h_{i+1,i} = \|\tilde{v}_{i+1}\|$, $v_{i+1} = \tilde{v}_{i+1}/h_{i+1,i}$;
    3.d $[h_{1,i}, h_{2,i}, \cdots, h_{i+1,i}]^T = \Omega_{i-1} \cdots \Omega_1 [h_{1,i}, h_{2,i}, \cdots, h_{i+1,i}]^T$;
    3.e Build Ω_i, $s_i = \dfrac{h_{i+1,i}}{\sqrt{|\tilde{h}_{i,i}|^2 + h_{i+1,i}^2}}$, $c_i = \dfrac{\tilde{h}_{i,i}}{\sqrt{|\tilde{h}_{i,i}|^2 + h_{i+1,i}^2}}$
    3.f $g \leftarrow \Omega_i g$;
    3.g if last element of g is sufficiently small go to 4; else i++.
4 Compute y as the solution of $H_{i,i}y = g_i$, where $H_{i,i}$ results from $H_{i+1,i}$ by deleting the last row, and g_i is obtained from g by deleting the last row.
5 $x \leftarrow x_0 + V_i y$

Bad conditioning of the matrix A could compound the robustness of the Arnoldi process. Therefore, use of a preconditioning matrix is most appropriate, especially when dealing with very large matrices. Preconditioning matrices and their utilization for improving convergence of the iterative solution process will be discussed in one of the following sections. For the purposes of this section it suffices to say that a preconditioner M^{-1} is an approximation to the inverse of A. When M^{-1} is available, we solve, instead, the following matrix equation:

$$M^{-1}Ax = M^{-1}b. \tag{4.65}$$

The changes required in the GMRES process are obvious. Step 1 in Algorithm (4.2) is changed to $r_0 = M^{-1}(b - Ax_0)$, while the matrix operation is Step 3.a becomes $\tilde{v}_{i+1} \leftarrow M^{-1}Av_i$.

The major drawback of GMRES is that CPU time and memory storage required per iteration both increase with the number of iterations. At each iteration, the new Arnoldi vector has to be made orthogonal to all previous Arnoldi vectors. Nevertheless, this cost may be acceptable if an effective preconditioner can be constructed to contain the number of iterations. The construction of such preconditioners using multigrid and multilevel methods is one of the key themes of this monograph and is discussed in detail later in this and subsequent chapters.

4.4 CONJUGATE GRADIENT METHOD

The conjugate gradient (CG) method is an effective method for the iterative solution of symmetric, positive definite systems [1]; however, it is also often used for the iterative solution of general symmetric systems. CG is the oldest and best known of the Krylov subspace-based iterative methods.

The method proceeds by generating sequences of approximate solutions, residuals corresponding to the approximate solutions, and search directions along which the next solution approximate is calculated and the new residual is computed. Although the length of these sequences increases with the number of iterations, only the very latest approximate solution, residual, and search direction are kept in memory. With regards to computational cost, in each iteration only two inner products and one matrix-vector product are computed.

The objective of the CG method is to find a vector $x \in x_0 + \mathcal{K}$,

$$x = x_0 + Vy, \tag{4.66}$$

such that the A-norm (4.5) of the error is minimized,

$$\min \|A^{-1}f - (x_0 + Vy)\|_A = \min \|A^{-1}r_0 - Vy\|_A. \tag{4.67}$$

Such a minimization requires that $A^{-1}r_0 - Vy$ is perpendicular to V with respect to an A-inner product; hence,

$$\begin{aligned} \langle A^{-1}r_0 - Vy, V \rangle_A = 0 &\Rightarrow \langle r_0 - AVy, V \rangle = 0 \\ &\Rightarrow r_0 - AVy \perp V. \end{aligned} \tag{4.68}$$

Contrasting this result to (4.53), it is immediately evident that GMRES minimizes the residual with respect to the Euclidean norm while CG minimizes the error with respect to the $A-$norm. Furthermore, considering the above result in the light of (4.48), we recognize CG as a projection method in which both the expansion and projection matrices are the same, $V = W$, both constructed from the same Krylov subspace K_m.

Multiplication of both sides of (4.56) by V_m^T yields

$$V_m^T AV_m = H_{m,m}. \tag{4.69}$$

If A is symmetric, then $H_{m,m}$ is also symmetric. In this case, the product AV_m is cast in the form

$$
AV_m = V_m T_m + \beta_{m+1} v_{m+1}, \quad T_m =
\begin{bmatrix}
\alpha_1 & \beta_2 & & & \\
\beta_2 & \alpha_2 & \beta_3 & & \\
& \beta_3 & \alpha_3 & \beta_3 & \\
& & \ddots & \ddots & \ddots \\
& & & \beta_m & \alpha_m
\end{bmatrix}.
\tag{4.70}
$$

This suggests that the construction of the orthonormal basis V_m of m-th Krylov subspace can be simplified through a *short recurrence* version of the Arnoldi algorithm, which is known as the Lanczos algorithm.

Algorithm (4.3): Lanczos Algorithm

1 $v_0 \leftarrow 0$, $\beta_1 = \|r_0\|$, $v_1 \leftarrow r_0/\beta_1$;
2 $i \leftarrow 1$;
3 for $i = 1, \cdots, m$
    3.a $\tilde{v}_{i+1} \leftarrow A v_i$;
    3.b $\alpha_i = \langle v_i, \tilde{v}_{i+1} \rangle$;
    3.c $\tilde{v}_{i+1} \leftarrow \tilde{v}_{i+1} - \beta_i v_{i-1} - \alpha_i v_i$;
    3.d $\beta_{i+1} = \|\tilde{v}_{i+1}\|$, $v_{i+1} = \tilde{v}_{i+1}/\beta_{i+1}$;

The above Lanczos algorithm generates the column vectors in V_m, referred to as *Lanczos vectors*, and the nonzero entries α_i, β_i, in T_m.

Since $W = V$, the approximate solution to the equation $Ax = f$ is obtained from (4.48) as follows:

$$
\begin{aligned}
V_m^T A V_m y = V_m^T r_0 &\Rightarrow y = T_m^{-1} \|r_0\| e_1 \\
&\Rightarrow x = x_0 + V_m y = x_0 + V_m T_m^{-1} \|r_0\| e_1,
\end{aligned}
\tag{4.71}
$$

where use was made of (4.70) and the fact that $V_m^T r_0 = \|r_0\| e_1$, which is due to Step 1 in the Lanczos algorithm. However, such a solution process is not efficient because all the Lanczos vectors in V_m must be stored and T_m has to be inverted at each iteration. This is where the CG method comes in to provide for a more efficient iterative solution process.

To explain, consider the equation that results after multiplying y by both sides of the matrix equation on the left of (4.70), and making use of the fact that $y = T_m^{-1}\|r_0\|e_1$ (see (4.71)). It is

$$
\begin{aligned}
AV_m y &= V_m T_m y + \beta_{m+1} v_{m+1} e_m^T y \\
&= V_m \|r_0\| e_1 + \beta_{m+1} v_{m+1} e_m^T T_m^{-1} \|r_0\| e_1.
\end{aligned}
\tag{4.72}
$$

In view of the fact that $V_m \|r_0\| e_1 = r_0$ and $\|r_0\| = \beta_1$, the above result may be cast in the following form:

$$
r_m = r_0 - AV_m y = -\beta_{m+1} \beta_1 v_{m+1} e_m^T T_m^{-1} e_1.
\tag{4.73}
$$

To proceed further, use will be made of the symmetric, tri-diagonal form of T_m to construct the following factorization:

$$
T_m = \underbrace{\begin{bmatrix}
1 & & & \\
\eta_2 & 1 & & \\
& \ddots & \ddots & \\
& & \eta_m & 1
\end{bmatrix}}_{L_m}
\underbrace{\begin{bmatrix}
\lambda_1 & & & \\
& \lambda_2 & & \\
& & \ddots & \\
& & & \lambda_m
\end{bmatrix}}_{\Lambda_m}
\underbrace{\begin{bmatrix}
1 & \eta_2 & & \\
& 1 & \eta_3 & \\
& & \ddots & \\
& & & 1
\end{bmatrix}}_{U_m},
\tag{4.74}
$$

where $U_m = L_m^T$, and
$$\eta_i = \beta_i/\lambda_{i-1}. \tag{4.75}$$
The inverse of T_m is obtained as $U_m^{-1}\Lambda_m^{-1}L_m^{-1}$, with the inverses of L_m and U_m given by

$$L_m^{-1} = \begin{bmatrix} 1 & & & & \\ -\eta_2 & 1 & & & \\ \eta_2\eta_3 & -\eta_3 & 1 & & \\ \vdots & \vdots & \vdots & \ddots & \\ (-1)^{m-1}\prod_{i=2}^m \eta_i & (-1)^{m-2}\prod_{i=3}^m \eta_i & \cdots & -\eta_m & 1 \end{bmatrix}, \tag{4.76}$$

$$U_m^{-1} = L_m^{-1^T}.$$

Use of the above result along with (4.75) in (4.73) yields

$$\begin{aligned}
r_m &= -\beta_{m+1}\beta_1 v_{m+1} e_m^T T_m^{-1} e_1 \\
&= -\beta_{m+1}\beta_1 v_{m+1} e_m^T U_m^{-1}\Lambda_m L_m^{-1} e_1 \\
&= (-1)^m \beta_{m+1}\beta_1 v_{m+1} \frac{1}{\lambda_m}\prod_{i=2}^m \eta_i \\
&= (-1)^m v_{m+1}\frac{\prod_{i=1}^{m+1}\beta_i}{\prod_{i=1}^m \lambda_i}.
\end{aligned} \tag{4.77}$$

This final result is most useful when interpreted in a matrix form that captures the sequence of residuals generated during the iteration process,

$$\underbrace{[r_0, r_1, \cdots, r_{m-1}]}_{R_m} = \underbrace{[v_1, v_2, \cdots, v_m]}_{V_m} \underbrace{\begin{bmatrix} \beta_1 & & & \\ & -\frac{\beta_1\beta_2}{\lambda_1} & & \\ & & \ddots & \\ & & & (-1)^{m-1}\frac{\prod_1^m \beta_i}{\prod_1^{m-1}\lambda_i} \end{bmatrix}}_{\Omega_m}. \tag{4.78}$$

In this form it is immediately evident that the residual vectors are orthogonal; hence, they constitute a scaled set of the Lanczos vectors. Thus one may use them in place of the vectors V_m in the equation for the approximate solution. This yields

$$\begin{aligned}
x &= x_0 + V_m y \\
&= x_0 + V_m T_m^{-1}\|r_0\|e_1 \\
&= x_0 + V_m U_m^{-1}\Lambda_m^{-1}L_m^{-1}\|r_0\|e_1 \\
&= x_0 + R_m\Omega_m^{-1}U_m^{-1}\Lambda_m^{-1}L_m^{-1}\|r_0\|e_1.
\end{aligned} \tag{4.79}$$

A more compact form of this result is obtained through the definition of the following matrices, U_m', P_m, and vector $\tilde{y}_m$.

$$U_m' = \Omega_m^{-1}U_m\Omega_m = \begin{bmatrix} 1 & -\eta_2^2 & & & \\ & 1 & -\eta_3^2 & & \\ & & \ddots & & \\ & & & 1 & -\eta_m^2 \\ & & & & 1 \end{bmatrix}, \tag{4.80}$$

$$P_m = [p_0, \ p_1, \ \cdots, \ p_{m-1}] = R_m U'^{-1}_m,$$

$$\tilde{y}_m = \Omega_m^{-1} \Lambda_m^{-1} L_m^{-1} \|r_0\| e_1 = \left[\frac{1}{\lambda_1}, \ \frac{1}{\lambda_2}, \ \cdots \ \frac{1}{\lambda_m} \right]^T. \tag{4.81}$$

In view of the above, (4.79) assumes the form,

$$x = x_0 + P_m \tilde{y}_m. \tag{4.82}$$

The column vectors in P_m are referred to as the *search vectors*. A slight manipulation of (4.81), yields a recursive equation for their construction,

$$R_m = P_m U'_m \ \Rightarrow \ p_i = r_i + \eta_i^2 p_{i-1} \quad (i = 1, 2, \cdots, m), \quad (p_0 = r_0). \tag{4.83}$$

In addition, recursive relations for both the approximate solution and the residual vector can be obtained from (4.82),

$$x_m = x_0 + P_m \tilde{y}_m \ \Rightarrow \ x_i = x_{i-1} + \frac{1}{\lambda_i} p_{i-1} \quad (i = 1, 2, \cdots, m),$$

$$r_m = f - A x_m \ \Rightarrow \ r_i = r_{i-1} - \frac{1}{\lambda_i} A p_{i-1} \quad (i = 1, 2, \cdots, m). \tag{4.84}$$

The relations in (4.83) and (4.84) can be combined to yield the following simple CG algorithm.

> **Algorithm (4.4): A simple CG .**
> 1 Initial guess x_0, $r_0 \leftarrow f - Ax_0$, and $p_0 \leftarrow r_0$;
> 2 do until convergence
>   2.a Build T_i using Algorithm 4.3;
>   2.b Factorize T_i for λ_i and η_i;
>   2.c $x_i = x_{i-1} + \frac{1}{\lambda_i} p_{i-1}$;
>   2.d $r_i = f - Ax_i$ and $p_i = r_i + \eta_i^2 p_{i-1}$.

Despite its simplicity (which stems from avoiding the storage of Lanczos vectors V_m), the above algorithm is hindered by the computational cost associated with the factorization of the matrices T_i. However, we can obtain the recursive coefficients without the factorization of T_i by making use of the orthogonality of the residual vectors (see (4.78)) and the fact that the column vectors of P_m are A−orthogonal. The latter is deduced from (4.80) and (4.84) as follows:

$$\begin{aligned} \langle AP_m, P_m \rangle &= P_m^T A P_m \\ &= U'^{-T}_m R_m^T A R_m U'^{-1}_m \qquad = U'^{-T}_m \Omega_m^T V_m^T A V_m \Omega_m U'^{-1}_m \\ &= \Omega_m^T U_m^{-T} V_m^T A V_m U_m^{-1} \Omega_m \quad = \Omega_m^T U_m^{-T} T_m U_m^{-1} \Omega_m \\ &= \Omega_m^T U_m^{-T} L_m \Lambda_m U_m U_m^{-1} \Omega_m = \Omega_m^T \Lambda_m \Omega_m. \end{aligned} \tag{4.85}$$

To see how these two orthogonality properties of R_m and P_m can be utilized for the development of recursive relations for x_i, r_i and p_i without the need to factor the matrix T_i, let us assume, in view of (4.83) and (4.84), that the desired recursive relations are as follows:

$$x_i = x_{i-1} + \alpha p_{i-1}, \tag{4.86}$$

$$r_i = r_{i-1} - \alpha A p_{i-1}, \tag{4.87}$$

$$p_i = r_i + \beta p_{i-1}. \tag{4.88}$$

The coefficients α and β are computed using the orthogonality properties of the vectors of R_m and P_m as shown next.

The inner-product of both sides of (4.87) with r_{i-1} yields the following formula for α,

$$\alpha = \frac{\langle r_{i-1}, r_{i-1} \rangle}{\langle r_{i-1}, Ap_{i-1} \rangle} = \frac{\langle r_{i-1}, r_{i-1} \rangle}{\langle p_{i-1} - \beta p_{i-2}, Ap_{i-1} \rangle} = \frac{\langle r_{i-1}, r_{i-1} \rangle}{\langle p_{i-1}, Ap_{i-1} \rangle}. \tag{4.89}$$

The inner-product of both sides of (4.88) with Ap_{i-1} yields the following formula for β,

$$\beta = -\frac{\langle r_i, Ap_{i-1} \rangle}{\langle p_{i-1}, Ap_{i-1} \rangle} = \frac{\langle r_i, r_i \rangle}{\langle r_{i-1}, r_{i-1} \rangle}, \tag{4.90}$$

where the results $Ap_{i-1} = -(r_i - r_{i-1})/\alpha$ from (4.87) and $\langle p_{i-1}, Ap_{i-1} \rangle = \langle r_{i-1}, r_{i-1} \rangle/\alpha$ from (4.89) were used. The complete CG algorithm is obtained as the combination of (4.86-4.90).

> **Algorithm (4.5): Conjugate Gradient Method.**
> ```
> 1 Initialize x₀, r₀ ← f − Ax₀, p₀ ← r₀
> 2 do until convergence
>   2.a α = ⟨r_{i-1},r_{i-1}⟩/⟨p_{i-1},Ap_{i-1}⟩ ;
>   2.b xᵢ = x_{i-1} + αp_{i-1}, rᵢ = r_{i-1} − αAp_{i-1};
>   2.c β = ⟨rᵢ,rᵢ⟩/⟨r_{i-1},r_{i-1}⟩ ;
>   2.d pᵢ = rᵢ + βp_{i-1};
>   2.e if rᵢ is small, converge and stop.
> ```

In the above algorithm each iteration involves the calculation of one matrix-vector product, Ap_{i-1}, and two inner products $\langle p_{i-1}, Ap_{i-1} \rangle$ and $\langle r_i, r_i \rangle$, since $\langle r_{i-1}, r_{i-1} \rangle$ is already available from the previous iteration.

When a symmetric preconditioner M is available the equation to be solved becomes

$$M^{-1}Ax = M^{-1}f. \tag{4.91}$$

Since the $M^{-1}A$ is, in general, not symmetric, the Euclidean inner product must be replaced by an M−inner product for the CG process to be applicable. This is easily seen by noting that it is

$$\langle M^{-1}Ax, y \rangle_M = (Ax)^T y = x^T M M^{-1} Ay = \langle x, M^{-1}Ay \rangle_M. \tag{4.92}$$

The resulting algorithm is the preconditioned CG algorithm [1].

> **Algorithm (4.6): Preconditioned Conjugate Gradient Algorithm (Version 1).**
> ```
> 1 Initialize x₀, r₀ ← (f − Ax₀), z₀ ← M⁻¹r₀, p₀ ← z₀;
> 2 do until convergence
>   2.a α = ⟨z_{i-1},z_{i-1}⟩_M/⟨p_{i-1},M⁻¹Ap_{i-1}⟩_M ;
>   2.b xᵢ = x_{i-1} + αp_{i-1}, rᵢ = r_{i-1} − αAp_{i-1},  zᵢ = M⁻¹rᵢ;
>   2.c β = ⟨zᵢ,zᵢ⟩_M/⟨z_{i-1},z_{i-1}⟩_M ;
>   2.d pᵢ = zᵢ + βp_{i-1};
>   2.e if rᵢ is small enough (convergence) stop.
> ```

Recognizing that $\langle z_i, z_i \rangle_M = z_i^T M z_i = \langle z_i, r_i \rangle$ and $\langle p_{i-1}, M^{-1}Ap_{i-1} \rangle_M = \langle p_{i-1}, Ap_{i-1} \rangle$, the above algorithm can be further simplified as follows:

> **Algorithm (4.7): Preconditioned Conjugate Gradient Method (Version 2).**
> ```
> 1 Initialize x₀, r₀ ← (f − Ax₀), z₀ ← M⁻¹r₀, p₀ ← z₀;
> ```

```
2 do until convergence
```
2.a $\alpha = \dfrac{\langle z_{i-1}, r_{i-1} \rangle}{\langle p_{i-1}, Ap_{i-1} \rangle}$;

2.b $x_i = x_{i-1} + \alpha p_{i-1}, \quad r_i = r_{i-1} - \alpha Ap_{i-1}, \quad z_i = M^{-1} r_i$;

2.c $\beta = \dfrac{\langle z_i, r_i \rangle}{\langle z_{i-1}, r_{i-1} \rangle}$;

2.d $p_i = z_i + \beta p_{i-1}$;

2.e if r_i is small enough (convergence) stop.

Compared to the CG Algorithm (4.5), each iteration of the preconditioned CG involves one additional solution of the *pseudo-residual equation* $M z_i = r_i$.

4.5 THE PRECONDITIONER MATRIX

A preconditioner matrix M^{-1} is an approximation of the inverse of the matrix A in the equation $Ax = f$. Thus the preconditioning operation $M^{-1}Ax = M^{-1}f$ may be interpreted as an operation to obtain an approximation of the solution x of the original matrix equation. In practice, the objective of preconditioning is to produce a new matrix, $M^{-1}A$, the iterative solution of which exhibits faster convergence than that of the original matrix A. The obvious requirements for an effective preconditioner are that it must be easy and computationally inexpensive to compute and apply, while at the same time it must be a good approximation to the inverse of A. Clearly, these are conflicting requirements. Thus the decision on the choice of a preconditioner comes from a compromise between the computational cost associated with its construction and application, and the improvement in convergence resulting from its use. In the following we review three commonly-used preconditioners, namely, the Jacobi preconditioner, the symmetric Gauss-Seidel preconditioner, and incomplete LU factorization preconditioner.

4.5.1 The Jacobi preconditioner

The Jacobi preconditioner is constructed through the iterative Jacobi process of (4.32). In view (4.32) the approximation after m steps is

$$x_m = R_J^m x_0 + \sum_{i=0}^{m-1} R_J^i D^{-1} f. \tag{4.93}$$

With $x_0 = 0$ the above becomes

$$x_m = \left(\sum_{i=0}^{m-1} R_J^i D^{-1} \right) f. \tag{4.94}$$

Then, the Jacobi preconditioner, M_J^{-1}, is defined by

$$M_J^{-1} = \sum_{i=0}^{m-1} R_J{}^i D^{-1}. \tag{4.95}$$

If A is symmetric, then each term in the sum is symmetric; thus the Jacobi preconditioner is symmetric.

The simplest preconditioner is obtained for $m = 1$, $M_J^{-1} = D^{-1}$. Hence, in this case the Jacobi preconditioner is, simply, the inverse of the diagonal matrix with elements the diagonal elements of A. The resulting matrix $D^{-1}A$ is interpreted as the scaling of each row of A by its diagonal element so as to make the diagonal elements of $D^{-1}A$ unity. This preconditoner is also referred to as the *diagonal scaling* preconditioner.

4.5.2 The symmetric Gauss-Seidel preconditioner

For the purposes of preconditioning of the system $Ax = f$, the matrix resulting from either the forward or the backward iteration of (4.37) or (4.38), respectively, can be used as a preconditioner. However if the matrix A is symmetric and the preconditioned CG is the desired iterative solver, the construction of a symmetric Gauss-Seidel (S-GS) preconditioner is needed. The way S-GS is constructed is described next.

Each S-GS iteration consists of one forward GS step and one backward GS step. Hence, we write

$$
\begin{aligned}
x_{k+\frac{1}{2}} &= (D - L)^{-1} U x_k + (D - L)^{-1} f &&\text{Forward GS,} \\
x_{k+1} &= (D - U)^{-1} L x_{k+\frac{1}{2}} + (D - U)^{-1} f &&\text{Backward GS.}
\end{aligned}
\tag{4.96}
$$

When combined into one, these two steps yield the following update equation:

$$
\begin{aligned}
x_{k+1} &= (D - U)^{-1} L (D - L)^{-1} U x_k + (D - U)^{-1} \left[L(D - L)^{-1} + I \right] f \\
&= \underbrace{(D - U)^{-1} L (D - L)^{-1} U}_{H_{S-GS}} \, x_k + \underbrace{(D - U)^{-1} D (D - L)^{-1}}_{R_{S-GS}} \, f.
\end{aligned}
\tag{4.97}
$$

With $x_0 = 0$, at step m the above equation yields

$$
x_m = \left(\sum_{i=0}^{m-1} H_{S-GS}^i R_{S-GS} \right) f.
\tag{4.98}
$$

Thus the S-GS preconditioner, M_{S-GS}^{-1}, is given by

$$
M_{S-GS}^{-1} = \sum_{i=0}^{m-1} H_{S-GS}^i R_{S-GS}.
\tag{4.99}
$$

Let us consider the ith term in the sum

$$
\begin{aligned}
H_{S-GS}^i R_{S-GS} &= \left[(D - U)^{-1} L (D - L)^{-1} U \right]^i (D - U)^{-1} D (D - L)^{-1} \\
&= (D - U)^{-1} \left[L(D - L)^{-1} U (D - U)^{-1} \right]^i D (D - L)^{-1}.
\end{aligned}
\tag{4.100}
$$

The proof of its symmetry is facilitated through the use of the following, easy to prove, identities:

$$
\begin{aligned}
D(D - U)^{-1} U &= U(D - U)^{-1} D, \\
D(D - L)^{-1} L &= L(D - L)^{-1} D.
\end{aligned}
\tag{4.101}
$$

Hence, with each one of the terms in the sum symmetric, the symmetry of M_{S-GS}^{-1} follows.

An alternative construction of a symmetric Gauss-Seidel preconditioner is possible through an m-step forward Gauss-Seidel iteration followed by an m-step backward one.

The relevant equations describing the construction process are

$$x_{k+1/2} = \underbrace{(D-L)^{-1}U}_{H_{f-GS}} x_k + \underbrace{(D-L)^{-1}b}_{R_{f-GS}}$$

$$\Rightarrow \ x_{m/2} = H_{f-GS}^m x_0 + \sum_{i=0}^{m-1} H_{f-GS}^i R_{f-GS} b,$$

$$x_{k+1/2} = \underbrace{(D-U)^{-1}L}_{H_{b-GS}} x_k + \underbrace{(D-U)^{-1}b}_{R_{b-GS}}$$

$$\Rightarrow \ x_m = H_{b-GS}^m x_{m/2} + \sum_{i=0}^{m-1} H_{b-GS}^i R_{f-GS} b. \tag{4.102}$$

Combination of the two m-step GS iterations with $x_0 = 0$ yields the following alternative S-GS preconditioner:

$$\tilde{M}_{S-GS}^{-1} = H_{b-GS}^m \sum_{i=0}^{m-1} H_{f-GS}^i R_{f-GS} + \sum_{i=0}^{m-1} H_{b-GS}^i R_{b-GS}. \tag{4.103}$$

The above expression for $\tilde{M}_{SGS}^{-1}$ may be recast in a form that reveals its symmetry,

$$\begin{aligned} \tilde{M}_{S-GS}^{-1} &= H_{b-GS}^m (I - H_{f-GS})^{-1}(I - H_{f-GS}^m)R_{f-GS} \\ &\quad + (I - H_{b-GS}^m)(I - H_{b-GS})^{-1}R_{b-GS} \\ &= H_{b-GS}^m A^{-1}(D-L)(I - H_{f-GS}^m)R_{f-GS} \\ &\quad + (I - H_{b-GS}^m)A^{-1}(D-U)R_{b-GS} \\ &= A^{-1} - \left[(D-U)^{-1}L\right]^m A^{-1} \left[U(D-L)^{-1}\right]^m. \end{aligned} \tag{4.104}$$

When $m = 1$, the S-GS preconditioning matrix for both cases considered above becomes $(D-U)^{-1}D(D-L)^{-1}$.

4.5.3 Incomplete LU factorization

As the name indicates incomplete LU factorization is obtained as an approximation of the complete LU factorization where some of the entries of the matrices L and U are "dropped" (hence, set to zero) in order to reduce the computational cost associated with the construction and use of the preconditioner. We begin the discussion of the process used for the construction of the incomplete LU (ILU) preconditioner by reviewing the LU factorization algorithm.

Consider a general matrix $A_n \in \mathbb{C}^{n \times n}$. Its LU factorization is

$$A_n = \begin{bmatrix} a_{11} & u^T \\ l & A_{n-1} \end{bmatrix} = \underbrace{\begin{bmatrix} a_{11} & 0 \\ l & L_{n-1} \end{bmatrix}}_{L_n} \underbrace{\begin{bmatrix} 1 & \frac{1}{a_{11}}u^T \\ 0 & U_{n-1} \end{bmatrix}}_{U_n}, \tag{4.105}$$

where

$$L_{n-1}U_{n-1} = A_{n-1} - \frac{1}{a_{11}}l\,u^T. \tag{4.106}$$

In view of the above relations, we can develop a recursive, memory-in-place LU factorization algorithm in which L_n is stored in the low-triangular and diagonal parts of A_n, while U_n is stored in its upper-triangular part. Since the diagonal elements of U_n are all equal to 1, they are not stored.

If A_n is symmetric, the matrix storage and computational cost can be reduced, approximately, by a factor of 2, by employing the Cholesky factorization,

$$A_n = C_n C_n^T, \tag{4.107}$$

where C_n is a lower-triangular matrix. The Cholesky factorization has the form

$$A_n = \begin{bmatrix} a_{11} & l^T \\ l & A_{n-1} \end{bmatrix} = \underbrace{\begin{bmatrix} \sqrt{a_{11}} & 0 \\ \frac{1}{\sqrt{a_{11}}}l & C_{n-1} \end{bmatrix}}_{C_n} \underbrace{\begin{bmatrix} \sqrt{a_{11}} & \frac{1}{\sqrt{a_{11}}}l^T \\ 0 & C_{n-1}^T \end{bmatrix}}_{C_n^T}, \tag{4.108}$$

where

$$C_{n-1}C_{n-1}^T = A_{n-1} - \frac{1}{a_{11}}l\,l^T. \tag{4.109}$$

In view of the above equations, we can develop a recursive, memory-in-place Cholesky factorization algorithm in which C_n is stored in the low-triangular and diagonal parts of A_n.

If A_n is a sparse matrix, the $l\,u^T/a_{11}$ or $l\,l^T/a_{11}$ in (4.106) and (4.109) can change the original zero-entry pattern and introduce new fill-ins in the next submatrix A_{n-1}. As the factorization algorithm proceeds, the new fill-ins eventually render the next submatrices dense, thus resulting in increased computational cost of the factorization process. Since the construction of the preconditioner does not require exact factorization, some fill-ins are dropped. The result is an incomplete LU factorization.

Depending on the criteria used for dropping the fill-ins, several variants of incomplete factorization schemes can be obtained. The simplest one is the zeroth level incomplete LU factorization, often referred to as ILU(0), where all the new fill-ins introduced by $l\,u^T/a_{11}$ or $l\,l^T/a_{11}$ in the next submatrix A_{n-1} or C_{n-1} are dropped. Thus the L and U matrices have the same non-zero patterns as A_n.

The accuracy of ILU(0) may be insufficient to yield an adequately fast rate of convergence of the iterative solver in which the constructed matrix is used as preconditioner. More accurate ILU schemes can be constructed by maintaining more fill-ins according to one of several schemes. For example, some ILU schemes use a threshold to decide on the "value" of a new fill-in. If the absolute value of the fill-in is below the threshold, the fill-in is dropped. A detailed discussion of numerous popular ILU schemes is given in [1].

Jacobi, symmetric GS and incomplete factorization preconditioners are effective for the finite element matrices obtained from the discretization of the elliptic-type partial differential equations that govern electrostatic and magnetostatic BVPs. However, they are less effective and often of unpredictable performance when used for the iterative solution of the finite-element matrices resulting from the discretization of the Helmholtz-type electric or magnetic partial differential equations that govern electrodynamic boundary value problems.

4.6 MULTIGRID PROCESS AND ITS USE AS A PRECONDITIONER

In our discussion of matrix preconditioning very little has been said about the impact of grid size on the convergence rate of the iterative process used for the solution of equations

$Ax = f$ and its preconditioned form $M^{-1}Ax = M^{-1}f$. However, there is strong theoretical evidence and numerical experience that the convergence of preconditioned iterative processes, including the class of preconditioned Krylov subspace methods, deteriorates with the dimension of the system, particularly when this increase in size is driven by the introduction of fine-size grids over portions of the computational domain.

In the context of the finite-element solution of electromagnetic BVPs the use of fine grids is driven by two factors. The first one is the presence of fine geometric features in the computation domain (e.g., thin slots in conducting plates, wires attached to platforms, thin dielectric layers), proper resolution of which during meshing is essential for solution accuracy. The second one concerns the utilization of a fine enough grid to contain *numerical dispersion error* and thus ensure the accuracy of numerical approximation of the wave interactions inside the computational domain. The numerical dispersion error is understood as the error in the numerical wave solution due to the fact that the numerical wave number of the sampled wave is different from the exact wavenumber [4]. For the case of linear elements it can be shown that the phase error in the finite element solution due to numerical dispersion is $\mathcal{O}[(h/\lambda)^2]$, where h is the grid size and λ is the wavelength. Hence, the smaller the grid size the smaller the phase error. In particular, it was shown in [4] that the larger the electrical size (i.e., the size in wavelengths) of the computational domain the smaller the grid size (or, equivalently, the higher the order of the finite element interpolation) must be to contain the numerical dispersion error below a desirable threshold.

Irrespective of which one of the aforementioned two factors is the dominant one, it is to be expected that the need for a very fine finite element grid will arise often in the electromagnetic analysis of structures of practical interest. Clearly, the finer the grid size the larger the dimension of the resulting finite element system. The consequences with regards to computational cost are obvious, namely, increased computer memory requirements and solution time. In the context of iterative methods, increased computation time manifests itself in terms of deterioration in convergence rate. Thus the development of iterative methods for which the convergence rate is independent of grid size is of particular importance. Multigrid methods were developed with this objective in mind.

The literature on multigrid methods is vast and continues to grow as new applications to both linear and nonlinear problems are being explored. Multigrid methods are loosely divided into two broad classes, namely, *geometric* and *algebraic* multigrid. While the former exploit the geometric structure of the finite difference or finite element grid used for the discretization of the partial differential equation of interest for the development of an fast iterative solution of the sparse matrix equation, the latter rely solely upon algebraic information contained in the system for achieving faster convergence. From the numerous references on multigrid methods in the mathematics literature we identify [5] and [6] as hands-on, intuitive, yet mathematically well-founded presentations of the key ideas behind multigrid methods. Reference [7] offers a more in-depth presentation of multigrid methods, including thorough discussions of the parallel implementation of multigrid methods and the algebraic multigrid. Up to date information on this rapidly involving field can be found at the website http://www.mgnet.org, sponsored by the National Science Foundation, Yale University, as well as numerous other universities and research organizations.

In this section, a brief overview is provided of the way multigrid can be used as an effective preconditioner for the iterative solution of finite element systems. Our presentation takes advantage of the illustrative power of geometric multigrid to motivate its use and highlight the key steps associated with its computer implementation. In subsequent chapters, a more in depth discussion of geometric multigrid is provided in the specific context of their application as effective preconditioners for the iterative solution of large sparse matrices resulting from the finite element approximation of electromagnetic BVPs.

4.6.1 Motivation for multigrid

For the purposes of this section, a one-dimensional model problem is used. More specifically, we will concern ourselves with the solution of the linear system resulting from the finite element approximation of the one-dimensional Helmholtz equation

$$\left(\frac{d^2}{dz^2} + \beta^2\right) E(z) = g(z), \quad 0 \le z \le 1 \tag{4.110}$$

with homogeneous Dirichlet boundary conditions $E(0) = E(1) = 0$, using linear elements. The forcing function, $g(z)$ will be assumed known; however, for the purposes of our discussion its exact form is not required.

Recognizing that $\sin(m\pi z)$ are the eigenfunctions of the operator $\mathcal{L} = \frac{d^2}{dz^2} + \beta^2$ over the domain $0 \le z \le 1$ with the indicated homogeneous Dirichlet boundary conditions, with corresponding eigenvalues $\lambda_m = \beta^2 - (m\pi)^2$, the analytic solution of (4.110) is easily found to be

$$E(z) = \sum_{m=1}^{\infty} \frac{g_m}{\beta^2 - (m\pi)^2} \sin(m\pi z),$$

$$g_m = 2 \int_0^1 g(z) \sin(m\pi z) dz, \quad m = 1, 2, \cdots. \tag{4.111}$$

The observation worth making here is that the analytic solution is cast as an infinite sum over the eigenfunctions of the operator $\mathcal{L}$. In a similar manner, the solution, x, to the linear system $Ax = f$ that results from the finite element approximation of (4.110) and, more importantly, the error, $e_k = x - x_k$ in its approximation, $\hat{x}_k$, obtained at step k of the iterative solution process, may be cast in terms of the eigenvectors of the iteration matrix. This, then, suggests that the convergence of the iterative process may be assessed in terms of the way the iterative process minimizes each one of the components in the eigenvector expansion of the error.

To explore this point further, consider the finite element system resulting from the approximation of (4.110) using linear elements in a uniform finite element grid, Ω_h, of N interior nodes. The grid size is $h = 1/(N+1)$. In view of the homogeneous Dirichlet boundary conditions at the two end nodes, node 0 at $z = 0$ and node $N+1$ at $z = 1$, the vector of unknowns, x, consists of the field value at the N interior nodes; hence, $x = [E_1, E_2, \cdots, E_N]^T$. The matrices A and f in the resulting finite element system, $Ax = f$, are given by

$$A = \begin{bmatrix} a & b & & & \\ b & a & b & & \\ & \ddots & \ddots & \ddots & \\ & & b & a & b \\ & & & b & a \end{bmatrix}, \quad f = h \begin{bmatrix} f_1 \\ f_2 \\ \vdots \\ f_{N-1} \\ f_N \end{bmatrix}, \tag{4.112}$$

where

$$a = \left(2 - \frac{2}{3}(\beta h)^2\right), \quad b = -\left(1 + \frac{1}{6}(\beta h)^2\right) \tag{4.113}$$

and

$$f_n = -\int_0^1 g(z)w_n(z)dz, \quad n = 1, 2, \ldots, N \tag{4.114}$$

with $w_n(z)$ denoting the nth linear expansion function associated with node n.

The eigenvalues and eigenvectors of A can be found in closed form. Let $v_n^{(h)} = [\sin\theta_n, \sin 2\theta_n, \ldots, \sin N\theta_n]^T$, where the superscript (h) is used to indicate the grid size of Ω_h. It is straightforward to show that

$$[A - (a + 2b\cos\theta_n)I]v_n^{(h)} = \sin((N+1)\theta_n)e_N, \tag{4.115}$$

where e_N is the Nth column of the identity matrix, I, of dimension N. It is immediately recognized that the vector $v_n^{(h)}$ will be an eigenvector of A with corresponding eigenvalue $\lambda_n = a + 2b\cos\theta_n$ for θ_n such that the right-hand side of (4.115) is zero. This is the case for

$$\theta_n = \frac{n\pi}{N+1}, \quad n = 1, 2, \ldots, N. \tag{4.116}$$

We conclude that the eigenvectors of A are

$$v_n^{(h)} = [\sin(n\pi z_1), \sin(n\pi z_2), \cdots, \sin(n\pi z_N)]^T, \quad n = 1, 2, \ldots, N, \tag{4.117}$$

where $z_i = i/(N+1)$, $i = 1, 2, \ldots, N$. The corresponding eigenvalues are given by

$$\lambda_n = 2\left[1 - \cos\left(\frac{n\pi}{N+1}\right)\right] - \frac{1}{3}(\beta h)^2\left[2 + \cos\left(\frac{n\pi}{N+1}\right)\right], \quad n = 1, 2, \ldots, N. \tag{4.118}$$

Depicted on the left columns of Fig. 4.1 and Fig. 4.2 are the nine eigenvectors of A for the case $N = 9$, where the circles are used to indicate that their values are associated with the nine interior nodes of the finite element grid, Ω_h, used for the approximation of (4.110). It is evident from the plots that the corresponding numerical modes, resulting from the finite interpolation over the domain $0 \le z \le 1$, may be split into two groups. The first group includes the first four modes (i.e., modes with $n < (N+1)/2$)). These modes are well-resolved by the grid, closely resembling the corresponding analytic eigenfunctions. They will be referred to as the *smooth modes*. The second group contains the modes with $n \ge (N+1)/2$. These modes are not well-resolved by the grid and will be referred to as the high-frequency or *oscillatory modes*.

Next, let us assume that the weighted Jacobi process will be used for the iterative solution of the finite element system. In view of (4.41), the error after k iterations is given by

$$e_k = (R_{J\omega})^k e_0. \tag{4.119}$$

Let us expand e_0 in terms of the eigenvectors of $R_{J\omega}$. In view of (4.36) and the fact that the elements of the diagonal matrix D are all equal, R_ω and A have the same eigenvectors. The corresponding eigenvalues, μ_n, $n = 1, 2, \ldots, N$, of $R_{J\omega}$ are readily obtained from those of A through the relation

$$\mu_n = 1 - \omega a^{-1}\lambda_n = (1 - \omega) + \omega\frac{1 + (\beta h)^2/6}{1 - (\beta h)^2/3}\cos\left(\frac{n\pi}{N+1}\right), \tag{4.120}$$

where use was made of (4.113). The eigenvector expansion of e_0 is then,

$$e_0 = \sum_{n=1}^N \zeta_n v_n^{(h)}. \tag{4.121}$$

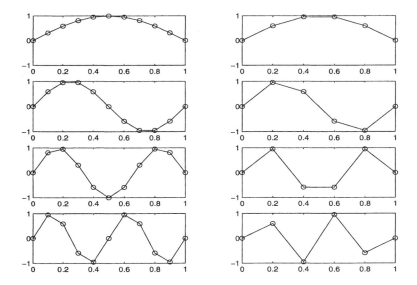

Figure 4.1 Left column (*from top to bottom*): Plots of the smooth eigenvectors (modes) of A for a uniform grid, Ω_h, with 9 interior nodes. Right column (*from top to bottom*): Projection of the five modes on a coarser grid, Ω_{2h} of grid size twice that of $G1$.

Use of (4.121) in (4.119) yields

$$e_k = (R_{J\omega})^k \sum_{n=1}^{N} \zeta_n v_n^{(h)} = \sum_{m=1}^{N} \zeta_n (R_{J\omega})^k v_n^{(h)} = \sum_{n=1}^{N} \zeta_n (\mu_n)^k v_n^{(h)}. \qquad (4.122)$$

This result indicates that the nth component in the eigenvector expansion of the error has been reduced by the factor $(\mu_n)^k$ after k iterations. As already stated earlier, convergence requires that $|\mu_n| < 1$, $n = 1, 2, \ldots, N$. More importantly, the result in (4.122) suggests that different components in the eigenvector expansion of the error are reduced at different rates and, hence, the different components in the eigenvector expansion of the finite element approximation will converge differently. In particular, the fastest converging components are the ones with the smallest eigenvalues.

The availability of the eigenvalues of $R_{J\omega}$ in closed form from (4.120) makes possible the quantification of the convergence rate of the classes of numerical modes identified above, namely, the smooth and oscillatory modes. To facilitate the discussion (4.120) is recast in the following form:

$$\mu_n = \left(1 + \omega \frac{(\beta h)^2/2}{1 - (\beta h)^2/3}\right) - 2\omega \frac{1 + (\beta h)^2/6}{1 - (\beta h)^2/3} \sin^2\left(\frac{n\pi}{2(N+1)}\right). \qquad (4.123)$$

For the oscillatory modes it is $n \geq (N+1)/2$; hence, $\sin^2\left(\frac{n\pi}{2(N+1)}\right) > 1/2$, and (4.123) yields in this case,

$$\mu_n < 1 - \omega, \quad n \geq (N+1)/2. \qquad (4.124)$$

This result suggests that if $0 < \omega < 1$ the error in the high-frequency modes can be made to decrease sufficiently fast. This result is confirmed by the data presented Table 4.1 for the eigenvalues of the nine modes calculated for $\omega = 0.5$ for different values of β. In all cases,

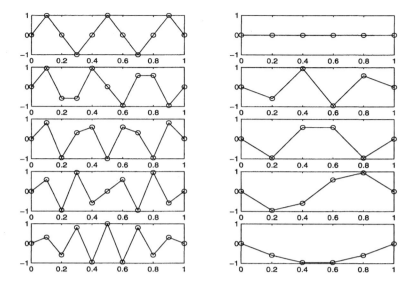

Figure 4.2   Left column (*from top to bottom*): Plots of the oscillatory eigenvectors of A for a uniform grid, Ω_h, with 9 interior nodes. Right column (*from top to bottom*): Projection of these modes on a coarser grid, Ω_{2h} of grid size twice that of Ω_h.

$\mu_n \leq 0.5$ for all high-frequency modes. Hence, the high-frequency components of the error undergo good reduction at each step of the iteration. Furthermore, in view of (4.124), the reduction rate is independent of the grid size h.

Table 4.1   Numerical Eigenvalues of $R_{J\omega}$ for Grid Ω_h, with $\omega = 0.5$

Mode	$\beta = 4/3$	$\beta = 2$	$\beta = 4$	$\beta = 6$
1	0.9798	0.9852	1.0157	1.0728
2	0.9081	0.9127	0.9387	0.9872
3	0.7965	0.7998	0.8187	0.8540
4	0.6559	0.6576	0.6676	0.6861
5	0.5000	0.5000	0.5000	0.5000
6	0.3441	0.3424	0.3324	0.3139
7	0.2035	0.2002	0.1813	0.1460
8	0.0919	0.0873	0.0613	0.0128
9	0.0202	0.0148	−0.0157	−0.0728

With regards to the rate at which the error in the smooth modes is reduced, the data in Table 4.1 suggest that the smoother the mode the slower the reduction rate. Furthermore, for the given choice of ω, depending of the value of β, the eigenvalue of the smoothest of the modes can become larger than 1.

To explore this point further, let us consider the smoothest (best resolved) numerical mode of $n = 1$. To begin with, it is useful to recall that the constant β appearing in the one-dimensional Helmholtz equation (4.110) is the wavenumber for the time-harmonic wave in the domain of interest. Its constant value for our purposes indicates a homogeneous domain.

Its expression in terms of the wavelength, λ, in the medium, $\beta = (2\pi)/\lambda$, is of particular relevance to this discussion. More specifically, it is noted that in the expressions (4.118) and (4.120) for the numerical eigenvalues of A and $R_{J\omega}$, respectively, the wavenumber enters through the term $\beta h = 2\pi(h/\lambda)$. Thus (βh) serves as a measure of how well the wavelength at the operating frequency is resolved by the finite element grid. For example, since $h = 0.1$ for the grid Ω_h, then (βh) assumes values of 0.133, 0.2, 0.4 and 0.6, respectively, for the values of β of 1.333, 2.0, 4.0 and 6.0, used for obtaining the data in Table 4.1. The smaller values of (βh) represent a well-resolved wavelength. We mention that in most practical applications of the finite element method for the modeling of wave phenomena, a wavelength resolution of at least ten degrees of freedom per wavelength is used. For linear elements, this is equivalent to a resolution of at least ten elements per wavelength. For the largest value of β considered ($\beta = 6.0$) corresponds to a wavelength of $\pi/3$, the grid size $h = 0.1$ amounts to a wavelength resolution of $\sim \lambda/10$. Hence, for all four cases considered in Table 4.1 the wavelength is well resolved by the finite element grid.

For $n = 1$ it is $\sin^2(\pi/2/(N+1)) = \sin^2(\pi/2/h) \simeq (\pi/2/h)^2$. Furthermore, term $p = (1/3)(\beta h)^2$ in the denominator of (4.123) is small enough for well-resolved wavelengths to allow for the approximation $(1 - p)^{-1} \simeq 1 + p$. Use of this approximation in (4.123) yields for $n = 1$ the following approximate expression for μ_1,

$$\mu_1 \simeq 1 + \frac{\omega}{2}\left(\beta^2 - \pi^2\right)h^2 + \mathcal{O}(h^4). \tag{4.125}$$

This result is consistent with the data presented in Table 4.1, indicating that, for $0 < \omega < 1$, the eigenvalue of the smoothest mode in the error remains close to 1 for well-resolved wavelengths on the grid, and it can even become greater than 1 for $\beta > \pi$. This leads us to the important result that the attenuation rate of smooth components of the error is slow, and that their reduction may even stall as wavelength resolution worsens. Multigrid attempts to rectify this situation by introducing a coarser grid on which the smooth modes appear as high-frequency ones. Thus, in view of (4.124), their iteration on the coarser grid is expected to enhance their attenuation rate and, hence, enhance solution convergence.

The multigrid idea may be explored further with the aid of Figs. 4.1 and 4.2. Let us define the coarse grid, Ω_{2h}, over the domain $0 \leq z \leq 1$, with element grid size $2h$. Hence, there are four interior nodes for Ω_{2h}, with coordinates $z_i^{(2h)} = i(2h), i = 1, 2, 3, 4$. With the nine interior nodes of Ω_h given by $z_j^{(h)} = jh, j = 1, 2, \ldots, 9$, it is $z_i^{(2h)} = z_{2i}^{(h)}, i = 1, 2, 3, 4$. The second column of Figs. 4.1 and 4.2 depicts the projection of the nine modes of the fine grid, Ω_h, onto the coarse grid, Ω_{2h}. Clearly, as seen in the right column of Fig. 4.1, the smooth modes of Ω_h become oscillatory on Ω_{2h}. More precisely, considering the elements of the nth mode, $v_n^{(h)}, 1 < n < (N+1)/2$, of Ω_h, associated with the even-numbered grid points, $2j, j = 1, 2, 3, 4$, we have

$$\left(v_n^{(h)}\right)_j = \sin\left(\frac{n\pi 2j}{N+1}\right) = \sin\left(\frac{n\pi j}{(N+1)/2}\right) = \left(v_n^{(2h)}\right)_j, \tag{4.126}$$

where $v_n^{(2h)}$ is recognized as the nth eigenvector, $1 < n < (N+1)/2$ of the Ω_{2h} grid. Furthermore, for the case of N odd considered in our example, the fine-grid mode $v_{(N+1)/2}^{(h)}$ is not represented on Ω_{2h}, as seen clearly by the top plot on the right column of Fig. 4.2. Finally, the fine-grid, high-frequency modes (i.e., the modes with $n > (N + 1)/2$) are not represented anymore on Ω_{2h}. More precisely, through the phenomenon of aliasing, these modes are misrepresented as smooth modes on Ω_{2h}, as depicted clearly in the right column of Fig. 4.2.

Returning to the smooth modes of the error on Ω_h, the observation that these modes are more oscillatory on the coarser grid Ω_{2h} suggests that, rather than continuing their iterative reduction on the fine grid, project them onto Ω_{2h} and attempt what will hopefully be a more expedient reduction on the coarser grid.

For example, for the model problem considered, we find from (4.123), with $\omega = 0.5$ and for $\beta = 4/3$, that the fine-grid, smooth modes $n = 3$ and $n = 4$, when projected onto Ω_{2h} have eigenvalues of 0.3399, and 0.0808, a vast improvement over their fine-grid values depicted in the first column of Table 4.1. Hence, a much faster attenuation is expected by iterating on them on Ω_{2h} rather than on Ω_h.

On the other hand, the smoothest, fine-grid modes $n = 1$ and $n = 2$, when projected onto Ω_{2h} have eigenvalues of 0.9192 and 0.6601, respectively, which, even though better than their fine-grid values, may not be sufficiently smaller than 1 to result in fast enough damping of the corresponding components of the error. The remedy to this should be obvious. It involves the introduction of an even coarser grid, Ω_{4h}, of grid size twice that of Ω_{2h}. From (4.123) (again, with $\omega = 0.5$ and for $\beta = 4/3$) we find that the eigenvalues of the two smoothest, fine-grid modes $n = 1$ and $n = 2$, when projected onto Ω_{4h} are 0.7764 and 0.2236, promising a faster damping compared to that obtained through their reduction on grid Ω_{2h}.

At this point the recursive nature of the multigrid process starts becoming evident. More specifically, the geometric multigrid process we have been discussing in the previous paragraphs calls for the definition of a hierarchy of grids of progressively decreasing density, starting with a finest grid, Ω_h, and ending with the coarsest grid Ω_H. The approximate solution is sought in an iterative manner through a process aimed at the expedient reduction of the error in an initial guess for the solution, by exploiting the fact that the damping rate for the smooth components of the error is larger on coarser grids. Clearly, such an iterative process requires *intergrid* transfer operators for moving back and forth between fine and coarse grids in the defined grid hierarchy. Along with these intergrid transfer operators, an effective *smoother* (i.e., a relaxation process (such as the weighted Jacobi process used above) is required to enable the expedient damping of the oscillatory components of the error at each grid. The detail procedure that leads to the algorithmic implementation of a multigrid process is discussed next.

4.6.2 The two-grid process

We begin with some notation. Let

$$A_h x_h = f_h \tag{4.127}$$

denote the finite element approximation of the boundary value problem of interest on a grid, Ω_h, of average grid size h. Let $\tilde{x}_h$ denote the approximate solution, and e_h and r_h the error and residual, respectively; hence,

$$\begin{aligned} e_h &= x_h - \tilde{x}_h, \\ r_h &= f_h - A_h \tilde{x}_h. \end{aligned} \tag{4.128}$$

From the above definitions it follows that the error satisfies the following residual equation:

$$A_h e_h = r_h. \tag{4.129}$$

In a similar manner, let x_H be the approximation on a coarser grid, Ω_H, with average grid size H. The finite element approximation of the boundary value problem on Ω_H has the form

$$A_H x_H = f_H. \tag{4.130}$$

Let $\tilde{x}_H$ denote the approximate solution of the system. Then the error e_H and the residual r_H, given by

$$\begin{aligned} e_H &= x_H - \tilde{x}_H, \\ r_H &= f_H - A_H \tilde{x}_H \end{aligned} \tag{4.131}$$

satisfy the residual equation

$$A_H e_H = r_H. \tag{4.132}$$

Equations (4.129) and (4.132) are most useful since they provide for the multigrid process to be performed directly in terms of the residual and the error vectors.

As mentioned in the last few paragraphs of the previous section, the purpose of the coarse grid is to enable for a more effective reduction of the smooth components of the error. Hence, once the residual is calculated on Ω_h, it must be transferred to the coarse grid for the smooth components of the error to be effectively processed there. This brings us to the introduction of one of the two intergrid transfer operators needed for going back and forth between the two grids, namely, the *restriction* operator. The restriction operator, I_h^H, restricts the residual r_h of the fine grid to the coarser grid. Hence, it is

$$r_H = I_h^H r_h. \tag{4.133}$$

With r_H known, let us assume that (4.132) can be solved for e_H. Then a *prolongation* or interpolation operator I_H^h is needed to interpolate the correction e_H back to the fine grid

$$e_h = I_H^h e_H. \tag{4.134}$$

In practice, both I_h^H and I_H^h are chosen to be linear operators.

The coarse-grid correction process described above completes with the new approximation, $\tilde{x}_h$, to the unknown vector, x_h, obtained through the correction operation

$$\tilde{x}_h \leftarrow \tilde{x}_h + e_h. \tag{4.135}$$

The following provides a compact summary of the algorithm.

Algorithm (4.8): $\tilde{x}_h \leftarrow$ **Coarse-Grid-Correction** $(\tilde{x}_h, f_h)$
1 $r_h = f_h - A_h \tilde{x}_h$;   (Compute residual on fine grid)
2 $r_H = I_h^H r_h$;   (Restrict residual to coarse grid)
3 $e_H = A_H^{-1} r_H$;   (Solve exactly on coarse grid)
4 $e_h = I_H^h e_H$;   (Prolongate correction to fine grid)
5 $\tilde{x}_h \leftarrow \tilde{x}_h + e_h$;   (Correct solution on fine grid)

In compact mathematical form, the above coarse-grid correction is written as follows:

$$\tilde{x}_h \leftarrow \tilde{x}_h + I_H^h A_H^{-1} I_h^H (f_h - A_h \tilde{x}_h). \tag{4.136}$$

It should be clear from the above process that for the intergrid transfer process to be effective the residual that is being transferred back and forth between the two grids must be smooth enough to provide for accurate interpolation. Hence, a smoothing process is needed prior to intergrid transfer, to damp sufficiently the high-frequency components of the error and thus improve the smoothness of the residual.

In the correction process described above, with the assumption that the residual equation is solved exactly on the coarse grid, such smoothing is not needed prior to the application of the prolongation operator. However, it is needed on the fine grid, prior to applying the restriction operator. As already stated in the previous section, a stationary iteration process, such as weighted Jacobi or Gauss-Seidel can be employed for this smoothing (or relaxation) process. For example, the combination of the symmetric GS of (4.103) with the coarse-grid correction of (4.136) leads to the following two-grid process for the approximate solution of $A_h x_h = f_h$:

> **Algorithm (4.9):** $\tilde{x}_h \leftarrow$ **Two-Grid Process**$(\tilde{x}_h, f_h)$.
> 1 $\tilde{x}_h \leftarrow H^m_{f-GS} \tilde{x}_h + \sum_{i=0}^{m-1} H^i_{f-GS} R_{f-GS} f_h$; (Pre-smoothing)
> 2 $r_h = f_h - A_h \tilde{x}$; (Compute residual on finegrid)
> 3 $r_H = I_h^H r_h$; (Restrict residual to coarse grid)
> 4 $e_H = A_H^{-1} r_H$; (Solve exactly on coarser grid)
> 5 $e_h = I_H^h e_H$; (Prolongate correction to fine grid)
> 6 $\tilde{x}_h \leftarrow \tilde{x}_h + e_h$; (Update solution)
> 7 $\tilde{x}_h \leftarrow H^m_{b-GS} \tilde{x}_h + \sum_{i=0}^{m-1} H^i_{b-GS} R_{b-GS} f_h$ (Post-smoothing)

Comparison of Algorithm (4.9) with Algorithm (4.8) reveals the coarse-grid correction step is inserted between the forward and the backward GS iterations. With the initial guess for the approximate solution $\tilde{x}_h$ set to zero, the mathematical statement of two-grid method is cast in the form

$$
\tilde{x}_h \leftarrow \left(H^m_{b-GS} \sum_{i=0}^{m-1} H^i_{f-GS} R_{f-GS} + \sum_{i=0}^{m-1} H_{b-GS}{}^i R_{b-GS} \right) f_h
$$
$$
+ \left(H^m_{b-GS} I_H^h A_H^{-1} I_h^H \left(I - A_h \sum_{i=0}^{m-1} H^i_{f-GS} R_{f-GS} \right) \right) f_h. \tag{4.137}
$$

The two-grid process (and, in general, a multigrid process utilizing a hierarchy of several grids) constitutes in itself a stand-alone iterative solver. However, it is also most useful as a convergence acceleration technique for use as a preconditioner for Krylov-subspace based iterative solvers such as GMRES or CG.

As we have already discussed, symmetry is required for a preconditioner that will be used in conjunction with CG. From (4.137), the two-grid preconditioner M_{2G}^{-1} is

$$
M_{2G}^{-1} = H^m_{b-GS} \sum_{i=0}^{m-1} H^i_{f-GS} R_{f-GS} + \sum_{i=0}^{m-1} H^i_{b-GS} R_{b-GS}
$$
$$
+ H^m_{b-GS} I_H^h A_H^{-1} I_h^H \left(I - A_h \sum_{i=0}^{m-1} H^i_{f-GS} R_{f-GS} \right). \tag{4.138}
$$

Compared with the symmetric Gauss-Seidel preconditioner M_{SGS}^{-1}, the coarse-grid correction introduces the extra third item in (4.138). The symmetry of the sum of the two terms in (4.138) has already been shown in (4.103). Thus it remains to show that the third term is symmetric also. Since it is

$$I - A_h \sum_{i=0}^{m-1} H_{f-GS}^i R_{f-GS} = I - A_h (I - H_{f-GS})^{-1} (I - H_{f-GS}^m) R_{f-GS}$$

$$= I - (D - L)(I - H_{f-GS}^m) R_{f-GS} \qquad (4.139)$$

$$= \left(U(D-L)^{-1} \right)^m,$$

the third term in (4.138) may be cast in the form

$$H_{b-GS}^m I_H^h A_H^{-1} I_h^H \left(I - A_h \sum_{i=0}^{m-1} H_{f-GS}^i R_{f-GS} \right) \qquad (4.140)$$

$$= \left((D-U)^{-1} L \right)^m I_H^h A_H^{-1} I_h^H \left(U(D-L)^{-1} \right)^m.$$

If the restriction operator I_h^H is the transpose of the prolongation operator I_H^h, that is, if

$$I_h^H = \left(I_h^H \right)^T, \qquad (4.141)$$

then the following relation holds between the coarse-grid matrix A_H and the fine-grid matrix A_h:

$$A_H = I_H^h A_h I_H^h. \qquad (4.142)$$

In this case the multigrid scheme is often referred to as being of the Galerkin type. Under this condition, it is straightforward to show that the matrix in (4.140) is symmetric, and thus the two-grid preconditioner M_{2G}^{-1} is symmetric.

4.6.3 The multigrid process

The extension of the two-grid process to a multigrid one is straightforward. Instead of solving exactly the coarse-grid residual equation exactly in Step 4 of Algorithm (4.9), the residual is restricted, following a smoothing process, onto a coarser grid. This process can be continued down to a grid coarse enough for the exact solution of the residual equation to be computationally efficient.

Consider a hierarchy of N grids, Ω_{ih}, $i = 1, 2, \ldots, N$, with average element size varying from h at the finest grid to Nh at the coarsest grid. We will often refer to such a multigrid process as an *N-level* process, with the highest level corresponding to the finest grid and the lowest level corresponding to the coarsest grid. The matrix equation at each level is

$$A_{ih} x_{ih} = f_{ih}, \quad i = 1, 2, \cdots, N, \qquad (4.143)$$

while e_{nh}, r_{nh}, are, respectively, the error and the residual at the nth level. The N-level multigrid process is described by the following algorithm.

Algorithm (4.10): $\tilde{x}_h \leftarrow$ **MultiGrid**$(f_h, n = 1)$.
1 if $n == N$, $\tilde{x}_{Nh} \leftarrow A_{Nh}^{-1} f_{Nh}$.
2 else
    2.1 $\tilde{x}_{nh} \leftarrow$ pre-smoothing using forward GS v_1 times
    2.2 $r_{nh} = f_{nh} - A_{nh} \tilde{x}_{nh}$
    2.3 $r_{(n+1)h} = I_{nh}^{(n+1)h} r_{nh}$, $e_{(n+1)h} \leftarrow 0$
    2.4 $e_{(n+1)h} \leftarrow$ MultiGrid$(r_{(n+1)h}, n + 1)$ α times
    2.5 $e_{nh} = I_{(n+1)h}^{nh} e_{(n+1)h}$

2.6 $\tilde{x}_{nh} \leftarrow \tilde{x}_{nh} + e_{nh}$

2.7 $\tilde{x}_{nh} \leftarrow$ post-smooth using backward GS v_2 times

The operators $I_{nh}^{(n+1)h}$ and $I_{(n+1)h}^{nh}$ are the intergrid transfer operators between grids Ω_{nh} and $\Omega_{(n+1)h}$. It is easy to show recursively that the multigrid preconditioner is symmetric. The parameter α represents the *shape* of the multigrid *cycle*. To elaborate, the transition from the finest grid down to the coarsest grid and back to the finest grid again is called a cycle. If $\alpha = 1$, Step 2.4 in Algorithm (4.10) is performed only once per iteration. This results in the *V-cycle* multigrid process, depicted in the left plot of Fig. 4.3. If $\alpha = 2$, Step 2.4 is performed twice per iteration. This results in the *W-cycle* multigrid process, depicted in the right plot of Fig. 4.4.

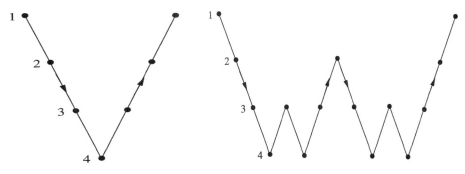

Figure 4.3 V-cycle: $N = 4, \alpha = 1$.   **Figure 4.4** W-cycle: $N = 4, \alpha = 2$.

The key to deriving an effective and robust multigrid algorithm is the selection of the appropriate relaxation processes (smoothers) and the construction of Galerkin-type intergrid transfer operators. Closely related to the former are the properties of the matrix A, and in particular its spectrum. For our model problem we chose the one-dimensional Helmholtz operator the finite element approximation of which may not be a positive-definite matrix, as it is immediately evident from its eigenvalues in (4.118). At this point, it is useful to point out that the positive-definiteness of its finite element matrix approximation is guaranteed only when $\beta = 0$. In this case, the Helmholtz operator reduces to the elliptic Laplace operator. While multigrid techniques have a proven record of success in the solution of finite-difference and finite-element approximation of elliptic BVPs, their application to the iterative solution of finite approximations of electrodynamic BVPs governed by Helmholtz-like operators has not been straightforward. However, recent progress in the understanding of the properties of the spectrum of discrete approximation to such problems have paved the way toward the establishment of effective multigrid-based methodologies and algorithms for the robust iterative solution of the associated matrix equations. These methodologies are discussed in detail in the remaining of the book.

REFERENCES

1. Y. Saad, *Iterative Methods for Sparse Linear Systems*, New York: SIAM, 2003.

2. R. Barrett, et al., *Templates for Iterative Solution of Linear Systems*, New York: SIAM, 1992.

3. G. H. Golub and C. F. Van Loan, *Matrix Computations*, 3rd ed., Baltimore: Johns Hopkins University Press, 1996.

4. R. Lee and A. Cangellaris, "A study of the discretization error in the finite element approximation of wave solutions," *IEEE Trans. Antennas Propagat.*, vol. AP-40, pp. 542-548, May 1992.

5. W. L. Briggs, V. E. Henson, and S. F. McCormick, *A Multigrid Tutorial*, 2nd ed., New York: SIAM, 2000.

6. S. McCormick, *Multigrid Methods*, vol 3 of SIAM Frontiers Series, Philadelphia:SIAM, 1987.

7. U. Trottenberg, C. O. Osterlee, and A. Schüller, *Multigrid*, London: Academic Press, 2001.

CHAPTER 5

NESTED MULTIGRID PRECONDITIONER

In this chapter, the multigrid preconditioning process is applied to the iterative solution of the finite element system resulting from the approximation of the scalar Helmholtz equation. To keep the presentation simple, only two-dimensional boundary value problems are considered. The simplicity of the two-dimensional framework is exploited for the explanation of the construction of the intergrid transfer operators. The experience and understanding gained from this discussion will be found useful in the discussion of the general three-dimensional vector case in Chapter 6. Another important aspect addressed in this chapter is the issue of the minimum acceptable wavelength resolution in the coarsest grid for overcoming numerical dispersion error and thus maintaining the solution accuracy needed for the numerical processing at finer grids.

5.1 WEAK STATEMENT OF THE TWO-DIMENSIONAL HELMHOLTZ EQUATION

The finite element solution of the two-dimensional (2D) electromagnetic scattering problem by an inhomogeneous cylindrical structure embedded in an unbounded homogeneous medium will be considered. Without loss of generality, the unbounded medium is taken to be free space. Let z be the direction in space along which geometry and electromagnetic fields exhibit no variation. Let Ω denote the planar computational domain on the plane $z = 0$ and let $\partial\Omega$ denote its boundary. The case considered here is that of a time-harmonic, transverse magnetic (TM$_z$) field excitation of angular frequency ω. With the time dependence $\exp(j\omega t)$ assumed and suppressed for simplicity the phasor form of the electric field is

$$\vec{E} = \hat{z}E_z, \tag{5.1}$$

while the magnetic field vector is expressed in terms of the electric field as follows:

$$\vec{H} = -\frac{1}{j\omega\mu}\nabla \times (\hat{z}E_z). \tag{5.2}$$

The governing scalar Helhmoltz equation is obtained from the substitution of the aforementioned expressions for the electric and magnetic fields into Ampère's law

$$\nabla \cdot \frac{1}{\mu_r}\nabla E_z + \omega^2\mu_0\epsilon_0\epsilon_r E_z = 0, \tag{5.3}$$

where μ_r and ϵ_r are, respectively, the position-dependent relative magnetic permeability and electric permittivity in the computational domain.

The above expressions are for the total electric and magnetic fields, which will constitute the unknowns of the two-dimensional boundary value problem (BVP). Two possible formulation of the weak statement of the problem are possible, namely, the total field formulation and the scattered field formulation. Their development is presented next.

5.1.1   Total field formulation

The weak statement of the problem is obtained by multiplying (5.3) with the nodal basis function, ψ, integrating over the computational domain Ω, and making use of the identity (A.5) and the two-dimensional form of the divergence theorem (A.14)

$$\iint_\Omega \nabla\psi \cdot \frac{1}{\mu_r}\nabla E_z \, ds - \omega^2\epsilon_0\mu_0 \iint_\Omega \psi\epsilon_r E_z \, ds - j\omega\mu_0 \int_{\partial\Omega} \psi\hat{z} \cdot \hat{n} \times \vec{H} \, dl = 0. \tag{5.4}$$

On the computational domain boundary, $\partial\Omega$, a first-order absorbing boundary condition (ABC) will be imposed. Clearly, this condition must be applied to the scattered (outgoing) field ($\vec{E}^{sc}, \vec{H}^{sc}$). Hence, with $\hat{n}$ denoting the outward pointing unit normal on $\partial\Omega$, the ABC condition has the form

$$\hat{n} \times \vec{H}^{sc} = \frac{1}{\eta_0}\hat{n} \times \hat{n} \times \vec{E}^{sc}, \tag{5.5}$$

where $\eta_0 = \sqrt{\mu_0/\epsilon_0}$ is the free-space intrinsic impedance. In view of the linearity of the problem, the scattered field can be defined as follows:

$$\vec{E}^{sc} = \vec{E} - \vec{E}^{inc}, \quad \vec{H}^{sc} = \vec{H} - \vec{H}^{inc}, \tag{5.6}$$

where the incident field ($\vec{E}^{inc}, \vec{H}^{inc}$) is defined to be the field in the absence of the inhomogeneous scatterer. With this definition, the ABC condition on $\partial\Omega$ may be recast in terms of the total fields as follows:

$$\hat{n} \times \vec{H} = \frac{1}{\eta}\hat{n} \times \hat{n} \times \vec{E} + \hat{n} \times \vec{H}^{inc} - \frac{1}{\eta}\hat{n} \times \hat{n} \times \vec{E}^{inc}. \tag{5.7}$$

Clearly, any set of consistent electric and magnetic fields that are known everywhere on the boundary $\partial\Omega$ may be used as excitation. However, for simplicity, we consider here one of the most common types of excitations, namely, a uniform, time-harmonic plane wave. For the TM$_z$ case considered the plane wave fields are given by

$$\vec{E}^{inc} = \hat{z}e^{-j\vec{k}_0 \cdot \vec{r}},$$

$$\vec{H}^{inc} = -\frac{1}{j\omega\mu}\nabla \times \vec{E}^{inc} = \hat{x}\frac{\sin\theta}{\eta_0}e^{-j\vec{k}_0 \cdot \vec{r}} - \hat{y}\frac{\cos\theta}{\eta_0}e^{-j\vec{k}_0 \cdot \vec{r}}, \tag{5.8}$$

where θ is the angle of incidence with respect to the x axis of the reference coordinate system and $k_0 = \omega\sqrt{\mu_0\epsilon_0}$ is the free space wavenumber. The wave vector $\vec{k}_0$ is given by

$$\vec{k}_0 = \omega\sqrt{\mu_0\epsilon_0}(\hat{x}\cos\theta + \hat{y}\sin\theta). \tag{5.9}$$

Substitution of (5.7) into (5.4) yields the weak form of the 2D scalar Helmholtz equation,

$$\iint_\Omega \nabla\psi \cdot \frac{1}{\mu_r}\nabla E_z \, ds - \omega^2\epsilon_0\mu_0 \iint_\Omega \psi\epsilon_r E_z \, ds + \frac{j\omega\mu_0}{\eta_0}\int_{\partial\Omega}\psi E_z \, dl$$
$$= \frac{j\omega\mu_0}{\eta_0}\int_{\partial\Omega}\psi\hat{z} \cdot (\eta_0\hat{n} \times \vec{H}^{inc} - \vec{E}^{inc}) \, dl. \tag{5.10}$$

Up to this point no specific mention has been made of the presence of material interfaces at which the electromagnetic properties of the material undergo abrupt changes. At such boundaries, the electromagnetic fields and/or their normal derivatives may exhibit discontinuities that must be properly accounted for in the development of the weak formulation. Pertinent to the enforcement of the appropriate boundary conditions at media interfaces is the boundary integral in (5.4). The way this term is used for the enforcement of boundary conditions is reviewed in the following.

The most commonly encountered types of such boundaries are those associated with the surface of a perfect electric conductor (PEC) and those at which the electric permittivity and/or the magnetic permeability exhibit a discontinuity. For the case of PEC boundaries, and for the TM$_z$ case under consideration, the pertinent boundary condition is $E_z = 0$. It is noted that the boundary of PEC scatterers is included as part of the enclosing boundary of the computational domain. Thus in the integrand of the boundary integral term of (5.4) the tangential magnetic field on the PEC is an unknown quantity. However, since the tangential total electric field (E_z) is zero at the finite element nodes on the PEC boundary, no testing is required for these nodes. In other words, since the total electric field is known on the boundary, the PEC boundary is excluded from the finite element statement of the problem by requiring the testing functions associated with the PEC boundary nodes to be zero.

With regards to the second type of material boundaries, it is immediately evident from the integrand of the boundary integral term of (5.4) that, in addition to the continuity of the tangential electric field at the nodes on the boundary, the continuity of the tangential magnetic field is naturally enforced. Thus, as long as the finite element grid is constructed in a manner such that the material interface is not crossed by any element edges, the appropriate electromagnetic boundary conditions at material interfaces are satisfied in the weak statement of the BVP.

5.1.2 Scattered field formulation

The development of the weak statement for the scattered electric formulation begins by using (5.6) in (5.3) to obtain the following form of the scalar Helmholtz equation for the scattered electric field

$$\nabla \cdot \frac{1}{\mu_r} \nabla E_z^{sc} + \omega^2 \mu_0 \epsilon_0 \epsilon_r E_z^{sc} = -\nabla \cdot \frac{1}{\mu_r} \nabla E_z^{inc} - \omega^2 \mu_0 \epsilon_0 \epsilon_r E_z^{inc}. \qquad (5.11)$$

The incident electric field satisfies the following scalar Helmholtz equation:

$$\nabla \cdot \nabla E_z^{inc} + \omega^2 \mu_0 \epsilon_0 E_z^{inc} = 0. \qquad (5.12)$$

Introducing this equation into the right-hand side of (5.11) yields

$$\nabla \cdot \frac{1}{\mu_r} \nabla E_z^{sc} + \omega^2 \mu_0 \epsilon_0 \epsilon_r E_z^{sc} = -\nabla \cdot \left(\frac{1}{\mu_r} - 1 \right) \nabla E_z^{inc} - \omega^2 \mu_0 \epsilon_0 \left(\epsilon_r - 1 \right) E_z^{inc}. \quad (5.13)$$

The weak statement of this equation is obtained by the integration of its weighted product by the scalar function ψ over the computational domain Ω, followed by the same integration-by-parts step used in the development of (5.4). This, combined with the application of the absorbing boundary condition (5.5) on the truncation boundary $\partial\Omega$, yields

$$\iint_\Omega \nabla\psi \cdot \frac{1}{\mu_r} \nabla E_z^{sc} \, ds - \omega^2 \epsilon_0 \mu_0 \iint_\Omega \psi \epsilon_r E_z^{sc} \, ds + \frac{j\omega\mu_0}{\eta} \int_{\partial\Omega} \psi E_z^{sc} \, dl$$

$$= \iint_\Omega \nabla\psi \cdot \left(\frac{1}{\mu_r} - 1 \right) \nabla E_z^{inc} \, ds - \omega^2 \epsilon_0 \mu_0 \iint_\Omega \psi \left(\epsilon_r - 1 \right) E_z^{inc} \, ds. \qquad (5.14)$$

The imposition of the appropriate electromagnetic boundary conditions at material interfaces in the context of (5.14) follows the process described in the last two paragraphs of the previous subsection. For the case of PEC boundaries the boundary condition for zero tangential electric field assumes the form $E_z^{sc} = -E_z^{inc}$. This essential boundary condition must be imposed in the selection of the expansion functions for the approximation of the scattered field.

This completes the development of the weak statement for the scattered field formulation. A direct comparison of (5.10) with (5.14) reveals the fact that, while in the total field formulation the excitation enters through the imposition of the absorbing boundary condition on the truncation boundary $\partial\Omega$, in the scattered field formulation the excitation appears in the form of equivalent current densities on the PEC boundaries and in the inhomogeneous regions of the computational domain.

5.2   DEVELOPMENT OF THE FINITE ELEMENT SYSTEM

The final step is the development of the linear system of equations that constitutes the finite element approximation of the two-dimensional BVP. A two-dimensional finite element mesh is assumed to be in place over the computational domain, with the edges of its elements tracing all material interfaces and the truncation boundary $\partial\Omega$. For our purposes, a triangular mesh is used. The unknown (scalar) total or scattered electric field is expanded in terms of the nodal basis functions, ϕ_i, discussed in Chapter 2

$$E_z \quad \text{or} \quad E_z^{sc} = \sum_{i=1}^{N_n - N_{Dn}} \phi_i e_i. \qquad (5.15)$$

The expansion coefficient e_i is associated with the $i-$th node in the mesh. N_n is the total number of nodes in the domain, while N_{Dn} is the number of nodes on the PEC boundary.

Application of Galerkin's method, where testing and expansion functions are taken to be the same, i.e., with $(\psi_i = \phi_i)$, in conjunction with the weak statements developed in the previous section, yields the finite element system,

$$Ax = f, \tag{5.16}$$

where A is the finite element matrix, x denotes the vector of the unknown expansion coefficients, and f is the discrete form of the excitation term. The finite element matrix is the same for both the total and the scattered field formulation, with elements given by

$$A_{i,j} = \iint_\Omega \nabla \phi_i \cdot \frac{1}{\mu_r} \nabla \phi_j \, ds - \omega^2 \epsilon_0 \mu_0 \iint_\Omega \phi_i \epsilon_r \phi_j \, ds + \frac{j\omega\mu_0}{\eta_0} \int_{\partial\Omega} \phi_i \phi_j \, dl. \tag{5.17}$$

The right-hand side vector, f, is the source vector and is different for the two formulations. For the total field formulation the elements of f are given by

$$f_i = \frac{j\omega\mu_0}{\eta_0} \int_{\partial\Omega} \phi_i \hat{z} \cdot \left(\eta_0 \hat{n} \times \vec{H}^{inc} - \vec{E}^{inc} \right) dl. \tag{5.18}$$

For the scattered field formulation, the incident field on the PEC boundary is approximated in terms of an expansion in the nodal basis functions associated with the N_{Dn} nodes of the finite element mesh as

$$E_z^{inc} = \sum_{i=N_n-N_{Dn}+1}^{N_n} \phi_i e_i^{inc}. \tag{5.19}$$

In view of the properties of the nodal basis functions ϕ_i, the expansion coefficients, e_i^{inc}, are simply the values of the incident electric field on the nodes on the PEC boundary. In view of (5.14), the forcing term vector will include an approximation of the induced current density in every element in the computational domain where the electromagnetic properties are different from those of the background medium. Thus the elements of f are given by

$$\begin{aligned} f_i = & \iint_\Omega \nabla \phi_i \cdot \left(\frac{1}{\mu_r} - 1 \right) \nabla E_z^{inc} \, ds - \omega^2 \epsilon_0 \mu_0 \iint_\Omega \phi_i \left(\epsilon_r - 1 \right) E_z^{inc} \, ds \\ & + \sum_{j=N_n-N_{Dn}+1}^{N_n} A_{i,j} e_j^{inc}, \end{aligned} \tag{5.20}$$

where $A_{i,j}$ is defined in (5.17).

This completes the development of the finite element system. In the following, we discuss its iterative solution using multigrid methods.

5.3 NESTED MULTIGRID PRECONDITIONER

Continuing from our discussion of the multigrid method in Section 4.6, a nested multigrid process will be presented in this section for the solution of the finite element system of (5.16). In a nested multigrid scheme, several grids of different granularity are utilized. The coarsest grid is generated first, and higher-level (finer) grids are subsequently built by subdividing each triangular element of the grid used in the previous level into four equal-area sub-triangles. This process is illustrated for the case of a two-level, nested multigrid method in Fig. 5.1.

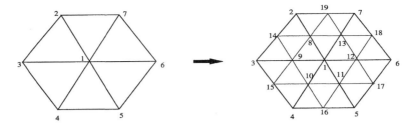

Figure 5.1   Coarse and fine grids used for a two-level, nested multigrid method.

Consider a hierarchy of N discretization levels, with the grid density varying from h at the finest grid level to Nh at the coarsest grid level. At each level, a finite element matrix equation is generated using the results of the previous section,

$$A^i x^{ih} = f^{ih}, \qquad i = 1, 2, \cdots, N. \tag{5.21}$$

Let $i = 1$ denote the finest level and $i = N$ the coarsest level. To obtain the solution of the finite element equation at level i, use is made of the solutions obtained in the coarser grids by means of the *coarse grid correction*. As discussed already in Section 4.6, coarse grid correction relaxes the matrix equation on grid i using a smoother, such as Gauss-Seidel, until convergence deteriorates. Next, its residual equation is projected and solved on the coarser grid $i + 1$. Finally, the solution is projected back to grid i, in order to correct the approximation first obtained there. The pseudo-code description of the process is the one given in Algorithm (4.10).

In the ideal situation, the number of levels N is dictated by our ability to solve the finite element equation at the coarsest grid directly through an LU factorization with reordering. Considering that the number of unknowns is reduced by a factor of 4 from each level to the next coarser level, this should always be possible for the case of scalar, 2D problems. For the rest of the levels, only a few relaxations need to be performed.

Finally, to improve the numerical robustness of the finite element solver, the multigrid technique is used as preconditioner in the iterative solution of the finite element system, using, for example, the conjugate gradient method. We will refer to this preconditioner as the *nested multigrid preconditioner*.

5.4   INTERGRID TRANSFER OPERATORS

In this section we develop the intergrid transfer operators used for projecting the finite element solution from one grid to the next. For the sake of simplicity, the two-level multigrid case of Fig. 5.1 is considered. The pertinent finite element matrix equations are

$$\begin{aligned} A^h x^h &= f^h &&\text{on the fine grid,} \\ A^{2h} x^{2h} &= f^{2h} &&\text{on the coarse grid.} \end{aligned} \tag{5.22}$$

We need to build the *restriction operator*, I_h^{2h}, which maps the residual of the fine grid solution onto the coarse grid, and the *interpolation operator*, I_{2h}^h, which maps the correction obtained in the coarse grid back onto the fine grid.

Let us consider the interpolation operator first. From Fig. 5.1, it is apparent that, once the solution on the coarse grid is calculated, the solution at the fine grid nodes is obtained in one of two ways. For those nodes of the fine grid that coincide with nodes of the coarse grid, a direct transfer of the solution is involved. Otherwise, the solution at a fine-grid node is obtained by averaging the values of the adjacent coarse-grid nodes. For example, with reference to Fig. 5.1, node 1 is a node on both grids. Thus it is

$$x_1^h = x_1^{2h}. \tag{5.23}$$

On the other hand, the value of node 8 in the fine grid is obtained through the equation

$$x_8^h = \frac{1}{2}(x_1^{2h} + x_2^{2h}). \tag{5.24}$$

The interpolation operator matrix I_{2h}^h can be constructed as a linear mapping between the two solution vectors, x^h and x^{2h},

$$x^h = I_{2h}^h x^{2h}. \tag{5.25}$$

Clearly, I_{2h}^h is a $N_n^h \times N_n^{2h}$ matrix, where N_n^h is the number of nodes in the fine grid, while N_n^{2h} is the number of nodes in the coarse grid. Each row of I_{2h}^h is associated with a node on the fine grid and has only one or two non-zero elements. The row associated with a fine-grid node that coincides with a node in the coarse grid has one non-zero element of value 1. The row associated with a fine-grid node that is placed at the middle point of a coarse-grid edge has two non-zero elements, both of value 0.5.

Let us consider I_{2h}^h from the point of view of the nodes in the coarse grid. Each column of I_{2h}^h is associated with a node in the coarse grid. The value of each entry of the column depends on the position of the corresponding fine-grid node relative to the coarse-grid node associated with the column. If the fine-grid node coincides with the coarse-grid node, the element value is 1. If the fine-grid node is at the middle point of a coarse-grid edge that has the coarse-grid node as one of its end points, the element value is 0.5. Otherwise, the element value is zero. For example, the first column of I_{2h}^h, associated with coarse-grid node 1, is

$$I_{2h}^h(:,1) = \begin{bmatrix} 1.0 & 0.5 & 0.5 & 0.5 & 0.5 & 0.5 & 0.5 \end{bmatrix}^T.$$

$$\begin{array}{ccccccc} \uparrow & & \uparrow & \uparrow & \uparrow & \uparrow & \uparrow & \uparrow \\ 1 & & 8 & 9 & 10 & 11 & 12 & 13 \end{array} \tag{5.26}$$

(In the above expression we have borrowed the Matlab® notation $M(:,j)$ to denote the column j of a matrix M.)

Next we consider the construction of the restriction operator. With Fig. 5.1 as reference, two sets of finite element, nodal basis functions are defined, one for the coarse grid,

$$\Phi^{2h} = \begin{bmatrix} \phi_1^{2h}, \phi_2^{2h}, \cdots, \phi_7^{2h} \end{bmatrix} \tag{5.27}$$

and one for the fine grid

$$\Phi^h = \begin{bmatrix} \phi_1^h, \phi_2^h, \cdots, \phi_{19}^h \end{bmatrix}. \tag{5.28}$$

Any nodal basis functions in the coarse grid can be written in terms of the ones in the fine grid, for example,

$$\phi_1^{2h} = \phi_1^h + \frac{1}{2}\left(\phi_8^h + \phi_9^h + \cdots + \phi_{12}^h + \phi_{13}^h\right). \tag{5.29}$$

Thus $\Phi^{2h} \subset \Phi^h$, and a linear relationship exists between the two sets of expansion functions.

It is not difficult to show that the matrix connecting the two sets of basis functions is I_{2h}^h; hence,

$$\Phi^{2h} = \Phi^h Q, \quad Q = I_{2h}^h. \tag{5.30}$$

Observe that the finite element matrix equation (5.10) has the form $A(\Phi^T, \Phi)x = f(\Phi^T)$, with a bilinear form on the left-hand side matrix and a linear form on the right-hand side vector. Starting with the finite element matrix equation on the fine grid,

$$A\left(\left(\Phi^h\right)^T, \Phi^h\right) x^h = f\left(\left(\Phi^h\right)^T\right), \tag{5.31}$$

let us substitute (5.25) and multiply on the left by $\left(I_{2h}^h\right)^T$. This yields

$$\left(I_{2h}^h\right)^T A\left(\left(\Phi^h\right)^T, \Phi^h\right) I_{2h}^h x^{2h} = \left(I_{2h}^h\right)^T f\left(\left(\Phi^h\right)^T\right). \tag{5.32}$$

The bilinear form of $A(\Phi^T, \Phi)$ combined with (5.30) allows us to rewrite the operator on the left-hand side of the above equation as follows:

$$
\begin{aligned}
\left(I_{2h}^h\right)^T A\left(\left(\Phi^h\right)^T, \Phi^h\right) I_{2h}^h x^{2h} &= A\left(\left(\Phi^h I_{2h}^h\right)^T, \Phi^h I_{2h}^h\right) x^{2h} \\
&= A\left(\left(\Phi^{2h}\right)^T, \Phi^{2h}\right) x^{2h}.
\end{aligned} \tag{5.33}
$$

Thus (5.31) becomes

$$A\left(\left(\Phi^{2h}\right)^T, \Phi^{2h}\right) x^{2h} = \left(I_{2h}^h\right)^T f\left(\Phi^h\right). \tag{5.34}$$

A comparison of the resulting equation with the finite element equation on the coarse grid (5.22), yields the following result:

$$f^{2h} = \left(I_{2h}^h\right)^T f^h. \tag{5.35}$$

This implies that the restriction operator I_h^{2h} is actually the transpose of the interpolation operator

$$I_h^{2h} = \left(I_{2h}^h\right)^T. \tag{5.36}$$

In the numerical implementation of the nested multigrid process it is not necessary to store the explicit matrix form of the operators I_{2h}^h and I_h^{2h}. Instead, the interpolation matrix-vector product $I_{2h}^h x^{2h}$ can be performed by traversing all of the nodes and edges in the coarse grid. The following pseudo-code describes the algorithm for interpolating the coarse grid vector p^{2h} to construct the fine-grid vector p^h.

> **Algorithm (5.1): Two-dimensional interpolation operator** $p^h \leftarrow I_{2h}^h p^{2h}$
> 1 for all nodes in the coarse grid
>   1.a $idx^{2h} \leftarrow$ index of the node in p^{2h};
>   1.b $idx^h \leftarrow$ index of the node in p^h;
>   1.c $p^h[idx^h] \leftarrow p^{2h}[idx^{2h}]$;
> 2 for all edges in the coarse grid
>   2.a $idx_1^{2h} \leftarrow$ index of one end-node in p^{2h};
>   2.b $idx_2^{2h} \leftarrow$ index of the other end-node in p^{2h};
>   2.c $idx^h \leftarrow$ index of the node in the middle of the edge in p^h;
>   2.d $p^h[idx^h] \leftarrow 0.5\left(p^{2h}[idx_1^{2h}] + p^{2h}[idx_2^{2h}]\right)$.

Similarly, the two-dimensional restriction operation of $I_h^{2h} p^{2h}$ can be performed by traversing all the nodes and edges in the coarse grid as described by the following algorithm.

Algorithm (5.2): Two-dimensional restriction operator $p^{2h} \leftarrow I_h^{2h} p^h$
```
1 for all nodes in coarse grid
   1.a  idx^2h ← index of the node in p^2h;
   1.b  idx^h ← index of the node in p^h;
   1.c  p^2h[idx^2h] ← p^h[idx^h];
2 for all edges in coarse grid
   2.a  idx_1^2h ← index of one end-node in p^2h;
   2.b  idx_2^2h ← index of the other end-node in p^2h;
   2.c  idx^h ← index of the node in the middle of the edge in p^h;
   2.d  p^2h[idx_1^2h] ← p^2h[idx_1^2h] + 0.5p^h[idx^h];
   2.e  p^2h[idx_2^2h] ← p^2h[idx_2^2h] + 0.5p^h[idx^h].
```

5.5 APPLICATIONS

In this section we present some results from the implementation of the nested multigrid process together with the conjugate gradient method for the iterative solution of finite element approximations of 2D electromagnetic scattering problems. The iterative solver will be referred to as the multigrid-preconditioned conjugate gradient (MGCG) solver. To provide for a reliable assessment of the accuracy of the solver, the sample problems considered possess analytic solutions. The stopping criterion for the iterative solver was taken to be $\|r\|_\infty = \text{stop\_tol} \|f\|_\infty$, where r is the residual and $\text{stop\_tol} = 10^{-6}$. Unless otherwise noted, this stopping criterion for the iterative solution is used throughout the book.

5.5.1 Plane wave scattering by a PEC cylinder

The first example problem is for plane wave scattering from a circular PEC cylinder of radius 0.5m. A circular boundary centered at the center of the PEC cylinder and of radius 1.0m was used to truncate the computational domain. For the coarsest grid, the average edge length is 5.4 cm and the maximum edge length is 8.43 cm. A three-level multigrid process was implemented. The number of unknowns on the three grids is 962, 3788 and 15032, respectively. As expected, the number of unknowns increases by a factor of four as we move from one grid to the next finer one.

The iterative solution was obtained using both the V-cycle and W-cycle processes of Section 4.6. In addition, the three preconditioners presented in Section 4.5, namely, the Jacobi, the symmetric Gauss-Seidel (SGS) and the zeroth-level incomplete LU factorization (ILU(0)), were used as smoothers. The performance of the various implementations, quantified in terms of number of iterations and CPU times, are listed in Table 5.1 for several excitation frequencies. In all cases excellent convergence is observed. The best smoother is the SGS, exhibiting the fastest convergence at all frequencies. A comparison of the V- and W-cycle multigrid schemes suggests the V-cycle algorithm as the most efficient one.

In order to compare the MGCG process with a commonly used iterative solver, the incomplete Cholesky conjugate gradient (ICCG) method was used to solve the problem at 1 GHz. The required number of iterations for convergence was 483 with a total CPU time of 25.36 seconds. Furthermore, it was observed that the convergence of ICCG deteriorated as the electrical size of the cylinder increased.

Another important observation that is made from the data in Table 5.1 is that the iteration number and CPU time increase with the excitation frequency. This is attributed to the wave attributes of the solution. To elaborate, for the coarse grid solution to be of any use to the development of the solution at a finer grid, the coarse grid must be fine enough to constrain the numerical dispersion error [1]-[5]. Otherwise, convergence is jeopardized. This was indeed confirmed by the numerical experiments.

Table 5.1   Test I. Scattering by a PEC Cylinder (15032 Unknowns, $v_1 = v_2 = 2$)

Freq (MHz)		900	1000	1100	1200	1300	1400
λ/h_{max}		3.95	3.56	3.24	2.97	2.74	2.54
λ/h_{avg}		6.17	5.55	5.05	4.63	4.27	3.97
V-cycle	Iteration	37	38	40	44	49	54
(Jacobi)	CPU (s)	8.65	8.82	9.26	10.16	11.22	12.30
W-cycle	Iteration	37	38	40	44	49	54
(Jacobi)	CPU (s)	10.18	10.32	10.81	11.93	13.23	14.50
V-cycle	Iteration	10	12	16	17	20	26
(S-GS)	CPU (s)	2.71	3.13	3.98	4.17	4.81	6.10
W-cycle	Iteration	10	12	16	17	20	26
(S-GS)	CPU (s)	3.12	3.50	4.60	4.84	5.63	7.07
V-cycle	Iteration	14	15	17	21	25	30
(ILU(0))	CPU (s)	5.19	5.50	6.15	7.48	8.78	10.39
W-cycle	Iteration	14	15	17	21	25	30
(ILU(0))	CPU (s)	6.02	6.37	7.12	8.72	10.23	12.17

To provide some specific guidelines for the effective application of the MGCG iterative process, let h_{max} denote the maximum edge length on the coarsest grid and h_{avg} the average edge length on the coarsest grid. For fast convergence, λ/h_{avg} should be larger than 4 or 5 at the maximum frequency in the bandwidth of interest. Considering that for the finest grid $\lambda/h_{avg} \sim 20$, a three-level multigrid should be sufficient for electromagnetic problems of reasonable electrical size. Use of larger number of levels does not seem to improve convergence and tends to increase the complexity and computational cost of the multigrid process.

Since SGS was found to be the most efficient and effective smoother, it will be adopted for all the multigrid preconditioning schemes discussed in later chapters.

What remains to be quantified in the effect of the number of pre- and post-processing operations, v_1 and v_2, respectively, on convergence. Table 5.2 summarizes the results of our study. The best choice for v_1 and v_2 is found to be 2 for two-dimensional scattering problems.

5.5.2   Plane wave scattering by dielectric cylinder

The second example concerns the plane wave scattering by a dielectric cylinder with relative permittivity $\epsilon_r = 9.0$. The radius of the cylinder is 0.5 m, and the radius of the circular truncation boundary is 1.0 m. Inside the dielectric cylinder the average edge length for the coarsest grid is 2.015 cm, while the maximum edge length is 3.151 cm. A three-level multigrid scheme was used in this case also. The number of unknowns for the three grids is

Table 5.2   Test I. Scattering by a PEC Cylinder (15032 Unknowns, V-cycle, SGS)

Freq (MHz)		1000	1200	1400
$v_1 = v_2 = 1$	Iteration	13	19	27
	CPU (s)	2.600	3.531	4.748
$v_1 = v_2 = 2$	Iteration	12	17	26
	CPU (s)	3.160	4.23	6.053
$v_1 = v_2 = 3$	Iteration	12	17	27
	CPU (s)	3.850	5.152	7.859
$v_1 = v_2 = 4$	Iteration	12	17	31
	CPU (s)	4.510	6.160	10.719
$v_1 = v_2 = 5$	Iteration	12	20	38
	CPU (s)	5.207	8.271	15.246

6289, 24945, and 99361, respectively. The results in Table 5.3 document the performance of the MGCG algorithm for this case. It is noted that at the highest frequency (800 MHz) it is $\lambda/h_{max} = 3.97$.

Table 5.3   Test II. Scattering by a Dielectric Cylinder (99361 Unknowns, $v_1 = v_2 = 2$)

Freq (MHz)		500	600	700	800
λ/h_{max}		6.33	5.29	4.533	3.97
λ/h_{avg}		9.27	8.26	7.09	6.2
V-cycle	Iteration	12	17	30	34
(SGS)	CPU (s)	22.908	30.574	50.344	56.484

5.5.3   Plane wave scattering by electrically large cylinders

Next, we return to the case of plane wave scattering by a PEC cylinder. However, for this study we consider excitation frequencies high enough for the cylinder to be electrically large. For example, with the cylinder radius kept at 0.5 m, the electrical length of its diameter is about thirteen wavelengths at 4.0 GHz. The radius of the circular truncation boundary was maintained at 1.0 m. The finite element meshes used were chosen such that the average edge length in the coarsest grid was 1.55 cm, while the maximum edge length is 2.50 cm; hence, it is $\lambda/h_{max} = 2.99$ at 4 GHz. A three-level multigrid method was used. The number of unknowns in the three grids are 11444, 45564, 181832, respectively. The results in Table 5.4 document the performance of the MGCG algorithm for frequencies ranging from 1.5 to 4.0 GHz.

Table 5.4 Test III. Scattering by a PEC Cylinder (181832 Unknowns, $v_1 = v_2 = 2$)

Freq (MHz)		1500	2000	2500	3000	3500	4000
λ/h_{max}		7.98	5.99	7.75	3.99	3.421	2.99
λ/h_{avg}		12.90	9.67	4.79	6.46	5.54	4.84
V-cycle	Iteration	7	9	16	25	42	73
(SGS)	CPU(s)	27.348	33.053	52.619	78.016	126.98	216.75

Like in the previous two test cases, the MGCG process exhibits robust and fast convergence. At this point it is worth mentioning that, in addition to its robustness, the MGCG process requires much less memory compared to other iterative methods with preconditioning. For example, the effectiveness of an ILU-type preconditioner depends on the number of fillings. Therefore, a good ILU preconditioner tends to require significant memory. However, in the case of the MCCG process, the extra matrices that must be generated and stored are the ones for the coarser grids, which are much smaller than that for the finest grid. For example, for the this example, which involved 181832 unknowns at the finest grid, the total memory required (for double-precision arithmetic) was 74 MB.

The plot in Fig. 5.2 illustrates the convergence performance of the three-level, V-cycle, MGCG solution for the case of the electrically-large PEC cylinder.

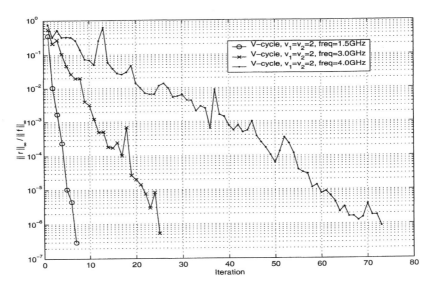

Figure 5.2 Convergence of the three-level V-cycle MCGC for scattering by a circular PEC cylinder.

These examples conclude this brief chapter, which was used to illustrate the basic ideas of the multigrid process in the context of the finite element solution of the scalar Helmholtz equation in two dimensions. For this case the elements involved are nodal elements. This provided for a straightforward and illustrative development of the intergrid transfer operators for the projection of the residual from one grid to its adjacent ones in the context of a hierarchy of nested grids.

The specific implementation considered involved the use of a three-level multigrid pre-conditioned, conjugate gradient (MGCG) iterative process. Despite the simplicity of the scalar two-dimensional problem, several important results were obtained, which will be shown to remain valid and relevant in the general case of multigrid-assisted iterative solution of vector electromagnetic problems in three dimensions.

First, it was found that, provided that the coarse grid discretization is able to resolve with reasonable accuracy the minimum wavelength of interest, MGCG exhibited robust convergence. More specifically, a resolution of 4 or 5 nodes per minimum wavelength at the coarsest grid was found to be sufficient. With such a choice, the three-level multigrid scheme in two dimensions yields $\sim$ 20 nodes per minimum wavelength resolution at the finest-grid level. Second, the simple V-cycle multigrid process one was found to be most robust with regards to convergence and most efficient with regards to CPU time. Finally, the numerical tests considered demonstrated that the MGCG process is more efficient in terms of memory usage compared to other preconditioned iterative techniques.

REFERENCES

1. R. Lee and A. C. Cangellaris "A study of discretization error in the finite element approximation of wave solutions," *IEEE Trans. Antennas Propagat.*, vol. 40, pp. 542-549, May 1992.

2. A. C. Cangellaris and R. Lee, "On the accuracy of numerical wave simulations based on finite methods," *Journal of Electromagnetic Waves and Applications*, vol. 6, pp. 1635-1653, Dec. 1992.

3. G. S. Warren and W. R. Scott, "Numerical Dispersion of Higher Order Nodal Elements in the Finite-Element Method," *IEEE Trans. Antenna Prop.*, vol. 44, no. 3, pp. 317-327, Mar. 1996.

4. R. Mullen and T. Belytschko, "Dispersion analysis of finite element semidiscretizations of the two-dimensional wave equation," *Int. J. Num. Methods Eng.*, vol. 18, pp. 11-29, 1982.

5. W. R. Scott, "Errors due to spatial discretization and numerical precision in the finite-element methods," *IEEE Trans. Antennas, Prop.*, vol. 42, pp. 1565-1570, Nov. 1994.

CHAPTER 6

NESTED MULTIGRID VECTOR AND SCALAR POTENTIAL PRECONDITIONER

In this chapter, the ideas of nested multigrid are extended to the development of robust preconditioners for the fast solution of the matrix equations resulting from the finite element discretization of the general, two- and three-dimensional, vector electrodynamic boundary value problem. In principle, one might argue that the three-dimensional vector Helmholtz equation may be considered as a system of three coupled scalar equations, one for each of the three components of the vector field quantity in a cartesian coordinate system. Thus the development of its finite element approximation can be achieved through the expansion of each component in terms of nodal basis functions. However, such an approach has been found to be problematic in the accurate handling of tangential field continuity at material interfaces and the singular behavior of fields at conducting edges and corners. In addition, such an approach has been found to produce spurious modes that contaminate the numerical solution and may render it useless [1]. These difficulties prompted the utilization of vector finite elements for the expansion of the electric and/or magnetic fields. The so-called tangentially-continuous (TV) vector spaces containing these vector basis functions are discussed in detail in Chapter 2. In this chapter, these spaces are utilized for the development of multigrid preconditioning techniques for the numerically stable and convergent iterative solution of the matrix equations resulting from the finite element approximation of three-dimensional, vector electromagnetic problems.

A detailed discussion of the various factors that impact the convergence of the iterative solution of the finite element matrix in the case of the electromagnetic problem is given in Chapter 3. The following paragraphs serve as a brief overview of these factors, and help us motivate some of the methodologies presented in this chapter toward the development of robust iterative finite element matrix solvers.

The primary reason for the slow convergence of the iterative solution of electromagnetic finite element matrices is the presence of non-zero gradient components in the vector expansion functions. These components are in the null space of the curl operator, thus rendering the stiffness matrix (matrix S in (3.124)) positive-semidefinite [2]. Furthermore, for a driven electromagnetic boundary value problem (BVP) at some (nonzero) frequency, their presence causes, through the matrix $-\omega^2 T$ in (3.124), the finite element matrix to have negative eigenvalues. Hence, the finite element matrix of the driven electromagnetic problem is indefinite, rendering the application of ILU preconditioners for its iterative solution ineffective and unreliable.

These spurious modes correspond to static fields and thus are not part of the electrodynamic solution we are trying to approximate. Thus one way to address the aforementioned convergence difficulty is through their elimination. This can be done through the introduction of an additional static field, which is used to annihilate the spurious static modes through the explicit enforcement of the divergence-free condition. A convenient mathematical framework for the implementation of this adjustment of the finite element solution is the vector and scalar potential formulation of the electromagnetic problem [3]. In this formulation, the electric field is computed in terms of its magnetic vector potential ($\vec{A}$) and electric scalar potential (V) contributions, according to the equation $\vec{E} = \vec{A} + \nabla V$. Tangentially continuous vector spaces are used for the expansion of $\vec{A}$, while scalar spaces are used for the expansion of V.

While the aforementioned ($\vec{A} - V$)-based process successfully eliminates the spurious static solutions, it fails to address the second source of poor convergence in the iterative solution of the finite element matrix, namely, the resolution of the low-frequency, physical eigenstates (modes) of the structure under consideration [4]. As already stated in our introductory discussion of multigrid methods in Chapter 4, the convergence of the iterative solution of the finite element matrix deteriorates or even stalls when the excitation frequency is close to eigenfrequencies of physical eigenstates of the structure that contribute to the electrodynamic solution and thus must be resolved by the finite element solution.

This difficulty can be address through the utilization of the multigrid process in the ($\vec{A} - V$)-based finite element solution of the electromagnetic problem. The result is what we shall call the nested multigrid vector and scalar potential (NMGAV) preconditioner. Based on our earlier discussion of multigrid methods, the NMGAV preconditioner uses the multigrid process to model accurately the low-frequency physical eigenmodes on coarser grids, transform them into finer grids, and, through the use of the smoothing operation on the vector and scalar potential formulation, enforce the elimination of any spurious gradient components from the electromagnetic solution.

The NMGAV process is very similar to the methodology presented in [5] and [6]. Aiming at the same objectives, this methodology uses the Helmholtz decomposition to remove the contribution of the spurious gradient components from the solution during each multigrid relaxation. The Helmholtz decomposition and the potential formulation are mathematically equivalent. However, an important advantage of the NMGAV method is that its mathematical and algorithmic framework are founded on the most popular vector finite element formulation of the electromagnetic problem, namely, the vector electric field formulation ($\vec{E}$ formulation). As elaborated in [3] (see also Chapter 3), the finite element matrices in the potential formulation can be constructed directly and in a very straightforward fashion from those of the field formulation. This attribute is utilized in this chapter to facilitate the understanding of how the application of the multigrid process enables the acceleration of the iterative numerical solution. In particular, it will become evident from the development of the proposed algorithm that NMGAV can be considered as the extension of the single-grid potential algorithm to a multigrid potential algorithm.

The chapter is organized as follows: First, the mathematical framework of the field and potential formulations and the associated finite element equations are considered for the case of two dimensional electromagnetic scattering by cylindrical targets in an unbounded domain. This facilitates a smooth transition from the two-dimensional scalar finite element multigrid process of Chapter 5 to the vector finite element one using triangular elements on a two-dimensional mesh. Finally, the extension of the methodology to the three-dimensional vector problem is presented, utilizing a multigrid process on tetrahedra.

6.1 TWO-DIMENSIONAL ELECTROMAGNETIC SCATTERING

Two-dimensional electromagnetic BVPs, involving geometries and excitations independent of one of the three coordinates in a cartesian coordinate system, can be cast in terms of the scalar Helmholtz equation for the single component of either the electric field, in the transverse magnetic case (TM), or the magnetic field, in the transverse electric (TE) case. For example, with the z axis taken to be the axis of invariance, the multigrid-assisted finite element solution of the TM_z scalar Helmholtz equation for the electric field E_z was considered in Chapter 5. For the TE_z case, a finite element solution can be developed in a similar manner for the scalar Helmholtz equation for H_z. However, as a prelude to the development of a multigrid preconditioner for the three-dimensional, vector electromagnetic BVP, the finite element problem considered next for the TE_z case uses the transverse vector electric field, $\vec{E} = \hat{x}E_x + \hat{y}E_y$, as the unknown field quantity.

6.1.1 Two-dimensional field formulation – TE_z case

With the time dependence $\exp(j\omega t)$ assumed and suppressed for simplicity, the vector Helmholtz equation for the electric field is

$$\nabla \times \frac{1}{\mu_r}\nabla \times \vec{E} - \omega^2\mu_0\epsilon_0\epsilon_r\vec{E} = 0, \tag{6.1}$$

where μ_r and ϵ_r are the position-dependent, relative magnetic permeability and electric permittivity, respectively. Multiplication of (6.8) by the vector basis function $\vec{w}$ and integration over the computational domain Ω yields

$$\iint_\Omega \nabla \times \vec{w} \cdot \frac{1}{\mu_r}\nabla \times \vec{E}\, ds - j\omega\mu_0\int_{\partial\Omega}\hat{n}\times\vec{H}\cdot\vec{w}\, dl - \omega^2\epsilon_0\mu_0\iint_\Omega \vec{w}\cdot\epsilon_r\vec{E}\, ds = 0. \tag{6.2}$$

Without loss of generality and for the sake of simplicity in the mathematical formulation of the weak statement of the problem, let us assume that the boundary, $\partial\Omega$, of the computational domain serves as a truncation boundary on which a first-order absorbing boundary condition (5.7) is imposed for the scattered field. Furthermore, the background homogeneous medium is assumed to be free space. Thus, in view of the linearity of the problem, with the total field defined as the superposition of the incident field, $\vec{E}^{inc}$, which is defined as the field that would exist everywhere in Ω in the absence of the scatterer, and the scattered field, $\vec{E}^{sc}$,

$$\vec{E} = \vec{E}^{inc} + \vec{E}^{sc} \tag{6.3}$$

the weak form of (6.2) becomes

$$
\iint_{\Omega} \nabla \times \vec{w} \cdot \frac{1}{\mu_r} \nabla \times \vec{E} ds + \frac{j\omega\mu_0}{\eta} \int_{\partial\Omega} (\hat{n} \times \vec{w}) \cdot \left(\hat{n} \times \vec{E} \right) dl
$$
$$
- \omega^2 \epsilon_0 \mu_0 \iint_{\Omega} \vec{w} \cdot \epsilon_r \vec{E} ds = \frac{j\omega\mu_0}{\eta} \int_{\partial\Omega} \hat{n} \times \vec{w} \cdot \left(\hat{n} \times \vec{E}^{inc} - \eta \vec{H}^{inc} \right) dl,
$$
(6.4)

where $\eta = \sqrt{\mu_0/\epsilon_0}$ is the intrinsic impedance of the free-space region outside $\partial\Omega$, and $\hat{n}$ is the unit normal vector on $\partial\Omega$. However, the possibility exists for portions of the boundary $\partial\Omega$ to be associated with surfaces on which impedance boundary conditions must be imposed. These involve the tangential component of the electric field and will be imposed via the second term in (6.2). Included in this case is the boundary condition of zero tangential electric field on any portions of the boundary of the computational domain that happen to be perfect electric conductors (PEC). On these portions the relevant boundary condition is

$$
\hat{n} \times \vec{E} = 0.
$$
(6.5)

It is imposed by requiring that the expansion functions for $\vec{E}$ satisfy (6.5) on these portions of the boundary. Equation (6.4), together with any boundary condition statements such as (6.5) for the tangential electric field value on $\partial\Omega$, constitute the weak statement for the total field formulation for $\vec{E}$.

A weak statement for the scattered electric field is also possible. Its development begins with the splitting of the total electric field into its scattered and incident field components in the vector Helmholtz equation,

$$
\nabla \times \frac{1}{\mu_r} \nabla \times \vec{E}^{sc} - \omega^2 \mu_0 \epsilon_0 \epsilon_r \vec{E}^{sc} = -\nabla \times \frac{1}{\mu_r} \nabla \times \vec{E}^{inc} + \omega^2 \mu_0 \epsilon_0 \epsilon_r \vec{E}^{inc}.
$$
(6.6)

However, according to its definition, the incident electric field satisfies the homogeneous, free-space Helmholtz equation

$$
\nabla \times \nabla \times \vec{E}^{inc} - \omega^2 \mu_0 \epsilon_0 \vec{E}^{inc} = 0.
$$
(6.7)

Addition of the left-hand side of the above equation to the right-hand side of (6.6) yields

$$
\nabla \times \frac{1}{\mu_r} \nabla \times \vec{E}^{sc} - \omega^2 \mu_0 \epsilon_0 \epsilon_r \vec{E}^{sc}
$$
$$
= -\nabla \times \left(\frac{1}{\mu_r} - 1 \right) \nabla \times \vec{E}^{inc}_z - \omega^2 \mu_0 \epsilon_0 \left(\epsilon_r - 1 \right) \vec{E}^{inc}.
$$
(6.8)

The weak statement is then obtained by multiplying (6.8) by the vector basis function $\vec{w}$ and integrating over Ω

$$
\iint_{\Omega} \nabla \times \vec{w} \cdot \frac{1}{\mu_r} \nabla \times \vec{E}^{sc} ds + \frac{j\omega\mu_0}{\eta} \int_{\partial\Omega} (\hat{n} \times \vec{w}) \cdot \left(\hat{n} \times \vec{E}^{sc} \right) dl
$$
$$
- \omega^2 \epsilon_0 \mu_0 \iint_{\Omega} \vec{w} \cdot \epsilon_r \vec{E}^{sc} ds = \iint_{\Omega} \nabla \times \vec{w} \cdot \left(\frac{1}{\mu_r} - 1 \right) \nabla \times \vec{E}^{inc} ds
$$
$$
- \omega^2 \epsilon_0 \mu_0 \iint_{\Omega} \vec{w} \cdot (\epsilon_r - 1) \vec{E}^{inc} ds,
$$
(6.9)

where the first-order absorbing boundary condition (5.5) has been imposed on the truncation boundary $\partial\Omega$. The splitting of the total field is used also in conjunction with any surface impedance boundary conditions present in the computational domain. For example, for the case of a PEC surface, the pertinent boundary condition assumes the form

$$\hat{n} \times \vec{E}^{sc} = -\hat{n} \times \vec{E}^{inc}. \tag{6.10}$$

The combination of (6.9) with equations of the form (6.10) pertinent to impedance boundary conditions, constitutes the weak statement for the scattered field formulation.

The two formulations are, in principle, equivalent. However, the manner in which the excitation enters in the weak statement for the two formulations is very different. While in the total field formulation the excitation is introduced through the surface boundary $\partial\Omega$, in the scattered field formulation it is effected in terms of what may be interpreted as distributed electric and magnetic current distributions over those portions of the computational domain with electromagnetic properties different than those of the background medium. Because of the numerical dispersion that results from the spatial discretization of the Helmholtz equation [7], this difference in the way the excitation is imposed may result in discrepancies in the finite element solutions obtained from the two formulations. Whether such a discrepancy is significant depends on several factors, among which average finite element grid size and the electrical size of the structure are the most important. Typically, fine grids that provide for sufficient wavelength resolution at the operating frequency help contain the impact of numerical dispersion error. Thus it is to be expected that the issue of sufficient wavelength resolution by the numerical grid is most pertinent to the choice of the coarsest grid that is acceptable in a nested multigrid process. This issue will be examined in some detail in later sections in this chapter.

6.1.2 The finite element matrix and its properties

To fix ideas, let us consider as excitation a time-harmonic, uniform plane wave with incident electric and magnetic fields given by

$$\begin{aligned}
\vec{E}^{inc} &= (-\hat{x} \sin\theta + \hat{y} \cos\theta)\eta e^{-j\vec{k}_0 \cdot \vec{r}} \\
\vec{H}^{inc} &= \hat{z} e^{-j\vec{k}_0 \cdot \vec{r}},
\end{aligned} \tag{6.11}$$

where θ is the angle of incidence, defining the wave vector $\vec{k}_0 = \omega\sqrt{\mu_0 \epsilon_0}\,(\hat{x}\cos\theta + \hat{y}\sin\theta)$. Let h be the average grid size of the finite element mesh used for the discretization of the computational domain. The total electric field is approximated as follows:

$$\vec{E} = \sum_{i=1}^{N_e - N_{De}} \vec{w}_i e_i, \tag{6.12}$$

where $\vec{w}_i$ is the edge element associated with the i-th edge and e_i is its corresponding expansion coefficient. N_e is the total number of edges in the computational domain, while N_{De} is the number of edges on the portion of the boundary on which the tangential electric field is specified (e.g., a PEC boundary). Application of Galerkin's method yields the following finite element system:

$$M_{EE} x_E = f_E, \tag{6.13}$$

where x_E is the vector of the expansion coefficients of $\vec{E}$, M_{EE} is the finite element matrix, and f_E is the forcing vector. Their entries are deduced from the integral terms in (6.4), which are repeated here for convenience

$$M_{EE,i,j} = \underbrace{\iint_\Omega \nabla \times \vec{w}_i \cdot \frac{1}{\mu_r} \nabla \times \vec{w}_j ds}_{S_{i,j}} + \underbrace{j\omega \frac{\mu_0}{\eta} \int_{\partial\Omega} \hat{n} \times \vec{w}_i \cdot \hat{n} \times \vec{w}_j dl}_{Z_{i,j}}$$

$$- \omega^2 \epsilon_0 \mu_0 \underbrace{\iint_\Omega \vec{w}_i \cdot \epsilon_r \vec{w}_j ds}_{T_{i,j}}, \qquad (6.14)$$

$$f_{E,i} = \frac{j\omega\mu_0}{\eta} \int_{\partial\Omega} \hat{n} \times \vec{w}_i \cdot \left(\hat{n} \times \vec{E}^{inc} - \eta \vec{H}^{inc} \right) dl.$$

Hence, the system matrix M_{EE} may also be written in terms of three matrices as follows:

$$M_{EE}(\omega) = \left(S + j\omega Z - \omega^2 T \right). \qquad (6.15)$$

As we have discussed in Chapter 3, S is positive indefinite. Its zero eigenvalues are attributed to the gradient components, which are contained in the edge elements and are in the null space of the curl operator. More specifically, the number of zero eigenvalues is equal to the total number of nodes excluding the ones on the Dirichlet portion of the boundary. The matrix Z is also positive-indefinite, since its nonzero entries are contributed only by those elements associated with the portion of the truncation boundary on which an absorbing boundary condition is imposed. Finally, T is a positive definite, provided $\epsilon_r > 0$ inside the computational domain.

The convergence of the iterative process depends on the spectrum of the matrix M_{EE}. Let us begin by examining the eigenvalues of M_{EE}, considering first the case where no radiation boundary is present. Thus the finite element matrix equation assumes the reduced form

$$M_{EE}(\omega)x = (S - \omega^2 T)x = f_E. \qquad (6.16)$$

As already discussed in Chapter 3, three sets of eigenmodes are identified as the ones who impact the convergence of the iterative solution of the above system [10]. *Type-A* eigenmodes are the ones associated with the eigenvectors of S with zero eigenvalues. As mentioned above, the number of these eigenmodes is equal to the number of vertices in the mesh, N_n, excluding the ones on the Dirichlet portion of the boundary, N_{Dn}. The corresponding eigenvectors of S are contained in the column vectors of the gradient matrix G, the size of which size is $(N_e - N_{De}) \times (N_n - N_{Dn})$. Clearly, the number of the zero eigenvalues of S is large, especially for electrically large domains.

Type-B eigenmodes are the approximations of those of the physical resonant modes of the electromagnetic domain under consideration that are resolved accurately by the finite element grid. These eigenmodes are obtained from the eigenvalue statement of (6.16), obtained by setting the forcing vector to zero

$$(S - \omega_i^2 T)x_i = 0. \qquad (6.17)$$

The number of these eigenmodes depends on the electrical size of the structure under consideration and the element size in the finite element grid. For a grid of average element size h and with the assumption that a resolution of ten elements per wavelength provides for an accurate approximation of the wave response on the grid, the structure eigenmodes that are approximated with good accuracy are within a frequency bandwidth with an approximate frequency upper bound of $\left((10h)\sqrt{\mu\epsilon} \right)^{-1}$.

Type-C eigenmodes are the finite element approximations of the high-frequency physical modes of the structure, with eigenfrequencies greater than $((10h)\sqrt{\mu\epsilon})^{-1}$. These modes are poorly resolved on the finite element grid; hence, there is some loss of accuracy in their approximation which is, nevertheless, acceptable since their contribution to the solution sought is not dominant. Furthermore, due to their coarse sampling by the grid, they appear highly oscillatory, lacking the smoothness exhibited by the finite element approximations of the well resolved, lower-frequency, physical modes.

Next we consider the eigenvalues of M_{EE} at a specific excitation frequency ω_0. The relevant matrix eigenvalue equation statement is

$$M_{EE}(\omega_0)\tilde{x}_i = \lambda_i \tilde{x}_i, \tag{6.18}$$

where $(\lambda_i, \tilde{x}_i)$ denotes one of the eigen-pairs of M_{EE}. We can use the generalized eigenpairs (ω_i^2, x_i) of (6.17) to establish bounds for the eigenvalues λ_i. We continue to assume that no absorbing boundary condition is imposed on the enclosing boundary and thus $Z = 0$ in (6.15). The Rayleigh quotient, $\rho(x_i)$, for $M_{EE}(\omega_0)$ is

$$\rho(x_i) = \frac{x_i^T (S - \omega_0^2 T) x_i}{x_i^T x_i} = (\omega_i^2 - \omega_0^2)\frac{x_i^T T x_i}{x_i^T x_i}. \tag{6.19}$$

Let τ_{min} and τ_{max} denote, respectively, the minimum and maximum eigenvalues of T. Use of the Courant-Fisher minmax theorem [4] yields

$$\mathrm{sgn}\lambda_i = \mathrm{sgn}(\omega_i^2 - \omega_0^2), \qquad |\omega_i^2 - \omega_0^2|\tau_{min} \le |\lambda_i| \le |\omega_i^2 - \omega_0^2|\tau_{max}. \tag{6.20}$$

However, for a fairly uniform finite element mesh, the eigenvalues of the mass matrix T will be clustered, and thus τ_{min} and τ_{max} are of commensurate value. Thus we conclude from (6.20) that the eigenvalues of $M_{EE}(\omega_0)$ are approximately obtained from those of S through a shift by ω_0^2.

This last result provides us with the ability to examine in a more concrete manner the impact of the aforementioned three types of eigenvalues on the convergence of its solution through an iterative solver. To begin with, the type-A eigenvalues of (6.17) appear, due to the negative shift by ω_0^2, as negative eigenvalues of (6.18). Thus the matrix $M_{EE}(\omega_0)$ becomes indefinite and ill-conditioned, especially for higher values of the excitation frequency ω_0. This, in turn, causes the poor convergence of the iterative solution of the finite element matrix. However, type-A eigenmodes are not solenoidal; hence, the associated electric fields do not satisfy the divergence-free equation,

$$\nabla \cdot \vec{D} = 0. \tag{6.21}$$

Consequently, the potential formulation that was proposed in [3] for improving the convergence of the iterative solution may be interpreted physically as the introduction of a charge distribution to cancel the spurious, non-solenoidal fields of these eigenmodes. This is done by explicitly imposing (6.21) in the weak statement of the problem.

Type-B eigenmodes of (6.17) are also shifted by ω_0^2. Depending on the value of ω_0, the eigenvalues of some of these eigenmodes may also become negative. This is another factor contributing to the poor convergence of iterative FEM solvers. To elaborate, as stated earlier, type-B eigenmodes are lower-eigenfrequency, physical modes that are well resolved by the finite element grid. If the frequency of the driven problem is close to the eigenfrequency of one of these modes, the FEM matrix is close to being singular and convergence of the iterative solution deteriorates. Furthermore, if the frequency of the driven problem is higher than the eigenfrequency of one of these modes, the negative shift contributed by the mass matrix T_h renders the FEM matrix indefinite. Thus convergence of the iterative solver deteriorates.

Since type-B eigenmodes are physical, solenoidal modes relevant to the electrodynamic solution, the remedy provided by the potential formulation through the explicit enforcement of (6.21) in the weak statement of the problem cannot be used to tackle the aforementioned numerical difficulties. This is where the multigrid method makes its contribution. More specifically, in the spirit of multigrid, a coarser grid is introduced to calculate in a stable and accurate manner all of the type-B modes that are significantly oversampled on the original grid. Subsequently, through an interpolation process, the computed modes are projected back onto the original grid.

Finally, the high-spatial-frequency, type-C modes must be considered. Of interest are those modes with wavelengths spanning a few elements of the finite element grid. Thus, even though not very well resolved by the grid, their contribution to the solution must be considered. As we already show in the discussion of multigrid methods in Section 4.6, the contribution of these modes to the finite element solution is captured easily through the use of an inexpensive local smoother, such as Jacobi, Gauss-Seidel, or incomplete LU factorization.

In summary, the discussion of the properties of the three types of eigenmodes of the finite element matrix suggests that an effective iterative solver can be founded on the combination of the vector and scalar potential formulation with the multigrid process. How this combination is effected is described in the following subsections.

6.1.3   Two-dimensional potential formulation

The potential formulation of the electromagnetic BVP and its finite element approximation was first described in [3]. For the purposes of our discussion, we consider the development of the weak form of the problem for the case of an unbounded domain, on the boundary of which the first-order absorbing boundary condition for the scattered electric field is imposed.

Adopting the notation of [3], the electrical field is written in terms of the vector magnetic and scalar electric potentials in the form $\vec{E} = \vec{A} + \nabla V$. Substitution of this representation into (6.4) yields the weak form of the equation

$$
\iint_\Omega \nabla \times \vec{w} \cdot \frac{1}{\mu_r} \nabla \times \vec{A} \, ds + \frac{j\omega\mu_0}{\eta} \int_{\partial\Omega} \hat{n} \times \vec{w} \cdot \hat{n} \times \left(\vec{A} + \nabla V \right) dl
$$
$$
- \omega^2 \epsilon_0 \mu_0 \iint_\Omega \vec{w} \cdot \epsilon_r \left(\vec{A} + \nabla V \right) ds \qquad (6.22)
$$
$$
= \frac{j\omega\mu_0}{\eta} \int_{\partial\Omega} \hat{n} \times \vec{w} \cdot \left(\hat{n} \times \vec{E}^{inc} - \eta \vec{H}^{inc} \right) dl.
$$

The presence of the ungauged scalar potential term ∇V in the above equation is exploited to annihilate the spurious DC components introduced by the type-A eigenmodes of the FEM matrix. This is done through the explicit imposition of (6.21) via an appropriate weak form. Multiplication of (6.21) with a scalar testing function ϕ and integration over Ω yields

$$
\iint_\Omega \phi \nabla \cdot \vec{D} \, ds = \iint_\Omega \left\{ \nabla \cdot \left[\phi \epsilon_r \left(\vec{A} + \nabla V \right) \right] - \nabla \phi \cdot \epsilon_r \left(\vec{A} + \nabla V \right) \right\} ds = 0. \quad (6.23)
$$

Use is made of Gauss' theorem to recast the first term in the above equation in terms of a surface integral over the enclosing boundary

$$\iint_\Omega \nabla \cdot \left(\phi \epsilon_r \vec{E} \right) ds = \int_{\partial\Omega} \phi \hat{n} \cdot \epsilon_r \vec{E} dl = \frac{1}{j\omega\epsilon_0} \int_{\partial\Omega} \phi \hat{n} \cdot \nabla \times \vec{H} dl. \tag{6.24}$$

Making use of well-known vector identities the last integral is re-written as follows:

$$\begin{aligned}
\frac{1}{j\omega\epsilon_0} \int_{\partial\Omega} & \phi \hat{n} \cdot \nabla \times \vec{H} dl = \frac{1}{j\omega\epsilon_0} \int_{\partial\Omega} \nabla\phi \cdot \hat{n} \times \vec{H} dl \\
&= \frac{1}{j\omega\epsilon_0} \int_{\partial\Omega} \nabla\phi \cdot \left(\frac{1}{\eta} \cdot \hat{n} \times \hat{n} \times \vec{E} + \hat{n} \times \vec{H}^{inc} - \frac{1}{\eta} \cdot \hat{n} \times \hat{n} \times \vec{E}^{inc} \right) dl \\
&= \frac{-1}{j\omega\epsilon_0\eta} \int_{\partial\Omega} \hat{n} \times \nabla\phi \cdot \hat{n} \times \vec{E} - \frac{1}{j\omega\epsilon_0} \int_{\partial\Omega} \hat{n} \times \nabla\phi \cdot \vec{H}^{inc} dl \\
&\quad + \frac{1}{j\omega\epsilon_0\eta} \int_{\partial\Omega} \hat{n} \times \nabla\phi \cdot \hat{n} \times \vec{E}^{inc} dl.
\end{aligned} \tag{6.25}$$

Substituting this result into (6.23) and multiplying by $\omega^2\epsilon_0\mu_0$ yields the following weak statement of the equation that describes the enforcement of the divergence-free requirement of the solution

$$\begin{aligned}
\frac{j\omega\mu_0}{\eta} \int_{\partial\Omega} \hat{n} \times \nabla\phi \cdot \hat{n} \times \left(\vec{A} + \nabla V \right) dl &- \omega^2\epsilon_0\mu_0 \iint_\Omega \nabla\phi \cdot \epsilon_r \\
\left(\vec{A} + \nabla V \right) ds &= \frac{j\omega\mu_0}{\eta} \int_{\partial\Omega} \hat{n} \times \nabla\phi \cdot \left(\hat{n} \times \vec{E}^{inc} - \eta\vec{H}^{inc} \right) dl.
\end{aligned} \tag{6.26}$$

Thus the weak statement of the electromagnetic BVP is given by the system of (6.22) and (6.26). The finite element approximation is obtained from this system through a Galerkin process. The resulting system and may be written in matrix form as follows:

$$\begin{bmatrix} M_{AA} & M_{AV} \\ M_{VA} & M_{VV} \end{bmatrix} \begin{bmatrix} x_A \\ x_V \end{bmatrix} = \begin{bmatrix} f_A \\ f_V \end{bmatrix}, \tag{6.27}$$

where

$$\begin{aligned}
M_{AA} &= M_{EE}, \quad M_{VA} = M_{AV}{}^T, \quad f_A = f_E \\
M_{AV i,j} &= \frac{j\omega\mu_0}{\eta} \int_{\partial\Omega} \hat{n} \times \vec{w}_i \cdot \hat{n} \times \nabla\phi_j \, dl - \omega^2\epsilon_0\mu_0 \iint_\Omega \vec{w}_i \cdot \epsilon_r\nabla\phi_j \, ds \\
M_{VV i,j} &= \frac{j\omega\mu_0}{\eta} \int_{\partial\Omega} \hat{n} \times \nabla\phi_i \cdot \hat{n} \times \nabla\phi_j \, dl - \omega^2\epsilon_0\mu_0 \iint_\Omega \nabla\phi_i \cdot \epsilon_r\nabla\phi_j \, ds \\
f_{V i} &= \frac{j\omega\mu_0}{\eta} \int_{\partial\Omega} \hat{n} \times \nabla\phi_i \cdot \left(\hat{n} \times \vec{E}^{inc} - \eta\vec{H}^{inc} \right) dl.
\end{aligned} \tag{6.28}$$

To further examine the properties of these matrices, let Φ denote the space spanned by the scalar, node element, basis functions ϕ, and $\vec{W}$ the space spanned by the vector, edge-element, basis functions $\vec{w}$. Recall from Chapter 2 that the set $\nabla\Phi$ (Whitney-0 form), formed by the gradients of the elements of Φ, is a subset of $\vec{W}$ (Whitney-1 form). As discussed in Chapter 2, Whitney-0 form is the lowest-order scalar space, containing the zero- and first-order scalar components; Whitney-1 form is the lowest-order, tangentially continuous vector space, containing the zero-order gradient components and first-order rotational components. Moreover, as elaborated in Chapter 2, this subset relationship holds not only for the two lowest-order spaces but also for higher-order spaces. Since $\nabla\Phi$ is a subset of $\vec{W}$, there exists a linear relationship between them, which is of the form

$$\nabla\Phi = \vec{W}\,G, \tag{6.29}$$

where G, the transition matrix between the two sets, was defined as the gradient matrix in Chapter 2. This relationship, combined with the bilinear form of the matrices of the finite element approximation, leads to the following equalities:

$$\begin{aligned}
M_{AV}(\vec{W}^T, \nabla\Phi) &= M_{AA}(\vec{W}^T, \vec{W})\,G, \\
M_{VA}(\nabla\Phi^T, \vec{W}) &= G^T\,M_{AA}(\vec{W}^T, \vec{W}), \\
M_{VV}(\nabla\Phi^T, \nabla\Phi) &= G^T\,M_{AA}(\vec{W}^T, \vec{W})\,G, \\
f_V(\nabla\Phi^T) &= G^T f_A(\vec{W}^T).
\end{aligned} \tag{6.30}$$

In view of these equalities and the fact that $M_{AA} = M_{EE}$, the finite element matrix of (6.27) can be written in terms of the finite element matrix M_{EE} of the $\vec{E}$ formulation [3]

$$\begin{aligned}
\begin{bmatrix} M_{AA} & M_{AV} \\ M_{VA} & M_{VV} \end{bmatrix} &= \begin{bmatrix} I \\ G^T \end{bmatrix} M_{EE} \begin{bmatrix} I & G \end{bmatrix}, \\
\begin{bmatrix} f_A \\ f_V \end{bmatrix} &= \begin{bmatrix} I \\ G^T \end{bmatrix} f_E,
\end{aligned} \tag{6.31}$$

where I is the identity matrix.

This result suggests that the preconditioner for an iterative solver for the finite element matrix for the $\vec{E}$-field formulation can be effected through the solution of the finite element system of the potential formulation. To explain, let us consider, step by step, the calculation of the pseudo-residual equation during the iterative solution of the finite element matrix of the $\vec{E}$ formulation. First, the equation

$$M_{EE}x_E = f_E \tag{6.32}$$

is used to solve for the expansion coefficients of $\vec{E}$. This is followed by a *correction* step, associated with the imposition of the weak form of the divergence-free requirement and involving the solution of the equation

$$M_{VV}x_V = G^T(f_A - M_{AA}x_A). \tag{6.33}$$

The solution of these two equations is effected either through an incomplete Cholesky factorization or through the Gauss-Seidel method. The potential solution is transformed back to the electric field solution through the matrix relationship between the unknown vectors in the two formulations

$$x_E = \begin{bmatrix} I & G \end{bmatrix} \begin{bmatrix} x_A \\ x_V \end{bmatrix}. \tag{6.34}$$

The above process can be summarized in terms of the following algorithm for the single-grid, vector and scalar potential ($SGAV$) preconditioner:

Algorithm (6.1): Single Grid Potential Preconditioner $x_E \leftarrow SGAV(f_E)$
1 $f_A \leftarrow f_E$, $x_A \leftarrow 0$

```
2 Pre-smooth M_AA x_A = f_A using forward GS v times.
3 f_V ← G^T (f_A - M_AA x_A), x_V ← 0
4 Pre-smooth M_VV x_V = f_V using forward GS v times.
5 Post-smooth M_VV x_V = f_V using backward GS v times.
6 f_A ← (f_A - M_AA G x_V).
7 Post-smooth M_AA x_A = f_A using backward GS v times.
8 x_E ← x_A + G x_V
```

Removing steps 3-6, and noting that $M_{AA} = M_{EE}$ and $f_A = f_E$, the above SGAV algorithm is simply the S-GS preconditioner for the $\vec{E}$-field formulation. The added steps 3-6 resolve the gradient components in the $\vec{E}$ field. The matrices G and G^T need not be generated explicitly. Rather, what is needed are only the matrix-vector-products involving these two matrices. Making use of the properties of the gradient matrix G, discussed in (2.125), the matrix-vector-product operations are performed using the following two algorithms. p_E is the vector of length the number of edges in the grid. p_V is the vector of length the number of nodes.

Algorithm (6.2): $p_V ← G^T p_E$
```
1 for all the edge (m, n | m < n)
   1.a idx_E ← index of edge (m, n) in p_E
   1.b idx_V ← index of node m in p_V
   1.c p_V[idx_V] += p_E[idx_E]
   1.d idx_V ← index of node n in p_V
   1.e p_V[idx_V] -= p_E[idx_E]
```

Algorithm (6.3): $p_E ← G p_V$
```
1 for all the edge (m, n | m < n)
   1.a idx_E ← index of edge (m, n) in p_E
   1.b idx_V ← index of node m in p_V
   1.c p_E[idx_E] += p_V[idx_V]
   1.d idx_V ← index of node n in p_V
   1.e p_E[idx_E] -= p_V[idx_V]
```

In summary, the aforementioned process enables us to deal effectively with the spurious DC modes and the loss of convergence caused by the presence of type-A eigenmodes. Instead of dealing with the $\vec{E}$-field formulation matrix M_{EE} in the solution of the pseudo-residual equation, the potential formulation is used to guarantee that the solution obtained does not contain any spurious, non-solenoidal fields.

However, a preconditioner based on the single-grid potential formulation is still not robust because of the convergence difficulties associated with the well-resolved by the finite element grid type-B eigenmodes. Thus it is necessary to enhance it through the implementation of a multigrid process, to enable the calculation of the type-B modes on coarser grids and thus improve the convergence properties of the iterative matrix solution. The way the multigrid process is implemented is described in the next two subsections.

6.1.4   Nested multigrid potential preconditioner

The nested multigrid process, where the coarse grid is generated first and higher-level (finer) grids are built by subdividing each triangular element of the previous level into four equal-area subtriangles, is the process we adopt for the development of the multigrid potential preconditioner [9].

Consider a hierarchy of N discretization levels, with average element size varying from Nh, at the lowest (coarsest grid) level, up to h at the highest (finest grid) level. At each level, a finite element matrix equation is generated,

$$M_{EE}^{ih} x_E^{ih} = f_E^{ih}, \qquad i = 1, 2, \cdots N. \tag{6.35}$$

The solution of the fine-grid matrix equation makes use of the solutions of the coarser-grids matrix equations using the *coarse grid correction* discussed in Chapter 4. Recall that the idea behind coarse grid correction is to relax the matrix equation on the fine grid using a smoother, such as Gauss-Seidel, until convergence deteriorates, and then restrict the residual down to the next coarser grid. Once the finite element solution has been obtained on the coarser grid, it is interpolated back onto the finer grid in order to correct the earlier approximation of the solution on this grid. The matrix equation on the lowest (coarsest) grid M_{EE}^{Nh} is solved directly by LU factorization.

The pseudo-code description of the nested multigrid potential (NMGAV) preconditioning algorithm is the following, typical V-cycle scheme:

Algorithm (6.4): Nested Multigrid Potential Preconditioner
$$x_E^h \leftarrow \text{NMGAV} \left(f_E^h, n = 1 \right)$$
1 If $n == N$, then solve $M_{EE}^{Nh} x_E^{Nh} = f_E^{Nh}$ exactly.
2 else
   2.a Pre-smooth v times on potential formulation.
      2.a.1 Pre-smooth $M_{EE}^{nh} x_E^{nh} = f_E^{nh}$ using forward GS v times
      2.a.2 $f_V^{nh} \leftarrow G^{nT} \left(f_E^{nh} - M_{EE}^{nh} x_E^{nh} \right),\ x_V^{nh} \leftarrow 0$
      2.a.3 Pre-smooth $M_{VV}^{nh} x_V^{nh} = f_V^{nh}$ using forward GS v times
      2.a.4 $x_E^{nh} \leftarrow x_E^{nh} + G^n x_V^{nh}$
   2.b $f_E^{(n+1)h} \leftarrow I_{nh}^{(n+1)h} (f_E^{nh} - M_{EE}^{nh} x_E^{nh})$
   2.c $x_E^{(n+1)h} \leftarrow \text{NMGAV} \left(f_E^{(n+1)h}, n + 1 \right)$
   2.d $x_E^{nh} \leftarrow x_E^{nh} + I_{(n+1)h}^{nh} x_E^{(n+1)h}$
   2.e Post-smooth v times on potential
      formulation.
      2.e.1 $f_V^{nh} \leftarrow G^{nT} (f_E^{nh} - M_{EE}^{nh} x_E^{nh}),\ x_V^{nh} \leftarrow 0$
      2.e.2 Post-smooth $M_{VV}^{nh} x_V^{nh} = f_V^{nh}$ using backward GS v times
      2.e.3 $f_E^{nh} \leftarrow \left(f_E^{nh} - M_{EE}^{nh} (G^n)^T x_V^{nh} \right)$
      2.e.4 Post-smooth $M_{EE}^{nh} x_E^{nh} = f_E^{nh}$ using backward GS v times
      2.e.5 $x_E^{nh} \leftarrow x_E^{nh} + G^n x_V^{nh}$

As already stated, this algorithm assumes a total of N nested grids, with level 1 representing the finest grid of average grid size h, and level N representing the coarsest one with average grid size Nh. G^n denotes the gradient matrix at the nth level. The matrix-vector-product operation with G^n and its transpose is discussed in Algorithm (6.2) and (6.3). The operator $I_{nh}^{(n+1)h}$ is the restriction operation that maps the residual of the n-th grid onto the $n+1$-st grid. $I_{(n+1)h}^{nh}$ is the interpolation operator that maps onto the n-th grid the correction obtained on the $n+1$-st grid. The construction of the two intergrid operators is discussed in the next section. The smoothing operation of 2.a and 2.e has to be done in a way similar to the single-level preconditioner, which is to use the potential formulation to eliminate the contribution caused by the nonphysical gradient eigenmodes.

Without Steps 2.a.4, 2.b-2.d, and 2.e.1, the NMGAV algorithm is identical to the SGAV algorithm. Steps 2.b-2.d constitute the coarse-grid-correction. Step 2.a.4 deflates the equations from potential formulation to field formulation so that the intergrid transfer operation is performed only for the field vector x_E^h. Step 2.e.1 inflates the equation from field formulation back to potential formulation so that the post-smoothing operation can be performed on the unknowns associated with the potential formulation.

A multigrid algorithm is discussed in [5]. Whereas NMGAV utilizes the vector-scalar potential formulation to remove the gradient components from the finite element solution, the process in [5] makes use of the Helmholtz decomposition to project the solution of the multigrid cycle onto the non-gradient space. Both processes are equivalent in the sense that they both render the solution divergence-free. The potential formulation used in NMGAV has the advantage that it can be most conveniently implemented in the numerous electro-dynamic FEM solvers that are based on the popular $\vec{E}$ formulation using edge elements. As evident from the development in the previous paragraphs, the implementation of the NMGAV algorithm relies entirely on the finite element matrices of the $\vec{E}$ formulation, with the only exception of the additional gradient matrix G used for the transformation between the field formulation and the potential formulation, which is easily constructed for edge elements as described in [10].

6.1.5 Two-dimensional intergrid transfer operators

In this section the construction of the intergrid transfer operators is presented. For the sake of simplicity, the development is presented for the case of a two-level multigrid with corresponding matrix equations

$$
\begin{aligned}
M_{EE}^h x_E^h &= f_E^h && \text{on the fine grid,} \\
M_{EE}^{2h} x_E^{2h} &= f_E^{2h} && \text{on the coarse grid.}
\end{aligned}
\tag{6.36}
$$

Any basis function in the coarse grid can be represented by the linear combination of basis functions in its nested fine grid. Let $\vec{W}^{2h}$ denote the vector space spanned by the vector edge-element basis functions in the coarse grid, and $\vec{W}^h$ denote the vector space spanned by the vector edge-element basis functions in the fine grid. The basis function in the coarse grid can be expressed as the linear combination of basis functions in the fine grid with appropriate coefficients. Thus the space $\vec{W}^{2h}$ is a subset of $\vec{W}^h$, and we write

$$
\vec{W}^{2h} = \vec{W}^h Q,
\tag{6.37}
$$

where Q is the transition matrix between the two sets of vector spaces. Since the approximation of electric field vector $\vec{E}$ is the same in both vector spaces,

$$
\vec{E}^h = \vec{E}^{2h} \quad \Rightarrow \quad \vec{W}^h x_E^h = \vec{W}^{2h} x_E^{2h}.
\tag{6.38}
$$

Substitution of (6.37) into (6.38) yields

$$
\vec{W}^h \left(x_E^h - Q \, x_E^{2h} \right) = 0 \quad \Rightarrow \quad x_E^h = Q \, x_E^{2h}.
\tag{6.39}
$$

Hence, the interpolation operator I_{2h}^h is the transition matrix Q.

Next, we consider the finite element matrix of the $\vec{E}$-formulation (6.13). It has the form

$$
M_{EE}^h (\vec{W}^{h^T}, \vec{W}^h) x_E^h = f_E^h (\vec{W}^{h^T}),
\tag{6.40}
$$

where the left-hand side is a bilinear form, whereas the right-hand side is a linear form. Substitution of (6.39) into this equation and multiplication on the left by Q^T yields

$$
Q^T M_{EE}^h (\vec{W}^{h^T}, \vec{W}^h) Q x^{2h} = Q^T f_E^h (\vec{W}^{h^T}).
\tag{6.41}
$$

The bilinear form of M_{EE}^h, combined with (6.37), allows us to rewrite the operator on the left-hand side of the above equation as follows:

$$Q^T M_{EE}^h(\vec{W}^{h^T}, \vec{W}^h)Q = M_{EE}^{2h}(\vec{W}^{2h^T}, \vec{W}^{2h}). \tag{6.42}$$

Thus (6.41) becomes

$$M_{EE}^{2h}(\vec{W}^{2h^T}, \vec{W}^{2h})x_E^{2h} = Q^T f_E^h(\vec{W}^{h^T}). \tag{6.43}$$

A comparison of the resulting equation with the finite element equation for the coarse grid (6.36), yields

$$f_E^{2h} = Q^T f_E^h. \tag{6.44}$$

Consequently, the restriction operator I_h^{2h} is Q^T. We conclude that the two intergrid operators are closely related with the transition matrix Q between the two sets of basis functions on the two grids.

Next, we discuss the construction of Q and its transpose. First consider the edge element $(1, 2)$ (associated with the edge pointing from node 1 to node 2) in the coarse grid triangle shown on the left side of Fig. 6.1. Shown on the right side of the same figure are the edge elements associated with the four triangles resulting from the coarse grid triangle through one level of grid refinement. The midpoints of the three edges of the coarse triangle provide the additional nodes used for the refinement. The weighting coefficients for these edge elements, used in their linear combination for the development of the coarse-grid edge element $(1, 2)$, are also shown.

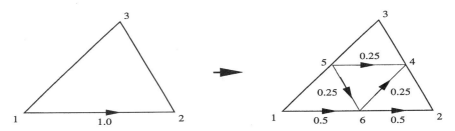

Figure 6.1   Transition between basis functions in coarse and fine grids; edge element $(1, 2)$.

Let $\vec{w}_{m,n}^{h/2h}$ denote the edge element pointing from node m to n, while the superscript h or $2h$ will be used to denote, respectively, an element on the fine or coarse grid. Referring to Fig. 6.1 we have

$$\vec{w}_{1,2}^{2h} = 0.5 \times \left(\vec{w}_{1,6}^h + \vec{w}_{6,2}^h\right) + 0.25 \times \left(\vec{w}_{5,6}^h + \vec{w}_{6,4}^h + \vec{w}_{5,4}^h\right). \tag{6.45}$$

Next we need to show that this transition equation holds true in each one of the four elements in the fine grid. Toward this objective, let us consider triangle $(4, 5, 6)$. It is

$$
\begin{aligned}
0.25 \times \left(\vec{w}_{5,6}^h + \vec{w}_{6,4}^h + \vec{w}_{5,4}^h\right) &= 0.25 \times (\lambda_5 \nabla \lambda_6 - \lambda_6 \nabla \lambda_5) \\
&\quad + 0.25 \times (\lambda_6 \nabla \lambda_4 - \lambda_4 \nabla \lambda_6) \\
&\quad + 0.25 \times (\lambda_5 \nabla \lambda_4 - \lambda_4 \nabla \lambda_5).
\end{aligned} \tag{6.46}
$$

From the geometric relationship between the triangles of the fine and coarse grids, the following identities hold in triangle $(4, 5, 6)$:

$$\begin{array}{lll}
\lambda_5 = 1 - 2\lambda_2 & \Rightarrow & \nabla\lambda_5 = -2\nabla\lambda_2 \\
\lambda_4 = 1 - 2\lambda_1 & \Rightarrow & \nabla\lambda_4 = -2\nabla\lambda_1 \\
\lambda_4 + \lambda_5 + \lambda_6 = 1 & \Rightarrow & \nabla\lambda_4 + \nabla\lambda_5 + \nabla\lambda_6 = 0.
\end{array} \qquad (6.47)$$

Making use of these identities, (6.46) yields

$$0.25 \times \left(\vec{w}_{5,6}^h + \vec{w}_{6,4}^h + \vec{w}_{5,4}^h\right) = \lambda_1\nabla\lambda_2 - \lambda_2\nabla\lambda_1 = \vec{w}_{1,2}^{2h}. \qquad (6.48)$$

Thus (6.45) is valid in triangle $(4, 5, 6)$. Its validity for the remaining three fine-grid triangles can be deduced in a similar manner.

Next we consider the edge element $(1, 3)$ in the coarse grid, depicted on the left in Fig. 6.2. The edge elements in the fine grid and their weighting coefficients that used for the construction of edge element $(1, 3)$ as a linear combination of fine-grid edge elements are shown on the right plot in Fig. 6.2. The pertinent equation is

$$\vec{w}_{1,3}^{2h} = 0.5 \times \left(\vec{w}_{1,5}^h + \vec{w}_{5,6}^h\right) + 0.25 \times \left(\vec{w}_{6,5}^h + \vec{w}_{5,4}^h + \vec{w}_{6,4}^h\right). \qquad (6.49)$$

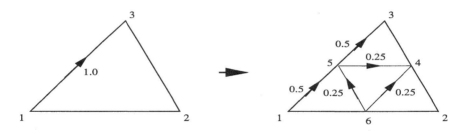

Figure 6.2   Transition between basis functions in coarse and fine grids; edge element $(1, 3)$.

Finally, we consider the coarse-grid edge element $(2, 3)$, shown on the left in Fig. 6.3. The edge elements in the fine grid and their weighting coefficients in their linear combination for the construction of edge element $(2, 3)$ are shown on the right in Fig. 6.3. The pertinent equation is

$$\vec{w}_{2,3}^{2h} = 0.5 \times \left(\vec{w}_{2,4}^h + \vec{w}_{4,3}^h\right) + 0.25 \times \left(\vec{w}_{6,4}^h + \vec{w}_{4,5}^h + \vec{w}_{6,5}^h\right). \qquad (6.50)$$

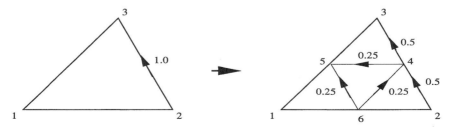

Figure 6.3   Transition between basis functions in coarse and fine grids; edge element $(2, 3)$.

The above discussion, along with the developed edge element basis functions transition relations, elucidates the way the transition matrix, and, hence, the intergrid-transfer matrices I_{2h}^h and I_h^{2h} are constructed in the case of triangular grids. However, we need not build the matrices explicitly. Rather, what is needed for the restriction and interpolation operations are matrix vector products involving the matrices I_{2h}^h and I_h^{2h}.

Let p^{2h} and p^h denote, respectively, vectors in the coarse and fine grids. The interpolation operation, effected through matrix I_{2h}^h, is summarized in the following algorithms:

Algorithm (6.5): Two-Dimensional Interpolation Operation $\mathbf{p^h \leftarrow I_{2h}^h p^{2h}}$.
```
1 for all edges (m,n) in coarse grid
  1.a find the middle point i of edge (m,n)
  1.b add_c2d(p^2h,p^h, m, n, m, i, 0.5)
  1.c add_c2d(p^2h,p^h, m, n, i, n, 0.5)
2 for all the triangles in coarse grid
  2.a store the three vertices in node[1..3]
  2.b store the middle point in three edge in node[4..6]
  2.c Edge-element (1,2)
    2.c.1 add_c2d(p^2h,p^h,node[1],node[2],node[5],node[6],0.25)
    2.c.2 add_c2d(p^2h,p^h,node[1],node[2],node[6],node[4],0.25)
    2.c.3 add_c2d(p^2h,p^h,node[1],node[2],node[5],node[4],0.25)
  2.d Edge-element (1,3)
    2.d.1 add_c2d(p^2h,p^h,node[1],node[3],node[6],node[5],0.25)
    2.d.2 add_c2d(p^2h,p^h,node[1],node[3],node[5],node[4],0.25)
    2.d.3 add_c2d(p^2h,p^h,node[1],node[3],node[6],node[4],0.25)
  2.e Edge-element (2,3)
    2.e.1 add_c2d(p^2h,p^h,node[2],node[3],node[6],node[4],0.25)
    2.e.2 add_c2d(p^2h,p^h,node[2],node[3],node[4],node[5],0.25)
    2.e.3 add_c2d(p^2h,p^h,node[2],node[3],node[6],node[5],0.25)
```

In this algorithm, all edges and all elements in the coarse grid are scanned. The function call "add_c2d" is given below.

Algorithm (6.6): add_c2d$(p^{2h}, p^h, m, n, i, j, coef)$
```
1 idx^2h ← index of edge (m,n) in p^2h
2 idx^h ← index of edge (i,j) in p^h
3 p^h[idx^h] += coef × p^2h[idx^2h]
```

The restriction operation effected through matrix I_h^{2h} is summarized in the following algorithm.

Algorithm (6.7): Two-Dimensional Restriction Operation $\mathbf{p^{2h} \leftarrow I_h^{2h} p^h}$
```
1 for all the edges (m,n,m < n) in coarse grid
  1.a find the middle point i of edge (m,n)
  1.b add_d2c(p^2h,p^h,m, n, m, i, 0.5)
  1.c add_d2c(p^2h,p^h,m, n, i, n, 0.5)
2 for all the triangles in coarse grid
  2.a store the three vertices in node[1..3]
  2.b Store the three middle points in three edge in node[4..6]
  2.c Edge-element (1,2)
    2.c.1 add_d2c(p^2h,p^h,node[1],node[2],node[5],node[6],0.25)
    2.c.2 add_d2c(p^2h,p^h,node[1],node[2],node[6],node[4],0.25)
    2.c.3 add_d2c(p^2h,p^h,node[1],node[2],node[5],node[4],0.25)
  2.d Edge-element (1,3)
    2.d.1 add_d2c(p^2h,p^h,node[1],node[3],node[6],node[5],0.25)
    2.d.2 add_d2c(p^2h,p^h,node[1],node[3],node[5],node[4],0.25)
    2.d.3 add_d2c(p^2h,p^h,node[1],node[3],node[6],node[4],0.25)
  2.e Edge-element (2,3)
    2.e.1 add_d2c(p^2h,p^h,node[2],node[3],node[6],node[4],0.25)
```

```
2.e.2 add_d2c(p^{2h},p^h,node[2],node[3],node[4],node[5],0.25)
2.e.3 add_d2c(p^{2h},p^h,node[2],node[3],node[6],node[5],0.25)
```

In this algorithm, all the edges and triangles in the coarse grid are scanned. The function call "add_d2c" is described below.

Algorithm (6.8): add_d2c($p^{2h}, p^h, m, n, i, j, coef$)
```
1 idx^{2h} ← index of edge (m,n) in p^{2h}
2 idx^h ← index of edge (i,j) in p^h
3 p^{2h}[idx^{2h}] += coef × p^h[idx^h]
```

When combining the relationships between node elements and edge elements of (5.30) and (6.37) on the two nested grids, we have the following diagram:

$$
\Phi^{2h} = \Phi^h Q_n \quad
\begin{array}{ccc}
\Phi^h & \xrightarrow{\nabla\Phi^h = \vec{W}^h G^h} & \vec{W}^h \\
\uparrow & & \uparrow \\
\Phi^{2h} & \xrightarrow{\nabla\Phi^{2h} = \vec{W}^{2h} G^{2h}} & \vec{W}^{2h},
\end{array}
\quad \vec{W}^{2h} = \vec{W}^h Q_e \qquad (6.51)
$$

where Q_n and Q_e are the intergrid transformation matrices between node elements and edge elements and G^h and G^{2h} are the gradient matrices on the fine and coarse grid. The scalar and vector potentials are expanded using the node and edge element spaces, respectively. The three-dimensional node and edge element spaces on the nested grids are also consistent with the diagram in (6.51).

6.1.6 Applications

The NMGAV process, described in the previous sections, is used as the preconditioner for the CG method to yield a nested multigrid $\vec{A}$-V preconditioned CG algorithm. In the following numerical examples, aimed at demonstrating the validity and effectiveness of the process, the number of pre-smoothing and post-smoothing operations is taken to be (unless stated otherwise) $v = 2$. These values are found to be optimal for two-dimensional applications.

Scattering by a PEC Cylinder We consider first the problem of TE$_z$ plane wave scattering from a perfectly electrical conducting cylinder of radius 0.5 m. The first-order ABC is imposed on a circular boundary of radius 1.0 m. The number of the edge unknowns in the coarsest grid is 2826. A three-level multigrid is implemented. Since the number of unknowns in the triangle-based, 2D nested multigrid increases by a factor of 4 from one grid to the next finer one, the number of unknowns in the finest grid is 44856.

The problem was solved using two potential preconditioning techniques, namely, the single-grid preconditioning of Algorithm (6.1) and the multigrid preconditioning of Algorithm (6.4). Their performance at various frequencies are compared in Table 6.1. Also included in the table are results obtained without making use of a preconditioner.

In Fig. 6.4, we plot the convergence performance of the three preconditioners at 1.5 GHZ. In Table (6.2) we compare the performance for different numbers of pre- and post-smoothing operations v. The two pre- and post- smoothing operations give the best convergence performance and smallest CPU time. Recall that for the iterative finite element solution of the 2D scalar wave equation discussed in the previous chapter, the optimum number of pre- and post- operations was found to be two also.

Table 6.1 TE$_z$ Plane Wave Scattering by a PEC Circular Cylinder (44856 Unknowns). Iterative Solver Convergence Performance

Freq. (GHz)		1.0	1.5	2.0
without	Iteration	5068	5281	4758
preconditioner	CPU (s)	894.6	932.04	838.46
SGAV	iter no.	664	786	878
	CPU (s)	705.9	918.78	933.85
NMGAV	iter no.	6	8	18
	CPU (s)	8.96	11.56	24.36

Table 6.2 TE$_z$ Plane Wave Scattering by a PEC Circular Cylinder (44856 Unknowns, V-cycle). Impact of Number of Smoothing Operations on Convergence

Freq (GHz)		1.0	1.5	2.0
$v = 1$	Iter	9	13	25
	Sec	10.75	15.01	28.09
$v = 2$	Iter	6	8	18
	Sec	8.96	11.56	24.36
$v = 3$	Iter	6	8	25
	Sec	10.10	13.14	39.32
$v = 4$	Iter	5	10	45
	Sec	9.67	18.11	76.87

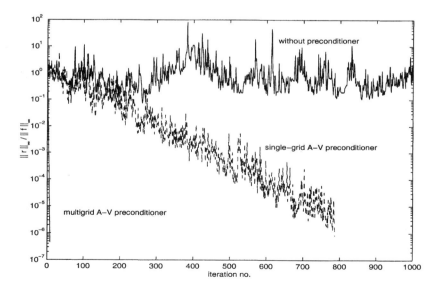

Figure 6.4 Iterative finite element solution convergence for TE$_z$ plane wave scattering by a circular PEC cylinder at 1.5 GHz. (After Zhu and Cangellaris [10], ©2002 IEEE.)

Relative residuals versus iteration numbers at various excitation frequencies are plotted in Fig. 6.5. An important observation is that both the number of iterations and the required CPU time for the NMGAV process increase with the excitation frequency. This is attributed to the wave nature of the fields. To elaborate, for the coarse grid solution to be of any use in the development of the solution on a finer grid, the coarse grid must be fine enough to constrain the numerical dispersion error [7] to within acceptable levels. Otherwise, convergence is jeopardized. This was indeed confirmed by the numerical experiments. Let $h_{\max}$ and h_{avg} denote the maximum and average edge length on the coarsest grid, respectively. For this test, $\lambda/h_{\max} = 1.78$ and $\lambda/h_{\text{avg}} = 2.78$ at 2.0 GHz. To guarantee fast convergence, $\lambda/h_{\text{avg}} >\sim 3$ at the maximum frequency in the bandwidth of interest. Considering that for the finest grid $\lambda/h_{\text{avg}} \sim 20$, a three- or four- level multigrid should be sufficient for accurate modeling. Use of a larger number of levels does not seem to improve convergence.

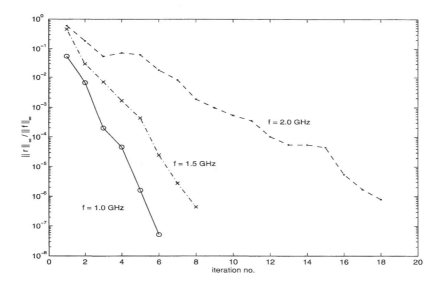

Figure 6.5   Convergence of NMGAV process for TE_z plane wave scattering by a PEC circular cylinder at various frequencies. (After Zhu and Cangellaris [10], ©2002 IEEE.)

Plane Wave Scattering by a Circular Dielectric Cylinder. The second numerical study considers TE_z plane wave scattering by a circular dielectric cylinder of radius 0.5m with relative permittivity $\epsilon_r = 9.0$. A circular boundary of radius 1.0m is used to truncate the computational domain and a first-order ABC is imposed on it. The number of unknowns in the coarsest grid is 18656. Once again, a three-level multigrid scheme is used. Thus the number of unknowns in the finest grid is 297248. The total memory required for the numerical solution is $\sim$98 MB.

The convergence performance of the NMGAV process is shown in Table 6.3. It is noted that at 1.4 GHz, $\lambda/h_{\max} = 2.27$ and $\lambda/h_{\text{avg}} = 3.54$. The calculated $|H_z|$ along the outer ABC boundary at 1.25 GHz is compared to the analytic solution in Fig. 6.6. Excellent agreement is observed.

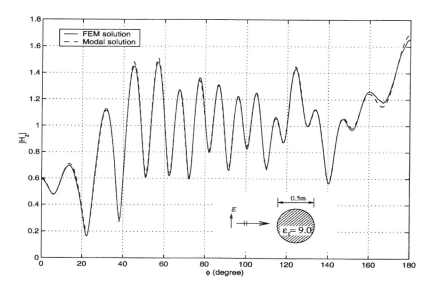

Figure 6.6 $|H_z|$ at $r = 1.0$ m (outer boundary) at 1.25 GHz for the case of TE$_z$ plane wave scattering by a circular dielectric cylinder. (After Zhu and Cangellaris [10], ©2002 IEEE.)

Table 6.3   TE$_z$ Plane Wave Scattering by a Dielectric Cylinder (297 248 Unknowns). Convergence Performance at Different Excitation Frequencies

Freq. (GHz)	1.0	1.25	1.5
λ/h_{avg}	4.96	3.97	3.31
λ/h_{max}	3.17	2.54	2.11
iteration	10	11	26
CPU (s)	96.96	106.082	236.23

Plane Wave Scattering by a PEC Inlet Structure.   The case of TE$_z$ plane wave scattering by the PEC inlet structure depicted in the insert of Fig. 6.7 is considered next. A first-order ABC is imposed on a circular boundary which is centered at the center of the rectangle formed by the inlet geometry and of radius 1.5 m. The plane wave is incident normally onto the open side of the inlet, as depicted in the figure.

A three-level multigrid scheme is used. The number of unknowns in the finest grid is 131500. The convergence performance of NMGAV is plotted in Fig. 6.7. With regards to wavelength resolution by the coarsest grid, at 2.0 GHz, $\lambda/h_{max} = 2.01$ and $\lambda/h_{avg} = 2.87$. What differentiates this example structure from the circular cylinders considered earlier is the occurrence of resonances inside the inlet. The convergence of the NMGAV process is not hindered by the presence of resonances.

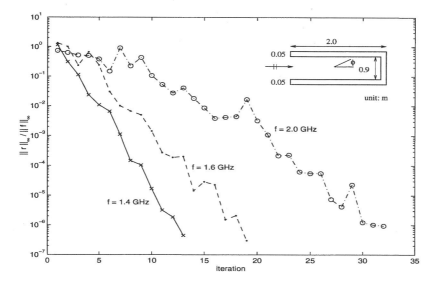

Figure 6.7   Convergence of the NMGAV process in the solution of TE$_z$ plane wave scattering by the PEC inlet structure. (After Zhu and Cangellaris [10], ©2002 IEEE.)

6.2 THREE-DIMENSIONAL ELECTROMAGNETIC SCATTERING

In this section we extend the NMGAV preconditioning process to applications involving electromagnetic field scattering by three-dimensional structures. As will soon become apparent, the NMGAV preconditioner for the three-dimensional vector Helmholtz equation is very similar to the one for the two-dimensional one except for the construction of the intergrid transfer operators.

6.2.1 Three-dimensional field formulation

We begin our discussion by reviewing the development of the weak statement and the associated finite element approximation of the electric field vector Helmholtz equation. As before, we refer to this as the $\vec{E}$-field formulation. Without loss of generality, the electromagnetic BVP of interest concerns a scatterer in an unbounded, homogeneous medium. The computational domain, Ω, includes all media exhibiting material inhomogeneities. For the purposes of this discussion all media are assumed to be linear and isotropic. However, the proposed methodology can be extended to the general case of anisotropic media in a fairly straightforward manner.

The pertinent vector equation for the electric field vector is

$$\nabla \times \frac{1}{\mu_r} \nabla \times \vec{E} - \omega^2 \mu_0 \epsilon_0 \epsilon_r \vec{E} = 0. \tag{6.52}$$

Multiplication of (6.52) by the vector basis function $\vec{w}$ and integration over the computational domain Ω, followed by a standard integration-by-parts step, yields

$$\iiint_\Omega \nabla \times \vec{w} \cdot \frac{1}{\mu_r} \nabla \times \vec{E} \, dv - j\omega\mu_0 \iint_{\partial\Omega} \hat{n} \times \vec{H} \cdot \vec{w} ds - \omega^2 \mu_0 \epsilon_0 \iiint_\Omega \vec{w} \cdot \epsilon_r \vec{E} dv = 0. \tag{6.53}$$

In deriving this result use was made of Faraday's law equation

$$\vec{H} = -\frac{1}{j\omega\mu_0\mu_r}\nabla \times \vec{E} \tag{6.54}$$

to introduce the tangential magnetic field in the integrand of the integral over the boundary $\partial\Omega$ of the computational domain. Except for portions associated with any closed perfectly conducting volumes present inside Ω, the rest of the computational domain boundary is, essentially, an artificial boundary used to truncate the domain. On this boundary, a first-order absorbing boundary condition is imposed for the outgoing radiated or scattered field, $(\vec{E}^{sc}, \vec{H}^{sc})$,

$$\hat{n} \times \vec{H}^{sc} = \frac{1}{\eta}\hat{n} \times \hat{n} \times \vec{E}^{sc}, \tag{6.55}$$

where $\hat{n}$ is the outwards-pointing unit normal on $\partial\Omega$, η is the intrinsic impedance of the homogeneous, background medium, and the superscript sc denotes the (outgoing) radiated or scattered fields. Without loss of generality, we choose the background medium to be free space; hence, $\eta = \sqrt{\mu_0/\epsilon_0}$. In view of the assumed linearity of the media, the total electromagnetic field can be written as the superposition of the incident field, $(\vec{E}^{inc}, \vec{H}^{inc})$, and the scattered field in the form

$$\vec{E} = \vec{E}^{inc} + \vec{E}^{sc}, \qquad \vec{H} = \vec{H}^{inc} + \vec{H}^{sc}. \tag{6.56}$$

Direct substitution in the first-order ABC yields

$$\hat{n} \times \vec{H} = \frac{1}{\eta} \cdot \hat{n} \times \hat{n} \times \vec{E} + \hat{n} \times \vec{H}^{inc} - \frac{1}{\eta} \cdot \hat{n} \times \hat{n} \times \vec{E}^{inc}. \tag{6.57}$$

Use of this result in the integrand of the boundary term in (6.53) yields

$$\iiint_\Omega \nabla \times \vec{w} \cdot \frac{1}{\mu_r}\nabla \times \vec{E}\,dv + \frac{j\omega\mu_0}{\eta}\iint_{\partial\Omega}(\hat{n} \times \vec{w}) \cdot (\hat{n} \times \vec{E})ds$$
$$-\omega^2\mu_0\epsilon_0\iiint_\Omega \vec{w} \cdot \epsilon_r\vec{E}dv = \frac{j\omega\mu_0}{\eta}\iint_{\partial\Omega}\hat{n} \times \vec{w} \cdot \left(\hat{n} \times \vec{E}^{inc} - \eta\vec{H}^{inc}\right)ds. \tag{6.58}$$

This is the weak statement of the BVP for the case where the unknown quantity is the total electric field. A weak formulation involving the scattered electric field is also possible to obtain, following the process used earlier for the two-dimensional case. The resulting equation is

$$\iiint_\Omega \nabla\vec{w} \cdot \frac{1}{\mu_r}\nabla \times \vec{E}^{sc}\,dv + \frac{j\omega\mu_0}{\eta}\iint_{\partial\Omega}\hat{n} \times \vec{w} \cdot \hat{n} \times \vec{E}^{sc}\,ds$$
$$-\omega^2\epsilon_0\mu_0\iiint_\Omega \vec{w} \cdot \epsilon_r\vec{E}^{sc}\,dv = \iiint_\Omega \nabla \times \vec{w} \cdot \left(\frac{1}{\mu_r} - 1\right)\nabla \times \vec{E}^{inc}\,dv \tag{6.59}$$
$$-\omega^2\epsilon_0\mu_0\iiint_\Omega \vec{w} \cdot (\epsilon_r - 1)\,\vec{E}^{inc}\,dv.$$

In the following, the development of the NMGAV preconditioner will be for the case of the total field formulation.

The finite element approximation of the BVP is obtained from (6.58) through a standard Galerkin's process. The resulting system has the form

$$M_{EE}x_E = f_E, \tag{6.60}$$

where the (i, j) element of the finite element matrix M_{EE} is given by

$$
M_{EEi,j} = \underbrace{\iiint_\Omega \nabla \times \vec{w}_i \cdot \frac{1}{\mu_r} \nabla \times \vec{w}_j \, dv}_{S_{i,j}} + \underbrace{jw \frac{\mu_0}{\eta} \iint_{\partial\Omega} \hat{n} \times \vec{w}_i \cdot \hat{n} \times \vec{w}_j \, ds}_{Z_{i,j}}
$$

$$
- \omega^2 \underbrace{\epsilon_0 \mu_0 \iiint_\Omega \vec{w}_i \cdot \epsilon_r \vec{w}_j \, dv}_{T_{i,j}} .
$$

(6.61)

The vector x_E contains the unknown expansion coefficients of $\vec{E}$, while the elements of the excitation vector f_E are given by

$$
f_{E,i} = \frac{j\omega\mu_0}{\eta} \iint_{\partial\Omega} \hat{n} \times \vec{w}_i \cdot \left(\hat{n} \times \vec{E}^{inc} - \eta \vec{H}^{inc} \right) ds.
$$

(6.62)

As already indicated in (6.61), the elements of the matrix M_{EE} may be broken down into three independent terms, thus allowing us to cast the matrix into the following form:

$$
M_{EE}(\omega) = (S + j\omega Z - \omega^2 T).
$$

(6.63)

It is immediately recognized that the finite element approximation for $\vec{E}$-field formulation of the three-dimensional electromagnetic problem is identical with the one for the two-dimensional case (6.15). The same is true for the potential formulation. Thus, in the next section only the key results are provided, since the process for their development is identical to the one presented in Section 6.1.2.

6.2.2 Three-dimensional potential formulation

Adopting the notation of [3] $\vec{E} = \vec{A} + \nabla V$ for the representation of the electric field vector in terms of the magnetic vector and electric scalar potentials, the following vector wave equation is obtained

$$
\nabla \times \frac{1}{\mu_r} \nabla \times \vec{A} - \omega^2 \mu_0 \epsilon_0 \epsilon_r \left(\vec{A} + \nabla V \right) = 0.
$$

(6.64)

This equation is supplemented by Gauss' law for the electric field, which, for the case of a source-free medium, has the form

$$
\nabla \cdot \epsilon_r \left(\vec{A} + \nabla V \right) = 0.
$$

(6.65)

Following the same process used in Section 6.2.1 for the two-dimensional case, the finite element approximation for the system of the above equations assumes the form

$$
\begin{bmatrix} M_{AA}^h & M_{AV}^h \\ M_{VA}^h & M_{VV}^h \end{bmatrix} \begin{bmatrix} x_A^h \\ x_V^h \end{bmatrix} = \begin{bmatrix} f_A^h \\ f_V^h \end{bmatrix},
$$

(6.66)

where the vectors x_A^h and x_V^h contain, respectively, the expansion coefficients in the approximations of the vector magnetic and scalar electric potentials, and the elements of the matrices and forcing vectors are given by

$$
\begin{aligned}
M^h_{AV\,ij} &= \frac{j\omega\mu_0}{\eta} \iint_{\partial\Omega} \hat{n} \times \vec{w}_i \cdot \hat{n} \times \nabla\phi_j ds - \omega^2\mu_0\epsilon_0 \iiint_\Omega \vec{w}_i \cdot \epsilon_r \nabla\phi_j dv \\
M^h_{VV\,ij} &= \frac{j\omega\mu_0}{\eta} \iint_{\partial\Omega} \hat{n} \times \nabla\phi_i \cdot \hat{n} \times \nabla\phi_j ds - \omega^2\epsilon_0\mu_0 \iiint_\Omega \nabla\phi_i \cdot \epsilon_r \nabla\phi_j dv \\
f^h_{V\,i} &= \frac{j\omega\mu_0}{\eta} \iint_{\partial\Omega} \hat{n} \times \nabla\phi_i \cdot \left(\hat{n} \times \vec{E}^{inc} - \eta\vec{H}^{inc} \right) ds \\
M^h_{AA} &= M^h_{EE}, \quad M^h_{VA} = (M^h_{AV})^T, \quad f^h_A = f^h_E.
\end{aligned}
\tag{6.67}
$$

Let Φ denote the space spanned by the scalar (nodal) basis functions ϕ, used for the approximation of V, and $\vec{W}$ the space spanned by the vector edge-element basis functions $\vec{w}$, used for the approximation of $\vec{A}$. In view of the properties of these spaces, which were discussed in detail in Chapter 2 and reviewed in Section 6.2.1, the finite element matrix in (6.66) may be cast in the following form:

$$
\begin{bmatrix} M_{AA} & M_{AV} \\ M_{VA} & M_{VV} \end{bmatrix} = \begin{bmatrix} I \\ G^T \end{bmatrix} [M_{EE}] \begin{bmatrix} I & G \end{bmatrix}.
\tag{6.68}
$$

In addition, it is

$$
\begin{bmatrix} f^h_A \\ f^h_V \end{bmatrix} = \begin{bmatrix} I \\ G^T \end{bmatrix} f^h_E,
\tag{6.69}
$$

where I is the identity matrix and the gradient matrix, G, is the linear transformation between the elements of the space $\vec{W}$ and the elements of the space $\nabla\Phi$

$$
\nabla\Phi = \vec{W}G.
\tag{6.70}
$$

Clearly, this equation is the analog of (6.29) for the three-dimensional case.

It is apparent that these matrix relations are identical to the ones obtained for the two-dimensional vector case. Thus the SGAV Algorithm 6.1 is directly applicable as a single-grid potential preconditioner for the three-dimensional case. Recall that the SGAV algorithm performs the smoothing operations on the potential formulation in order to remove the spurious, non-solenoidal fields associated with the type A and type C eigenmodes of the finite element system. The multigrid process needed for addressing the convergence difficulties associated with type B eigenmodes is discussed in the next two sections.

6.2.3 Nested multigrid potential preconditioner

For the purposes of implementing a geometric (nested) multigrid method, a coarse (lowest-level) grid is constructed first, with higher-level (finer) grids built subsequently by subdividing each tetrahedron of the previous level into eight equal-volume sub-tetrahedra. For an N-level algorithm, a hierarchy of N discretization levels is introduced with average element size varying from h, at the finest-grid level, to Nh at the coarsest-grid level. At each level, a finite element matrix equation is generated. The solution of the fine-grid, pseudo-residual equation makes use of the solutions on coarser grids. More specifically, the matrix equation is relaxed on the finer grid until convergence deteriorates. At that point the residual is restricted down to the coarser grid. Once the finite element equation is solved on the coarser grid, the solution is interpolated back to the finer grid in order to correct the earlier approximation on this grid. The matrix equation on the coarsest grid is solved directly by LU factorization.

It should be evident that the pseudo code description of the nested multigrid potential (NMGAV) preconditioning process for the three-dimensional case is identical to the one for the two-dimensional case given in Algorithm 6.4. Recall that the purpose of the subroutine "pre- and post-smooth(x_E^h, f_E^h)" is to project the problem from the $\vec{E}$-field formulation to the potential formulation, relax the potential solution using Gauss-Seidel method, and then project the solution back to the $\vec{E}$-field formulation.

The only difference between the two-dimensional and the three-dimensional case is in the construction and the resulting forms of the restriction operator, $I_{nh}^{(n+1)h}$, and the interpolation operator, $I_{(n+1)h}^{n}{}^{h}$. The construction of these operators for the case of a three-dimensional finite element grid of tetrahedra is discussed in the next section.

6.2.4   Three-dimensional intergrid transfer operators

The development of the three-dimensional intergrid transfer operators assumes a two-level multigrid with finite element equations on the two grids given by (6.36). All elements in the grids are assumed to be tetrahedra. Let $\vec{W}^{2h}$ denote the set of the vector edge-element basis functions in the coarse grid and $\vec{W}^{h}$ the set of the vector edge-element basis functions in the fine grid. Since $\vec{W}^{2h}$ is a subset of $\vec{W}^{h}$, a linear mapping exists between elements of the two sets. The matrix form, Q, of this mapping is built locally on an element-by-element basis, following the process discussed later in this section. The matrix form of the mapping between the vector containing the elements of $\vec{W}^{2h}$ and the vector containing the elements of $\vec{W}^{h}$ is given by

$$\vec{W}^{2h} = \vec{W}^{h}Q, \qquad (6.71)$$

where the notation $\vec{W}^{2h}$ and $\vec{W}^{h}$ has been used for the two row vectors containing the elements of the two spaces. Since both spaces are used for the expansion of the electric field vector, the same arguments used in 6.1.5 for the case of two-dimensional vector multigrid can be used to conclude that the matrix form of the interpolation operator I_{2h}^{h} is nothing else but the matrix Q.

We proceed by noting the bilinear form of the finite element matrix (6.60) and the linear form of the forcing term on the right-hand side. Thus following exactly the same procedure used for the two-dimensional case (described by equations (6.38)-(6.44) we deduce that the restriction operator I_{h}^{2h} equals Q^{T}. Thus the three-dimensional intergrid transfer operators I_{2h}^{h} and I_{h}^{2h} are both obtained directly from the transformation matrix Q between the basis functions of the coarse and the fine grids.

The construction of Q commences with the partition of one tetrahedron in the coarse grid into eight equal-volume sub-tetrahedra [11]. Toward this, we introduce first the middle points of the six sides and connect them as depicted on the left of Fig. 6.8.

Next, we remove the four "outermost" sub-tetrahedra, each one of which contains one of the four vertices of the tetrahedron. The result is the solid depicted on the right-hand side of Fig. 6.8, which can be considered as the combination of two pyramids. The base shared by the two pyramids can be anyone of the three quadrilateral surfaces defined in terms of their vertices as $(5, 7, 10, 8)$, $(5, 6, 10, 9)$, or $(6, 7, 9, 8)$ in Fig. 6.8. The combination of the two pyramids can be further split into four sub-tetrahedra using anyone of the three different choices depicted in Fig. 6.9. The choice of the partition is dictated by the quality of the resulting sub-tetrahedra. The quality of a tetrahedron is, in turn, determined by the perturbation of the shape of the tetrahedron from the ideal case of an equilateral one; the smaller the perturbation, the better its quality. Maintaining high-quality tetrahedra during the development of finer grids is of paramount importance for ensuring that the

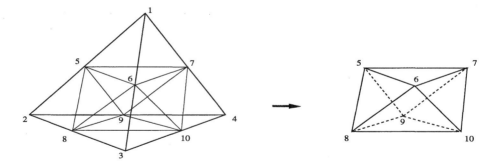

Figure 6.8 Decomposition of a coarse-grid tetrahedron into fine-grid tetrahedra.

finite element matrices remain well conditioned. The way this is effected in a systematic and effective manner during mesh generation remains a topic of on-going investigation. A simple heuristic that we would like to propose is for the choice of the sub-tetrahedra decomposition in Fig. 6.9 to be guided by the length of the new edge which is introduced. More specifically, from the three possible choices, the edge of length closest to the average length of the edges of the two pyramids should be selected. In the following, without loss of generality, we discuss the construction of the transformation matrix for the case of the first partition in Fig. 6.9 where edge $(6, 9)$ is added.

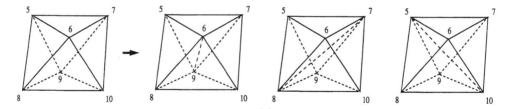

Figure 6.9 Splitting of two pyramids sharing a common base into four sub-tetrahedra.

We begin by considering the linear combination of the basis functions in the fine grid for the development of each basis function in the coarse grid. We consider, first, the edge element $(1, 2)$ shown on the left in Fig. 6.10. The edge element can be expressed as the linear combination of the edge elements in the nested (fine) grid marked by arrows in the right plot of Fig. 6.10. The pertinent expansion coefficients for these elements are also indicated on the figure. The linear combination is

$$\vec{w}_{1,2} = 0.5 \times (\vec{w}_{1,5} + \vec{w}_{5,2}) + 0.25 \times (\vec{w}_{6,5} + \vec{w}_{5,8} + \vec{w}_{6,8})$$
$$+ 0.25 \times (\vec{w}_{7,5} + \vec{w}_{5,9} + \vec{w}_{7,9}) + 0.25 \times \vec{w}_{6,9}. \tag{6.72}$$

Use of the parentheses allows us to identify four groups of fine-grid edge elements contributing to this representation. The first group consists of the two fine-grid edge elements that are associated with edge (1,2) of the coarse grid. The second and third groups consist each of three fine-grid edge elements on faces (1,2,3) and (1,2,4), respectively, of the coarse grid. Finally, the fourth group consists of only one edge element, the one formed by the middle points of edges (2,4) and (1,3) of the coarse-grid tetrahedron.

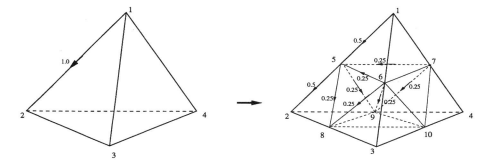

Figure 6.10   Edge-element $(1, 2)$.

The validity of 6.72 can be shown by examining its correctness in each one of the eight sub-tetrahedra in the finer grid. For example, for sub-tetrahedron $(5, 6, 8, 9)$, we have

$$0.25 \times (\underbrace{\lambda_6 \nabla \lambda_5 - \lambda_5 \nabla \lambda_6}_{\vec{w}_{6,5}} + \underbrace{\lambda_6 \nabla \lambda_8 - \lambda_8 \nabla \lambda_6}_{\vec{w}_{6,8}} + \underbrace{\lambda_6 \nabla \lambda_9 - \lambda_9 \nabla \lambda_6}_{\vec{w}_{6,9}} +$$
$$\underbrace{\lambda_5 \nabla \lambda_9 - \lambda_9 \nabla \lambda_5}_{\vec{w}_{5,9}} + \underbrace{\lambda_5 \nabla \lambda_8 - \lambda_8 \nabla \lambda_5}_{\vec{w}_{5,8}}). \tag{6.73}$$

The barycentric coordinates associated with the four vertices (5,6,8,9) of the fine-grid tetrahedron are related to those of the four vertices (1,2,3,4) on the coarser grid as follows:

$$\lambda_5 = 1 - 2\lambda_3 - 2\lambda_4, \qquad \lambda_6 = 1 - 2\lambda_2 - 2\lambda_4,$$
$$\lambda_8 = 1 - 2\lambda_1 - 2\lambda_4, \qquad \lambda_9 = 1 - 2\lambda_1 - 2\lambda_3. \tag{6.74}$$

Substitution of the above relationships into (6.73) leads immediately to the result that the expression in (6.73) is equal to $\vec{w}_{1,2}$ in sub-tetrahedron $(5, 6, 8, 9)$. A similar process is used to show that (6.72) is valid in the remaining seven sub-tetrahedra.

Next we consider edge element $(1, 3)$. With the aid of Fig. 6.11, the edge element can be expressed as the linear combination of the edge elements in the nested (fine) grid as follows:

$$\vec{w}_{1,3} = 0.5 \times (\vec{w}_{1,6} + \vec{w}_{6,3}) + 0.25 \times (\vec{w}_{5,6} + \vec{w}_{6,8} + \vec{w}_{5,8})$$
$$+ 0.25 \times (\vec{w}_{7,6} + \vec{w}_{6,10} + \vec{w}_{7,10}) . \tag{6.75}$$

The validity of this equation is shown following the same procedure used for proving (6.72).

In a similar manner, we have for edge element $(1, 4)$ (see Fig. 6.12),

$$\vec{w}_{1,4} = 0.5 \times (\vec{w}_{1,7} + \vec{w}_{7,4}) + 0.25 \times (\vec{w}_{5,7} + \vec{w}_{7,9} + \vec{w}_{5,9})$$
$$+ 0.25 \times (\vec{w}_{6,7} + \vec{w}_{7,10} + \vec{w}_{6,10}) + 0.25 \times \vec{w}_{6,9}. \tag{6.76}$$

For edge element $(2, 3)$ (see Fig. 6.13) it is

$$\vec{w}_{2,3} = 0.5 \times (\vec{w}_{2,8} + \vec{w}_{8,3}) + 0.25 \times (\vec{w}_{5,8} + \vec{w}_{8,6} + \vec{w}_{5,6})$$
$$+ 0.25 \times (\vec{w}_{9,8} + \vec{w}_{8,10} + \vec{w}_{9,10}) + 0.25 \times \vec{w}_{9,6}. \tag{6.77}$$

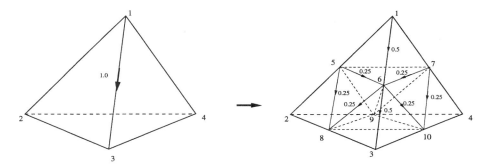

Figure 6.11   Edge element $(1, 3)$.

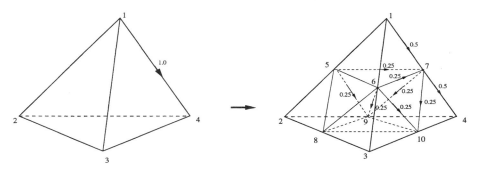

Figure 6.12   Edge element $(1, 4)$.

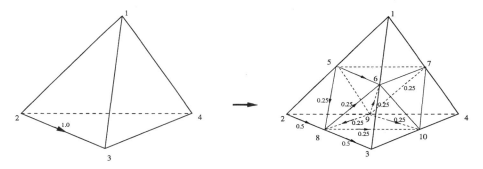

Figure 6.13   Edge element $(2, 3)$.

For edge element $(2, 4)$ (see Fig. 6.14) it is

$$\vec{w}_{2,4} = 0.5 \times (\vec{w}_{2,9} + \vec{w}_{9,4}) + 0.25 \times (\vec{w}_{5,9} + \vec{w}_{9,7} + \vec{w}_{5,7})$$
$$+ 0.25 \times (\vec{w}_{8,9} + \vec{w}_{9,10} + \vec{w}_{8,10}) . \tag{6.78}$$

Finally, for edge element $(1, 3)$ (see Fig. 6.15) we have

$$\vec{w}_{3,4} = 0.5 \times (\vec{w}_{3,10} + \vec{w}_{10,4}) + 0.25 \times (\vec{w}_{8,10} + \vec{w}_{10,9} + \vec{w}_{8,9})$$
$$+ 0.25 \times (\vec{w}_{6,10} + \vec{w}_{10,7} + \vec{w}_{6,7}) + 0.25 \times \vec{w}_{6,9} . \tag{6.79}$$

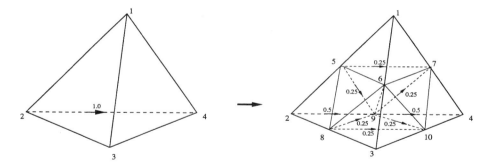

Figure 6.14   Edge element $(2, 4)$.

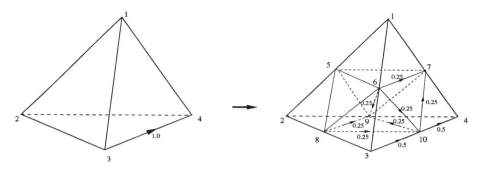

Figure 6.15   Edge element $(3, 4)$.

The derived relationships are employed to build the transformation matrix Q and its transpose Q^T between the two grids. However, Q and Q^T, which are, respectively, I_h^{2h} and I_{2h}^h, are not needed explicitly for the implementation of the multigrid process. Rather, what is needed in Algorithm 6.4 are the matrix-vector products involving I_{2h}^h and I_h^{2h}. Let p^{2h} and p^h denote two vectors in the coarse and fine grid, respectively. The matrix-vector-product operation with I_{2h}^h is described in the following algorithm.

Algorithm (6.9): Three-Dimensional Interpolation Operation $p^h \leftarrow I_{2h}^h p^{2h}$
```
1 for all the edges (m,n) in coarse grid
  1.a i ← the middle point of edge (m,n) in fine grid
  1.b add_c2d(p^2h, p^h, m, n, m, i, 0.5)
  1.c add_c2d(p^2h, p^h, m, n, i, n, 0.5)
2 for all faces in coarse grid
  2.a Store the three vertices in node[1..3]
  2.b Store the three middle points of the three edges in node[4..6]
  2.c Edge element (1,2)
    2.c.1 add_c2d(p^2h, p^h,node[1],node[2],node[5],node[6],0.25)
    2.c.2 add_c2d(p^2h, p^h,node[1],node[2],node[6],node[4],0.25)
    2.c.3 add_c2d(p^2h, p^h,node[1],node[2],node[5],node[4],0.25)
  2.d Edge element (1,3)
    2.d.1 add_c2d(p^2h, p^h,node[1],node[3],node[6],node[5],0.25)
    2.d.2 add_c2d(p^2h, p^h,node[1],node[3],node[5],node[4],0.25)
    2.d.3 add_c2d(p^2h, p^h,node[1],node[3],node[6],node[4],0.25)
  2.e Edge element (2,3)
    2.e.1 add_c2d(p^2h, p^h,node[2],node[3],node[6],node[4],0.25)
    2.e.2 add_c2d(p^2h, p^h,node[2],node[3],node[4],node[5],0.25)
    2.e.3 add_c2d(p^2h, p^h,node[2],node[3],node[6],node[5],0.25)
```

```
3 for all tetra in coarse grid
  3.a find the new edge (i,j) inside the tetra:
      i is the middle point of edge (m,n);
      j is the middle point of edge (p,q).
  3.b add_c2d(p^{2h}, p^h, m, p, i, j, 0.25)
  3.c add_c2d(p^{2h}, p^h, m, q, i, j, 0.25)
  3.d add_c2d(p^{2h}, p^h, n, p, i, j, 0.25)
  3.e add_c2d(p^{2h}, p^h, n, q, i, j, 0.25)
```

The process "add_c2d" is defined in Algorithm 6.6. The edge elements in the nested (fine) grid can be categorized into three groups. Group 1 includes edge elements associated with the edges of the coarse grid. Group 2 includes edges elements on the facets of the coarse grid. Group 3 includes edge elements with fine-grid edges in the interior of the coarse-grid tetrahedron. The above algorithm handles these three groups of edge elements one group at a time. Step 1 is for Group 1. Step 2 is for Group 2, where the pertinent node indices are marked in Figs. 6.1-6.3. Finally, Step 3 is for Group 3, where the pertinent node indices are depicted in Fig. 6.16.

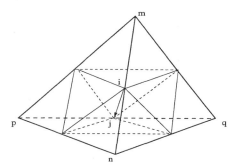

Figure 6.16   An interior edge (i, j) for a coarse-grid tetrahedron.

The matrix-vector-product operation with I_h^{2h} is described in the following algorithm.

Algorithm (6.10): Three-Dimensional Restriction Operation $p^{2h} \leftarrow I_h^{2h} p^h$.
```
1 for all the edges (m,n) in coarse grid
  1.a i ← the middle point of edge (m,n) in fine grid
  1.b add_d2c(p^{2h},p^h, m, n, m, i, 0.5)
  1.c add_d2c(p^{2h},p^h, m, n, i, n, 0.5)
2 for all facets in coarse grid
  2.a Store the three vertices in node[1:3]
  2.b Store the three middle points of the three edges in node[4:6]
  2.c Edge element (1,2)
    2.c.1 add_d2c(p^{2h},p^h,node[1],node[2],node[5],node[6],0.25)
    2.c.2 add_d2c(p^{2h},p^h,node[1],node[2],node[6],node[4],0.25)
    2.c.3 add_d2c(p^{2h},p^h,node[1],node[2],node[5],node[4],0.25)
  2.d Edge element (1,3)
    2.d.1 add_d2c(p^{2h},p^h,node[1],node[3],node[6],node[5],0.25)
    2.d.2 add_d2c(p^{2h},p^h,node[1],node[3],node[5],node[4],0.25)
    2.d.3 add_d2c(p^{2h},p^h,node[1],node[3],node[6],node[4],0.25)
  2.e Edge element (2,3)
    2.e.1 add_d2c(p^{2h},p^h,node[2],node[3],node[6],node[4],0.25)
    2.e.2 add_d2c(p^{2h},p^h,node[2],node[3],node[4],node[5],0.25)
    2.e.3 add_d2c(p^{2h},p^h,node[2],node[3],node[6],node[5],0.25)
3 for all tetra in coarse grid
  3.a find the newly-added edge (i,j) inside the tetra:
      i is the middle point of edge (m,n);
```

```
      j is the middle point of edge (p,q).
3.b add_d2c(p^{2h},p^h, m, p, i, j, 0.25)
3.c add_d2c(p^{2h},p^h, m, q, i, j, 0.25)
3.d add_d2c(p^{2h},p^h, n, p, i, j, 0.25)
3.e add_d2c(p^{2h},p^h, n, q, i, j, 0.25)
```

where the process "add_d2c" is defined in Algorithm 6.8.

This completes the NMGAV algorithm that is used as a multigrid preconditioner in the iterative solution of the three-dimensional electromagnetic BVP (6.60), with the potential formulation implemented to suppress the non-solenoidal spurious field error occurring during the iteration process.

6.2.5   Grid truncation via a boundary integral operator

Thus far our discussion of the finite element approximation for an electromagnetic BVP in an unbounded domain assumed the use of an absorbing boundary condition on a truncation boundary, which was placed sufficiently far away from the scatterer to guarantee the outgoing character of the scattered fields. Such an approach becomes inaccurate and unreliable when the truncation boundary is in close proximity to the scatterer. The placement of the truncation boundary as close as possible to the scatterer is prompted by the desire to minimize the computational and memory overhead associated with the finite element discretization of "empty space." Toward this objective hybrid schemes have been proposed, which combine the finite element approximation of the interior with a boundary integral equation (BIE) statement that is consistent with the physics of the exterior and involves the tangential electric and magnetic fields on the truncation boundary [12]-[21]. In the following we review the basics of the mathematical formalism of such hybrid FEM/BIE methods, in preparation of the discussion of a preconditioning scheme for its iterative solution.

With Ω denoting the volume of the interior region and $\partial\Omega$ denoting, for the purposes of this discussion, the surface of the truncation boundary, the weak statement of the vector Helmoltz equation for the electric field is

$$\iiint_\Omega \nabla \times \vec{w} \cdot \frac{1}{\mu_r} \nabla \times \vec{E}\, dv - j\omega\mu_0 \iint_{\partial\Omega} \hat{n} \times \vec{H} \cdot \vec{w}\, ds$$
$$- \omega^2 \epsilon_0 \mu_0 \iiint_\Omega \vec{w} \cdot \epsilon_r \vec{E}\, ds = 0. \tag{6.80}$$

Let us assume that the electric field vector, $\vec{E}$, is expanded in terms of edge elements, while for the magnetic field vector, $\vec{H}$, on $\partial\Omega$, the surface RWG basis functions are used for its approximation (see Chapter 2.2.3 for a discussion of the RWG functions and their relation to edge elements)

$$\vec{H} = \sum_{i=1}^{N_s} \hat{n} \times \vec{w}_i h_i. \tag{6.81}$$

In the above expansion $\hat{n}$ is the outward-pointing unit normal on $\partial\Omega$, N_s is the number of edges on $\partial\Omega$, $\vec{w}_i$ is the edge element associated with the i-th edge, and $\hat{n} \times \vec{w}_i$ is the associated RWG basis function.

Following a standard Galerkin's process, the finite element statement obtained from (6.80) assumes the form

$$(S - \omega_0^2 T)x_E + j\omega\mu_0 Z x_H = 0, \tag{6.82}$$

where S and T are defined in (6.61),

$$Z_{ij} = \iint_{\partial\Omega} \hat{n} \times \vec{w}_i \cdot \hat{n} \times \vec{w}_j ds \tag{6.83}$$

and x_E and x_H are column vectors containing the electric field expansion coefficients and surface magnetic field expansion coefficients, respectively.

One of the possible forms of the boundary integral statement for the calculation of the electric field in the exterior is in terms of the so-called electric field integral equation (EFIE) formulation,

$$\begin{aligned}
\vec{E}(\vec{r}) = \vec{E}^{inc}(\vec{r}) &+ \iint_{\partial\Omega} \left(\hat{n}' \times \vec{E}(\vec{r}') \right) \cdot \left(\nabla' \times \bar{\bar{G}}(\vec{r}, \vec{r}') \right) ds' \\
&+ \iint_{\partial\Omega} \left(\hat{n}' \times \nabla' \times \vec{E}(\vec{r}') \right) \cdot \bar{\bar{G}}(\vec{r}, \vec{r}') \, ds',
\end{aligned} \tag{6.84}$$

where $\bar{\bar{G}}(\vec{r}, \vec{r}')$ is the dyadic Green's function for the exterior domain. Without loss of generality we assume that the exterior domain is free space. Thus it is

$$\bar{\bar{G}}(\vec{r}, \vec{r}') = \left(\bar{\bar{I}} + \frac{1}{k_0^2} \nabla\nabla \right) G(\vec{r}, \vec{r}'), \tag{6.85}$$

where $\bar{\bar{I}} = \hat{x}\hat{x} + \hat{y}\hat{y} + \hat{z}\hat{z}$, $k_0 = \omega\sqrt{\mu_0 \epsilon_0}$, and

$$G(\vec{r}, \vec{r}') = \frac{e^{-jk_0|\vec{r}-\vec{r}'|}}{4\pi|\vec{r} - \vec{r}'|} \tag{6.86}$$

is the Green's function for the scalar Helmholtz equation in free space. Noting that it is

$$\nabla' \times \bar{\bar{G}}(\vec{r}, \vec{r}') = \nabla' G(\vec{r}, \vec{r}') \times \bar{\bar{I}} \tag{6.87}$$

(6.84) assumes the form

$$\begin{aligned}
\vec{E}(\vec{r}) = \vec{E}^{inc}(\vec{r}) &+ \iint_{\partial\Omega} \left(\hat{n}' \times \vec{E}(\vec{r}') \right) \times \nabla' G(\vec{r}, \vec{r}') ds' \\
&+ \iint_{\partial\Omega} \left(\hat{n}' \times \nabla' \times \vec{E}(\vec{r}') \right) \cdot \left(\bar{\bar{I}} G(\vec{r}, \vec{r}') + \frac{1}{k_0^2} \nabla\nabla G(\vec{r}, \vec{r}') \right) ds'.
\end{aligned} \tag{6.88}$$

Application of the surface divergence theorem leads to the following alternative form of the above expression:

$$\begin{aligned}
\vec{E}(\vec{r}) = \vec{E}^{inc}(\vec{r}) &+ \iint_{\partial\Omega} \left(\hat{n}' \times \vec{E}(\vec{r}') \right) \times \nabla' G(\vec{r}, \vec{r}') ds' \\
&+ \iint_{\partial\Omega} \left(\hat{n}' \times \nabla' \times \vec{E}(\vec{r}') \right) G(\vec{r}, \vec{r}') ds' \\
&+ \iint_{\partial\Omega} \frac{1}{k_0^2} \left[\nabla' \cdot \left(\hat{n}' \times \nabla' \times \vec{E}(\vec{r}') \right) \right] \nabla G(\vec{r}, \vec{r}') ds'.
\end{aligned} \tag{6.89}$$

The tangential component of the above equation on $\partial\Omega$ is readily obtained as follows:

$$\begin{aligned}
\hat{n} \times \vec{E}(\vec{r}) = \hat{n} \times \vec{E}^{inc}(\vec{r}) &+ \hat{n} \times \iint_{\partial\Omega} \left(\hat{n}' \times \vec{E}(\vec{r}') \right) \times \nabla' G(\vec{r}, \vec{r}') ds' \\
- j\omega\mu_0 \hat{n} \times &\iint_{\partial\Omega} \left[\hat{n}' \times \vec{H}(\vec{r}') G(\vec{r}, \vec{r}') + \frac{1}{k_0^2} \nabla' \cdot \hat{n}' \times \vec{H}(\vec{r}') \nabla G(\vec{r}, \vec{r}') \right] ds'.
\end{aligned} \tag{6.90}$$

Multiplication of the above equation by $\hat{n} \times \vec{w}_i$ and integration over $\partial\Omega$ yields

$$
\iint_{\partial\Omega} \hat{n} \times \vec{w}_i \cdot \hat{n} \times \vec{E} ds - \iint_{\partial\Omega} \iint_{\partial\Omega} \vec{w}_i \cdot \left(\hat{n}' \times \vec{E}(r') \right) \times \nabla' G(\vec{r}, \vec{r}') ds' ds
$$

$$
+ j\omega\mu_0 \iint_{\partial\Omega} \iint_{\partial\Omega} \vec{w}_i \cdot \left[\hat{n}' \times \vec{H}(\vec{r}') G(\vec{r}, \vec{r}') + \frac{1}{k_0^2} \nabla' \cdot \hat{n}' \times \vec{H}(\vec{r}') \nabla G(\vec{r}, \vec{r}') \right] ds' ds
$$

$$
= \iint_{\partial\Omega} \hat{n} \times \vec{w}_i \cdot \hat{n} \times \vec{E}^{inc} ds.
$$

$$(6.91)$$

The matrix form of the above integral equation statement is obtained by expanding the tangential electric field in terms of surface edge elements, utilizing (6.81), and performing the testing described by the above equations for each one of the surface edge elements $\hat{n} \times \vec{w}_i$. This yields the following matrix equation:

$$
P x_{E_s} + Q x_H = b, \tag{6.92}
$$

where the vectors of unknowns x_H was defined earlier in conjunction with the expansion in (6.81) and x_{E_s} is a vector containing the subset of electric field unknown coefficients associated with the edges on $\partial\Omega$. The elements of P and Q are given by

$$
P_{i,j} = \iint_{\partial\Omega} \hat{n} \times \vec{w}_i \cdot \hat{n} \times \vec{w}_j ds - \iint_{\partial\Omega} \iint_{\partial\Omega} \vec{w}_i(\vec{r}) \cdot (\hat{n}' \times \vec{w}_j(\vec{r}')) \times \nabla' G \, ds' ds
$$

$$
Q_{i,j} = j\omega\mu_0 \iint_{\partial\Omega} \iint_{\partial\Omega} \vec{w}_i(\vec{r}) \cdot \left[\vec{w}_j(\vec{r}') G + \frac{1}{k_0^2} \nabla' \cdot \vec{w}_j(\vec{r}') \nabla G \right] ds' ds \qquad (6.93)
$$

$$
b_i = \iint_{\partial\Omega} \hat{n} \times \vec{w}_i \cdot \hat{n} \times \vec{E}^{inc} ds.
$$

It should be apparent that, due to the global form of $G(\vec{r}, \vec{r}')$ and its derivatives, the matrices P and Q are dense. Equations (6.82) and (6.92) are combined into the following matrix statement of the discrete approximation of the unbounded electromagnetic BVP:

$$
\begin{bmatrix} S - \omega^2 T & j\omega\mu_0 Z \\ \hat{P} & Q \end{bmatrix} \begin{bmatrix} x_E \\ x_H \end{bmatrix} = \begin{bmatrix} 0 \\ b \end{bmatrix}. \tag{6.94}
$$

In the above equation the matrix $\hat{P}$ is obtained from P in (6.92) by adding the appropriate number of zero columns to adjust its column dimension to the length of x_E. The effective preconditioner for the solution of this matrix equation is the topic of the next section.

6.2.6 Approximate boundary integral equation preconditioner

The finite-element/boundary-integral (FE-BI) matrix in (6.94) contains a dense submatrix associated with the integral equation interactions of the electric field and magnetic field unknowns on $\partial\Omega$. Even with the application of the multigrid process the dense submatrix is still present in the coarsest-grid version of (6.94), rendering its factorization computationally expensive. It was shown in [22] that the eigenvalue spectrum of the finite element matrix of the same problem but with an absorbing boundary condition (ABC) used on $\partial\Omega$ is very similar to that of the FE-BI matrix. This, then, suggests that the finite element with ABC truncation (FE-ABC) matrix is a good candidate for use as an approximate preconditioner for the FE-BI matrix equation.

In order to implement such a preconditioner we need, first, to recast the FE-ABC matrix in a form consistent with the set of unknowns involved in the FE-BI equation. Assuming that the first-order ABC of (5.7) will be used on $\partial\Omega$, its weak statement is obtained in a straightforward manner as follows:

$$
\iint_{\partial\Omega} \vec{w}_i \cdot \hat{n} \times \vec{H} ds = -\frac{1}{\eta} \iint_{\partial\Omega} \hat{n} \times \vec{w}_i \cdot \hat{n} \times \vec{E} ds
$$
$$
+ \underbrace{\frac{1}{\eta} \iint_{\partial\Omega} \hat{n} \times \vec{w}_i \cdot \left(\hat{n} \times \vec{E}^{inc} - \eta \vec{H}^{inc} \right) ds}_{f_E} . \tag{6.95}
$$

From this equation a matrix statement is readily obtained through the expansion of $\vec{E}$ in terms of surface edge elements and the substitution of (6.81)

$$
\frac{1}{\eta} Z x_E - Z x_H = f_E, \tag{6.96}
$$

where f_E is defined in (6.62). Combining this equation with the portion of the finite element matrix of (6.82) associated with the interior of the computational domain yields the FE-ABC matrix equation

$$
\begin{bmatrix} S - \omega^2 T & j\omega\mu_0 Z \\ \frac{1}{\eta} Z & -Z \end{bmatrix} \begin{bmatrix} x_E \\ x_H \end{bmatrix} = \begin{bmatrix} 0 \\ f_E \end{bmatrix}. \tag{6.97}
$$

It is immediately evident that the elimination of the tangential magnetic field unknowns, x_H, on $\partial\Omega$ reduced this equation to the FE-ABC equation of (6.60).

Use of the FE-ABC matrix of (6.97) as a preconditioner for the FE-BI equation entails the utilization of the following preconditioned FE-BI equation:

$$
\begin{bmatrix} S - \omega^2 T & j\omega\mu_0 Z \\ \frac{1}{\eta} Z & -Z \end{bmatrix}^{-1} \begin{bmatrix} S - \omega^2 T & j\omega Z \\ P & Q \end{bmatrix} \begin{bmatrix} x_E \\ x_H \end{bmatrix}
$$
$$
= \begin{bmatrix} S - \omega^2 T & j\omega\mu_0 Z \\ \frac{1}{\eta} Z & -Z \end{bmatrix}^{-1} \begin{bmatrix} 0 \\ b \end{bmatrix}. \tag{6.98}
$$

Thus each iteration involves the solution of an FE-ABC equation

$$
\begin{bmatrix} S - \omega^2 T & j\omega\mu_0 Z \\ \frac{1}{\eta} Z & -Z \end{bmatrix} \begin{bmatrix} z_E \\ z_H \end{bmatrix} = \begin{bmatrix} r_E \\ r_H \end{bmatrix}. \tag{6.99}
$$

Its solution is carried out in a two-step process as follows: First we solve for z_E through the familiar equation

$$
\left(S - \omega^2 T + \frac{j\omega\mu_0}{\eta} Z \right) x_E = r_E + j\omega\mu_0 r_H. \tag{6.100}
$$

Depending on the size of the matrix, its solution is obtained either directly or iteratively using NMGAV as preconditioner. Once z_E is computed, z_H is readily obtained through the equation

$$
z_H = \frac{1}{\eta} z_E - Z^{-1} r_H. \tag{6.101}
$$

The matrix Z is symmetric positive-definite, involving only the unknowns on the boundary; hence, it can be solved efficiently using ICCG.

Finally, let us observe that since the FE-ABC matrix is only used as the preconditioner for the FE-BIE equation, it is not necessary to solve (6.100) with high accuracy. Instead, it suffices to use the solution obtained from the NMGAV preconditioner as its approximate solution. We shall refer to this preconditioning process for the FE-BIE equation as approximate BIE preconditionining. Its algorithmic implementation is summarized below.

Algorithm 6.11 Approximate BIE preconditioning
1 Form the FE-ABC matrix of (6.99)
2 Solve approximately using NMGAV preconditioner
$$\left(S - \omega^2 T + \frac{j\omega\mu_0}{\eta} Z\right) z_E = r_E + j\omega\mu_0 r_H$$
3 $z_H \leftarrow \frac{1}{\eta} e - Z^{-1} r_H$.

6.2.7 Applications

To demonstrate the effectiveness of the multigrid preconditioning method discussed in this section let us consider its application to the extensively studied problem of time-harmonic plane wave scattering by a perfectly conducting cube in free space. The frequency of the sinusoidally varying, incident plane wave is 300 MHz. Three different cubes were considered, of side lengths 1.5λ, 3.0λ, and 5.0λ, where λ is the free-space wavelength at the excitation frequency. The ABC is imposed on a cubic boundary placed 0.25λ away from the sides of the cube. A three-level multigrid is implemented.

Table 6.4 provides numerical computation-related information for the three cases. More specifically, information is provided for the number of unknowns in each level, the average edge length in the finest grid, the number of iterations, and the total CPU time and memory requirements. For all three cases, three pre- and post-smoothing operations were used. For all three sizes convergence was achieved after eleven iterations.

Table 6.4   Numerical Computation Data for Scattering by PEC Cubes

Cube	h_{avg}/λ	Unknowns	Iter.	CPU(s)	Mem(MB)
$1.5\lambda_0$	13.4	2563/19497/152030	11	60.52	38
3.0λ	14.3	10 464/81725/645758	11	267.98	162
5.0λ	13.6	22306/174953/1385686	11	532.68	349

Table 6.5   Convergence Performance Comparison (1.5λ Cube)

CG-based solver	Iterations	CPU (s)
No preconditioner	2 509	1592.49
SGAV	77	347.47
NMGAV	11	60.52

In Table 6.5 a comparison is given of the convergence performance of the CG-based iterative finite element solution of scattering by the 1.5λ cube, carried out a) without preconditioning, b) with a single-grid potential (SGAV) preconditioning, and c) with nested multigrid potential (NMGAV) preconditioning. The impact of the NMGAV process on improving convergence is evident.

The superior performance of the NMGAV-preconditioner is illustrated in a more visual manner through the convergence curves in Fig. 6.17. It is noted here that, due to its slow convergence, the curve for the iterative CG-based solution without preconditioning was truncated at iteration step 400.

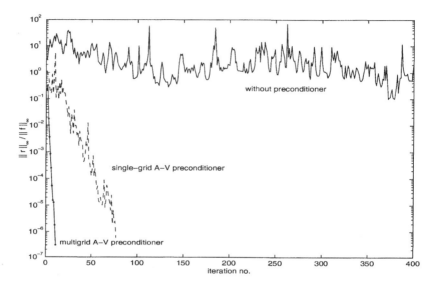

Figure 6.17   Convergence behavior of three CG-based iterative solvers. (After Yu and Cangellaris [24], ©2002 AGU.)

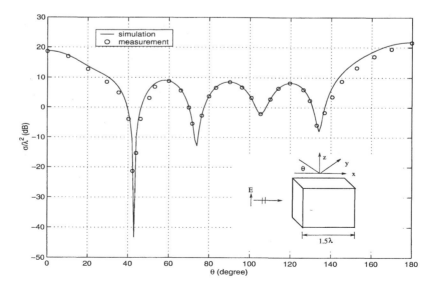

Figure 6.18   Bistatic radar cross-section for the 1.5λ PEC cube. (After Yu and Cangellaris [24], ©2002 AGU.)

To examine the accuracy of the finite element solution, the generated bistatic scattering radar cross-sections for the 1.5λ cubes are plotted in Figs. 6.18 and compared with measured data obtained from [23]. Excellent agreement is observed.

For the coarse-grid solution to be of use in the development of the solution on a finer grid, the spatial resolution of the electromagnetic fields provided by the coarse grid must be sufficient to constrain the numerical dispersion error [7], [8]. For the three cube sizes considered, the average grid size on the coarsest grids is ~ 3.62 pts/λ. In order to test the robustness of the NMGAV algorithm, the excitation frequency is increased to 450 MHz, thus reducing the coarsest grid resolution to $h_{avg}/\lambda = 2.42$. In this case 31 iterations are needed for solution convergence for all three cube sizes.

6.3 FINITE ELEMENT MODELING OF PASSIVE MICROWAVE COMPONENTS

Next, we turn our attention to the application of the multigrid potential preconditioners to the iterative solution of finite element approximations of BVPs associated with the electromagnetic modeling of passive microwave components and planar integrated circuits. This class of applications provides us with the opportunity to discuss in some detail the special case of excitation of an electromagnetic structure via the modal fields associated with the uniform waveguiding structures through which the structure connects with the rest of the microwave network.

Our discussion begins with the development of the weak statement of the electromagnetic BVP. Special attention is given to the surface integrals associated with the waveguide ports and the appropriate imposition of the modal excitation in a way that facilitates the direct generation of the modal scattering-parameter matrix for the structure. This is followed by the application of the multigrid potential preconditioner for the iterative solution of the resulting finite element system [25]. The section concludes with representative examples from the application of the iterative solver to the analysis of typical, planar microwave circuits.

6.3.1 Electromagnetic ports and associated boundary condition

Consider the generic microwave device shown in Fig. 6.19. Without loss of generality, the materials in the interior of the volume Ω are assumed to be linear, isotropic, and position-dependent. The boundary $\partial\Omega$ is split into two portions. Portion S_0 is meant to represent a truncation boundary between Ω and an unbounded domain that, for the purposes of this discussion, is assumed to be homogeneous. For example, for the case of a planar microstrip circuit, the unbounded medium is the homogeneous medium (e.g., air) that constitutes the microstrip superstrate. For the purposes of this discussion, the first-order absorbing boundary condition (5.7) is imposed on S_0. The remaining of $\partial\Omega$ consists of the union of the surfaces S_i, $i = 1, 2, \ldots, p$, each one of which is assumed to coincide with the cross-section of a uniform waveguide. We will refer to each one of the surfaces S_i as a *port*.

Returning to the development of the finite element approximation of the electromagnetic BVP inside Ω, the weak statement for the vector Helmholtz equation for the electric field

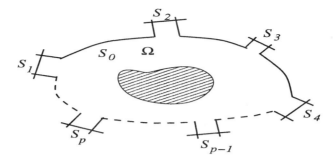

Figure 6.19   A multiport electromagnetic structure.

vector is repeated here for convenience

$$\iiint_{\Omega} \nabla \times \vec{w} \cdot \nabla \times \vec{E} \, dv - j\omega\mu_0 \iint_{\partial\Omega} \hat{n} \times \vec{H} \cdot \vec{w} ds$$
$$- \omega^2 \mu_0 \epsilon_0 \iiint_{\Omega} \vec{w} \cdot \epsilon_r \vec{E} dv = 0. \tag{6.102}$$

Let us assume that the structure is excited through port k. Furthermore, the port boundaries S_i are chosen in such a manner that only the dominant mode (understood to be the waveguide mode with the lowest cutoff frequency) of the associated waveguide is propagating at the excitation frequency. This restriction is not necessary and will be removed in the next section. Thus for each port $i \neq k$ it is

$$\hat{n} \times \vec{H}_i = \frac{1}{Z_i} \hat{n} \times \hat{n} \times \vec{E}_i, \qquad i \neq k \tag{6.103}$$

where $\hat{n}$ is the outward-pointing, unit normal on S_i, $\vec{E}_i$ and $\vec{H}_i$ are the transverse parts of the dominant mode at port i, and Z_i is the associated modal impedance [27]. As far as the excitation port, S_k, is concerned, the above equation relates the transverse parts of the reflected modal electric and magnetic fields. Thus we have

$$\hat{n} \times (\vec{H}_k - \vec{H}_k^{inc}) = \frac{1}{Z_k} \hat{n} \times \hat{n} \times (\vec{E}_k - \vec{E}_k^{inc}) \tag{6.104}$$

where the superscript inc denotes the incident modal electric and magnetic fields. Since the excitation is assumed to be through the dominant mode in the waveguide connected at port k pointing into the structure, it is

$$(-\hat{n}) \times \vec{H}_k^{inc} = \frac{1}{Z_k} (-\hat{n}) \times (-\hat{n}) \times \vec{E}_k^{inc}. \tag{6.105}$$

Substitution of this relation into (6.104) yields the following boundary condition on S_k

$$\hat{n} \times \vec{H}_k = \frac{1}{Z_k} \hat{n} \times \hat{n} \times \vec{E}_k - \frac{2}{Z_k} \hat{n} \times \hat{n} \times \vec{E}_k^{inc}. \tag{6.106}$$

Use of (6.103) and (6.106) into (6.102) results in the following weak form

$$\iiint_{\Omega} \nabla \times \vec{w} \cdot \nabla \times \vec{E}\, dv + \frac{j\omega\mu_0}{\eta} \iint_{S_0} \hat{n} \times \vec{w} \cdot \hat{n} \times \vec{E}\, ds$$

$$+ \sum_{i=1}^{p} \frac{j\omega\mu_0}{Z_i} \iint_{S_i} \hat{n} \times \vec{w} \cdot \hat{n} \times \vec{E}\, ds - \omega^2 \mu_0 \epsilon_0 \iiint_{\Omega} \vec{w} \cdot \epsilon_r \vec{E}\, dv \qquad (6.107)$$

$$= \frac{2j\omega\mu_0}{Z_k} \iint_{S_k} \hat{n} \times \vec{w} \cdot \hat{n} \times \vec{E}_k^{inc}\, ds.$$

We note that the subscript i has been dropped from the field quantities in the integrand of the port integral term on the left-hand side of the above equation. The reason for this is that in the subsequent development of the finite element approximation the electric field everywhere inside the computational domain (including the port surfaces) is expanded in terms of edge elements. However, consistent with our earliest assumption, the port boundaries are placed sufficiently far away from the junctions of the associated waveguides with the electromagnetic device for the dominant modes only to contribute to the field distribution over the port boundaries. Thus the finite element approximation of the electric field on each port boundary, S_i, is expected to satisfy with acceptable accuracy the modal impedance boundary condition (6.103).

Through a standard Galerkin's process the finite element matrix equation for the electric field formulation is obtained

$$M_{EE} x_E = f_E, \qquad (6.108)$$

where the elements of the matrix M_{EE} are identical to those for the scattering problem except for the fact that, in addition to an ABC boundary, port boundaries are also present in this case. However, their contribution to the elements of M_{EE} through the matrix Z in (6.61), are identical to those for the ABC boundary except for the fact that the unbounded medium intrinsic impedance η is now replaced by the modal impedance of the dominant mode at each port.

At this point, and prior to proceeding with the development of the weak form and the finite element matrix for the $A - V$ formulation, it is appropriate to comment on the desirable output from the finite element analysis of the multiport electromagnetic device. As already mentioned earlier, multiport electromagnetic devices are most commonly described in terms of one of several network parameters matrices, namely, the impedance, admittance, or scattering-parameter (S-parameter) matrix. For our purposes it suffices to consider the S-parameter matrix only, since the impedance and admittance matrices can be readily obtained from the S-parameter matrix using well-known relationships [26].

Referring to the modal field description of the field propagation in and out of each one of the waveguide ports, the ijth element, S_{ij}, $i, j = 1, 2, \ldots, p$ of the S-parameter matrix can be defined as the transmission coefficient between ports i and j. More specifically, under the assumption that only the dominant mode is propagating through each port at the operating frequency, S_{ij} is understood to be the amplitude of the dominant mode transmitted through port i when port j is excited by its dominant mode with amplitude 1. Clearly, the diagonal elements, S_{ii}, $i = 1, 2, \ldots, p$, of the S-parameter matrix are the reflection coefficients of the ports.

In general, different types of waveguides will be connected to the p ports of the electromagnetic device. Therefore, in order to account for the different modal impedances for the dominant modes at the p ports, the so-called *generalized* S-parameter matrix is most convenient for the network description of the electromagnetic device [26]. To elaborate, let

$A_k^{(+)}$, $B_k^{(-)}$ be, respectively, the amplitudes of the incident and reflected electric field of the fundamental mode at port k and Z_k its associated modal impedance. Then the *normalized incident wave*, a_k, and *normalized reflected wave*, b_k, are defined as follows:

$$a_k = \frac{A_k^{(+)}}{\sqrt{Z_k}}, \quad b_k = \frac{A_k^{(-)}}{\sqrt{Z_k}}. \tag{6.109}$$

It is noted that for a propagating mode (above cutoff) in a lossless waveguide the modal impedance is real. Furthermore, it is evident from the above definition that the square of the magnitude of the normalized incident and reflected waves are proportional (to within a constant) to the time-average incident and reflected power, respectively, at the port. Using these definitions for the normalized waves at each port, the generalized scattering parameter S_{ij}, quantifying electromagnetic power transmission from port j to port i, is computed as follows:

$$S_{ij} = \frac{b_i}{a_j}\big|_{a_k=0,k\neq j} = \frac{A_i^{(-)}\sqrt{Z_j}}{A_j^{(+)}\sqrt{Z_i}}\big|_{A_k^{(+)}=0,k\neq j}. \tag{6.110}$$

From the above definition of the generalized scattering parameters it is evident that the computation of the S-parameter matrix can be effected one column at a time by exciting, sequentially, each one of the ports by the associated dominant mode with amplitude 1 and extracting from the resulting electric field distribution over each port the amplitude of the transmitted dominant mode. This extraction process is facilitated by the orthogonality of the eigenmodes of a uniform waveguide [27]. This orthogonality is discussed in more detail in Chapter 9 in conjunction with the eigen-analysis of uniform electromagnetic waveguides. At this point it suffices to recall the fact that, through an appropriate normalization of the electric and magnetic field distributions of the eigenmodes, the following mode orthogonality relationship is useful to the computation of the scattering parameters:

$$\iint_{S_i} \hat{n} \cdot \left(\vec{e}_{i,m} \times \vec{h}_{i,n}\right) = \delta_{mn}, \quad (i = 1, 2, \cdots, p). \tag{6.111}$$

In this expression $\vec{e}_{i,m}$, $\vec{h}_{i,m}$ denote the normalized transverse parts (tangential to the port surface) of the electric and magnetic field for mode m of the waveguide connected at port S_i, [27].

The transverse field distributions for the dominant mode ($\vec{e}_{i,1}$ and $\vec{h}_{i,1}$) are assumed to be known at each one of the p ports, and are used both for the excitation at each port

$$\vec{E}_k^{inc} = \vec{e}_{k,1} \tag{6.112}$$

and for the calculation of the scattering parameters through the following integrals over the port surfaces

$$S_{ik} = \iint_{S_i} \hat{n} \cdot \left(\vec{E} \times \vec{h}_{i,1}\right) ds \quad i \neq k. \tag{6.113}$$

As far as the reflection coefficient at the excitation port k is concerned, expressing the reflected transverse electric field on S_k in terms of the total and incident fields and making use of the orthogonality relationship (6.111), yields the following result:

$$\begin{aligned} S_{kk} &= \iint_{S_k} \hat{n} \cdot \left[\left(\vec{E} - \vec{E}_k^{inc}\right) \times \vec{h}_{k,1}\right] ds = \iint_{S_k} \hat{n} \cdot \left[\left(\vec{E} - \vec{e}_{k,1}\right) \times \vec{h}_{k,1}\right] ds \\ &= \iint_{S_k} \hat{n} \cdot \left(\vec{E} \times \vec{h}_{k,1}\right) ds - 1.0. \end{aligned} \tag{6.114}$$

This completes the discussion of the extraction of the generalized S-parameter matrix for the electromagnetic multiport. Even though the extracted S-parameters are referenced to the modal impedances of the dominant modes at the ports, their re-normalization in terms of a common reference impedance is straightforward in view of (6.109) and (6.110) [26]. Once again, we emphasize that this process assumes that only the dominant (fundamental) mode is propagating, and thus present, at the port cross-section. This is equivalent to saying that the extraction process is accurate only when the waveguide ports are placed at a distant sufficiently far away from the junction between the waveguide and the electromagnetic device for all higher-order, non-propagating modes to have subsided sufficiently. The general case where no restrictions are imposed on the number of modes contributing to the field distribution over the port cross-section is discussed in the next subsection.

We conclude this subsection with the derivation of the weak statement for the $A - V$ formulation of this boundary value problem. Similar to the electromagnetic scattering problem case, we substitute $\vec{E} = \vec{A} + \nabla V$ into (6.107) to obtain one of the two equations that constitute the desired weak statement

$$
\iiint_\Omega \nabla \times \vec{w} \cdot \nabla \times \vec{A}\, dv + \frac{j\omega\mu_0}{\eta} \iint_{S_0} \hat{n} \times \vec{w} \cdot \hat{n} \times (\vec{A} + \nabla V)\, ds
$$

$$
+ \sum_{i=1}^{p} \frac{j\omega\mu_0}{Z_i} \iint_{S_i} \hat{n} \times \vec{w} \cdot \hat{n} \times (\vec{A} + \nabla V)\, ds \tag{6.115}
$$

$$
- \omega^2 \mu_0 \epsilon_0 \iiint_\Omega \vec{w} \cdot \epsilon_r (\vec{A} + \nabla V)\, dv = \frac{2j\omega\mu_0}{Z_k} \iint_{\partial\Omega} \hat{n} \times \vec{w} \cdot \hat{n} \times \vec{E}_k^{inc}\, ds.
$$

The second equation is the weak form of $\nabla \cdot \vec{D} = 0$. Using the same integration-by-parts process as in the case of the electromagnetic scattering problem we obtain

$$
\frac{j\omega\mu_0}{\eta} \iint_{S_0} \hat{n} \times \nabla\phi \cdot \hat{n} \times (\vec{A} + \nabla V)\, ds + \sum_{i=1}^{p} \frac{j\omega\mu_0}{Z_i} \iint_{S_i} \hat{n} \times \nabla\phi \cdot \hat{n} \times (\vec{A} + \nabla V)\, ds
$$

$$
- \omega^2 \mu_0 \epsilon_0 \iiint_\Omega \nabla\phi \cdot \epsilon_r (\vec{A} + \nabla V)\, dv = \frac{2j\omega\mu_0}{Z_k} \iint_{S_k} \hat{n} \times \nabla\phi \cdot \hat{n} \times \vec{E}_k^{inc}\, ds.
$$

$$
\tag{6.116}
$$

A Galerkin's process yields from (6.115) and (6.116) the finite element matrix for the $A-V$- formulation. The resulting matrix has the exact same form as (6.66). Thus the NMGAV Algorithm 6.4 is readily applicable for the construction of a multigrid preconditioner for this problem.

6.3.2 Transfinite-element boundary truncation

The transfinite element (TFE) method was proposed in [29]-[31] as an effective and accurate means for the truncation of uniform waveguide sections attached to an otherwise three-dimensional electromagnetic structure. In this section, we will review the TFE method and discuss its adaptation for use in conjunction with the $A - V$ formulation-based multigrid preconditioner.

For the purposes of this discussion, and without loss of generality, we will assume that only one uniform waveguide is connected with the three-dimensional volume Ω. Thus only one physical waveguide port, S_1 is present. Referring to the geometry of Fig. 6.19, the computational domain boundary $\partial\Omega$ is, simply, $\partial\Omega = S_0 \bigcup S_1$. Furthermore, it is assumed that through an eigenvalue analysis an appropriate subset of N_m waveguide eigenmodes, $\left(\vec{E}_i, \vec{H}_i\right)$, $i = 1, 2, \ldots, N_m$, along with their eigenvalues and modal impedances, have been obtained. It is assumed that N_m is sufficient for the truncated sum

$$\sum_{i=1}^{N_m} \alpha_i \vec{E}_i \tag{6.117}$$

to provide for a sufficiently accurate approximation of the electric field on the port boundary S_1. (The reader is referred to the classical electromagnetic theory texts [27] and [28] for the detailed discussion of the utilization of the eigenmodes of uniform electromagnetic waveguides as a basis for the representation of arbitrary field distributions inside the waveguide.) Furthermore, it is assumed that the transverse electric and magnetic field mode profiles, $(\vec{e}_i, \vec{h}_i)$, $i = 1, 2, \ldots, N_m$, are normalized according to (6.111).

Consider, next, the weak statement (6.102) of the E-field formulation. Assuming a tetrahedron element-based discretization of the computational domain, the electric field is expanded as follows:

$$\vec{E} = \sum_{i=1}^{N_e - N_{pe}} \vec{w}_i E_i + \sum_{i=1}^{N_m} \alpha_i \vec{e}_i + \vec{e}_k, \tag{6.118}$$

where N_e is number of edges in the mesh excluding those on PEC surfaces, N_{pe} is the number of edges on the port boundary S_1, and α_i is the coefficient for the i-th mode. Thus the tangential electric field on the port boundary is solely determined by the second term on the right-hand side of (6.118). Notice also that we have explicitly identified in (6.118) the excitation for the problem, taken to be the k-th mode with amplitude 1. It follows immediately from the definition of the normalized modal fields in the waveguide that the tangential magnetic field at the port boundary can be expanded in terms of the transverse parts of the modal magnetic fields as follows:

$$\vec{H} = \sum_{i=1}^{N_m} \alpha_i \vec{h}_i - \vec{h}_k, \tag{6.119}$$

where the minus sign in front of $\vec{h}_k$ is due to the fact that the incident (excitation) wave propagates into the computational domain, while our definition for the normalized modal electric and magnetic fields was based on the direction of propagation along the waveguide axis taken to be along the outward-pointing, unit normal on the port boundary.

A simplified, alternative form of (6.118) and (6.119) is obtained by absorbing the incident electric field mode into the modal summation; hence, $\alpha_k \leftarrow \alpha_k + 1.0$, and the two revised expansions become

$$\vec{E} = \sum_{i=1}^{N_e - N_{pe}} \vec{w}_i E_i + \sum_{i=1}^{N_m} \alpha_i \vec{e}_i \tag{6.120}$$

and

$$\vec{H} = \sum_{i=1}^{N_m} \alpha_i \vec{h}_i - 2\vec{h}_k. \tag{6.121}$$

Substitution of (6.120) and (6.121) into (6.102), followed by the imposition of a first-order ABC on S_0 and the application of Galerkin's process yields the following finite element equation for $j = 1, 2, \ldots, N_e - N_{pe}$:

$$
\iiint_\Omega \nabla \times \vec{w}_j \cdot \nabla \times \left(\sum_{i=1}^{N_e - N_{pe}} \vec{w}_i E_i + \sum_{i=1}^{N_m} \alpha_i \vec{e}_i \right) dv
$$

$$
+ \frac{j\omega\mu_0}{\eta} \iint_{S_0} \hat{n} \times \vec{w}_j \cdot \hat{n} \times \left(\sum_{i=1}^{N_e - N_{pe}} \vec{w}_i E_i + \sum_{i=1}^{N_m} \alpha_i \vec{e}_i \right) ds
$$

$$
- j\omega\mu_0 \iint_{S_1} \hat{n} \times \left(\sum_{i=1}^{N_m} \alpha_i \vec{h}_i - 2\vec{h}_k \right) \cdot \vec{w}_j ds
$$
$$
\tag{6.122}
$$

$$
- \omega^2 \mu_0 \epsilon_0 \iiint_\Omega \vec{w}_j \cdot \epsilon_r \left(\sum_{i}^{N_e - N_{pe}} \vec{w}_i E_i + \sum_{i=1}^{N_m} \alpha_i \vec{e}_i \right) dv = 0.
$$

However, the surface integral on S_1 vanishes, since edge elements on S_1 have been excluded from the expansion of the electric field vector, and thus $\hat{n} \times \vec{w}_j = 0$ on S_1. Combining the $N_e - N_{pe}$ equations above in matrix form, we have

$$
\left(S - \omega^2 T + j\omega Z \right) x_E + P x_\alpha = 0, \tag{6.123}
$$

where S, T and Z are defined in (6.61), the vector x_E contains the expansion coefficients for the $N_e - N_{pe}$ edge elements (excluding the ones on PEC boundaries), and x_α,

$$
x_\alpha = [\alpha_1, \alpha_2, \cdots, \alpha_{N_m}]^T \tag{6.124}
$$

is the vector containing the expansion coefficients for the modal field representation of the tangential electric field on port boundary S_1. The elements of the matrix P are given by

$$
P_{i,j} = \iiint_\Omega \nabla \times \vec{w}_i \cdot \nabla \times \vec{e}_j \, dv + \frac{j\omega\mu_0}{\eta} \iint_{S_0} \hat{n} \times \vec{w} \cdot \hat{n} \times \vec{e}_i ds
$$
$$
- \omega^2 \mu_0 \epsilon_0 \iint_\Omega \vec{w} \cdot \epsilon_r \vec{e}_j dv. \tag{6.125}
$$

In a similar manner, substitution of (6.120) and (6.121) into (6.102), followed by multiplication with $\vec{e}_j$, $j = 1, 2, \ldots, N_m$, and integration over the computational domain yields

$$
\iiint_\Omega \nabla \times \vec{e}_j \cdot \nabla \times \left(\sum_{i=1}^{N_e - N_{pe}} \vec{w}_i E_i + \sum_{i=1}^{N_m} \alpha_i \vec{e}_i \right) dv
$$

$$
+ \frac{j\omega\mu_0}{\eta} \iint_{S_0} \hat{n} \times \vec{e}_j \cdot \hat{n} \times \left(\sum_{i=1}^{N_e - N_{pe}} \vec{w}_i E_i + \sum_{i=1}^{N_m} \alpha_i \vec{e}_i \right) ds
$$

$$
- j\omega\mu_0 \iint_{S_1} \hat{n} \times \left(\sum_{i=1}^{N_m} \alpha_i \vec{h}_i - 2\vec{h}_k \right) \cdot \vec{e}_j ds
$$
$$
\tag{6.126}
$$

$$
- \omega^2 \mu_0 \epsilon_0 \iiint_\Omega \vec{e}_j \cdot \epsilon_r \left(\sum_{i=1}^{N_e - N_{pe}} \vec{w}_i E_i + \sum_{i=1}^{N_m} \alpha_i \vec{e}_i \right) dv = 0.
$$

Considering the surface integral over S_1 and making use of the orthogonality property of the modes (6.111),only the term associated with $\vec{h}_j$ is nonzero. In matrix form the above equations become

$$P^T x_E + (R + j\omega\mu_0 I) x_\alpha = f, \tag{6.127}$$

where P is defined in (6.125), I is identity matrix, and the elements of R are given by

$$R_{i,j} = \iiint_\Omega \nabla \times \vec{e}_i \cdot \nabla \times \vec{e}_j \, dv + \frac{j\omega\mu_0}{\eta} \iint_{S_0} \hat{n} \times \vec{e}_i \cdot \hat{n} \times \vec{e}_j ds$$
$$- \omega^2 \mu_0 \epsilon_0 \iiint_\Omega \vec{e}_i \cdot \epsilon_r \vec{e}_j \, dv. \tag{6.128}$$

With regards to the excitation vector f, all elements are zero except for the element in the row associated with excitation mode k, which has the value

$$f_k = 2j\omega\mu_0. \tag{6.129}$$

Finally, the finite element matrix statement of the problem is obtained by combining (6.123) and (6.127),

$$\begin{bmatrix} (S + j\omega Z - \omega^2 T) & P \\ P^T & R + j\omega\mu_0 I \end{bmatrix} \begin{bmatrix} x_E \\ x_\alpha \end{bmatrix} = \begin{bmatrix} 0 \\ f \end{bmatrix}. \tag{6.130}$$

Let us now explain the details associated with the numerical computation of the elements of matrices P and R. The pertinent integrals require expressions for the tangential parts of the modal electric fields and their curls. Expanding the tangential parts of the modal electric fields in terms of the N_{pe} edge elements on the port boundary S_1, we have

$$\vec{e}_i = \sum_{n=N_e-N_{pe}+1}^{N_e} \vec{w}_n b_n^{(i)}, \quad i = 1, 2, \ldots, N_m \tag{6.131}$$

where $b_n^{(i)}$, $n = 1, 2, \ldots, N_m$, are the expansion coefficients in the edge-element expansion of the i-th modal field. Since the transverse modal electric fields are known, these coefficients are readily computed. The matrix form expression of the above equation, accounting for all N_m modes, is

$$\left[\vec{e}_1, \vec{e}_2, \cdots, \vec{e}^{N_m} \right] = \left[\vec{w}_{N_e-N_{pe}+1}, \vec{w}_{N_e-N_{pe}+2}, \cdots, \vec{w}_{N_e} \right] \underbrace{\left[b^1, b^2, \cdots, b^{N_m} \right]}_{B}, \tag{6.132}$$

where the vector b^i contains the expansion coefficients in the edge-element expansion of the i-th mode

$$b^i = \begin{bmatrix} b_{N_e-N_{pe}+1}^{(i)} & b_{N_e-N_{pe}+2}^{(i)} & \cdots & b_{N_e}^{(i)} \end{bmatrix}^T. \tag{6.133}$$

At this point it is useful to recognize that we have reinstated the edge elements on S_1 as expansion functions for the unknown electric field. Thus the space W of all edge elements used in the finite element approximation is decomposed into two subsets as follows:

$$W = \underbrace{\left[\vec{w}_1, \cdots, \vec{w}_{N_e-N_{pe}} \right]}_{W_I} \oplus \underbrace{\left[\vec{w}_{N_e-N_{pe}+1}, \cdots, \vec{w}_{N_e} \right]}_{W_P}, \tag{6.134}$$

where W_I denotes the subset of the edge elements in the computational domain excluding those on PEC surfaces and port S_1, and W_P denotes the subset of the edge elements on S_1.

With this classification, if we were to ignore for a minute the boundary condition on S_1, the finite element matrix can be written in the following compact form due to the bilinear form of S, Z and T

$$\begin{bmatrix} \langle W_I^T, W_I \rangle & \langle W_I^T, W_P \rangle \\ \langle W_P^T, W_I \rangle & \langle W_P^T, W_P \rangle \end{bmatrix}, \tag{6.135}$$

where the notation W_q, $q \in \{I, P\}$, is used to denote the row vector containing the elements of subset W_q, and

$$\langle W_s^T, W_q \rangle = \left[S(W_s^T, W_q)) + j\omega Z(W_s^T, W_q) - \omega^2 T(W_s^T, W_q) \right]. \tag{6.136}$$

In view of the above, (6.125), (6.128) and (6.132) lead to the following expressions for the matrices P and R:

$$\begin{aligned} P &= \langle W_I^T, W_P \rangle B \\ R &= B^T \langle W_P^T, W_P \rangle B. \end{aligned} \tag{6.137}$$

The explicit forms for the elements of P and R are obtained by substituting (6.132) into (6.125) and (6.128), respectively,

$$P_{i,j} = \sum_{m=N_e-N_{pe}+1}^{N_e} \left(\begin{array}{l} \iiint_\Omega \nabla \times \vec{w}_i \cdot \nabla \times \vec{w}_m \, dv \\ + \dfrac{j\omega\mu_0}{\eta} \iint_{S_0} \hat{n} \times \vec{w}_i \cdot \hat{n} \times \vec{w}_m ds \\ - \omega^2 \mu_0 \epsilon_0 \iint_\Omega \vec{w} \cdot \epsilon_r \vec{w}_m \, dv \end{array} \right) b_m^{(j)}$$

$$R_{i,j} = \sum_{m=N_e-N_{pe}+1}^{N_e} \sum_{n=N_e-N_{pe}+1}^{N_e} b_m^{(i)} \left(\begin{array}{l} \iiint_\Omega \nabla \times \vec{w}_m \cdot \nabla \times \vec{w}_n \, dv \\ + \dfrac{j\omega\mu_0}{\eta} \iint_{S_0} \hat{n} \times \vec{w}_m \cdot \hat{n} \times \vec{w}_n ds \\ - \omega^2 \mu_0 \epsilon_0 \iint_\Omega \vec{w}_m \cdot \epsilon_r \vec{w}_n \, dv \end{array} \right) b_n^{(j)}. \tag{6.138}$$

Use of (6.137) in (6.130) yields the following form of the finite element matrix for the case where the TFE process is used to truncate a waveguide port,

$$\begin{bmatrix} \langle W_I^T, W_I \rangle & \langle W_I^T, W_P \rangle B \\ B^T \langle W_P^T, W_I \rangle & B^T \langle W_P^T, W_P \rangle B + j\omega\mu_0 I \end{bmatrix} \begin{bmatrix} x_E \\ x_\alpha \end{bmatrix} = \begin{bmatrix} 0 \\ f \end{bmatrix}. \tag{6.139}$$

The solution of the above matrix equation yields directly the electric field solution in the interior of the electromagnetic device in terms of the vector of the edge-element coefficients x_E, along with the reflection coefficients for the modal fields at the waveguide port S_1. More specifically, the i-th element of the vector x_α equals the reflection coefficient of the i-th mode for $i \neq k$, where k is the excitation mode. As far as the reflection coefficient for the excitation mode is concerned, it is obtained by subtracting 1 from the k-th element of x_α.

While the aforementioned discussion was for the simple case of a single-port device, the methodology is readily extendable to the case where multiple waveguide ports are present. At this point it is important to recognize the fact that, depending on the operating frequency, more than one waveguide modes may be propagating modes. This, in turn, implies that the electromagnetic characterization of the structures requires the computation of reflection and transmission coefficient for all propagating modes through which the device exchanges energy with the rest of the microwave circuit. In other words, the S-parameter matrix, defined earlier on the basis of the number of physical ports and the dominant mode at each port, must be extended to allow for multiple propagating modes to be present at each port [32]. The orthogonality of the modes at each port provides for a systematic methodology for the definition of such an extended scattering-parameter matrix, which utilizes *modal* ports and their associated incident and reflected normalized modal waves to quantify the reflection and transmission properties of the electromagnetic multiport. It should be evident that the TFE method is, by construction, the most suitable methodology for the direct generation of the S-parameter matrix in this case.

6.3.3   Nested multigrid potential TFE preconditioner

In order to apply the nested multigrid potential preconditioner 6.4 to the iterative solution of the TFE matrix equation, we must first develop the potential formulation for the TFE truncation scheme, along with the transformation between the E-field formulation and $A-V$ potential formulation of the finite element approximations of the problem. In addition, expressions for the intergrid transfer operators for the additional matrices associated with the TFE truncation must be constructed.

Let us begin by re-writing (6.120) in terms of expansion functions associated with the magnetic vector and electric scalar potential contributions to the electric field vector

$$\vec{E} = \sum_{i=1}^{N_e - N_{pe}} \vec{w}_i a_i + \sum_{i=1}^{N_n - N_{pn}} \nabla \phi_i v_i + \sum_{i=1}^{N_m} \alpha_i \vec{e}_i + \vec{e}_k. \tag{6.140}$$

In the above expression N_n denotes the number of nodes in the computational domain excluding those on PEC surfaces, while N_{pn} is the number of nodes on the port boundary S_1. The finite element matrix equation for the $A-V$ formulation follows the standard Galerkin's process discussed earlier. More specifically, two of the necessary matrix equations in the system are obtained from (6.102), with (6.140) substituted for the electric field and the resulting equations tested first by edge elements $\vec{w}_i, i = 1, 2, \ldots, N_e - N_{pe}$, and then by the tangential modal electric fields $\vec{e}_i, i = 1, 2, \ldots, N_m$. Using the notation from the previous subsection, and making use of the relation between the gradient of node elements and edge elements (described by the gradient matrix G) the pertinent equations may be cast in the following compact form:

$$\begin{bmatrix} \langle W_I^T, W_I \rangle & \langle W_I^T, W_P \rangle B & \langle W_I^T, W_I \rangle G \\ B^T \langle W_P^T, W_I \rangle & B^T \langle W_P^T, W_P \rangle B + j\omega\mu_0 I & B^T \langle W_P^T, W_I \rangle G \end{bmatrix} \begin{bmatrix} x_A \\ x_\alpha \\ x_V \end{bmatrix} = \begin{bmatrix} 0 \\ f \end{bmatrix}.$$
$$\tag{6.141}$$

These equations must be augmented by the system of equations obtained from the explicit enforcement of the divergence-free constraint on the solution, $\nabla \cdot \vec{D} = 0$. The pertinent weak statement is

$$\frac{1}{j\omega\epsilon_0} \iint_{S_0} \nabla\phi \cdot \hat{n} \times \vec{H} ds + \frac{1}{j\omega\epsilon_0} \iint_{S_1} \nabla\phi \cdot \hat{n} \times \vec{H} ds - \iiint_{\Omega} \nabla\phi \cdot \epsilon_r \vec{E} dv = 0. \tag{6.142}$$

Since the node elements associated with the nodes on the port boundary S_1 are not included in the expansion of the electric field, the surface integral over S_1 is zero. Substitution of

(6.140) and (6.121) into the above equation, followed by multiplication by $\omega^2 \epsilon_0 \mu_0$ and a standard Galerkin's process, yields the following matrix equation:

$$\begin{bmatrix} G^T \langle W_I^T, W_I \rangle & G^T \langle W_I^T, W_P \rangle B & G^T \langle W_I^T, W_I \rangle G \end{bmatrix} \begin{bmatrix} x_A \\ x_\alpha \\ x_V \end{bmatrix} = 0, \tag{6.143}$$

where, once again, use was made of the relationship between the gradient of the nodal elements and edge elements.

The final form of the finite element matrix for the $A - V$ formulation of the problem is obtained by combining (6.141) and (6.143),

$$\begin{bmatrix} \langle W_I^T, W_I \rangle & \langle W_I^T, W_P \rangle B & \langle W_I^T, W_I \rangle G \\ B^T \langle W_P^T, W_I \rangle & B^T \langle W_P^T, W_I \rangle B + j\omega\mu_0 I & B^T \langle W_P^T, W_I \rangle G \\ G^T \langle W_I^T, W_I \rangle & G^T \langle W_I^T, W_P \rangle B & G^T \langle W_I^T, W_I \rangle G \end{bmatrix} \begin{bmatrix} x_A \\ x_\alpha \\ x_V \end{bmatrix} = \begin{bmatrix} 0 \\ f \\ 0 \end{bmatrix}. \tag{6.144}$$

It is straightforward to check that the finite element matrix in the above equation can be expressed in terms of the one for the E-field formulation of the previous subsection as follows:

$$\begin{bmatrix} \langle W_I^T, W_I \rangle & \langle W_I^T, W_P \rangle B & \langle W_I^T, W_I \rangle G \\ B^T \langle W_P^T, W_I \rangle & B^T \langle W_P^T, W_I \rangle B + j\omega\mu_0 I & B^T \langle W_P^T, W_I \rangle G \\ G^T \langle W_I^T, W_I \rangle & G^T \langle W_I^T, W_P \rangle B & G^T \langle W_I^T, W_I \rangle G \end{bmatrix}$$
$$= \begin{bmatrix} I & \\ & I \\ G^T & \end{bmatrix} \begin{bmatrix} \langle W_I^T, W_I \rangle & \langle W_I^T, W_P \rangle B \\ B^T \langle W_P^T, W_I \rangle & B^T \langle W_P^T, W_I \rangle B + j\omega\mu_0 I \end{bmatrix} \begin{bmatrix} I & & G \\ & I & \end{bmatrix}. \tag{6.145}$$

This result serves as the transformation equation between the two formulations.

The final task involves the development of the intergrid transfer operators. Consider two sets of grids, one is the coarse grid and the other is its nested, fine grid. Let W_I^{2h} denote the space spanned by the edge elements of the coarse grid, excluding those elements on a PEC boundary and the port boundary S_1. Similarly, let W_I^h denote the space spanned by the edge elements of the nested grid, again excluding elements on S_1 and on PEC boundaries. The elements of the two spaces are related by the transformation matrix Q, discussed in detail in Sections 6.1.5 and 6.2.4

$$W_I^{2h} = W_I^h Q. \tag{6.146}$$

Thus the following relationship exists between the coarse and nested grid finite element matrices for the E-field formulation

$$\begin{bmatrix} \langle W_I^{2h^T}, W_I^{2h} \rangle & \langle W_I^{2h^T}, W_P^h \rangle B \\ B^T \langle W_P^{h^T}, W_I^{2h} \rangle & B^T \langle W_P^{h^T}, W_I^h \rangle B + j\omega\mu_0 I \end{bmatrix}$$
$$= \underbrace{\begin{bmatrix} Q^T & \\ & I \end{bmatrix}}_{I_h^{2h}} \begin{bmatrix} \langle W_I^{h^T}, W_I^h \rangle & \langle W_I^{h^T}, W_P^h \rangle B \\ B^T \langle W_P^{h^T}, W_I^h \rangle & B^T \langle W_P^{h^T}, W_I^h \rangle B + j\omega\mu_0 I \end{bmatrix} \underbrace{\begin{bmatrix} Q & \\ & I \end{bmatrix}}_{I_{2h}^h}, \tag{6.147}$$

where the intergrid transfer operators I_h^{2h} and I_{2h}^h have been indicated. From the above expression it is evident that the edge-elements on the port boundary S_1 are fixed to be those of the fine grid. This way, a single edge-element expansion of the tangential modal fields is used in the multigrid process. With the intergrid transfer operators available, the multigrid potential preconditioning is performed using Algorithm 6.4.

6.3.4 Applications

We present next two examples from the application of the multigrid potential precondition-ing process to the extraction of the scattering parameters of two planar microwave circuits. Both circuits are of the microstrip type. In both cases the frequency bandwidth over which the scattering parameters are computed is such that only the fundamental (quasi-TEM) mode is propagating in the input and output microstrip lines. In addition, the microstrip ports are placed sufficiently far away from the planar device under analysis for the quasi-TEM microstrip mode to suffice for the description of the transverse electromagnetic field distribution over the port boundary.

The first device is the microstrip low-pass filter depicted in the insert of Fig. 6.20. This filter was also studied in [33]. The dimensions of the filter are shown in the filter geometry in the insert of Fig. 6.20. A truncation boundary, on which a first-order ABC is imposed, is placed 4 mm away from the top and the two ends of the 20.32-mm long microstrip segment, which is perpendicular to the input and output 10-mm long sections. The average element size in the fine grid is 0.79 mm, and the number of unknowns is 60534. The average element size in the coarse grid is 1.57 mm and the number of unknowns is 7823.

Shown in Fig. 6.20 are S-parameter magnitude plots for the low-pass filter. Their com-parison with those in [33] reveals very good agreement. The convergence plots in Fig. 6.21

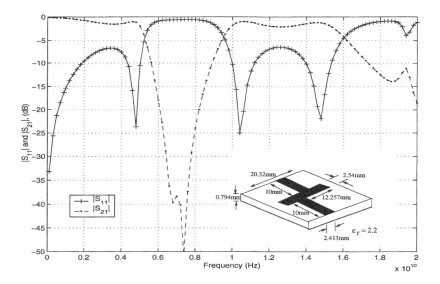

Figure 6.20   S-parameter magnitude plots for a low-pass filter. (After Zhu and Cangellaris [25], ©2001 IEEE.)

depict the performance of the preconditioned iterative solver at low (1 GHz), medium (10 GHz), and high (20 GHz) frequencies. On the PC platform used for the analysis the total CPU time to achieve convergence at 1, 10, and 20 GHz was, respectively, 36.10, 44.87, and 53.66 seconds. The required memory was 20 MB.

To further examine the convergence and solution efficiency of the iterative solver, the number of iterations was recorded for each one of the indicated frequency points on the curves of Fig. 6.20. On the average, 12 iterations and 45 seconds of CPU time were needed for convergence. In [33], the same problem was analyzed with a finite element grid of 33352

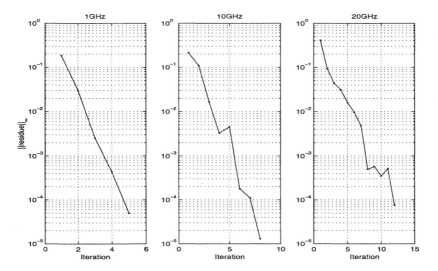

Figure 6.21 Convergence behavior of the preconditioned iterative solver for the microstrip filter depicted in the insert of Fig. 6.20. (After Zhu and Cangellaris [25], ©2001 IEEE.)

unknowns. The solution required 20 minutes per frequency point in the lower frequency range and 10 minutes per frequency point in the higher frequency range. Clearly, the nested multigrid potential preconditioned solver offers significant improvement in convergence and computational efficiency. Finally, to further demonstrate the robustness of the preconditioner, the excitation frequency was set at 30 GHz. At this frequency the average element size on the coarsest grid is 3.26 points per wavelength, approaching the lowest acceptable limit for avoiding numerical solution degradation due to numerical dispersion. In this case the solver converged in 25 iterations.

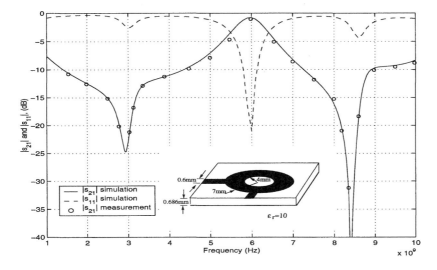

Figure 6.22 S-parameter magnitude plots for an annular microstrip resonator. (After Zhu and Cangellaris [25], ©2001 IEEE.)

The second microwave device considered is the annular microstrip resonator shown in the insert of Fig. 6.22. The structure was truncated at the top and the two sides where no microstrip ports are located with a truncation boundary on which a first-order ABC was imposed. The distance of the top truncation boundary from the dielectric interface was 2 mm. The computational domain used was $18 \times 18 \times 2.635$ mm^3. The average element size in the fine grid is 0.70 mm, while the number of unknown is 27840. For the coarse grid the element size is 1.4 mm and the number of unknowns is 3521.

The magnitudes of the scattering parameters for this device are plotted in Fig. 6.22. Shown in the figure are also data measurement data for the transmission coefficient obtained from [34]. Very good agreement is observed. The convergence plots in Fig. 6.23 depict the performance of the preconditioned iterative solver at low (1 GHz), medium (5 GHz), and high (10 GHz) frequencies. On the PC used for the analysis the total CPU time to achieve convergence at these three frequencies was, respectively, 14.01, 15.16, and 18.75 seconds. On the average, 11 iterations and 16 seconds of CPU time were needed for convergence.

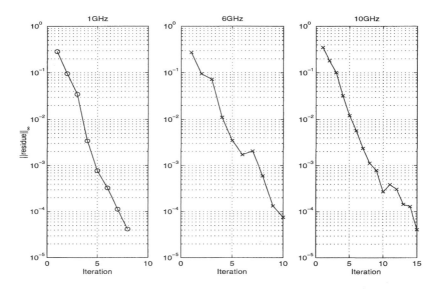

Figure 6.23 Convergence behavior of the preconditioned iterative solver for the microstrip filter depicted in the insert of Fig. 6.22. (After Zhu and Cangellaris [25], ©2001 IEEE.)

6.4 SYMMETRY OF THE NESTED MULTIGRID POTENTIAL PRECONDITIONER

We conclude this chapter with the examination of the symmetry of the nested multigrid preconditioner. As we have already mentioned, the symmetry of the preconditioner is important since it enables the use of the memory-efficient CG or minimum residual (MINRES) iterative solvers. With regards to the smoothing operations, Gauss-Seidel provides for an effective and low-cost smoother. We saw in Section 4.6.2 that the use of the symmetric Gauss-Seidel (SGS) preconditioner of 4.5.2 for the smoothing operations in the multigrid process provided for a symmetric multigrid preconditioner. However, in the case of the nested multigrid potential preconditioner, the smoothing operations are performed on the vector-scalar potential formulation. In the following we examine the symmetry of the nested multigrid potential preconditioner described in Algorithm 6.4. We consider one set of nested grids, a fine grid and a coarse grid, which are denoted by the superscript h and H, respectively. The multigrid case involving a nested hierarchy of grids can be proved by induction.

6.4.1 Potential smoothing operators

We begin with the pre-smoother in Step 2.a of Algorithm 6.4. The smoothing operation can be considered as a two-step process. Step 1 involves m forward GS smoothing operations on the field or the vector potential part with zero initial guess $x_E = 0$,

$$M_{EE}^h x_E = f_E \quad \Rightarrow \quad x_E \leftarrow \sum_{i=0}^{m-1} H_{EE,f-GS}^i R_{EE,f-GS} f_E, \quad (6.148)$$

where

$$H_{EE,f-GS} = (D_{EE} - L_{EE})^{-1} U_{EE}, \quad R_{EE,f-GS} = (D_{EE} - L_{EE})^{-1}. \quad (6.149)$$

The matrices L_{EE}, D_{EE}, and U_{EE} are, respectively, the strict low-triangular, diagonal, and strict upper-triangular matrices of M_{EE}^h. After the smoothing on the vector potential part of the solution, the residual is mapped onto the scalar potential part. Step 2 involves the smoothing on the scalar potential part with zero initial guess $x_V = 0$

$$M_{VV} x_V = G^T (f_E - M_{EE}^h x_E) \quad \Rightarrow$$

$$x_V \leftarrow \sum_{i=0}^{m-1} H_{VV,f-GS}^i R_{VV,f-GS} G^T (f_E - M_{EE}^h x_E), \quad (6.150)$$

where

$$H_{VV,f-GS} = (D_{VV} - L_{VV})^{-1} U_{VV}, \quad R_{VV,f-GS} = (D_{VV} - L_{VV})^{-1}. \quad (6.151)$$

The matrices L_{VV}, D_{VV}, and U_{VV} are, respectively, the strict low-triangular, diagonal, and strict upper-triangular matrices of M_{VV}^h. Combining the solutions of (6.148) and (6.150), we have

$$x_E \leftarrow x_E + G x_V = \underbrace{\left[\underline{R}_{EE} + G \underline{R}_{VV} G^T \left(I - M_{EE}^h \underline{R}_{EE} \right) \right]}_{R} f_E, \quad (6.152)$$

where

$$\underline{R}_{EE} = \sum_{i=0}^{m-1} H_{EE,f-GS}^i R_{EE,f-GS}, \quad \underline{R}_{VV} = \sum_{i=0}^{m-1} H_{VV,f-GS}^i R_{VV,f-GS}. \quad (6.153)$$

Next we consider the post-smoothing in Step 2.e of Algorithm 6.4. The post-smoothing process follows the reverse of the pre-smoothing one. Thus Step 2.e.2 involves m backward GS smoothing operations on the scalar potential part with zero initial guess $x_V = 0$

$$M_{VV}^h x_V = G^T (f_E - M_{EE}^h x_E) \quad \Rightarrow$$

$$x_V \leftarrow \sum_{i=0}^{m-1} H_{VV,b-GS}^i R_{VV,b-GS} G^T (f_E - M_{EE}^h x_E), \quad (6.154)$$

where

$$H_{VV,b-GS} = (D_{VV} - U_{VV})^{-1} L_{VV}, \quad R_{VV,b-GS} = (D_{VV} - U_{VV})^{-1}. \quad (6.155)$$

After the backward smoothing operation on the scalar potential part, the new residual in the vector potential part is calculated in Step 2.e.3. Then, Step 2.e.4 involves m backward GS smoothing operations on the vector potential part with nonzero initial guess obtained from Step 2.c

$$M_{EE}^h x_E = f_E - M_{EE}^h G x_V \quad \Rightarrow$$

$$x_E \leftarrow H_{EE,b-GS}^m x_E + \sum_{i=0}^{m-1} H_{EE,b-GS}^i R_{EE,b-GS} (f_E - M_{EE}^h G x_V), \quad (6.156)$$

where

$$H_{EE,b-GS} = (D_{EE} - U_{EE})^{-1} L_{EE}, \quad R_{EE,f-GS} = (D_{EE} - U_{EE})^{-1}. \quad (6.157)$$

In Step 2.e.5, the solutions for the vector and scalar potentials are combined. Substitution of x_v in (6.154) into (6.156) yields

$$\begin{aligned}
x_E &\leftarrow x_E + G x_V \\
&= H_{EE,b-GS}^m x_E + \overline{R}_{EE} \left[f_E - M_{EE}^h G \overline{R}_{VV} G^T (f_E - M_{EE}^h x_E) \right] \\
&\quad + G \overline{R}_{VV} G^T (f_E - M_{EE}^h x_E) \\
&= \underbrace{\left[H_{EE,b-GS}^m - (I - \overline{R}_{EE} M_{EE}^h) G \overline{R}_{VV} G^T M_{EE}^h \right]}_{\overline{H}} x_E \\
&\quad + \underbrace{\left[\overline{R}_{EE} + (I - \overline{R}_{EE} M_{EE}^h) G \overline{R}_{VV} G^T \right]}_{\overline{R}} f_E,
\end{aligned} \quad (6.158)$$

where it is

$$\overline{R}_{EE} = \sum_{i=0}^{m-1} H_{EE,b-GS}^i R_{EE,b-GS}, \quad \overline{R}_{VV} = \sum_{i=0}^{m-1} H_{VV,b-GS}^i R_{VV,b-GS}. \quad (6.159)$$

6.4.2 Symmetric nested two-grid potential preconditioner

The results of the previous subsection allow us to write the nested two-grid potential preconditioner in matrix form as follows:

$$x_E \leftarrow \underbrace{\left[\overline{H}\underline{R} + \overline{R} + \overline{H} I_H^h M_{EE}^{H}{}^{-1} I_h^H (I - M_{EE}^h \underline{R}) \right]}_{P} f_E. \quad (6.160)$$

Thus, in order for the process to be symmetric, the preconditioning matrix P must be symmetric. Starting with the first two terms, substitution of (6.152) and (6.158) yields

$$\begin{aligned}
\overline{H}\underline{R} + \overline{R} &= H_{EE,b-GS}^m \underline{R}_{EE} + \overline{R}_{EE} + \left[H_{EE,b-GS}^m G \overline{R}_{VV} G^T \left(I - M_{EE}^h \underline{R}_{EE} \right) \right] \\
&\quad + \left(I - \overline{R}_{EE} M_{EE}^h \right) G \overline{R}_{VV} G^T \left(I - M_{EE}^h \underline{R}_{EE} \right) \\
&\quad - \left(I - \overline{R}_{EE} M_{EE}^h \right) G \overline{R}_{VV} G^T M_{EE}^h G \underline{R}_{VV} G^T \left(I - M_{EE}^h \underline{R}_{EE} \right).
\end{aligned} \quad (6.161)$$

The first two terms $\overline{H}^m_{EE} \underline{R}_{EE} + \overline{R}_{EE}$ are symmetric as proven in (4.103). Using the following, easy-to-prove identities

$$\overline{R}_{EE} = \underline{R}^T_{EE}, \quad \overline{R}_{VV} = \underline{R}^T_{VV} \tag{6.162}$$

we obtain the proof of the symmetry of $\overline{H}\underline{R} + \overline{R}$. With regards to the term $I - M^h_{EE}$, we have

$$I - M^h_{EE}\underline{R} = \left(I - M^h_{EE}\underline{R}_{EE}\right) - M^h_{EE}G\underline{R}_{VV}G^T\left(I - M^h_{EE}\underline{R}_{EE}\right). \tag{6.163}$$

However, it was shown in (4.139) that it is

$$I - M^h_{EE}\underline{R}_{EE} = \left(U_{EE}(D_{EE} - L_{EE})^{-1}\right)^m. \tag{6.164}$$

Substitution of this result into (6.163) yields

$$I - M^h_{EE}\underline{R} = \left(I - M^h_{EE}G\underline{R}_{VV}G^T\right)\left(U_{EE}(D_{EE} - L_{EE})^{-1}\right)^m. \tag{6.165}$$

Furthermore, since

$$\begin{aligned}
I - \overline{R}_{EE}M^h_{EE} &= I - \left(\sum_{i=0}^{m-1} H^i_{EE,b-GS} R_{EE,b-GS}\right)M^h_{EE} \\
&= I - \left(I - H^m_{EE,b-GS}\right)\left(I - H_{EE,b-GS}\right)^{-1} R_{EE,b-GS}M^h_{EE} \\
&= I - \left(I - H^m_{EE,b-GS}\right)\left(I - H_{EE,b-GS}\right)^{-1}\left(I - H_{EE,b-GS}\right) \\
&= H^m_{EE,b-GS} = \left((D_{EE} - L_{EE})^{-1}U_{EE}\right)^m,
\end{aligned} \tag{6.166}$$

we have, from (6.158),

$$\overline{H} = \left[(D_{EE} - L_{EE})^{-1}U_{EE}\right]^m \left(I - G\overline{R}_{VV}G^T M^h_{EE}\right) = \left[I - M^h_{EE}\underline{R}\right]^T, \tag{6.167}$$

where use was made of (6.164). Hence, the third term in P is symmetric. This completes the proof of the symmetry of nested multigrid potential preconditioner.

In summary, what we have established in this chapter is the robustness of a preconditioning scheme based on the combination of the multigrid process with the vector-scalar potential formulation of the electromagnetic BVP for the iterative solution of the matrix obtained from the finite element approximation of full-wave vector electromagnetic problems. The resulting iterative solver is effective for the analysis of both scattering and guided-wave problems. The NMGAV preconditioner leads to expedient convergence and high accuracy of the iterative solution, provided that the discretization at the coarsest grid can resolve the minimum wavelength of interest with reasonable accuracy. In particular, the example studies indicated that an average element size of $\lambda/3$ on the coarsest grid suffices for fast convergence to an accurate solution.

REFERENCES

1. M. L. Barton and Z. J. Cendes, "New vector finite elements for three-dimensional magnetic field computation," *J. Appl. Phys.*, vol. 61, pp. 3919-3921, 1987.

2. R. Dyczij-Edlinger and O. Biro, "A joint vector and scalar potential formulation for driven high frequency problems using hybrid edge and nodal finite elements," *IEEE Trans. Microwave Theory Tech.*, vol. 44, pp. 15-23, Jan. 1996.

3. R. Dyczij-Edlinger, G. Peng, and J. F. Lee "A fast vector-potential method using tangentially continuous vector finite elements," *IEEE Trans. Microwave Theory Tech.* vol. 46, pp. 863-868, Jun. 1998.

4. R. Dyczij-Edlinger, G. Peng, and J. F. Lee, "Stability conditions for using TVFEMs to solve Maxwell equations in the frequency domain," *Int. J. Numer. Model.: Electron. Netw. Devices Fields*, vol. 13, pp. 245-260, 2000.

5. R. Beck and R. Hiptmair, "Multilevel solution of the time-harmonic Maxwell's equations based on edge elements," *Int. J. Numer. Methods Eng.*, vol. 45, pp. 901-920, 1999.

6. R. Hiptmair, "Multigrid method for Maxwell's equations," *SIAM J. Numer. Anal.*, vol. 36, no. 1, pp. 204-225, 1999.

7. R. Lee and A. C. Cangellaris, "A study of discretization error in the finite element approximation of wave solutions," *IEEE Trans. Antennas Propagat.*, vol. 40, pp. 542-549, May 1992.

8. A. F. Peterson and D. R. Wilton, "Curl-conforming mixed-order edge elements for discretizing the 2D and 3D vector Helmholtz equation," in *Finite Element Software for Microwave Engineering*, T. Itoh, G. Pelosi and P.P. Silvester, Eds. New York: John Wiley & Sons, 1996, pp. 101-125.

9. J. P. Webb and S. McFee, "Nested tetrahedra finite elements for h-adaption," *IEEE Trans. Magnetics*, vol. 35, pp. 1338-1341, May 1999.

10. Y. Zhu and A. C. Cangellaris "Nested multigrid vector and scalar potential finite element method for fast computation of two-dimensional electromagnetic scattering," *IEEE Trans. Antennas Propagat.*, Dec 2002. Vol. 50, 1850-1858.

11. J. P. Webb and S. McFee, "Nested tetrahedral finite elements for h-adaption," *IEEE Trans. Magnetics*, vol. 15, pp. 1338-1341, May 1999.

12. S. P. Marin, "Computing scattering amplitudes for arbitrary cylinders under incident plane waves," *IEEE Trans. Antennas Propagat.*, vol. AP-30, pp. 1045-1049, Nov. 1982.

13. K. D. Paulsen, D. R. Lynch and J. W. Strohbehn, "Three-dimensional finite, boundary, and hybrid element solutions of the Maxwell equations for lossy dilectric media," *IEEE Trans. Microwave Theory Tech.*, vol. 36, pp. 682-693, Apr. 1988.

14. J. M. Jin and V. V. Liepa, "A note on hybrid finite element method for solving scattering problems," *IEEE Trans. Antennas Propagat.*, vol. 36, pp. 1486-1490, Oct. 1988.

15. Z. Gong and A. W. Glisson, "A hybrid equation approach for the solution of electromagnetic scattering problems involving two-dimensional inhomogeneous dielectric cylinders," *IEEE Trans. Antennas Propagat.*, vol. 38, pp. 60-68, Jan. 1990.

16. X. Yuan, "Three-dimensional electromagnetic scattering from inhomogeneous objects by the hybrid moment and finite element method," *IEEE Trans. Microwave Theory Tech.*, vol. 38, pp. 1053-1058, Aug. 1990.

17. J. M. Jin and J. L. Volakis, "A hybrid finite element method for scattering and radiation by two- and three-dimensional structures," *IEEE Trans. Antennas Propagat.*, vol. 39, pp. 1598-1604. Nov. 1991.

18. S. S. Bindiganavale and J. L. Volakis, "A hybrid FE-FMM technique for electromagnetic scattering," *IEEE Trans. Antennas Propagat.*, vol. 45, pp. 180-183, Jan. 1997.

19. J. L. Volakis, A. Chatterjee, and J. Gong, "A class of hybrid finite element methods for electromagnetics: A review," *J. Electromagn. Waves Applications*, vol. 8, no. 9/10, pp. 1095-1124, 1994.

20. X. Q. Sheng, J. M. Jin, J. Song. C. C. Liu, and W. C. Chew, "On the formulation of hybrid finite-element and boundary integral methods for 3-D scattering," *IEEE Trans. Antennas Propagat.*, vol. 46, no. 3, pp. 303-313, Mar. 1998.

21. N. Lu and J. M. Jin, "Application of fast multipole method to finite-element boundary integral solution of scattering problems," *IEEE Trans. Antennas Propagat.*, vol. 44, no. 6, pp. 781-791, Jun. 1996.

22. J. Liu and J. M. Jin, "A highly effective preconditioner for solving the finite element-boundary integral matrix equation of 3-D scattering," *IEEE Trans. Antennas Propaga.*, vol. 50, no. 9, Sep. 2002.

23. M. C. Cote, M. B. Woodworth, and A. D. Yaghjian, "Scattering from the perfectly conducting cube," *IEEE Trans. Antennas Propagat.*, vol. 36, pp. 1321-1329, Sep. 1988.

24. Y. Zhu and A. C. Cangellaris, "Nested multigrid vector and scalar potential finite element method for three-dimensional time-harmonic electromagnetic analysis," *Radio Science*, vol. 37, May 2002.

25. Y. Zhu and A. C. Cangellaris, "Robust multigrid preconditioner for fast finite element modeling of microwave devices," *IEEE Microw. Wireless Compon. Letters*, vol. 11, pp. 416-418, Oct. 2001.

26. D. M. Pozar, *Microwave Engineering*, 2nd ed., New York: John Wiley & Sons Inc., 1998.

27. R. F. Harrington, *Time-Harmonic Electromagnetic Fields*, IEEE Press Series on Electromagnetic Wave Theory, New York: Wiley Interscience, John Wiley & Sons Inc., 2001.

28. R. E. Collin, *Field Theory of Guided Waves*, 2nd ed., IEEE Press Series on Electromagnetic Wave Theory, New York: Wiley Interscience, John Wiley & Sons Inc., 1990.

29. Z. J. Cendes and J. F. Lee, "The transfinite element method for modeling MMIC devices," *IEEE Trans. Microwave Theory Tech.*, vol 36, no. 12, pp. 1639-1649, Dec. 1988.

30. D-K. Sun, J. F. Lee, and Z. Cendes, "The transfinite element time-domain method," *IEEE Antennas and Propagation Society Symposium*, 2003 IEEE, vol. 1, p.693, 22-27 Jun. 2003.

31. D. Crawford and Z. Cendes, "Domain decomposition via the transfinite element method," *IEEE Antennas and Propagation Society Symposium*, 2004 IEEE, vol. 1, pp. 347-350, 20-25 Jun. 2004.

32. A. Morini and T. Rozzi, "On the definition of the generalized scattering matrix of a lossless multiport," *IEEE Trans. Microwave Theory Tech.*, vol. 49, no. 1, pp. 160-165, Jan. 2001.

33. A. C. Polycarpou, P. A. Tirkas, and C. A. Balanis, "The finite-element method for modeling circuits and interconnects for electromagnetic packaging," *IEEE Trans. Microwave Theory Tech.*, vol. 45, pp. 1868-1874, Oct. 1997.

34. G. Dínzeo, F. Giannini, and R. Sorrentino, "Wide-band equivalent circuits of microwave planar networks," *IEEE Trans. Microwave Theory Tech.*, vol. 10, pp. 1107-1113, Oct. 1980.

CHAPTER 7

HIERARCHICAL MULTILEVEL AND HYBRID POTENTIAL PRECONDITIONERS

In the previous chapter it was shown how the combination of the potential formulation of the electromagnetic problem with a nested multigrid technique led to the development of an effective preconditioner, which provided for the iterative solution of finite element approximations of two- and three-dimensional problems with very good accuracy and fast convergence. We referred to this preconditioner as the nested multigrid vector-scalar $(\vec{A} - V)$ potential preconditioner (NMGAV). Since the geometric nested multigrid process constructs the finer-level grids by dividing each tetrahedron in the coarser grid into eight equal-volume sub-tetrahedra, it can be considered as an h-refinement finite element method. While h-refinement schemes are very effective when applied to problems in which the fields exhibit rapid variation over small distances, p-refinement is typically found to be more effective and computationally more efficient in providing wave solution accuracy and numerical error dispersion control in computational domains with regions where the electromagnetic field variation is fairly smooth. We recall that p refinement provides enhanced field resolution through higher-order polynomial interpolation over a fixed-size, finite element grid[1]. Thus the first objective in this chapter is the development of a hierarchical, multilevel, potential preconditioner, which uses a fixed finite element grid and the set of hierarchical basis function spaces $\mathcal{H}^p(curl)$, discussed in Chapter 2, for the development of a robust and fast-converging, iterative solver for finite element approximations of electromagnetic boundary value problems.

One of the computational difficulties of the hierarchical multilevel potential preconditioner is the direct solution of the finite element matrix obtained from the approximation of the problem on the $\mathcal{H}^0(curl)$ space. This limits the application of this preconditioner to problems involving electromagnetic structures of moderate electrical size. To overcome this shortcoming, a hybrid preconditioner can be developed, which uses the hierarchical multilevel scheme on top of a nested multigrid scheme. More specifically, the nested multigrid scheme is utilized for the solution of the finite element matrix at the $\mathcal{H}^0(curl)$ level of the multilevel scheme.

The chapter is organized as follows: First, the finite element approximations for the field and potential formulations of the electromagnetic problem are developed using the high-order $\mathcal{H}^p(curl)$ spaces. Through this development, those attributes of the potential formulation that improve the convergence of the iterative matrix solution process are highlighted. This is followed by the presentation of the hierarchical multilevel and hybrid potential preconditioners. The chapter concludes with examples from the application of the proposed preconditioners to the electromagnetic analysis of various passive electromagnetic devices. These examples help demonstrate the convergence attributes of the resulting iterative solvers.

7.1 HIGHER-ORDER FIELD FORMULATION

The electromagnetic analysis of a passive, multiport electromagnetic device of the generic type depicted in Fig. 6.19 will be used as the vehicle for the development of the multilevel and the hybrid multigrid/multilevel preconditioners. The media within the computational domain are assumed to be linear, isotropic, time-invariant, and non-magnetic. A first-order absorbing boundary condition is imposed on the truncation boundary S_0, while single-mode port boundary conditions are utilized for the truncation at the port boundaries S_i, $i = 1, 2, \ldots, p$. Thus, with the time-harmonic variation of $\exp(j\omega t)$ assumed and suppressed for simplicity, the weak form of the driven problem for the electric-field (E-field) formulation is

$$
\iiint_\Omega \nabla \times \vec{w} \cdot \nabla \times \vec{E}\, dv + \frac{j\omega\mu_0}{\eta} \iint_{S_0} \hat{n} \times \vec{w} \cdot \hat{n} \times \vec{E}\, ds
$$
$$
+ \sum_{i=1}^{p} \frac{j\omega\mu_0}{Z_i} \iint_{S_i} \hat{n} \times \vec{w} \cdot \hat{n} \times \vec{E}\, ds - \omega^2 \mu_0 \epsilon_0 \iiint_\Omega \vec{w} \cdot \epsilon_r \vec{E}\, dv \qquad (7.1)
$$
$$
= \frac{2j\omega\mu_0}{Z_k} \iint_{S_k} \hat{n} \times \vec{w} \cdot \hat{n} \times \vec{e}_k\, ds,
$$

where Z_i is the modal impedance for the dominant propagating mode at the i-th port, and $\vec{e}_k$ is the incident modal electric field at the k-th (driven) port.

Following a standard Galerkin's testing scheme, the testing functions, $\vec{w}$, and the expansion functions in the approximation of $\vec{E}$ are taken to be the basis functions in the $\mathcal{H}^p(curl)$ subspace of the TV space as discussed in Chapter 2. Recall that $\mathcal{H}^p(curl)$ is the space for which both its basis functions and their *curl* are complete to the p-th degree vector polynomial functions as explained in Chapter 2. The lowest (zeroth) order space is $\mathcal{H}^0(curl)$, with basis functions the so-called edge elements. Let the i-th edge of a tetrahedron be defined as (m, n), where m, n are the associated vertices. Then it is

$$
\mathcal{H}^0(curl) = \text{span}\left\{ \vec{w}_{e,i} = \lambda_m \nabla \lambda_n - \lambda_n \nabla \lambda_m \mid \text{edge } i\ (m, n) \right\}. \qquad (7.2)
$$

Expansion of the electric field using edge elements yields the finite element matrix in (6.108). The first-order space can be written as

$$
\mathcal{H}^1(curl) = \mathcal{H}^0(curl) \oplus \nabla W_{s,e}^2, \oplus R_{tv,f}^2 \qquad (7.3)
$$

where $W_{s,e}^2$ represents the second-order, edge-type scalar subspace, the gradient of which is added to make the basis functions in $\mathcal{H}^1(curl)$ complete to first-degree polynomials. For the i-th edge (m, n), there is one basis function in $W_{s,e}^2$

$$W_{s,e}^2 = \text{span} \left\{ w_{v,i} = \lambda_m \lambda_n \mid \text{edge } i \ (m, n) \right\}. \tag{7.4}$$

The space $R_{tv,f}^2$ denotes the second-order, facet-type, non-gradient TV subspace, the addition of which makes the curl of the basis functions in $\mathcal{H}^1(curl)$ complete to first-degree polynomials. For the i-th facet (m, p, q) there are two basis functions in $R_{tv,f}^2$ as

$$R_{tv,f}^2 = \text{span} \left\{ \begin{matrix} \vec{w}_{f,i,1} = 4\lambda_m(\lambda_n \nabla \lambda_p - \lambda_p \nabla \lambda_n) \\ \vec{w}_{f,i,2} = 4\lambda_n(\lambda_p \nabla \lambda_m - \lambda_m \nabla \lambda_p) \end{matrix} \middle| \text{ face } i \ (m, n, p) \right\}. \tag{7.5}$$

Thus the expansion of the electric field vector in $\mathcal{H}^1(curl)$ assumes the form

$$\vec{E} = \sum_{i=1}^{N_e} (\vec{w}_{e,i} e_{e,i} + \nabla w_{v,i} e_{v,i}) + \sum_{i=1}^{N_f} (\vec{w}_{f,i,1} e_{f,i,1} + \vec{w}_{f,i,2} e_{f,i,2}), \tag{7.6}$$

where $e_{e,i}$, $e_{v,i}$, $e_{f,i,1}$ and $e_{f,i,2}$ are the expansion coefficients for the corresponding basis functions. Since the basis functions on PEC surfaces are excluded, N_e and N_f are, respectively, the number of edges and facets in the domain excluding any edges and facets on PEC boundaries.

Substitution of (7.6) into (7.1), followed by Galerkin's testing, yields the finite element matrix for the second-order E-field formulation. Following the notation used in (6.136), the finite element matrix equation is written as

$$\begin{bmatrix} \langle W_f^T, W_f \rangle & \langle W_f^T, \nabla W_v \rangle & \langle W_f^T, W_e \rangle \\ \langle \nabla W_v^T, W_f \rangle & \langle \nabla W_v^T, \nabla W_v \rangle & \langle \nabla W_v^T, W_e \rangle \\ \langle W_e^T, W_f \rangle & \langle W_e^T, \nabla W_v \rangle & \langle W_e^T, W_e \rangle \end{bmatrix} \begin{bmatrix} x_{E,f} \\ x_{E,v} \\ x_{E,e} \end{bmatrix} = \begin{bmatrix} f_{E,f} \\ f_{E,v} \\ f_{E,e} \end{bmatrix}, \tag{7.7}$$

where the elements of the (row vector) W_e are the functions $\vec{w}_{e,i}$, those of W_v are the functions $w_{v,i}$, and those of W_f are the functions $\vec{w}_{f,i,1}$ and $\vec{w}_{f,i,2}$. The row vector ∇W_v contains the gradients of the elements of W_v. The vector $x_{E,e}$ contains the expansion coefficients $e_{e,i}$, $x_{E,v}$ contains the expansion coefficients $e_{v,i}$, and $x_{E,f}$ contains the expansion coefficients $e_{f,i,1}$ and $e_{f,i,2}$ in (7.6). The specific forms of the integrals associated with the entries in each one of the submatrices in the above equation are easily deduced through a direct comparison with (7.1).

It is useful to recast the above matrix equation in a form that separates the zeroth-order and first-order contributions to the electric field expansion. Thus we write

$$\begin{bmatrix} M_{EE}^{11} & M_{EE}^{10} \\ M_{EE}^{01} & M_{EE}^{00} \end{bmatrix} \begin{bmatrix} x_E^1 \\ x_E^0 \end{bmatrix} = \begin{bmatrix} f_E^1 \\ f_E^0 \end{bmatrix}, \tag{7.8}$$

where

$$x_E^0 = x_{E,e}, \quad x_E^1 = \begin{bmatrix} x_{E,f} \\ x_{E,v} \end{bmatrix}, \quad f_E^0 = x_{E,e}, \quad f_E^1 = \begin{bmatrix} f_{E,f} \\ f_{E,v} \end{bmatrix}. \tag{7.9}$$

The specific forms of the entries of the matrices M_{EE}^{ij}, $i, j \in \{1, 0\}$, are easily deduced from (7.7).

The reasons for the slow convergence of iterative solvers, such as CG, when applied to the solution of the above matrix have been discussed in detail in the previous chapter and will not be elaborated here. Loosely speaking, for the purpose of offering a brief summary of these reasons, we may state that the slow convergence is due to the presence of spurious zero-frequency (static) modes, as well as some modes with eigenfrequencies below the excitation frequency. In the finite element approximation of the driven problem, these two classes of eigenmodes manifest themselves as eigemnodes of zero and/or negative eigenvalues, rendering the system matrix indefinite and, thus, ill-conditioned. Thus common preconditioners, such as incomplete Cholesky factorization, tend to perform very poorly.

An effective way to address the preconditioning of (7.8) was proposed in [2] and [3], making use of the additive or multiplicative Schwartz preconditioner. The additive Schwartz preconditioner solves the following residual equation

$$
\begin{bmatrix} z_E^1 \\ z_E^0 \end{bmatrix} = \begin{bmatrix} \tilde{M}_{EE}^{11^{-1}} & 0 \\ 0 & M_{EE}^{00^{-1}} \end{bmatrix} \begin{bmatrix} r_E^1 \\ r_E^0 \end{bmatrix},
\tag{7.10}
$$

where $M_{EE}^{00^{-1}}$ is the direct factorization of M_{EE}^{00}, and $\tilde{M}_{EE}^{00^{-1}}$ is the incomplete Cholesky factorization of M_{EE}^{11}. Since it solves the diagonal blocks of (7.8), the additive Schwartz preconditioner resembles the Jacobi process. In the following, the Schwartz preconditioner will be applied to the iterative solution of the matrix resulting from the finite element approximation of the potential formulation [6].

7.2   HIGHER-ORDER POTENTIAL FORMULATION

The potential formulation of the weak statement of the electromagnetic BVP provides for the elimination of the spurious DC modes through the explicit imposition of the divergence-free constraint on the numerical solution. This, combined with a multilevel scheme for the selective solution of eigenmodes of different spatial frequencies at different resolution levels, results in a robust iterative process for the solution of the finite element approximation of the electromagnetic boundary value problem.

We begin the development with the derivation of the finite element matrix for the potential formulation in a hierarchical set of expansion functions associated with a fixed finite element grid. The process is similar to that of the previous section for the E-field formulation. Thus we start with the weak statement of (7.8) with $\vec{E}$ replaced by its expression in terms of $\vec{A}$ and V,

$$
\iiint_\Omega \nabla \times \vec{w} \cdot \nabla \times \vec{A}\, dv + \frac{j\omega\mu_0}{\eta} \iint_{S_0} \hat{n} \times \vec{w} \cdot \hat{n} \times (\vec{A} + \nabla V)\, ds
$$
$$
+ \sum_{i=1}^{N_m} \frac{j\omega\mu_0}{Z_i} \iint_{S_i} \hat{n} \times \vec{w} \cdot \hat{n} \times (\vec{A} + \nabla V)\, ds
\tag{7.11}
$$
$$
- \omega^2 \mu_0 \epsilon_0 \iiint_\Omega \vec{w} \cdot \epsilon_r (\vec{A} + \nabla V)\, dv = \frac{2j\omega\mu_0}{Z_k} \iint_{S_k} \hat{n} \times \vec{w} \cdot \hat{n} \times \vec{e}_k ds.
$$

This equation must be augmented by the weak form of Gauss' law $\nabla \cdot \vec{D} = 0$ which, as shown in detail in the previous chapter, is readily found to be

$$\frac{j\omega\mu_0}{\eta} \iint_{S_0} \hat{n} \times \nabla\phi \cdot \hat{n} \times (\vec{A} + \nabla V)ds$$

$$+ \sum_{i=1}^{N_m} \frac{j\omega\mu_0}{Z_i} \iint_{S_i} \hat{n} \times \nabla\phi \cdot \hat{n} \times (\vec{A} + \nabla V)ds \tag{7.12}$$

$$- \omega^2\mu_0\epsilon_0 \cdot \iiint_{\Omega} \nabla\phi \cdot \epsilon_r(\vec{A} + \nabla V)dv = \frac{2j\omega\mu_0}{Z_k} \iint_{S_k} \hat{n} \times \nabla\phi \cdot \hat{n} \times \vec{e}_k ds.$$

The "coarse" approximation of the unknown potentials makes use of the basis functions of $\mathcal{H}^0(curl)$ for the expansion of $\vec{A}$, and the basis functions of the first-order, nodal elements, scalar subspace $W_{s,n}^1$,

$$W_{s,n}^1 = \text{span}\{\lambda_i \mid \text{node } i\} \tag{7.13}$$

for the expansion of V. Thus we have

$$\vec{w} \in \mathcal{H}^0(curl), \qquad \phi \in W_{s,n}^1. \tag{7.14}$$

Let us refer to this as the $\mathcal{H}^0(curl)$ expansion.

The hierarchical refinement, which will be referred to as the $\mathcal{H}^1(curl)$ expansion, is effected through the augmentation of the spaces above as follows:

$$\vec{w} \in \mathcal{H}^0(curl) \oplus R_{tv,f}^2, \qquad \phi \in W_{s,n}^1 \oplus W_{s,e}^2. \tag{7.15}$$

That is, the basis functions in $\mathcal{H}^1(curl)$ are decomposed into two parts. The part containing the non-gradient subspaces, $\mathcal{H}^0(curl)$ and $R_{tv,f}^2$, is used to expand the vector potential, while the part containing the gradient subspaces, $W_{s,n}^1$ and $W_{s,e}^2$, is used to expand the scalar potential. Thus we have

$$\vec{A} = \sum_{i=1}^{N_e} \vec{w}_{e,i} a_{e,i} + \sum_{i=1}^{N_f} (\vec{w}_{f,i,1} a_{f,i,1} + \vec{w}_{f,i,2} a_{f,i,2})$$

$$V = \sum_{i=1}^{N_n} w_{n,i} v_{n,i} + \sum_{i=1}^{N_e} w_{v,i} v_{v,i}, \tag{7.16}$$

where $v_{n,i}$, $v_{v,i}$, $a_{e,i}$, $a_{f,i,1}$, and $a_{f,i,2}$ are the expansion coefficient for the corresponding basis functions. Since the basis functions on any PEC surfaces present in the domain are excluded from the above expansions, N_n, N_e and N_f represent, respectively, the number of thoses nodes, edges, and facets in the domain which are not located on the PEC surfaces. Substitution of (7.16) into (7.11) and (7.12) and application of Galerkin's method yields the second-order potential formulation matrix for the weak statement in (7.1). Following the notation used in (6.136), this finite element matrix equation is written as

$$\begin{bmatrix} \langle W_f^T, W_f \rangle & \langle W_f^T, \nabla W_v \rangle & \langle W_f^T, W_e \rangle & \langle W_f^T, \nabla W_n \rangle \\ \langle \nabla W_v^T, W_f \rangle & \langle \nabla W_v^T, \nabla W_v \rangle & \langle \nabla W_v^T, W_e \rangle & \langle \nabla W_v^T, \nabla W_n \rangle \\ \langle W_e^T, W_f \rangle & \langle W_e^T, \nabla W_v \rangle & \langle W_e^T, W_e \rangle & \langle W_e^T, \nabla W_n \rangle \\ \langle \nabla W_n^T, W_f \rangle & \langle \nabla W_n^T, \nabla W_v \rangle & \langle \nabla W_n^T, W_e \rangle & \langle \nabla W_n^T, \nabla W_n \rangle \end{bmatrix} \begin{bmatrix} x_{A,f} \\ x_{V,v} \\ x_{A,e} \\ x_{V,n} \end{bmatrix} = \begin{bmatrix} f_{A,f} \\ f_{V,v} \\ f_{A,e} \\ f_{V,n} \end{bmatrix},$$
$$\tag{7.17}$$

where the row vector W_n contains the nodal basis functions and the row vector ∇W_n their gradients. The column vector $x_{A,e}$ contains the expansion coefficients $a_{e,i}$, while the column vector $x_{A,f}$ contains the expansion coefficients $a_{f,i,1}$ and $a_{f,i,2}$. The vectors $x_{V,n}$ and $x_{V,v}$ contain, respectively, the coefficients $v_{n,i}$ and $v_{v,i}$. The specific forms of the entries in each submatrices are easily deduced from the direct comparison of the terms in each row of the above matrix equation with the integral terms in the weak statements (7.11) and (7.12).

The hierarchical field and potential formulations are equivalent. Their equivalency can be shown through careful examination of (7.7) and (7.17). Because of the bilinear form of the system matrices and the linear form of the right-hand sides, we may write

$$
\begin{bmatrix}
\langle W_f^T, W_f \rangle & \langle W_f^T, \nabla W_v \rangle & \langle W_f^T, W_e \rangle & \langle W_f^T, \nabla W_n \rangle \\
\langle \nabla W_v^T, W_f \rangle & \langle \nabla W_v^T, \nabla W_v \rangle & \langle \nabla W_v^T, W_e \rangle & \langle \nabla W_v^T, \nabla W_n \rangle \\
\langle W_e^T, W_f \rangle & \langle W_e^T, \nabla W_v \rangle & \langle W_e^T, W_e \rangle & \langle W_e^T, \nabla W_n \rangle \\
\langle \nabla W_n^T, W_f \rangle & \langle \nabla W_n^T, \nabla W_v \rangle & \langle \nabla W_n^T, W_e \rangle & \langle \nabla W_n^T, \nabla W_n \rangle
\end{bmatrix}
$$

$$
=
\begin{bmatrix}
I & & & \\
& I & & \\
& & I & \\
& & & G^T
\end{bmatrix}
\begin{bmatrix}
\langle W_f^T, W_f \rangle & \langle W_f^T, \nabla W_v \rangle & \langle W_f^T, W_e \rangle \\
\langle \nabla W_v^T, W_f \rangle & \langle \nabla W_v^T, \nabla W_v \rangle & \langle \nabla W_v^T, W_e \rangle \\
\langle W_e^T, W_f \rangle & \langle W_e^T, \nabla W_v \rangle & \langle W_e^T, W_e \rangle
\end{bmatrix}
\begin{bmatrix}
I & & \\
& I & \\
& I & G
\end{bmatrix}
$$

(7.18)

and

$$
\begin{bmatrix}
x_{E,f} \\
x_{E,v} \\
x_{E,e}
\end{bmatrix}
=
\begin{bmatrix}
I & & \\
& I & \\
& I & G
\end{bmatrix}
\begin{bmatrix}
x_{A,f} \\
x_{V,v} \\
x_{A,e} \\
x_{V,n}
\end{bmatrix}
,
\qquad
\begin{bmatrix}
f_{A,f} \\
f_{V,v} \\
f_{A,e} \\
f_{V,n}
\end{bmatrix}
=
\begin{bmatrix}
I & & \\
& I & \\
& & I \\
& & G^T
\end{bmatrix}
\begin{bmatrix}
f_{E,f} \\
f_{E,v} \\
f_{E,e}
\end{bmatrix}.
$$

(7.19)

In the above, I denotes the identity matrix. Also, use was made of the fact that $\nabla W_n = W_e G$.

Using the second relation in (7.19), the forcing vector for the field formulation can be mapped onto the one for the potential formulation. Similarly, the first relation in (7.19) can be used to obtain from the solution of the potential formulation the one of the field formulation. Hence, the transformation between the two formulations is straightforward and enables the construction of a robust preconditioner for the iterative solution of (7.8) free from spurious modes.

Let us re-write (7.17) in the following compact hierarchical form:

$$
\begin{bmatrix}
M_{AV}^{11} & M_{AV}^{10} \\
M_{AV}^{01} & M_{AV}^{00}
\end{bmatrix}
\begin{bmatrix}
x_{AV}^1 \\
x_{AV}^0
\end{bmatrix}
=
\begin{bmatrix}
f_{AV}^1 \\
f_{AV}^0
\end{bmatrix},
$$

(7.20)

where

$$
x_{AV}^0 =
\begin{bmatrix}
x_{A,e} \\
x_{V,n}
\end{bmatrix},
\qquad
x_{AV}^1 =
\begin{bmatrix}
x_{A,f} \\
x_{V,v}
\end{bmatrix},
\qquad
f_{AV}^0 =
\begin{bmatrix}
f_{A,e} \\
f_{V,n}
\end{bmatrix},
\qquad
f_{AV}^1 =
\begin{bmatrix}
f_{A,f} \\
f_{V,v}
\end{bmatrix}
$$

(7.21)

and the specific forms of the entries of the matrices M_{AV}^{ij}, $i, j \in \{0, 1\}$, are easily deduced through a direct comparison with (7.17). Then the following matrix relations can be written between the matrices, forcing vectors and unknown vectors of the compact form of the field formulation (7.8) and the compact form of the potential formulation (7.20)

$$
\begin{bmatrix}
M_{AV}^{11} & M_{AV}^{10} \\
M_{AV}^{01} & M_{AV}^{00}
\end{bmatrix}
=
\begin{bmatrix}
I & \\
& \begin{bmatrix} I \\ G^T \end{bmatrix}
\end{bmatrix}
\begin{bmatrix}
M_{EE}^{11} & M_{EE}^{10} \\
M_{EE}^{01} & M_{EE}^{00}
\end{bmatrix}
\begin{bmatrix}
I & \\
& \begin{bmatrix} I & G \end{bmatrix}
\end{bmatrix}
$$

(7.22)

and

$$\begin{bmatrix} x_E^1 \\ x_E^0 \end{bmatrix} = \begin{bmatrix} I & \\ & [I \ \ G] \end{bmatrix} \begin{bmatrix} x_{AV}^1 \\ x_{AV}^0 \end{bmatrix}, \ \begin{bmatrix} f_{AV}^1 \\ f_{AV}^0 \end{bmatrix} = \begin{bmatrix} I & \\ & \begin{bmatrix} I \\ G^T \end{bmatrix} \end{bmatrix} \begin{bmatrix} f_E^1 \\ f_E^0 \end{bmatrix}. \tag{7.23}$$

With these results the stage is set for the discussion of the hierarchical multilevel potential preconditioner.

7.3 HIERARCHICAL MULTILEVEL POTENTIAL PRECONDITIONER

Consider the calculation of the pseudo-residual equation in each step of the iterative solution of (7.8),

$$\begin{bmatrix} M_{EE}^{11} & M_{EE}^{10} \\ M_{EE}^{01} & M_{EE}^{00} \end{bmatrix} \begin{bmatrix} z_E^1 \\ z_E^0 \end{bmatrix} = \begin{bmatrix} r_E^1 \\ r_E^0 \end{bmatrix}. \tag{7.24}$$

The pre-smoothing operation can be performed in two steps. Step 1 involves the approximate solution for the higher-order electric field unknowns, z_E^1, from the equation

$$M_{EE}^{11} z_E^1 = r_E^1 - M_{EE}^{10} z_E^0. \tag{7.25}$$

Step 2 is more involved, consisting of several sub-steps. First, the residual is mapped onto the potential formulation through the equation

$$r_{AV}^0 = \begin{bmatrix} I \\ G^T \end{bmatrix} \left(r_E^0 - M_{EE}^{01} z_E^1 - M_{EE}^{00} z_E^0 \right). \tag{7.26}$$

Next, the approximate solution for z_{AV}^0 in the lowest-order potential formulation is obtained from

$$M_{AV}^{00} z_{AV}^0 = r_{AV}^0. \tag{7.27}$$

Finally, the solution is mapped back onto the field formulation through the operation

$$z_E^0 \leftarrow z_E^0 + \begin{bmatrix} I & G \end{bmatrix} z_{AV}^0. \tag{7.28}$$

The post-smoothing is the reverse of the above pre-smoothing operation, where Step 2 is performed before Step 1. Both the approximate solution of (7.25) and (7.27) are effected either through an incomplete Cholesky factorization or through the Gauss-Seidel method. Clearly, since the smoothing operation on the lowest-order space is done in the potential formulation, the spurious DC modes are suppressed in the complete second-order gradient space. The algorithmic description of the process is as follows:

Algorithm (7.1): Potential Multiplicative Schwarz Smoother
```
1 Pre-smooth(z_E, r_E).
  1.a Pre-smooth M_EE^11 z_E^1 = r_E^1 - M_EE^10 z_E^0 using forward GS v times
  1.b Pre-smooth M_AV^00 z_AV^0 = [ I ; G^T ] (r_E^0 - M_EE^01 z_E^1 - M_EE^00 z_E^0)
      using forward GS v times
  1.c z_E^0 ← z_E^0 + [ I  G ] z_AV^0
2 Post-smooth(z_E, r_E)
  2.a Post-smooth M_AV^00 z_AV^0 = [ I ; G^T ] (r_E^0 - M_EE^01 z_E^1 - M_EE^00 z_E^0)
      using backward GS v times
  2.b z_E^0 ← z_E^0 + [ I  G ] z_AV^0.
  2.c Post-smooth M_EE^11 z_E^1 = r_E^1 - M_EE^10 z_E^0 using backward GS v times
```

The above algorithm describes the smoothing operations for the potential multiplicative Schwartz process. Smoothing operations for the field formulation, which constitute the $\vec{E}$-field Schwartz process, also involve two steps. In Step 1 the approximate solution for the higher-order electric field unknowns, z_E^1, is obtained from the equation

$$M_{EE}^{11} z_E^1 = r_E^1 - M_{EE}^{10} z_E^0. \tag{7.29}$$

Step 2 involves the approximate solution for the lowest-order electric field unknowns, z_E^0, through the equation

$$M_{EE}^{00} z_E^1 = r_E^0 - M_{EE}^{01} z_E^1. \tag{7.30}$$

The post-smoothing is the reverse of the pre-smoothing operation, with Step 2 performed before Step 1. The algorithmic description of the field multiplicative Schwartz process is given below.

Algorithm (7.2): Field Multiplicative Schwartz Smoother
```
1 Pre-smooth(z_E, r_E)
    1.a Pre-smooth M^11_EE z^1_E = r^1_E - M^10_EE z^0_E using forward GS v_1 times
    1.b Pre-smooth M^00_EE z^0_E = r^0_E - M^01_EE z^1_E using forward GS v_1 times
2 Post-smooth(z_E, r_E)
    2.a Post-smooth M^00_EE z^0_E = r^0_E - M^01_EE z^1_E using backward GS v_2 times
    2.b Post-smooth M^11_EE z^1_E = r^1_E - M^10_EE z^0_E using backward GS v_2 times
```

As already mentioned, the single-level potential preconditioner, consisting only of the pre- and post-smoothing processes, cannot tackle the ill-conditioning caused by low-frequency modes. To overcome this difficulty and improve convergence, the above single-level smoothing process is combined with multilevel techniques. This extension of the single-level preconditioner to a multilevel one is described in the following typical multilevel pseudo-code.

Algorithm (7.3): Hierarchical Multilevel Potential Preconditioner
```
              z_E ← Hie-ML(r_E, i)
1 z_E ← 0
2 if i == 0, then solve M^00_EE z_E = r_E.                    (lowest level)
  else
    2.a Pre-smooth(z_E, r_E) for v times
    2.b r^{i-1}_E ← I^{i-1}_i (r_E - M_EE z_E)
    2.c z^{i-1}_Z ← Hie-ML(r^{i-1}_E, i-1)
    2.d z_E ← z_E + I^i_{i-1} z^{i-1}_E
    2.e Post-smooth(z_E, r_E) for v times
```

Removal of Steps 2.b–2.d results in the single-level potential preconditioner. The operator I_i^{i-1} is the restriction operator that maps the residual of the i-th level matrix equation down to the $(i-1)$th level. I_{i-1}^i is the interpolation operator that interpolates the correction obtained on the $(i-1)$th level back to the ith level. The use of the hierarchical basis functions facilitates greatly the implementation of these operations.

7.4 HYBRID MULTILEVEL/MULTIGRID POTENTIAL PRECONDITIONER

The nested multigrid potential solver uses lowest-order basis functions on a set of nested grids, while the hierarchical multilevel potential solver uses a set of hierarchical basis functions on a fixed grid. The hierarchical multilevel solver can tackle numerical dispersion error more effectively. On the other hand, the nested multigrid preconditioner requires the direct factorization on a matrix much smaller than that for the hierarchical multilevel solver. Clearly, a combination of the merits of the two preconditioners is desirable since it will lead to improved computational efficiency and solution accuracy.

This can be effected through a hybrid multigrid/multilevel preconditioning process, which uses hierarchical multilevel techniques on top of a nested multigrid process [7]. The pseudo code, Hbd-MLMG(z_E, r_E, i, j), of this hybrid preconditioner, the flowchart of which is shown in Fig. 7.1, is given below. The indices i and j indicate, respectively, the level numbers for the multilevel and the multigrid processes.

Algorithm (7.4): Hybrid Multileve/Multgrid Potential Preconditioner

$$z_E \leftarrow \texttt{Hbd-MLMG}(r_E,\ i,\ j)$$

```
1 zE ← 0
2 if i == 0                                          (nested multigrid)
  2.a if j == 0, then solve MEEzE = rE              (coarsest grid)
  2.b else
    2.b.1 Pre-smooth(zE, rE) for v times
    2.b.2 r_E^{2h} ← I_h^{2h}(rE − MEEzE)
    2.b.3 z_E^{2h} ← Hbd-MLMG(r_E^{2h}, 0, j − 1)
    2.b.4 zE ← zE + I_{2h}^h z_E^{2h}
    2.b.5 Post-Smooth(zE, rE) for v times
3 else                                               (hierarchical multilevel)
  3.a Pre-Smooth(zE, rE) for v times
  3.b r_E^{i−1} ← I_i^{i−1}(rE − MEEzE)
  3.c z_E^{i−1} ← Hbd-MLMG(r_E^{i−1}, i − 1, j )
  3.d zE ← zE + I_{i−1}^i z_E^{i−1}
  3.e Post-Smooth(zE, rE) for v times
```

From both the flowchart description of the process and its algorithmic implementation it is evident that the nested multigrid preconditioning is placed within the solution of the $\mathcal{H}^0(curl)$ matrix equation of the hierarchical multilevel preconditioner. In this manner, the dimension of the matrix that needs to be factored is reduced further. The smoothing operations in both the nested and the hierarchical preconditioning processes are performed in the potential formulation.

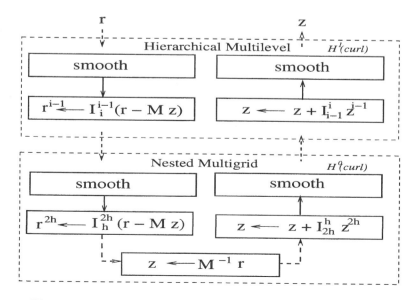

Figure 7.1   Flowchart of the hybrid multilevel/multigrid preconditioner.

I_i^{i-1} and I_{i-1}^i denote the interlevel operators that map the residual and the correction between two adjacent levels. Since hierarchical basis functions are used, the construction of the two operators is trivial. Finally, we recall that the matrices I_h^{2h} and I_{2h}^h are the intergrid matrix operators, which provide for the transformation between the two sets of the basis functions of two adjacent nested grids. Their properties and their construction are described in Chapters 5 and 6.

7.5 NUMERICAL EXPERIMENTS

In this section we present some results from the application of the hierarchical multilevel and the hybrid multilevel/multigrid preconditioners to the solution of several electromagnetic BVPs. The objective is to highlight the beneficial impact of these preconditioners on numerical solution accuracy and convergence speed of the iterative solution process. The hierarchical multilevel preconditioners were implemented within a CG-based iterative solver framework. In all cases the number of pre-smoothing and post-smoothing operations was taken to be $v = 3$.

Section of a Microstrip Line. The first application considers the electromagnetic analysis of the simple microstrip line structure depicted in the insert of the top figure in Fig. 7.2. First-order absorbing boundary conditions were imposed at the top boundary and the two side boundaries (along the microstrip axis) of the computational domain. While the cross section of the microstrip waveguide is kept unchanged, its length was increased from 200 mils to 1200 mils for the purpose of examining the dependence of the convergence performance of the iterative solution on electrical size.

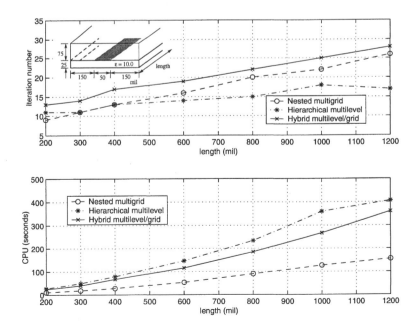

Figure 7.2 Number of iterations and CPU time versus the length of the microstrip line. (After Zhu and Cangellaris [7], ©2002 IEEE.)

Let us compare the performance of the three different preconditioning methodologies we have considered, namely, nested multigrid, hierarchical multilevel, and hybrid multilevel-multigrid. Figure 7.2 depicts the number of iterations and the required CPU time versus the length of the microstrip line. With the operating frequency set at 20 GHz, the electrical size of the 200-mils-long section is $\sim 1.5\lambda$, while the electrical size of the 1200-mils section is $\sim 10.0\lambda$. This results in an increase in the number of unknowns from 24936 to 160962 when using $\mathcal{H}^1(curl)$. The average spatial resolution in the coarsest grid is ~ 3.5 points/λ.

The hybrid preconditioner uses a two-level hierarchical and a two-grid nested multigrid process. The hierarchical multilevel preconditioner uses two sets of basis functions, $\mathcal{H}^1(curl)$ and $\mathcal{H}^0(curl)$. In the case of the hierarchical multilevel preconditioner, a direct matrix solution of the pseudo-residual equation is required at the level $\mathcal{H}^0(curl)$. In contrast, in the case of the hybrid preconditioner the dimension of the pseudo-residual matrix equation that must be solved at each iteration is further reduced through a projection onto a coarser grid. With regards to the flowchart of Fig. 7.1, the hierarchical preconditioner corresponds to the upper block, while the two-grid nested multigrid process (which solves the problem in $\mathcal{H}^0(curl)$ using the two nested grids) corresponds to the lower block in the flowchart.

From Fig. 7.2 it is clear that both the number of iterations and the CPU time increase with the length of the microstrip line. This is attributed to the numerical dispersion error which is known to increase with the electrical length of the structure under analysis. The hierarchical multilevel preconditioner exhibits the best convergence performance. Its superiority to the hybrid preconditioner is attributed to the fact that it solves the $\mathcal{H}^0(curl)$ matrix directly. In contrast, in the case of the hybrid preconditioner, the projection of the $\mathcal{H}^0(curl)$ matrix onto a coarser grid is solved, resulting in a larger error. However, this improvement in convergence comes at the cost of more CPU time. More specifically, for the 1200-mils-long microstrip line, there are 160962 and 30209 unknowns, respectively, at the $\mathcal{H}^1(curl)$ and $\mathcal{H}^0(curl)$ levels. In the case of the hierarchical multilevel preconditioner the $\mathcal{H}^0(curl)$ matrix of size 30209 must be factored, whereas in the case of the hybrid preconditioner the projection of the $\mathcal{H}^0(curl)$ matrix down to the coarser grid must be factored. The dimension of this reduced matrix is only 3930. This result in reduced solution time and reduced memory requirements. More specifically, for the case of the 1200-mils-long line, the memory requirement for the hybrid preconditioner was 60 MB while that for the hierarchical multilevel preconditioner was 112 MB.

In summary, the hybrid multilevel/multigrid preconditioner offers improved computational efficiency over the hierarchical multilevel preconditioner at the cost of slower convergence.

A Rectangular Waveguide Filter. The geometry of the waveguide filter is depicted in the insert of Fig. 7.3. Measured data for the scattering parameters of this filter were provided in [8]. The finite element grid used provides for an average spatial sampling rate of about 7.80 points/λ at 18 GHz. The number of unknowns at the $\mathcal{H}^1(curl)$ level is 42526. The calculated scattering parameters, obtained through the use of the multilevel preconditioner, are seen to be in very good agreement with the results obtained from measurements.

To demonstrate the superiority of the multilevel preconditioner, the convergence performance of three CG-based solvers, namely, one without preconditioning, a single-level potential preconditioner (SLAV), and a multilevel potential preconditioner (MLAV), at 15 GHz is described in terms of the numbers in Table 7.1 and the pictorial description in Fig. 7.4(a). The single-level potential preconditioner is obtained by neglecting Steps 3.b-3.d in the multilevel/multigrid Algorithm 7.4. Clearly, its performance is better than that of the un-preconditioned CG because it renders the search vectors divergence-free. The convergence performance is further improved with the implementation of the multilevel potential preconditioner, which provides for the effective handling of the physical low-frequency modes.

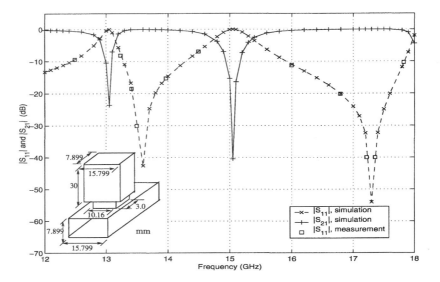

Figure 7.3   Magnitude of scattering parameters for a waveguide filter. Measured data obtained from [8]. (After Zhu and Cangellaris [6], ©2002 IEEE.)

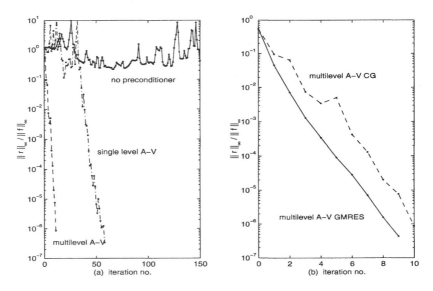

Figure 7.4   (a) Convergence comparison of three CG solvers. (b) Convergence comparison of multilevel preconditioned CG and multilevel preconditioned GMRES. (After Zhu and Cangellaris [6], ©2002 IEEE.)

Table 7.1 Convergence Comparison of Three CG-based solvers at 15.0 GHz

	Without Preconditioner	SLAV	MLAV
CPU time (s)	1339.22	108.01	34.23
Iteration	4394	58	11

In place of CG a different iterative solver, for example, GMRES, may be used. Fig. 7.4(b) provides a comparison of the convergence performance of the two solvers with multilevel potential preconditioning used in both cases. The data are from the solution of the waveguide filter at 15 GHz.

Over the frequency range from $12-18$ GHz, it takes less than 15 iterations for the pre-conditioned solver to converge at each frequency point. More specifically, the number of iterations was 8, 10 and 13, respectively, at 12, 15, and 18 GHz. The required CPU times were 32.91 s, 34.23 s, and 39.65 s, respectively.

Microstrip Low-pass Filter. The next example considers the application of the multilevel potential preconditioners to the iterative solution of the low-pass microstrip filter shown in the insert of Fig. 7.5. A computational domain of size $345 \times 375 \times 150$ mil^3 was used. At the top side of the domain a first-order absorbing boundary condition was imposed at a distance of 75 mils above the air/dielectric interface. Absorbing boundary conditions were also used at the side boundaries (parallel to the top and bottom sides of the footprint of the filter depicted in Fig. 7.5). The finite element grid used had an average spatial sampling rate of 8.5 points/λ. The number of unknowns at the $\mathcal{H}^1(curl)$ level was 88708.

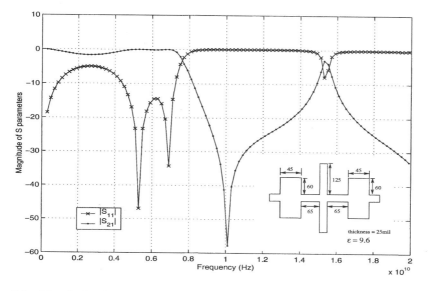

Figure 7.5 Magnitude of the scattering parameters of the depicted low-pass filter. (After Zhu and Cangellaris [6], ©2002 IEEE.)

This filter was also analyzed in [9]. The calculated magnitudes of the scattering parameters, depicted in Fig. 7.5, are in good agreement with those in [9]. It is noted that the mesh used for our solution is rather uniform throughout the domain. Use of a nonuniform mesh, utilizing a higher density in the immediate vicinity of the metallization, should result in improved solution accuracy. The convergence of the multilevel potential preconditioned iterative solver is excellent. The required CPU times for the solution at 0.1 GHz, 10.0 GHz, and 20.0 GHz are 190.14 s, 209.44 s, and 219.29 s, respectively. Over the frequency range from 0.1-20.0GHz, it takes less than 18 iterations at each frequency point.

To test the robustness of the potential preconditioners, the operating frequency was set at 40.0 GHz. At this frequency, the average grid size is 4.2 points/λ. The convergence behavior of the three CG-based iterative solvers, depicted in Fig. 7.6, demonstrates clearly the beneficial impact of the single-level and multilevel potential preconditioners.

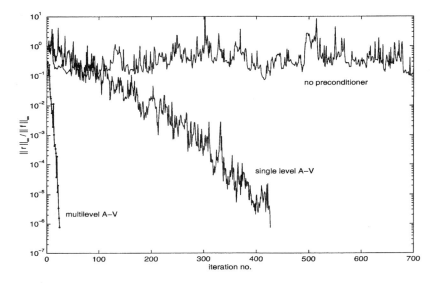

Figure 7.6   Convergence comparison of three CG-based iterative solvers applied to the solution of the microstrip filter depicted in the insert of Fig. 7.5. (After Zhu and Cangellaris [6], ©2002 IEEE.)

Band-stop Rectangular Waveguide Filter. The final example to be presented is for the band-stop waveguide filter depicted in the insert of Fig. 7.7. The cross-sectional dimensions for all waveguides are 22.86 mm and 10.16 mm. The distance between resonators is 19.63 mm. The length of each one of the top two resonators is 16.54 mm, while the length of the bottom resonator is 16.94 mm. All three irises have the same width of 3.05 mm. The length of the two irises in the top two resonators is 12.22 mm. The width of the iris in the bottom resonator is 11.63 mm. Waveguide walls were assumed to be infinitesimally thin.

The iterative solver utilized a two-level/two-grid hybrid preconditioner. The number of unknowns at $\mathcal{H}^1(curl)$ and $\mathcal{H}^0(curl)$ are 152062 and 26375, respectively. The average coarsest grid resolution is ~ 5.1 points/λ at 12.4 GHz. The dimension of the matrix at the coarsest grid that is solved at each iteration is 2784.

Measured data for the scattering parameters of this filter were provided in [8] and are used here for comparison purposes. The calculated scattering parameters are in very good agreement with the measured data. The convergence behavior of the iterative solver at three

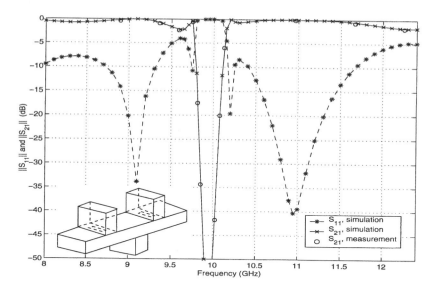

Figure 7.7 Magnitude of the scattering parameters of a rectangular waveguide band-stop filter. (After Zhu and Cangellaris [7], ©2002 IEEE.)

frequencies is shown in Fig. 7.8. On the average, the required CPU time is about 150 s, while the total memory requirement is 53 MB.

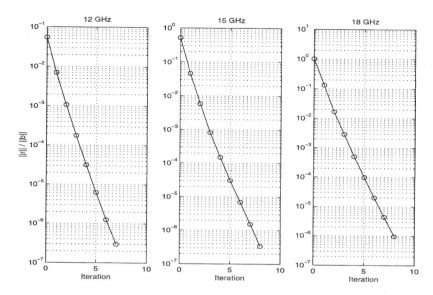

Figure 7.8 Convergence performance of the iterative solver with hybrid preconditioning for the waveguide filter depicted in Fig. 7.7. (After Zhu and Cangellaris [7], ©2002 IEEE.)

7.6 SYMMETRIC HIERARCHICAL MULTILEVEL POTENTIAL PRECONDITIONER

Because of its efficiency and robustness the conjugate gradient method is the preferred choice in the development of an iterative solver. However, its use requires a symmetric preconditioner. In this section we will demonstrate the symmetry of our implementation of the hierarchical multilevel preconditioner for the special case of a two-level scheme. The symmetry of the multilevel one follows by induction.

The relevant equation is the following pseudo-residual matrix system

$$
\begin{bmatrix} M_{EE}^{11} & M_{EE}^{10} \\ M_{EE}^{01} & M_{EE}^{00} \end{bmatrix} \begin{bmatrix} z_E^1 \\ z_E^0 \end{bmatrix} = \begin{bmatrix} r_E^1 \\ r_E^0 \end{bmatrix},
\tag{7.31}
$$

where now the superscripts 1 and 0 in the z and r vectors are used to indicate the basis function spaces, $\mathcal{H}^1(curl)$ and in $\mathcal{H}^0(curl)$, respectively, in which the vector is represented. The superscripts used in the earlier sections to indicate the iteration at which the vectors z and r are computed, are suppressed for simplicity.

7.6.1 Potential smoothing operations

To examine the symmetry of the hierarchical multilevel preconditioning process we begin with the potential smoothing operations. The smoothing operation can be considered as a two-step process. Step 1 involves the m smoothing operations in the $\mathcal{H}^1(curl)$ space with initial $z_E^0 = 0$ and $z_E^1 = 0$,

$$
M_{EE}^{11} z_E^1 = r_E^1 - M_{EE}^{10} z_E^0, \quad \Rightarrow \quad z_E^1 \leftarrow \underline{R}_{11} r_E^1,
\tag{7.32}
$$

where

$$
\underline{H}_{11} = \left(D_{EE}^{11} - L_{EE}^{11} \right)^{-1} U_{EE}^{11}, \quad \underline{R}_{11} = \sum_{i=0}^{m-1} \underline{H}_{11}^i \left(D_{EE}^{11} - L_{EE}^{11} \right)^{-1}.
\tag{7.33}
$$

The matrices D_{EE}^{11}, L_{EE}^{11}, and U_{EE}^{11} are, respectively, the diagonal, low-triangular, and upper-triangular parts of M_{EE}^{11}.

Subsequently, the m additional smoothing operations continue with initial $z_E^0 = 0$ in the $\mathcal{H}^0(curl)$ space as discussed in Chapter 6,

$$
M_{EE}^{00} z_E^0 = r_E^0 - M_{EE}^{01} z_E^1 \quad \Rightarrow \quad z_E^0 \leftarrow \underline{R}(r_E^0 - M_{EE}^{01} z_E^1),
\tag{7.34}
$$

where $\underline{R}$ are defined in (6.152). Equations (7.32) and (7.34) are combined into the following matrix form

$$
\begin{bmatrix} z_E^1 \\ z_E^0 \end{bmatrix} \leftarrow \underbrace{\begin{bmatrix} \underline{R}_{11} & \\ -\underline{R} M_{EE}^{01} \underline{R}_{11} & \underline{R} \end{bmatrix}}_{\underline{R}} \begin{bmatrix} r_E^1 \\ r_E^0 \end{bmatrix}.
\tag{7.35}
$$

The post-smoothing process proceeds in a sequence that is the reverse of the pre-smoothing one presented above. Thus m smoothing operations are performed first in the $\mathcal{H}^0(curl)$ space,

$$
M_{EE}^{00} z_E^0 = r_E^0 - M_{EE}^{01} z_E^1 \quad \Rightarrow \quad z_E^0 \leftarrow \overline{H} z_E^0 + \overline{R}(r_E^0 - M_{EE}^{01} z_E^1),
\tag{7.36}
$$

where $\overline{H}$ and $\overline{R}$ are defined in (6.158). Subsequently, m smoothing operations are performed in the $\mathcal{H}^1(curl)$ space,

$$M_{EE}^{11} z_E^1 = r_E^1 - M_{EE}^{10} z_E^0 \quad \Rightarrow \quad z_E^1 \leftarrow \overline{H}_{11}^m z_E^1 + \overline{R}_{11}(r_E^1 - M_{EE}^{10} z_E^0), \qquad (7.37)$$

where

$$\overline{H}_{11} = \left(D_{EE}^{11} - U_{EE}^{11}\right)^{-1} L_{EE}^{11}, \quad \overline{R}_{11} = \sum_{i=0}^{m-1} \overline{H}_{11}^i \left(D_{EE}^{11} - U_{EE}^{11}\right)^{-1}. \qquad (7.38)$$

The combination of (7.36) and (7.37) into one matrix equation yields

$$\begin{bmatrix} z_E^1 \\ z_E^0 \end{bmatrix} = \underbrace{\begin{bmatrix} \overline{H}_{11}^m + \overline{R}_{11} M_{EE}^{10} \overline{R} M_{EE}^{01} & -\overline{R}_{11} M_{EE}^{10} \overline{H} \\ -\overline{R} M_{EE}^{01} & \overline{H} \end{bmatrix}}_{\overline{\mathbf{H}}} \begin{bmatrix} z_E^1 \\ z_E^0 \end{bmatrix} + \underbrace{\begin{bmatrix} \overline{R}_{11} & -\overline{R}_{11} M_{EE}^{10} \overline{R} \\ & \overline{R} \end{bmatrix}}_{\overline{\mathbf{R}}} \begin{bmatrix} r_E^1 \\ r_E^0 \end{bmatrix}.$$

$$(7.39)$$

7.6.2 Symmetric hierarchical two-level potential preconditioner

The equations derived in the previous two subsections are combined to yield the following form for the hierarchical multilevel potential preconditioning matrix P

$$P = \overline{\mathbf{H}}\underline{\mathbf{R}} + \overline{\mathbf{R}} + \overline{\mathbf{H}} I_0^1 M_{EE}^{00}{}^{-1} I_1^0 \left(I - \begin{bmatrix} M_{EE}^{11} & M_{EE}^{10} \\ M_{EE}^{01} & M_{EE}^{00} \end{bmatrix} \underline{\mathbf{R}} \right). \qquad (7.40)$$

The matrices I_0^1 and I_1^0 are the inter-level transfer operators. Let us note, first, that it is

$$I - \begin{bmatrix} M_{EE}^{11} & M_{EE}^{10} \\ M_{EE}^{01} & M_{EE}^{00} \end{bmatrix} \underline{\mathbf{R}} = \begin{bmatrix} I - M_{EE}^{11} \underline{R}_{11} + M_{EE}^{10} \underline{R} M_{EE}^{01} \underline{R}_{11} & -M_{EE}^{10} \underline{R} \\ -(I - M_{EE}^{00} \underline{R}) M_{EE}^{01} \underline{R}_{11} & I - M_{EE}^{00} \underline{R}. \end{bmatrix}. \qquad (7.41)$$

Making use of (6.167) and the relations

$$\overline{R}_{11} = \underline{R}_{11}^T, \quad \overline{R} = \underline{R}^T \qquad (7.42)$$

we conclude that the result in the above equation is equal to the transpose of $\overline{\mathbf{H}}$ in (7.39). As far as the first two terms in (7.40) are concerned, we have

$$\overline{\mathbf{H}}\underline{\mathbf{R}} + \overline{\mathbf{R}} = \begin{bmatrix} \overline{H}_{11}^m \underline{R}_{11} + \overline{R}_{11} + \overline{R}_{11} M_{EE}^{10}(\overline{H}\underline{R} + \overline{R}) M_{EE}^{01} \underline{R}_{11} & -\overline{R}_{11} M_{EE}^{10}(\overline{H}\underline{R} + \overline{R}) \\ -(\overline{R} + \overline{H}\underline{R}) M_{EE}^{01} \underline{R}_{11} & \overline{H}\underline{R} + \overline{R} \end{bmatrix}.$$

In the above equation the symmetry of $\overline{H}\underline{R} + \overline{R}$ and $\overline{H}_{11}^m \underline{R}_{11} + \overline{R}_{11}$ was proven in (6.161) and (4.103), respectively.

In view of the above results the symmetry of the matrix $\overline{\mathbf{H}}\underline{\mathbf{R}} + \overline{\mathbf{R}}$ is deduced. The symmetry of our implementation of the nested multigrid potential preconditioner was proven in the last chapter. It follows that the hybrid multigrid/multilevel preconditioner is also symmetric.

7.7 KEY ATTRIBUTES OF MULTIGRID AND MULTILEVEL POTENTIAL PRECONDITIONERS

At this point we have concluded the presentation of methodologies for the development of symmetric, multigrid and multilevel potential preconditioners for use in the iterative solution of finite element approximations of electromagnetic BVPs. The key attributes of these preconditioners are briefly reviewed and contrasted in the following paragraphs.

The hierarchical, multilevel potential preconditioner combines the vector and scalar potential formulation of the electromagnetic BVP with a hierarchical multilevel finite element expansion to avoid the ill-conditioning of the finite element matrix and, thus, provide for a fast converging iterative solution. Compared to the nested multigrid preconditioner, only one grid is used in this case, with the refinement performed through the utilization of higher-order expansion functions in the elements of the single grid.

The hierarchical multilevel preconditioner is most useful when dealing with the finite element modeling of electromagnetic interactions inside domains that are electrically large and exhibit fairly smooth variation in their electromagnetic properties. For such domains, suppressing numerical dispersion error at sufficiently low levels is most important for solution accuracy. The p-refinement enabled by the hierarchical multilevel finite element expansion of the unknown fields serves this objective well. However, the multilevel preconditioner requires the direct factorization of the finite element matrix at the lowest ($\mathcal{H}^0(curl)$) level. Thus, for a given computing platform, the dimension of this matrix is the one with the decisive impact on the size of the problem that can be handled efficiently using this preconditioning scheme.

A way to overcome this potential bottleneck is through the utilization of the hybrid multigrid/multilevel potential preconditioner, which results from employing a nested multigrid process at the lowest ($\mathcal{H}^0(curl)$) level of the multilevel preconditioner. This helps reduce the dimension of the matrix that is factored directly during the iterative process, thus extending the applicability of the multilevel preconditioner to the solution of electrically-large structures. This hybridization, which may be thought of as an adaptive hp-refinement process, comes at the cost of slower convergence compared to the case where the multilevel preconditioner is used alone. This is due to the error introduced by the lower resolution of the field distribution in the coarser grids utilized by the nested multigrid process.

Intuitively one expects that the resolution at the coarsest grid should be adequate to describe the field variation in space. Since for structures with smoothly-varying electromagnetic material properties the characteristic length of such variation is the wavelength at the operating frequency, a resolution of at least three points per wavelength appears to be a reasonable lowest limit, consistent with the Nyquist sampling theorem. Our numerical experiments support this conjecture.

This prompts us to make the recommendation that, for structures such that the finite element size is dictated by the operating wavelength and not the minimum feature size, the coarsest grid used in the hybrid multigrid/multilevel potential preconditioner should be such that, on the average, the element size should not exceed one third of the wavelength.

REFERENCES

1. J. P. Webb and B. Forghani, "Hierarchical scalar and vector tetrahedra," *IEEE Trans. Magnetics*, vol. 29, pp. 1495-1498, May 1993.

2. G. Peng, "Multigrid preconditioning in solving time-harmonic wave propagation problems using tangential vector finite elements," Ph.D. dissertation, Worcester Polytechnic Institute, Worcester, MA, 1997.

3. G. Peng, R. Dyczij-Edlinger, and J. F. Lee, "Hierarchical methods for solving matrix equations from TVFEM for microwave components," *IEEE Trans. Magnetics*, vol. 35, pp. 1474-1477, May 1999.

4. I. Tsukerman and A. Plaks, "Hierarchical basis multilevel preconditioners for 3D magnetostatic problems," *IEEE Trans. Magnetics*, vol. 35, pp. 1143-1147, May 1999.

5. C. Geuzaine, B. Meys, P. Dular and W. Legros, "Convergence of high order curl-conforming finite elements," *IEEE Trans. Magnetics*, vol. 35, pp. 1442-1445, May 1999.

6. Y. Zhu and A. Cangellaris, "Hierarchical multilevel potential preconditioner for fast finite element modeling of microwave devices," *IEEE Trans. Microwave Theory Tech.*, vol. 50, pp. 1984-1989, Aug. 2002.

7. Y. Zhu and A. C. Cangellaris, "Hybrid multilevel/multigrid potential preconditioner for fast finite element modeling", *IEEE Microwave Wireless Compon. Lett.*, Volume: 12, Issue: 8, pp. 290-292, Aug. 2002.

8. T. Sieverding and F. Arndt, "Field theoretic CAD of open or aperture matched T-junction coupled rectangular waveguide structures," *IEEE Trans. Microwave Theory Tech.*, vol. 40, pp. 353-363, Feb. 1992.

9. J. E. Bracken, D. K. Sun, and Z. J. Cendes, "S-domain methods for simultaneous time and frequency characterization of electromagnetic devices," *IEEE Trans. Microwave Theory Tech.*, vol. 42, pp. 1277-1290, Sep. 1998.

CHAPTER 8

KRYLOV-SUBSPACE BASED EIGENVALUE ANALYSIS

In this chapter, the machinery used for the numerical computation of eigenvalues and eigen-vectors of large sparse matrices is reviewed. Our presentation is neither comprehensive nor complete. Rather, it concentrates on some of the concepts, methodologies and algorithms that are most suitable and effective in dealing with the types of sparse matrices resulting from the finite element approximation of electromagnetic eigenvalue problems. For a more in-depth discussion of the material presented herein the reader is referred to comprehensive texts in the applied computational mathematics literature, such as [1]-[5].

A basic review of the definitions and fundamental results pertinent to eigenvalue analysis and eigen-decomposition of linear matrices was given in the first section in Chapter 4. It is recommended that the reader reviews this background material to become familiar with the notation and nomenclature used in this chapter.

8.1 SUBSPACE ITERATION

The simplest algorithm for solving large sparse-matrix eigenvalue problems is the subspaces iteration algorithm. It can be viewed as the block generalization of the power method [1]. The following pseudo-code describes the key steps of the algorithm for obtaining n eigenvectors of A.

Algorithm(8.1): Subspace Iteration for n Eigenvectors
```
1 Choose initial vectors  X₀ = [x₁,x₂,···,xₙ];
2 Iterate from  k = 1:
  2.a Compute  Xₖ ← AXₖ₋₁;
  2.b QR factorization of  Xₖ:   Xₖ = QRₖ;
  2.c  Xₖ ← Q;
  2.d If  ‖Xₖ − Xₖ₋₁‖ < ρ,  convergence is achieved and goto (3);
  2.e  k ← k + 1;
3 Eigen-decomposition of  Rₖ:   Rₖ = YΛY⁻¹;
4 Eigenvalues are in  Λ; eigenvectors are  V = XₖY.
```

The QR factorization in Step 2.b can be performed using the Gram-Schmidt process to orthonormalize the n column vectors of X_k and render R_k upper triangular. The pseudocode for the process is given below.

Algorithm(8.2): Gram-Schmidt orthogonalization
```
1 Rₖ(1,1) = ‖x₁‖ and  q₁ = x₁/Rₖ(1,1);
2 for  i = 2,···,n
  2.a  qᵢ = xᵢ;
  2.b for  j = 1,···,i − 1,  Rₖ(j,i) = qⱼᵀqᵢ and  qᵢ = qᵢ − Rₖ(j,i)qⱼ;
  2.c  Rₖ(i,i) = ‖qᵢ‖ and  qᵢ = qᵢ/Rₖ(i,i);
3 Q = [q₁,q₂,···,qₙ].
```

Upon convergence of the subspace iteration algorithm we have

$$AX_K = X_K R_K, \tag{8.1}$$

where K is the total number of iterations, X_K contains n orthogonomal vectors and R_K is triangular of dimension n. Comparing the result with the Schur decomposition (see Chapter 4), it is apparent that it constitutes an incomplete Schur decomposition of A. The n column vectors in X_K span an invariant space with respect to multiplication with A; hence, if $v \in X_K$, then $Av \in X_K$.

The eigen-decomposition of R_K is $Y\Lambda Y^{-1}$. Its substitution into (8.1) yields the first n eigenvalues and their associated eigenvectors,

$$AV = V\Lambda, \quad V = X_K Y. \tag{8.2}$$

From the process described in Step 2 of Algorithm (8.1) it is clear that the recursive equation of the algorithm is

$$AX_{k-1} = X_k R_k. \tag{8.3}$$

Hence, by induction, we conclude that, at step k of the process, it is

$$X_k R_k R_{k-1} \cdots R_1 = A^k X_0. \tag{8.4}$$

Thus we conclude that the algorithm builds the orthonormal basis for $A^k X_0$. Use of the QR factorization helps improve convergence by enhancing the difference (i.e., the linear independence) between the column vectors in X_k, which is compromised by the influence of the first few dominant eigenvectors as the number of iterations increases. In the ideal case, the column vectors of X_k converge, for sufficiently large values of k, to the Schur vectors associated with the first n dominant eigenvalues of A, $|\lambda_1| > |\lambda_2| > \cdots |\lambda_n|$.

The convergence of the subspace iteration algorithm depends on how distinct are the first n eigenvalues from the rest. In general, its convergence performance is rather poor, especially when a large number of eigenvalues are sought. Thus, it is of limited use, and more effective processes are utilized instead for the construction of subspaces of A spanned by its first n dominant eigenvectors. A class of such processes is described next.

8.2 METHODS BASED ON KRYLOV SUBSPACE PROJECTION

Krylov subspace projection methods were introduced in Section 4.2 in conjunction with the development of iterative methods for the solution of large sparse matrices. A projection method involves the approximation of the exact eigenvector v by a vector $\tilde{v}$ that belongs to the *expansion subspace*, $\mathcal{K}$, by imposing the constraint that the residual vector $A\tilde{v} - \tilde{\lambda}\tilde{v}$ is orthogonal to the *projection subspace*, $\mathcal{L}$. Thus, for the approximate eigenpair $(\tilde{\lambda}, \tilde{v})$, we write

$$A\tilde{v} - \tilde{\lambda}\tilde{v} \perp \mathcal{L}, \qquad \tilde{v} \in \mathcal{K}. \tag{8.5}$$

Assume that the orthonormal basis $\{v_1, v_2, \cdots, v_n\}$ for $\mathcal{K}$ is available, and let V be the matrix with column vectors the elements of the orthonormal basis. Similarly, let $\{w_1, w_2, \cdots, w_n\}$ constitute the orthonormal basis for $\mathcal{L}$, and let W be a matrix that has the elements of the basis as its columns. The approximate eigenvector $\tilde{v}$ may be expanded as follows:

$$\tilde{v} = Vy. \tag{8.6}$$

Then, in view of (8.5), we conclude that the vector y and the scalar $\tilde{\lambda}$ satisfy the following relation:

$$B_n y = \tilde{\lambda} y, \tag{8.7}$$

where $B_n = W^T A V$ is a square matrix of dimension n. Thus the original eigenvalue problem is projected onto an eigenvalue problem of smaller dimension. The eigendecomposition of B_n can be carried out using direct methods available in standard matrix software libraries such LAPACK [6]. The desired eigenvalues, $\tilde{\lambda}$, are selected from the n eigenvalues of B_n, and their corresponding eigenvectors of A are obtained from those of B_n through the operation $\tilde{v} = Vy$.

Krylov subspace projection methods use a Krylov subspace as the expansion subspace for approximate eigenvectors. In contrast with the subspace iteration, the dimension of the subspace used for the approximation of the eigenvectors increases with the number of iterations, and all vectors in the generated Krylov subspace are utilized. These attributes will become clear through the following discussion of two of the most popular members in the class of Krylov subspace projection methods, namely, the Arnoldi and the Lanczos algorithms.

8.2.1 Arnoldi algorithm

The Arnoldi process of Algorithm (4.1) can transform a matrix A into an *upper Hessenberg* matrix [3] which is often referred to as *Hessenberg reduction*. Subsequently, through the Schur decomposition of the Hessenberg matrix one obtains the eigenvalues of the matrix A.

The important point here is that the Arnoldi algorithm can reduce a matrix to an upper Hessenberg matrix whose eigenvalues yields good approximations to some of the eigenvalues of the original matrix even *well before* the completion of the full Hessenberg reduction. Hence, the Arnoldi process becomes a good candidate for approximating some of the eigenvalues of a large sparse matrix.

Starting with the incomplete Hessenberg reduction which is carried out using the Arnoldi process, from Section (4.2) we have

$$AQ_m = Q_m H_m + \tilde{q}_{m+1} e_m^T, \tag{8.8}$$

where we recall that e_m is a vector of length m with its mth element of value 1 and all the rest of its elements zero. Let $(\tilde{\lambda}_i, y_i)$ denote one of the m eigenpairs of H_m, i.e., $H_m y_i = \tilde{\lambda}_i y_i$. Then, multiplication of both sides of (8.8) on the right by y_i yields the residual of the approximate eigenpair $(\tilde{\lambda}_i, \tilde{v}_i = Q_m y_i)$ for A,

$$A\tilde{v}_i - \tilde{\lambda}_i \tilde{v}_i = \tilde{q}_{m+1} e_m^T y_i, \qquad \tilde{v}_i = Q_m y_i. \tag{8.9}$$

This statement shows that the approximate eigenvector $\tilde{v}_i$ is in the space spanned by the column vectors of Q_m. The residual $\tilde{q}_{m+1} e_m^T y_i$ is perpendicular to this space since $\tilde{q}_{m+1}$ is orthogonal to the column vectors of Q_m. The relative convergence tolerance, ρ_i, for the ith eigenpair is

$$\rho_i = \frac{\|A\tilde{v}_i - \tilde{\lambda}_i \tilde{v}_i\|}{\|\tilde{\lambda}_i \tilde{v}_i\|} = \frac{\|AQ_m y_i - \tilde{\lambda}_i Q_m y_i\|}{\|\tilde{\lambda}_i Q_m y_i\|} = \frac{\|\tilde{q}_{m+1} e_m^T y_i\|}{\|\tilde{\lambda}_i Q_m y_i\|}. \tag{8.10}$$

Combination of Algorithm (4.1) for Hessenberg reduction with the above tolerance estimation yields the complete Arnoldi algorithm for the approximation of n eigenpairs of matrix A.

Algorithm (8.3): Arnoldi algorithm for n eigenpairs of matrix A

```
1 q₁ ← r₀/‖r₀‖;                                  (Normalize an arbitrary vector)
2 for m = 1,...
   2.a q̃ₘ₊₁ ← Aqₘ;
   2.b for i = 1,...,m                           (Normalize with Arnoldi vectors)
      2.b.1 Hₘ(i,m) ← ⟨qᵢ, q̃ₘ₊₁⟩ = qᵢᵀq̃ₘ₊₁;
      2.b.2 q̃ₘ₊₁ ← q̃ₘ₊₁ − Hₘ(i,m)qᵢ;
   2.c Compute m eigenpairs (yᵢ, λ̃ᵢ) of Hₘ (i = 1,...,m);
   2.d for i = 1,...m, compute ρᵢ = ‖eₘᵀyᵢ‖‖q̃ₘ₊₁‖/‖λ̃ᵢQₘyᵢ‖;
   2.e If n eigepairs (λ̃ᵢ, ṽᵢ = Qₘyᵢ) have converged, goto 3;
   2.f Hₘ(m+1,m) = ‖q̃ₘ₊₁‖, qₘ₊₁ = qₘ₊₁/Hₘ(m+1,m);
3 End
```

The above algorithm computes simultaneously approximations to the first n dominant eigenvalues with largest magnitude $|\tilde{\lambda}_1| > |\tilde{\lambda}_2| \cdots > |\tilde{\lambda}_n|$. However, the convergence rates are different for different eigenvalues, with the eigenvalues of large magnitude converging faster. Clearly, once an eigenvalue has converged, its repeated calculation through the Arnoldi process is not needed. Thus, it is common to extract the converged eigenvalues using *deflation*. Deflation techniques will be addresses in the next section.

The computational cost of the Arnoldi algorithm increases with the number of iterations since every new Arnoldi vector is orthonormalized to the previous vectors. One way to alleviate the increasing computational complexity is by *restarting* the algorithm. More specifically, after a run of the algorithm that yields M Arnoldi vectors, we compute the approximate eigenvectors and use their linear combination as an initial vector for the next run. The pseudo-code description of the restarted Arnoldi algorithm is given below.

Algorithm (8.4): Restarted Arnoldi algorithm for n eigenpairs of matrix A

```
1 q₁ ← r₀/‖r₀‖;                                  (Normalize an arbitrary vector)
2 for m = 1,...,M
   2.a q̃ₘ₊₁ ← Aqₘ;
   2.b for i = 1,...,m                           (Normalize with Arnoldi vectors)
      2.b.1 Hₘ(i,m) ← ⟨qᵢ, q̃ₘ₊₁⟩ = qᵢᵀq̃ₘ₊₁;
      2.b.2 q̃ₘ₊₁ ← q̃ₘ₊₁ − Hₘ(i,m)qᵢ;
   2.c Compute m eigenpairs (yᵢ, λ̃ᵢ) of Hₘ (i = 1,···,m);
   2.d for i = 1···m, compute ρᵢ = ‖eₘᵀyᵢ‖‖q̃ₘ₊₁‖/‖λ̃ᵢQₘyᵢ‖;
   2.e If n eigenpairs (λ̃ᵢ, ṽᵢ = Qₘyᵢ) have converged, goto 4;
```

```
2.f  Hm(m + 1, m) = ‖q̃m+1‖,  qm+1 = qm+1/Hm(m + 1, m);
3  q1 = Σⁿᵢ₌₁ δᵢṽᵢ.  Normalize q1 and goto 2;
4  End
```

In Step 3, the initial vector q_1 is taken as the linear combination of the desired (yet unconverged) eigenvectors where δ_i are weighting coefficients. The simplest choice for the values of δ_i is to set them all equal to 1. Upon convergence, the initial vector is the linear combination of the desired eigenvectors. It is noted that in the restarted Arnoldi process the number of Arnoldi vectors and the size of the Hessenberg matrix do not exceed the value M. For more complicated restarted algorithms, the reader is referred to [7].

8.2.2 Lanczos algorithm

The Lanczos algorithm can be viewed as a simplification of the Arnoldi process for the particular case when the matrix A is symmetric. As shown in Section 4.4, when A is symmetric its Hessenberg reduction matrix H_m is a tri-diagonal matrix T_m.

$$AQ_m = Q_m \underbrace{\begin{bmatrix} \alpha_1 & \beta_2 & & & \\ \beta_2 & \alpha_2 & \beta_3 & & \\ & \ddots & \ddots & \ddots & \\ & & & \beta_m & \alpha_m \end{bmatrix}}_{T_m} + \tilde{q}_{m+1} e_m^T. \tag{8.11}$$

Algorithm (4.3) generates the Lanczos vectors in Q_m and T_m. The criterion for convergence is the same as (8.10). Thus, the Lanczos algorithm is easily obtained through a slight modification of the Arnoldi one presented in the previous subsection.

```
Algorithm (8.5): Lanczos algorithm for n eigenpairs of a symmetric matrix A
1  q0 ← 0,  β1 = ‖r0‖,  q1 ← r0/β1;                        (r0 is an arbitrary vector)
2  for  m = 1, 2, ...
   2.a  q̃m+1 ← Aqm;
   2.b  αm ← ⟨qm, q̃m+1⟩ = qmᵀq̃m+1;
   2.c  q̃m+1 ← q̃m+1 − βmqm−1 − αmqm;
   2.d  Compute m eigenpairs (λi, yi) of Tm,  (j = 1, ··· , m);
   2.e  for i = 1, ··· , m, compute ρi = ‖emᵀyi‖‖q̃m+1‖/‖λ̄iQmyi‖;
   2.f  If n eigenpairs (λ̄i, ṽi = Qmyi) have converged, goto 3;
   2.g  βm+1 = ‖q̃m+1‖,  qm+1 = q̃m+1/βm+1;
3  End
```

Because each Lanczos vector is only orthogonalized explicitly to the previous one, the orthogonality among the Lanczos vectors in Q_m is gradually lost due to roundoff error. This loss of orthogonality can be overcome either by *selective orthogonalization* of the generated vectors or by restarting the Lanczos process.

Selective orthogonalization is a computationally more efficient alternative than full orthogonalization. It is motivated by the fact that the most desired properties of the Lanczos algorithm are preserved provided that the generated eigenvectors are (approximately) orthogonal to half the machine precision. Clearly, the algorithmic implementation of such a process requires monitoring of the deterioration of orthogonality during the Lanczos process. Effective means through recurrences for such monitoring have been proposed and are discussed, along with the algorithmic implementation of selective orthogonalization in [5].

Shown below is the pseudo-code implementation of the restarted Lanczos algorithm. Like the restarted Arnoldi algorithm the restarting of the Lanczos process has the advantage of limiting the number of the generated Lanczos vectors before orthogonality is compromised.

Algorithm (8.6): Restarted Lanczos algorithm for n eigenvectors of a symmetric matrix A

1 $q_0 \leftarrow 0$, $\beta_1 = \|r_0\|$, $q_1 \leftarrow r_0/\beta_1$;   ($r_0$ is an arbitrary vector)
2 for $m = 1, 2, \ldots, M$
   2.a $\tilde{q}_{m+1} \leftarrow Aq_m$;
   2.b $\alpha_m \leftarrow \langle q_m, \tilde{q}_{m+1} \rangle = q_m^T \tilde{q}_{m+1}$;
   2.c $\tilde{q}_{m+1} \leftarrow \tilde{q}_{m+1} - \beta_m q_{m-1} - \alpha_m q_m$;
   2.d Compute m eigenpairs (λ_i, y_i) of T_m, $(j = 1, \cdots, m)$;
   2.e for $i = 1, \cdots, m$, compute $\rho_i = \|e_m^T y_i\| \|\tilde{q}_{m+1}\| / \|\tilde{\lambda}_i q_m y_i\|$;
   2.f If n eigenpairs $(\bar{v}_i = Q_m y_i, \tilde{\lambda}_i)$ have converged, goto 4;
   2.g $\beta_{m+1} = \|\tilde{q}_{m+1}\|$, $q_{m+1} = \tilde{q}_{m+1}/\beta_{m+1}$;
3 $r_0 \leftarrow \sum_{i=1}^{n} \delta_i \bar{v}_i$ and goto 1;
4 End

In Step 3, the initial vector r_0 is taken as the linear combination of the desired (yet un-converged) eigenvectors. The number of Lanczos vectors in Q_m is limited to M.

8.3 DEFLATION TECHNIQUES

As already mentioned above, when computing multiple eigenvalues simultaneously using the Arnoldi or the Lanczos algorithms, they converge at different rates. In general, eigenvalues with larger magnitude converge faster. Once an eigenvalue and its associated eigenvector have converged, it makes sense to extract it from the iteration process since its dominance slows down the convergence of the remaining eigenvalues and penalizes the computational cost of the process. *Deflation* is the process through which the extraction of the converged eigenvectors is achieved.

Given a matrix A, suppose that we have computed the eigenpair (λ_1, v_1), where λ_1 is the eigenvalue of largest magnitude. Deflation techniques amount to modifications to the original matrix A aimed at extracting the eigenvalue λ_1 while keeping all the remaining eigenvalues and their associated eigenvectors unchanged. In this manner, the subsequent application of the eigensolver to the modified matrix leads to the computation of the eigenpair (λ_2, v_2), where λ_2 has the largest magnitude among the remaining eigenvalues.

8.3.1 Deflation techniques for symmetric matrices

Let $\Lambda_{k-1} = \text{diag}(\lambda_1, \lambda_2, \cdots, \lambda_{k-1})$, be the diagonal matrix containing the first $k-1$ eigenvalues with corresponding eigenvectors the columns of the matrix V_{k-1}, $V_{k-1} = [v_1, v_2, \cdots, v_{k-1}]$. Hence, it is

$$AV_{k-1} = V_{k-1}\Lambda_{k-1}. \tag{8.12}$$

Furthermore, let us assume that the eigenvectors are orthonormal,

$$V_{k-1}^T V_{k-1} = I_{k-1}, \tag{8.13}$$

where I_{k-1} is the identity matrix of dimension $k-1$. Let r_0 be a vector orthogonal to all columns of V_{k-1}; hence, $V_{k-1}^T r_0 = 0$. However, in view of (8.12), this result may be cast in the following form:

$$V_{k-1}^T r_0 = (\Lambda_{k-1})^{-1}(AV_{k-1})^T r_0 = 0 \Rightarrow V_{k-1}^T A r_0 = 0. \tag{8.14}$$

By induction, it follows that the vectors $Ar_0, A^2 r_0, \ldots$ are orthogonal to V_{k-1}. Hence, it follows that if r_0 is chosen orthogonal to V_{k-1} the Krylov subspace generated from r_0 is also orthogonal to V_{k-1}

$$r_0 \perp V_{k-1} \Rightarrow [r_0, Ar_0, A^2 r_0, \cdots] \perp V_{k-1}. \tag{8.15}$$

If it were not for roundoff error, according to the above result the orthogonality of the initial vector, r_0, in the Krylov expansion space to V_{k-1} suffices for the Lanczos Algorithm (8.6) to compute correctly the next eigenvalue different from the ones that have already converged. However, because of finite numerical precision in the calculations, the orthogonality of the generated new Lanczos vectors to those in V_{k-1} deteriorates gradually, leading to lose of linear independence and, thus, inability to compute correctly the next eigenvalue. Therefore, deflation is needed, in the sense that orthogonality is explicitly enforced in the Lanczos process.

Algorithm (8.7): Restarted deflated Lanczos algorithm for n eigenvectors of a symmetric matrix A
1 $V_0 \leftarrow 0$;
2 for $k = 1, 2, \cdots, n$
  2.a $r_0 = (I - V_{k-1}V_{k-1}^T)r_0$;
  2.b $q_0 = 0$, $\beta_1 = \|r_0\|$, $q_1 = r_0/\beta_1$;
  2.c for $m = 1, 2, \cdots, M$
    2.c.1 $\tilde{q}_{m+1} \leftarrow (I - V_{k-1}V_{k-1}^T)Aq_m$;
    2.c.2 $\alpha_m \leftarrow \langle q_m, \tilde{q}_{m+1}\rangle = q_m^T \tilde{q}_{m+1}$;
    2.c.3 $\tilde{q}_{m+1} \leftarrow \tilde{q}_{m+1} - \beta_m q_{m-1} - \alpha_m q_m$;
    2.c.4 Compute m eigenpairs (λ_i, y_i) of T_m, $(i = 1, \cdots, m)$;
    2.c.5 For $i = 1, \cdots, m$, compute $\rho_i = \|e_m^T y_i\|\|\tilde{q}_{m+1}\|/\|\tilde{\lambda}_i Q_m y_i\|$;
    2.c.6 If one eigenpair $(\tilde{\lambda}_i, \tilde{v}_i = Q_m y_i)$ converge, goto 2.e;
    2.c.7 $\beta_{m+1} = \|\tilde{q}_{m+1}\|$, $q_{m+1} = \tilde{q}_{m+1}/\beta_{m+1}$;
  2.d $r_0 \leftarrow \tilde{v}_i$ and goto 2.a;
  2.e Normalize $\tilde{v}_i$ and $V_k \leftarrow [V_{k-1}, \tilde{v}_i]$.

To show that in the above algorithm the Lanczos vectors in Q_m starting from r_0 are explicitly forced to be orthogonal to the converged eigenvectors in V_{k-1}, we first note that Step 2.a forces the initial vector r_0 to be orthogonal to the space spanned by the columns of V_{k-1}. Thus, in view of Step 2.b, q_0 is also orthogonal to V_{k-1}. Step 2.c handles the enforcement of orthogonality, proved by induction as follows.

Assume $V_{k-1} \perp q_{1,2,\cdots,m}$, then Step 2.c.3 yields

$$\tilde{q}_{m+1} = (I - V_{k-1}V_{k-1}^T)Aq_m - \beta_m q_{m-1} - \alpha_m q_m \Rightarrow V_{k-1}^T \tilde{q}_{m+1} = 0; \tag{8.16}$$

hence, $V_{k-1} \perp q_{m+1}$.

Since the Krylov subspace Q_m is orthogonal to the converged eigenvectors in V_{k-1}, the algorithm will capture the next dominant eigenvalue and its associated eigenvector. The algorithm computes the eigenvalues one at a time instead of all at once as in Algorithm (8.6). Also, the converged eigenvectors are accumulated in V_k one by one.

Besides explicitly enforcing the orthogonality of the constructed Krylov subspace with the converged eigenvectors, the algorithm can be understood as effecting a modification to the original matrix A through the following matrix manipulation

$$\tilde{A} = (I - V_{k-1}V_{k-1}^T)A(I - V_{k-1}V_{k-1}^T) = A - V_{k-1}\Lambda_{k-1}V_{k-1}^T \tag{8.17}$$

where use was made of (8.12). It follows immediately from the above result that any eigenvector v of A, such that $v \in V_{k-1}$, is also an eigenvector of $\tilde{A}$ with zero eigenvalue. On the other hand, if $v \notin V_{k-1}$, then v is still an eigenvector of $\tilde{A}$ with the same eigenvalue as A.

In summary, deflation may be described as a process through which the original matrix A is modified in such a manner that, a) the resulting matrix has the same eigenvectors as A; b) all converged eigenvectors appear in the modified matrix with eigenvalue zero; c) the remaining eigenvectors still to be determined have the same eigenvalues as in A.

8.3.2   Deflation techniques for non-symmetric matrices

Deflation techniques are also applicable to non-symmetric matrices. In this case, the added complexity stems from the fact that the development of the pertinent techniques requires the use of both right and left eigenvectors.

Let us begin with the assumption that the first $k-1$ eigenvalues and their corresponding left and right eigenvectors have been computed. Using the notation of Section 4.1, we define the following three matrices. $\Lambda_{k-1} = \text{diag}(\lambda_1, \lambda_2, \cdots, \lambda_{k-1})$, contains the $k-1$ eigenvalues. $U_{k-1} = [u_1, u_2, \cdots, u_{k-1}]$ contains the corresponding k left eigenvectors. $V_{k-1} = [v_1, v_2, \cdots, v_{k-1}]$ contains the corresponding right eigenvectors. Hence, it is

$$AV_{k-1} = V_{k-1}\Lambda_{k-1}, \quad A^T U_{k-1} = U_{k-1}\Lambda_{k-1}, \quad U_{k-1}^T V_{k-1} = I_{k-1}, \qquad (8.18)$$

where I_{k-1} is the identity matrix of size $k-1$.

In order to compute the k-th eigenvalue and its corresponding left and right eigenvectors, we modify the original matrix A as follows:

$$\tilde{A} = (I - V_{k-1}U_{k-1}^T)A(I - V_{k-1}U_{k-1}^T). \qquad (8.19)$$

Use of (8.18) into (8.19) yields the following simplified form

$$\tilde{A} = A - V_{k-1}\Lambda_{k-1}U_{k-1}^T. \qquad (8.20)$$

It follows immediately from this result that $\tilde{A}$ has the same eigenvectors as A. However, the eigenvalues of the first $k-1$ eigenvectors are now zero, while the eigenvalues of the remaining eigenvectors are unchanged. Finally, we mention that the deflation technique for symmetric matrices can be viewed as the special case of (8.19) with $U_{k-1} = V_{k-1}$.

The above deflation technique requires the computation of both left and right eigenvectors, thus doubling the associated computational cost. A more economical alternative is considered next.

Suppose that we are interested in computing the Schur-vectors instead of the eigenvectors, and we already computed the first $k-1$ dominant eigenvalues and their Schur vectors. Let $W_{k-1} = [w_1, w_2, \cdots, w_{k-1}]$ be the unitary matrix with columns the $k-1$ Schur vectors. Hence, it is (see Section 4.1)

$$AW_{k-1} = W_{k-1}R_{k-1}, \qquad (8.21)$$

where R_{k-1} is an upper triangular matrix of size $k-1$. (The change in the notation of Section 4.1 for the Schur vectors from Q to W is aimed at avoiding confusion with the matrix Q used in this chapter for storing the Arnoldi vectors.) The complete Schur decomposition of A is

$$A = \begin{bmatrix} W_{k-1} & W_{N-k+1} \end{bmatrix} \begin{bmatrix} R_{k-1} & R_{k-1,N-k+1} \\ 0 & R_{N-k+1} \end{bmatrix} \begin{bmatrix} W_{k-1}^T \\ W_{N-k+1}^T \end{bmatrix}. \qquad (8.22)$$

Considering a deflation-related modification of the original matrix A as follows:

$$\tilde{A} = (I - W_{k-1}W_{k-1}^T)A(I - W_{k-1}W_{k-1}^T). \tag{8.23}$$

Substitution of (8.22) into the above equation yields the Schur-decomposition of $\tilde{A}$

$$\tilde{A} = \begin{bmatrix} W_{k-1} & W_{N-k+1} \end{bmatrix} \begin{bmatrix} 0 & 0 \\ 0 & R_{N-k+1} \end{bmatrix} \begin{bmatrix} W_{k-1}^T \\ W_{N-k+1}^T \end{bmatrix}. \tag{8.24}$$

It is now apparent that the modified matrix $\tilde{A}$ has the first $k-1$ Schur vectors of A as eigenvectors with zero eigenvalue. If we continue applying the Arnoldi algorithm to $\tilde{A}$, then the next dominant eigenvalue in R_{N-k+1} will be computed along with its associated Schur-vector in W_{N-k+1}. The process continues until the first n dominant Schur vectors in W_n have been computed. This is followed by the formation of the upper-triangular matrix, R_n,

$$R_n = W_n^T A W_n \tag{8.25}$$

and its subsequent, complete eigen-decomposition,

$$R_n = Y_n \Lambda_n Y_n^{-1}. \tag{8.26}$$

Since R_n is an upper-triangular matrix its eigen-decomposition is quite expedient. The eigenvalue matrix Λ_n is simply obtained as the diagonal of R_n; the matrix V_n containing the desired eigenvectors is obtained as $V_n = W_n Y_n$.

The aforementioned deflation process, combined with the restarted Arnoldi algorithm yields the following deflated restarted Arnoldi process for the calculation of n eigenpairs of the non-symmetric matrix A.

Algorithm (8.8): Deflated restarted Arnoldi algorithm for n eigenpairs of a non-symmetric matrix A

1 $W_0 \leftarrow 0$;
2 for $k = 1, 2, \cdots, n$
  2.a $r_0 = (I - W_{k-1}W_{k-1}^T)r_0$;
  2.b $q_1 \leftarrow r_0/\|r_0\|$;
  2.c for $m = 1, \cdots, M$
    2.c.1 $\tilde{q}_{m+1} \leftarrow (I - W_{k-1}W_{k-1}^T)Aq_m$;
    2.c.2 for $i = 1, \cdots, m$
      $H_m(i,m) \leftarrow \langle q_i, \tilde{q}_{m+1}\rangle = q_i^T \tilde{q}_{m+1}$ and $\tilde{q}_{m+1} \leftarrow \tilde{q}_{m+1} - H_m(i,m)q_i$;
    2.c.3 Compute m eigenpairs $(\tilde{\lambda}_i, y_i)$ of H_m $(i = 1, \cdots, m)$;
    2.c.4 Compute $\rho_i = \|e_m^T y_i\|\|\tilde{q}_{m+1}\|/\|\tilde{\lambda}_i Q_m y_i\|$, $(i = 1, \cdots, m)$;
    2.c.5 If one eigenpair $(\lambda_i, \tilde{v}_i = Q_m y_i)$ converges, goto 2.e;
    2.c.6 $H_m(m+1,m) = \|\tilde{q}_{m+1}\|$, $q_{m+1} = q_{m+1}/H_m(m+1,m)$;
  2.d $r_0 \leftarrow \tilde{v}_i$ and goto 2.a;
  2.e Normalize $\tilde{v}_i$, and $W_k \leftarrow [W_{k-1}, \tilde{v}_i]$;
3 Compute $R_n = W_n^T A W_n$;
4 Eigen-decomposition of $R_n = Y_n \Lambda_n Y_n^{-1}$;
5 n eigenvalues are in Λ_n; eigenvectors are in $V_n = W_n Y_n$.

The above algorithm forces the Arnoldi vectors to be orthogonal to the converged Schur vectors in W_k in Step 2.c.1. The first n dominant Schur vectors are computed one by one, by making use of the previously converged Schur vectors to deflate the original matrix. Once the first n dominant Schur vectors have been obtained, their linear combination through multiplication with the matrix of the eigenvectors of R_n yields the desired n eigenvectors of A.

We conclude by noting the similarity between the deflated Arnoldi algorithm (8.8) and the deflated Lanczos algorithm (8.7). Since for a symmetric matrix, the Schur vectors are also its eigenvectors, once the Schur vectors are obtained, the extra steps (Steps 3 and 4) in Algorithm (8.8) are not needed for the derivation of the n dominant eigenvectors.

8.4   NON-STANDARD EIGENVALUE PROBLEMS

Up till now, we discussed methodologies and deflation techniques for the computation of a set of the dominant eigenvalues for standard matrix eigenvalue problems of the form $Ax = \lambda x$. However, it is quite often in practice that finite element approximations of electromagnetic eigenvalue problems result in either the "generalized" matrix eigenvalue statement of

$$Ax = \lambda Bx, \tag{8.27}$$

or the "quadratic" form

$$\left(\lambda^2 M + \lambda C + K\right) x = 0. \tag{8.28}$$

In this section we discuss techniques for converting the above non-standard matrix eigenvalue problems to standard form so that the algorithms presented in previous sections may be used for their solution.

8.4.1   Generalized eigenvalue problems

We begin the discussion of methodologies used for handling eigenvalue problems of the form (8.27) with some pertinent nomenclature from the mathematics literature.

Given the two square matrices A and B, both of dimension N, the set of all matrices of the form $A - zB$, $z \in \mathbb{C}$, is called a *pencil*. The eigenvalues of the pencil are elements of the set $\lambda(A, B)$, defined as follows:

$$\lambda(A, B) = \{\lambda \in \mathbb{C} : \det(A - \lambda B) = 0\}. \tag{8.29}$$

The vector $v \neq 0$ satisfying the equation $Av = \lambda Bv$, where $\lambda \in \lambda(A, B)$, is the eigenvector of $A - \lambda B$.

The generalized eigenvalue problem has N eigenvalues if and only if the rank of B is N. This is the most commonly encountered situation in the context of finite element approximations of electromagnetic boundary value problems.

In most cases of electromagnetic eigenvalue problems of practical interest the matrices A and B exhibit special properties (e.g., they are either symmetric or Hermitian matrices). These special properties can be exploited for the development of efficient and robust methods for their solution. Prior to presenting these methods we will review four results which will be found most useful in our subsequent discussions.

Theorem I: If A and B are symmetric matrices, and v_i and v_j are two distinct eigenvectors with distinct eigenvalues, then v_i and v_j are *B-orthogonal*, that is, $\langle v_i, Bv_j \rangle = 0$.

The proof is straightforward and begins with the eigenvalue equation statements for the two eigenpairs,

$$Av_i = \lambda_i Bv_i, \qquad Av_j = \lambda_j Bv_j. \tag{8.30}$$

Multiplication of the first equation by v_j^T, followed by matrix transposition yields

$$v_i^T Av_j = \lambda_i v_i^T Bv_j, \tag{8.31}$$

which, in view of the second equation in (8.30), becomes

$$(\lambda_i - \lambda_j)v_i^T B v_j = 0. \tag{8.32}$$

If $\lambda_i \neq \lambda_j$, then $v_i^T B v_j = 0$. For the degenerate case where $\lambda_i = \lambda_j$, the Gram-Schmidt orthogonalization process may be used to construct a linear combination of v_i and v_j which is B-orthogonal. ∎

Hence, we conclude that for symmetric generalized eigenvalue problems, two distinct eigenvectors are B-orthogonal. Furthermore, through proper scaling, the eigenvectors can be made orthonormal. That is,

$$V^T B V = I, \qquad V^T A V = \Lambda \tag{8.33}$$

where V is the matrix with columns the eigenvectors and Λ is a diagonal matrix with elements the corresponding eigenvalues.

Theorem II: If A and B are Hermitian matrices, and v_i and v_j are two eigenvectors with corresponding eigenvalues, λ_i and λ_j, respectively, then the following are true:

1. For real eigenvalue λ_i, the B-norm of its corresponding eigenvector v_i is nonzero, that is, $v_i^H B v_i \neq 0$.

2. For complex eigenvalue λ_i, the B-norm of its corresponding eigenvector v_i is zero, that is, $v_i^H B v_i = 0$.

3. For any complex eigenvalue λ_i, there exists another eigenvalue λ_j, such that $\lambda_j = \lambda_i^*$, and their corresponding eigenvectors satisfy $v_i^H B v_j \neq 0$.

4. For any two complex eigenvalues λ_i and λ_j, such that $\lambda_j \neq \lambda_i^*$, their corresponding eigenvectors satisfy $v_i^H B v_j = 0$.

The proof starts from he eigenvalue equation statements for the two eigenpairs

$$A v_i = \lambda_i B v_i, \qquad A v_j = \lambda_j B v_j. \tag{8.34}$$

Multiplication of the first equation by v_j^H, followed by matrix conjugate transposition yields

$$v_i^H A v_j = \lambda_i^* v_i^H B v_j. \tag{8.35}$$

In view of the second equation in (8.34) this becomes

$$(\lambda_i^* - \lambda_j)v_i^H B v_j = 0. \tag{8.36}$$

Results 1 and 2 follows immediately from the above equation with $i = j$. Results 3 and 4 follows with $i \neq j$. ∎

If the Hermitian matrix B is positive-definite, that is, $v_i^H B v_i > 0 \; \forall v_i$ and $v_i \neq 0$, then it follows from (8.36) that there are only real eigenvalues and their eigenvectors are orthogonal.

Theorem III: If A is symmetric and B is skew-symmetric (that is, $B^T = -B$), or if A is skew-symmetric and B is symmetric, then the following are true:

1. For each nonzero eigenvalue, λ_i, the B-norm of its corresponding eigenvector, v_i, is zero, that is, $v_i^T B v_i = 0$.

2. For each nonzero eigenvalue, λ_i, there exists another eigenvalue, λ_j, such that $\lambda_i = -\lambda_j$ and their corresponding eigenvectors satisfy the relation $v_i^T B v_j \neq 0$.

3. For any two eigenvalues λ_i and λ_j, such that $\lambda_j \neq -\lambda_i$, their corresponding eigenvectors satisfy $v_i^T B v_j = 0$.

The proof is straightforward. Once again, we start with the eigenvalue equation statements for the two eigenpairs,

$$Av_i = \lambda_i B v_i, \qquad Av_j = \lambda_j B v_j. \tag{8.37}$$

Multiplication of the first equation by v_j^T, followed by matrix transposition yields,

$$v_i^T A v_j = -\lambda_i v_i^T B v_j. \tag{8.38}$$

In view of the second equation in (8.40), the above equation is cast in the form

$$(\lambda_i + \lambda_j)v_i^T B v_j = 0. \tag{8.39}$$

Result 1 in Theorem III follows immediately from the above equation with $i = j$. Results 2 and 3 follows with $i \neq j$. ∎

Theorem IV: If A is Hermitian and B is skew-Hermitian (that is, $B^H = -B$), or if A is skew-Hermitian and B is Hermitian, then the following are true.

1. For pure imaginary eigenvalue, λ_i, the B-norm of its corresponding eigenvector, v_i, is non-zero, that is, $v_i^H B v_i \neq 0$.

2. For complex eigenvalue, λ_i, the B-norm of its corresponding eigenvector, v_i, is zero, that is, $v_i^H B v_i = 0$.

3. For any complex eigenvalue λ_i, there exists another eigenvalue λ_j, such that $\lambda_j = -\lambda_i^*$, and their corresponding eigenvectors satisfy $v_i^H B v_j \neq 0$.

4. For any two distinct complex eigenvalues λ_i and λ_j, such that $\lambda_j \neq -\lambda_i^*$, their corresponding eigenvectors satisfy $v_i^H B v_j = 0$.

The proof starts with the eigenvalue equation statements for the two eigenpairs,

$$Av_i = \lambda_i B v_i, \qquad Av_j = \lambda_j B v_j. \tag{8.40}$$

Multiplication of the first equation by v_j^H, followed by matrix complex transposition yields,

$$v_i^H A v_j = -\lambda_i^* v_i^H B v_j. \tag{8.41}$$

In view of the second equation in (8.40), the above equation is cast in the form

$$(\lambda_i^* + \lambda_j)v_i^H B v_j = 0. \tag{8.42}$$

Results 1 and 2 in Theorem IV follow immediately from the above equation with $i = j$. Results 3 and 4 follow with $i \neq j$. ∎

If the Hermitian matrix B is positive-definite, it follows from (8.42) that there are only pure imaginary eigenvalues, and their eigenvectors are orthogonal.

8.4.1.1 Lanczos Algorithm for Generalized Symmetric Eigenvalue Problems

The generalized eigenvalue problem may be cast in the standard form $B^{-1}Av = \lambda v$. If both A and B are symmetric, the matrix $B^{-1}A$ is symmetric only if B^{-1} and A commute, that is, $B^{-1}A = AB^{-1}$. However, the matrix $B^{-1}A$ is self-adjoint with respect the B-inner product,

$$\langle x, y \rangle_B = x^T B y. \tag{8.43}$$

This is shown in a straightforward manner as follows:

$$\langle x, B^{-1}Ay \rangle_B = x^T Ay = x^T AB^{-1}By = \langle B^{-1}Ax, y \rangle_B. \tag{8.44}$$

Thus, the Lanczos Algorithms (8.5) and (8.6) are still applicable with the matrix A replaced by $B^{-1}A$ and the Euclidean inner product replaced by the B-inner product. Hence, the generated Lanczos vectors in Q_m satisfy the relations

$$\begin{aligned}
\langle Q_m, Q_m \rangle_B &= Q_m^T B Q_m = I_m, \\
\langle Q_m, B^{-1}AQ_m \rangle_B &= Q_m^T A Q_m = T_m,
\end{aligned} \tag{8.45}$$

where I_m is the identity matrix of dimension m, and T_m is a tri-diagonal matrix of dimension m. The B-inner product involves multiplication by B. Step 2.b in both Algorithms (8.5) and (8.6) becomes

$$\alpha_m = \langle q_m, \tilde{q}_{m+1} \rangle_B = \langle q_m, B^{-1}Aq_m \rangle_B = q_m^T Aq_m. \tag{8.46}$$

Similarly, the B-norm calculation in Step 2.g can also be simplified by observing that $\tilde{q}_{m+1}$ is B-orthogonal to the vectors q_m and q_{m-1}. Thus it is

$$\begin{aligned}
\|\tilde{q}_{m+1}\|^2 &= \langle \tilde{q}_{m+1}, \tilde{q}_{m+1} \rangle_B \\
&= \langle \tilde{q}_{m+1}, B^{-1}Aq_m \rangle_B - \alpha_m \langle \tilde{q}_{m+1}, q_m \rangle_B - \beta_m \langle \tilde{q}_{m+1}, q_{m-1} \rangle_B \\
&= \tilde{q}_{m+1}^T Aq_m.
\end{aligned} \tag{8.47}$$

It can be seen from (8.46) and (8.47) that an extra vector is needed for Av_i. Finally, the modified Lanczos algorithm for the generalized eigenvalue problem when both A and B are symmetric is given below.

Algorithm (8.9): Lanczos algorithm for n eigenvectors of $Av = \lambda Bv$ with A and B symmetric

```
1 q₀ ← 0,  β₁ = ‖r₀‖,  q₁ ← r₀/β₁;  (r₀ is an arbitrary vector)
2 for m = 1, 2, ···
   2.a w ← Aqₘ;
   2.b αₘ ← ⟨qₘ, w⟩ = qₘᵀw;
   2.c q̃ₘ₊₁ ← B⁻¹w − βₘqₘ₋₁ − αₘqₘ;
   2.d Compute the eigenpairs (λᵢ, yᵢ) of Tₘ, (i = 1, ···, m);
   2.e for (i = 1, ···, m), compute ρᵢ = ‖eₘᵀyᵢ‖‖q̃ₘ₊₁‖/‖λ̃ᵢQₘyᵢ‖;
   2.f If n eigenpairs (λ̃ᵢ, ṽᵢ = Qₘyᵢ) converge, goto 3;
   2.g βₘ₊₁ = √(q̃ₘ₊₁ᵀw),  qₘ₊₁ = q̃ₘ₊₁/βₘ₊₁;
3 End.
```

Note that in Step 2.c the inverse of B is required. This operation is computationally costly for large matrices. We will discuss this issue further in later chapters and in the context of specific applications of this algorithm to electromagnetic eigenvalue problems.

A restarted version of the above algorithm can be developed from Algorithm (8.6) in a straightforward fashion.

8.4.1.2 Deflation Technique for Generalized Symmetric Eigenvalue Problems

The development of a deflation process for generalized symmetric eigenvalue problems follows closely the one in Section 8.4.1 for the standard problem. The primary difference is that in place of the standard inner product, the B-inner product of (8.43) is used.

Let $\Lambda_{k-1} = \text{diag}(\lambda_1, \lambda_2, \cdots, \lambda_{k-1})$, be the diagonal matrix containing the (already computed) first $k-1$ eigenvalues with corresponding eigenvectors the columns of the matrix V_{k-1}, $V_{k-1} = [v_1, v_2, \cdots, v_{k-1}]$. Hence, it is

$$AV_{k-1} = BV_{k-1}\Lambda_{k-1}. \tag{8.48}$$

Furthermore, let us assume that the eigenvectors have been orthonormalized,

$$V_{k-1}^T BV_{k-1} = I_{k-1}. \tag{8.49}$$

Let r_0 be a vector orthogonal to all the columns in V_{k-1}; hence, $V_{k-1}^T Br_0 = 0$. In view of (8.48), this result may be cast in the following form:

$$\begin{aligned}
\Lambda_{k-1}V_{k-1}^T Br_0 = (AV_{k-1})^T r_0 = 0 \\
\Rightarrow V_{k-1}^T B(B^{-1}Ar_0) = \langle V_{k-1}, B^{-1}Ar_0\rangle_B = 0.
\end{aligned} \tag{8.50}$$

By induction, it follows that the vectors $(B^{-1}A)r_0, (B^{-1}A)^2 r_0, \ldots$ are orthogonal to V_{k-1}. Hence, it follows that if r_0 is chosen B-orthogonal to V_{k-1} the Krylov subspace generated from r_0 is also B-orthogonal to V_{k-1}

$$r_0 \perp V_{k-1} \Rightarrow \left[r_0, (B^{-1}A)r_0, (B^{-1}A)^2 r_0, \cdots\right] \perp V_{k-1}. \tag{8.51}$$

If it were not for roundoff error, according to this result, the orthogonality of the initial vector, r_0, in the Krylov expansion space to V_{k-1} suffices for the Lanczos Algorithm (8.9) to compute correctly the next eigenvalue different from the ones that have already converged. However, because of finite numerical precision in the calculations, the orthogonality of the generated new Lanczos vectors to those in V_{k-1} deteriorates gradually, leading to lose of linear independence and, thus, inability to compute correctly the next eigenvalue. Therefore, deflation is needed, in the sense that orthogonality is explicitly enforced in the Lanczos process.

Algorithm (8.10): Restarted deflated Lanczos algorithm for n eigenvectors of $Av = \lambda Bv$ with A and B symmetric

1 $V_0 \leftarrow 0$;
2 for $k = 1, 2, \cdots, n$
  2.a $r_0 = (I - V_{k-1}V_{k-1}^T B)r_0$;
  2.b $q_0 = 0$, $\beta_1 = \|r_0\|$, $q_1 = r_0/\beta_1$;
  2.c for $m = 1, 2, \cdots, M$
    2.c.1 $w \leftarrow (I - BV_{k-1}V_{k-1}^T)Aq_m$;
    2.c.2 $\alpha_m \leftarrow \langle q_m, w\rangle = q_m^T w$;
    2.c.3 $\tilde{q}_{m+1} \leftarrow B^{-1}w - \beta_m q_{m-1} - \alpha_m q_m$;
    2.c.4 Compute the eigenpairs $(\tilde{\lambda}_i, y_i)$ of T_m, $(i = 1, \cdots, m)$;
    2.c.5 Compute $\rho_i = \|e_m^T y_i\|\|\tilde{q}_{m+1}\|/\|\tilde{\lambda}_i Q_m y_i\|$, $(i = 1, \cdots, m)$;
    2.c.6 If one eigenpair $(\lambda_i, \tilde{v}_i = Q_m y_i)$ converge, goto 2.e;
    2.c.7 $\beta_{m+1} = \sqrt{\tilde{q}_{m+1}^T w}$, $q_{m+1} = \tilde{q}_{m+1}/\beta_{m+1}$;
  2.d $r_0 \leftarrow \tilde{v}_i$ and goto 2.a;
  2.e B-Normalization: $v_i = \tilde{v}_i/\sqrt{\tilde{v}_i^T B\tilde{v}_i}$ and $V_k \leftarrow [V_{k-1}, v_i]$.

Step 2.a forces the initial vector r_0 to be B-orthogonal to the previous $k-1$ converged eigenvectors in V_{k-1}. Theoretically, it is sufficient to render the Lanczos vectors in Q_m also B-orthogonal to V_{k-1}. However, due to roundoff, Step 2.c.1 is necessary to enforce orthogonality explicitly.

Step 2.c.3 yields

$$\tilde{q}_{m+1} = B^{-1}(I_{k-1} - BV_{k-1}V_{k-1}^T)Aq_m - \beta_m q_{m-1} - \alpha_m q_m. \tag{8.52}$$

It is shown next that, if all previous Lanczos vectors are B-orthogonal to V_{k-1}, then $\tilde{q}_{m+1}$ is also B-orthogonal to V_{k-1}. This follows immediately from (8.52) making use of the result $V_{k-1}^T BV^{k-1} = I_{k-1}$,

$$V_{k-1}^T B\tilde{q}_{m+1} = V_{k-1}^T(I_{k-1} - BV_{k-1}V_{k-1}^T)Aq_m = 0. \tag{8.53}$$

Besides explicitly enforcing the orthogonality of the constructed Krylov subspace with the converged eigenvectors, the algorithm can be understood as effecting a modification to the original matrix A through the following matrix manipulation:

$$\tilde{A} = (I - BV_{k-1}V_{k-1}^T)A(I - V_{k-1}V_{k-1}^T B) = A - BV_{k-1}\Lambda_{k-1}V_{k-1}^T B, \tag{8.54}$$

where use was made of (8.48) and (8.49). It follows immediately from the above result that any eigenvector v of $A - \lambda B$, such that $v \in V_{k-1}$, is also an eigenvector of $\tilde{A} - \lambda B$ with zero eigenvalue. On the other hand, if $v \notin V_{k-1}$, then v is still an eigenvector of $\tilde{A}$ with the same eigenvalue as A.

In summary, in a manner identical to the one discussed earlier in the context of the standard eigenvalue problem, deflation may be described as a process through which the original matrix A of the eigenvalue problem $A - \lambda B$ is modified in such a manner that, a) the resulting matrix, $\tilde{A}$ is such that the modified eigenvalue problem, $\tilde{A} - \lambda B$, has the same eigenvectors as $A - \lambda B$; b) all converged eigenvectors appear in the modified matrix with eigenvalue zero; c) the remaining eigenvectors still to be determined have the same eigenvalues as in $A - \lambda B$.

8.4.2 Quadratic eigenvalue problems

Quadratic eigenvalue problems, understood to mean that the eigenvalue and its square are explicitly present in the eigenvalue statement, are encountered in electromagnetic applications in conjunction with the modal analysis of lossy structures. The general statement of the finite element approximation of such problems has the form

$$\left(\lambda^2 M + \lambda C + K\right) x = 0. \tag{8.55}$$

The most common way of dealing with quadratic eigenvalue problems above problem is to transform it into a linear generalized eigenvalue problem. For example, introducing the vector $\tilde{x} = \lambda x$, (8.55) may be cast in the form

$$\underbrace{\begin{bmatrix} 0 & I \\ -K & -C \end{bmatrix}}_{A} \begin{bmatrix} x \\ \lambda x \end{bmatrix} = \lambda \underbrace{\begin{bmatrix} I & \\ & M \end{bmatrix}}_{B} \begin{bmatrix} x \\ \lambda x \end{bmatrix}. \tag{8.56}$$

Depending on the properties of the matrices K, C, M, alternative forms may offer computational advantages. For example, if the matrices are symmetric,the following form results in a symmetric generalized eigenvalue problem

$$\underbrace{\begin{bmatrix} 0 & K \\ K & C \end{bmatrix}}_{A} \begin{bmatrix} x \\ \lambda x \end{bmatrix} = \lambda \underbrace{\begin{bmatrix} K & \\ & -M \end{bmatrix}}_{B} \begin{bmatrix} x \\ \lambda x \end{bmatrix}. \tag{8.57}$$

In the context of electromagnetic problems, such quadratic eigenvalue statements result when the governing differential equation statements of the eigenvalue problem are based on the vector Helmholtz equations for the electric or the magnetic field. An alternative approach to obtaining a linear eigenvalue problem is to work directly with Maxwell's curl equations for the electric and the magnetic flux density. In this manner, making use of the field-flux formulation of Chapter 3 for the development of the finite element approximation of these equations, a linear generalized eigenvalue problem is obtained. A specific application of such an approach is presented in Chapter 10 in conjunction with the modal analysis of lossy electromagnetic resonators, and in Chapter 11 in conjunction with the development of model order reduction methodologies for finite element approximation of electromagnetic BVPs.

8.5 SHIFT-AND-INVERT PRECONDITIONER

The algorithms introduced in the previous sections are aimed at the computation of the eigenvalues with the largest magnitude. However, it is often the case that these are not the desired eigenvalues. For example, in the eigenmodal analysis of electromagnetic cavities the finite element approximations of the associated continuous eigenvalue problem possess three classes of eigenvalues. The first class, which will be referred to as Type-A eigenvalues, are the zero eigenvalues associated with the presence of spurious gradient modes. These modes are of no practical interest. The same is true for the eigenvalues with the largest magnitude, that is, the high-frequency eigenvalues. These eigenvalues, which will be referred to as Type-C eigenvalues, correspond to poorly resolved modes in the finite element approximation. Finally, the Type-B eigenvalues are the ones of practical interest, associated with physical, well-resolved eigenvectors that capture the spatial distribution of the fields for the mode(s) involved in the practical application of the cavity. Since the magnitudes of Type-B eigenvalues are neither the smallest nor the largest in the spectrum, a slight modification of the eigenvalue equation statement is needed for the algorithms presented in this Chapter to be applicable for their computation. A simple and effective method, referred to as *shift-and-invert preconditioner*, is given below.

Consider the following transformation to the standard eigenvalue problem

$$Ax = \lambda x \Rightarrow (A - \lambda_0 I)^{-1} x = \frac{1}{\lambda - \lambda_0} x, \tag{8.58}$$

where λ_0 is referred to as the *expansion point*. The eigenvalues closest to λ_0 are shifted to the ones with largest magnitude. Hence, the aforementioned algorithms can be applied for their expedient computation. For example, if A is symmetric, $A - \lambda_0 I$ is also symmetric; hence, the Lanczos algorithm is applicable to the modified eigenvalue problem.

For the case of generalized symmetric eigenvalues problems, the transformation assumes the form

$$Ax = \lambda Bx \Rightarrow Bx = \frac{1}{\lambda - \lambda_0}(A - \lambda_0 B)x. \tag{8.59}$$

The eigenvalues closest to λ_0 are shifted to the ones with largest magnitude. Hence, Algorithm (8.10) is applicable for the calculation of the eigenvalues and eigenvectors of A in the vicinity of λ_0.

For generalized non-symmetric eigenvalue problems the shift-and-invert transformation is

$$Ax = \lambda Bx \Rightarrow (A - \lambda_0 B)^{-1} Bx = \frac{1}{\lambda - \lambda_0} x. \tag{8.60}$$

The resulting eigenvalue problem can be solved numerically using the Arnoldi process of Algorithm (8.8).

REFERENCES

1. Y. Saad, *Numerical Methods for Large Eigenvalue Problems*, New York: Halsted Press, 1992.

2. J. Cullum and R. Willoughby, *Lanczos Algorithms for Large Symmetric Eigenvalue Computations*, Basel: Birkhause, 1985.

3. G. H. Golub and C. F. Van Loan, *Matrix Computations*, 3rd ed., Baltimores: Johns Hopkins University Press, 1996.

4. B. N. Parlett, *The Symmetric Eigenvalue Problem*, Englewood Cliffs, NJ: Prentice-Hall, 1980.

5. Z. Bai, J. Demmel, J. Dongarra, A. Ruhe, and H. van der Vorst (eds.), *Templates for the Solution of Algebraic Eigenvalue Problems – A Practical Guide*, Philadelphia: SIAM, 2000.

6. E. Anderson, Z. Bai, C. Bischof, S. Blackford, D. Demmel, J. Dongarra, J. Du Croz, A. Greenbaum, S. Hammarling, A. McKenney, D. Sorensen, *LAPACK Users' Guide*, Third edition, Philadelphia: SIAM, 1999.

7. R. B. Lehoucq, D. C. Sorensen, and C. Yang, *ARPACK Users' Guide: Solution of Large-Scale Eigenvalue Problems with Implicitly Restarted Arnoldi Methods*, Philadelphia: SIAM, 1998.

CHAPTER 9

TWO-DIMENSIONAL EIGENVALUE ANALYSIS OF WAVEGUIDES

Advances in the design of integrated electronic and optical circuits have necessitated the utilization of complex waveguiding structures, of both metallic and dielectric guiding cores, which comply with the planar nature of integrated circuits. Furthermore, it is often the case that more than one guiding cores are utilized, either to facilitate the design of controlled power splitting and power combining, or simply to increase the bandwidth of the waveguide medium. By changing the material properties of both the waveguide core media and the integrating substrate, the electromagnetic properties of the resulting waveguiding systems (e.g., propagation constant, coupling efficiency, and signal dispersion) can be altered.

The design of such multiple, coupled waveguides relies on the availability of a versatile electromagnetic field eigensolver, capable of handling the geometric and material complexity involved, including material anisotropy and inhomogeneity. The consequence of such material complexity is that the eigenmodes of the structure are *hybrid*, that is, they cannot be reduced into the simpler sets of transverse magnetic or transverse electric modes, encountered in simpler waveguiding structures such as homogeneous or partially-filled metallic waveguides. Therefore the electromagnetic analysis of such waveguides needs to be based on the complete system of Maxwell's equations with the only simplification that the solutions sought exhibit a spatial dependence of the form $\exp(-\gamma w)$, where γ is the propagation constant (eigenvalue) to be determined and w is the spatial variable in the direction parallel to the axis of the waveguiding system.

Finite element-based eigensolvers are most suitable for the modal analysis of waveguiding systems exhibiting significant geometric and material complexity. This chapter discusses the application of the finite element method to the numerical solution of the two-dimensional electromagnetic eigenvalue problem pertinent to waveguiding structures of constant cross-sectional geometry along their axis. Several possible formulations are considered, and their attributes in the context of numerical efficiency and robustness of the associated finite element eigen-problem are discussed.

9.1 FEM FORMULATIONS OF THE TWO-DIMENSIONAL EIGENVALUE PROBLEM

Consider the lossy, inhomogeneous, anisotropic waveguide of arbitrary cross-section in the $x - y$ plane, depicted in Fig. 9.1.

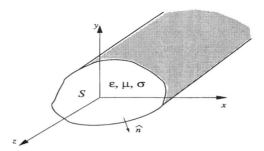

Figure 9.1 Generic geometry of a uniform electromagnetic waveguide.

The cross-sectional geometry is assumed to remain constant along the axis z of the waveguide. Over the years, several finite element formulations have been proposed for waveguide eigenmode analysis [1]-[22]. They can be classified into the following four types:

1. **Longitudinal-Field Formulations** solve for the longitudinal components of both electric and magnetic fields, $\vec{E}_z$ and $\vec{H}_z$, as proposed in [1]-[5].

2. **Transverse-Field Formulations** solve for the transverse components of either the electric field, $\vec{E}_t$, or the magnetic field, $\vec{H}_t$, as proposed in [7] and [8].

3. **Transverse-Longitudinal-Field Formulations** solve for all three components of one of the fields, $\vec{E}$ or $\vec{H}$, as proposed in [9]-[14].

4. **Transverse-Transverse-Field Formulations** solve for the transverse components of both the electric and the magnetic field, $\vec{E}_t$ and $\vec{H}_t$, as proposed in [15] and [21].

The longitudinal-field (LF) formulations usually result in an eigenvalue problem with frequency being the eigenvalue to be determined for a given value of the propagation constant. This is contrary to the most commonly used statement of a waveguide eigenvalue problem, where the propagation constant is the eigenvalue to be determined for a given value of frequency. The LF formulation suffers from the presence of spurious, non-physical eigenmodes which do not satisfy the divergence-free condition $\nabla \cdot \vec{D} = 0$. We will not discuss LF formulations in this chapter. The interested reader may consult references [1]-[5].

The transverse-field (TF) formulation proposed in [7] has the advantage that it solves only for the transverse field components and results in generalized eigenvalues problems with the propagation constant being the eigenvalue to be determined. However, in its mathematical statement both the curl and the divergence operator are imposed on the transverse part of the electric field. This prevents the use of tangentially-continuous vector (TV) basis functions for the expansion of the transverse electric field, since the divergence of edge elements is identically zero. The use of other types of expansion functions complicates the imposition of electromagnetic boundary conditions at material interfaces and is prone to the generation of spurious modes.

To alleviate this difficulty, the technique proposed in [11]-[14] introduces the longitudinal component, $\hat{z}E_z$, in the mathematical statement of the eigenvalue problem. In this manner, the transverse field is expanded in terms of the TV basis functions, while the longitudinal field is expanded in terms of scalar basis functions. The resulting finite element approximation is not free from spurious modes; hence, an appropriate process must be established for their removal during the solution of the eigenvalue problem.

When the material in the waveguide is uniaxially anisotropic, the mode propagating in the $+z$ direction is degenerate with the mode propagating in the $-z$ direction; hence, the eigenvalue to be obtained is, actually, γ^2. In the general case of material anisotropy the two modes are not degenerate. In this case the mathematical statement of the eigenvalue problem contains both the propagation constant and its square. Hence, a quadratic eigenvalue problem must be solved. As discussed in Chapter 8, the quadratic eigenvalue problem may be recast in the form of an equivalent generalized linear eigenvalue problem by doubling the dimension of the unknown eigenvector. One possible formulation, namely, the transverse-transverse-field (TTF) formulation, for the case of the electromagnetic eigenvalue problem was proposed in [15]. It solves for both the transverse electric and magnetic fields, $\vec{E}_t$ and $\vec{H}_t$, over the cross-section of waveguide. When the material is uniaxial, the mode propagating in the $+z$ direction is degenerate with the mode propagating in the $-z$ direction [23], and the size of eigen-system can be reduced to solve an eigen-equation of γ^2.

In the following, starting from Maxwell's equations, we develop the general mathematical framework for the two-dimensional (2D) vector eigenvalue problem for the waveguiding structure of Fig. 9.1. Subsequently, in Sections 9.2 and 9.3 the TLF and the TTF formulations are discussed in detail, along with the development of their finite element approximations and algorithms for their solution.

9.1.1 Mathematical statement of the 2D vector eigenvalue problem

With reference to the generic geometry depicted in Fig. 9.1, the geometry and material composition of the waveguide is assumed to be uniform along its longitudinal z-axis. The permittivity and permeability tensors of the anisotropic medium are assumed to be of the form

$$\bar{\epsilon} = \begin{bmatrix} \bar{\epsilon}_{tt} & \bar{\epsilon}_{tz} \\ \bar{\epsilon}_{zt} & \epsilon_{zz} \end{bmatrix}, \qquad \bar{\mu} = \begin{bmatrix} \bar{\mu}_{tt} & \bar{\mu}_{tz} \\ \bar{\mu}_{zt} & \mu_{zz} \end{bmatrix}, \tag{9.1}$$

where the subscripts tt, tz, tz, and zz are used to indicate the 2×2, 2×1, 1×2, and 1×1 tensors, respectively, entering in the statement of the permittivity and permeability tensors. Unless stated otherwise, we will assume throughout our discussion that the tz and zt tensors are zero; hence, the above expressions simplify as follows:

$$\bar{\epsilon} = \begin{bmatrix} \bar{\epsilon}_{tt} & 0 \\ 0 & \epsilon_{zz} \end{bmatrix}, \qquad \bar{\mu} = \begin{bmatrix} \bar{\mu}_{tt} & 0 \\ 0 & \mu_{zz} \end{bmatrix}. \tag{9.2}$$

Under time-harmonic conditions the electric and magnetic field in the waveguide satisfy the following system of Maxwell's equations

$$
\begin{cases}
\nabla \times \vec{E} = -j\omega\bar{\mu} \cdot \vec{H} \\
\nabla \times \vec{H} = j\omega\bar{\epsilon} \cdot \vec{E} \\
\nabla \cdot \vec{D} = 0 \\
\nabla \cdot \vec{B} = 0.
\end{cases}
\tag{9.3}
$$

Any loss in the media, associated with the presence of a finite conductivity, σ, is taken into account by allowing the permittivity to be complex; hence, $\epsilon + \sigma/(j\omega) \to \epsilon$.

As discussed in Chapter 3, even though the last two of the equations in (9.3) may be deduced from the two curl equations by operating on both sides by the divergence operator, the unique definition of the electromagnetic field quantities requires statements for both the curl and the divergence of the vector fields in the domain of interest. This issue will manifest itself in our discussion of the presence of spurious (non-physical) modes in the finite element approximation of the eigenvalue problem and the techniques that must be used for their elimination.

By canceling the magnetic field from the two curl equations we obtain

$$
\nabla \times \bar{\mu}^{-1} \cdot \nabla \times \vec{E} - \omega^2 \bar{\epsilon} \cdot \vec{E} = 0.
\tag{9.4}
$$

Let S denote the two-dimensional domain associated with the cross-section of the waveguide. On its enclosing boundary, ∂S, the following types of boundary conditions are imposed. A perfect electric conductor (PEC) boundary condition is imposed on the portion ∂S_1 of the boundary; a perfect magnetic conductor (PMC) condition is imposed on the portion ∂S_2 of the boundary; an impedance boundary condition is imposed on the portion ∂S_3 of the boundary. Hence, the associated mathematical forms are given by

$$
\begin{cases}
\hat{n} \times \vec{E} = 0, & \text{on } \partial S_1 \\
\hat{n} \times \vec{H} = 0, & \text{on } \partial S_2 \\
\hat{n} \times \hat{n} \times \vec{E} = Z_s \hat{n} \times \vec{H} & \text{on } \partial S_3,
\end{cases}
\tag{9.5}
$$

where Z_s is the surface impedance on ∂S_3, and the unit normal, $\hat{n}$, is taken to be in the outward direction as indicated in Fig. 9.1.

Of interest to the eigenmode analysis of a uniform waveguide are solutions to (9.3) of the form

$$
\begin{aligned}
\vec{E} &= \left[\vec{E}_t(x,y) + \hat{z}E_z(x,y) \right] e^{-\gamma z}, \\
\vec{H} &= \left[\vec{H}_t(x,y) + \hat{z}H_z(x,y) \right] e^{-\gamma z},
\end{aligned}
\tag{9.6}
$$

where the subscripts t and z denote, respectively, the transverse and longitudinal parts of the fields. In view of the defined variation, $e^{-\gamma z}$, of the fields along z, we write

$$
\nabla = \nabla_t - \gamma\hat{z},
\tag{9.7}
$$

where $\nabla_t = \hat{x}\frac{\partial}{\partial x} + \hat{y}\frac{\partial}{\partial y}$. In view of (9.7) and (9.6) we have

$$
\begin{aligned}
\nabla \times \vec{E} &= (\nabla_t - \gamma\hat{z}) \times \left(\vec{E}_t + \hat{z}E_z \right) \\
&= \nabla_t \times \vec{E}_t + \nabla_t \times \hat{z}E_z - \gamma\hat{z} \times \vec{E}_t \\
&= \nabla_t \times \vec{E}_t - \hat{z} \times \left(\nabla_t E_z + \gamma\vec{E}_t \right),
\end{aligned}
\tag{9.8}
$$

where use was made of the vector identity (A.7). Use of the above result allows the calculation of $\nabla \times \left(\bar{\mu}^{-1} \cdot \nabla \times \vec{E} \right)$ as follows

$$
\nabla \times \left(\bar{\mu}^{-1} \cdot \nabla \times \vec{E} \right) = (\nabla_t - \gamma \hat{z}) \times \left[\mu_{zz}^{-1} \nabla_t \times \vec{E}_t - \bar{\mu}_{tt}^{-1} \cdot \hat{z} \times \left(\nabla_t E_z + \gamma \vec{E}_t \right) \right]
$$

$$
= \nabla_t \times \left(\mu_{zz}^{-1} \nabla_t \times \vec{E}_t \right) - \nabla_t \times \left(\bar{\mu}_{tt}^{-1} \cdot \hat{z} \times \nabla_t E_z \right) - \gamma \nabla_t \times \left(\bar{\mu}_{tt}^{-1} \cdot \hat{z} \times \vec{E}_t \right)
$$

$$
+ \gamma \hat{z} \times \left(\bar{\mu}_{tt}^{-1} \cdot \hat{z} \times \nabla_t E_z \right) + \gamma^2 \hat{z} \times \left(\bar{\mu}_{tt}^{-1} \cdot \hat{z} \times \vec{E}_t \right).
$$

$$(9.9)$$

Substitution of the above result into (9.4) yields two equations as follows: The first equation, which will be referred to as the *transverse equation*, is

$$
\nabla_t \times \left(\mu_{zz}^{-1} \nabla_t \times \vec{E}_t \right) + \gamma \hat{z} \times \left(\bar{\mu}_{tt}^{-1} \cdot \hat{z} \times \nabla_t E_z \right)
$$

$$
+ \gamma^2 \hat{z} \times \left(\bar{\mu}_{tt}^{-1} \cdot \hat{z} \times \vec{E}_t \right) - \omega^2 \bar{\epsilon}_{tt} \cdot \vec{E}_t = 0.
$$

$$(9.10)$$

The second, which will be referred to as the *longitudinal equation*, is

$$
-\nabla_t \times \left(\bar{\mu}_{tt}^{-1} \cdot \hat{z} \times \nabla_t E_z \right) - \gamma \nabla_t \times \left(\bar{\mu}_{tt}^{-1} \cdot \hat{z} \times \vec{E}_t \right) - \omega^2 \epsilon_{zz} E_z \hat{z} = 0. \qquad (9.11)
$$

The divergence-free condition for the electric flux density assumes the form

$$
\nabla \cdot \left(\bar{\epsilon} \cdot \vec{E} \right) = 0 \quad \Rightarrow \quad \nabla_t \cdot \left(\bar{\epsilon}_{tt} \cdot \vec{E}_t \right) - \gamma \epsilon_{zz} E_z = 0. \qquad (9.12)
$$

Since the three equations (9.10-9.12) are not independent, (e.g., applying $\nabla_t \cdot$ on (9.10) and adding to it (9.11) multiplied by $-\gamma \hat{z}$, yields (9.12)), we may follow one of two possible schemes for the development of the eigenvalue problem. One approach would be to use (9.12) to eliminate E_z in (9.10), and then solve an eigenvalue problem in terms of $\vec{E}_t$ only. Obviously, this is the TF formulation. The other is to use either one of (9.11), (9.12) together with (9.10) to solve an eigenvalue problem for all three components of the electric field. This is the TLF formulation.

Both formulations will be considered in the next two sections. However, let us first develop expressions for the boundary conditions, cast in terms of the transverse and longitudinal components of the electric field. We begin by noting that the magnetic field, $\vec{H}$, may be written in terms of the electric field components as follows:

$$
\vec{H} = -\frac{1}{j\omega\mu} \nabla \times \vec{E} = \underbrace{-\frac{\mu_{zz}^{-1}}{j\omega} \nabla_t \times \vec{E}_t}_{\hat{z} H_z} + \underbrace{\frac{\bar{\mu}_{tt}^{-1}}{j\omega} \cdot \hat{z} \times \left(\nabla_t E_z + \gamma \vec{E}_t \right)}_{\vec{H}_t}. \qquad (9.13)
$$

Thus the boundary conditions on the PEC and PMC portions of the boundary, respectively, assume the forms

$$
\hat{n} \times \vec{E} = 0 \Rightarrow \begin{cases} \hat{n} \times \vec{E}_t = 0 \\ E_z = 0 \end{cases} \qquad \text{(on } \partial S_1)
$$

$$
\hat{n} \times \vec{H} = 0 \Rightarrow \begin{cases} \nabla_t \times \vec{E}_t = 0 \\ \hat{n} \times \left[\bar{\mu}_{tt}^{-1} \cdot \hat{z} \times \left(\nabla_t E_z + \gamma \vec{E}_t \right) \right] = 0 \end{cases} \qquad \text{(on } \partial S_2).
$$

$$(9.14)$$

The impedance boundary condition on ∂S_3 becomes

$$\frac{1}{Z_s}\left(\hat{n}\times\hat{n}\times\vec{E}\right)=\hat{n}\times\vec{H} \Rightarrow \begin{cases} \dfrac{1}{Z_s}\left(\hat{n}\times\hat{n}\times\vec{E}_t\right)=\hat{n}\times\hat{z}H_z \\[2mm] -\dfrac{1}{Z_s}\hat{z}E_z=\hat{n}\times\vec{H}_t. \end{cases} \tag{9.15}$$

Substitution of (9.13) into the above result yields the final form of the impedance boundary condition

$$\frac{1}{Z_s}\left(\hat{n}\times\hat{n}\times\vec{E}\right)=\hat{n}\times\vec{H} \Rightarrow \begin{cases} \dfrac{j\omega}{Z_s}\hat{n}\times\vec{E}_t=-\mu_{zz}^{-1}\nabla_t\times\vec{E}_t \\[3mm] -\dfrac{j\omega}{Z_s}\hat{z}E_z=\hat{n}\times\left(\bar{\mu}_{tt}^{-1}\cdot\hat{z}\times\left(\nabla_t E_z+\gamma\vec{E}_t\right)\right). \end{cases} \tag{9.16}$$

9.2 TRANSVERSE-FIELD METHODS

As already mentioned, in the transverse-field (TF) formulation the eigenvalue problem is cast in terms of the transverse part of the electric field, $\vec{E}_t$, only. One way to obtain an eigenvalue matrix equation in terms of $\vec{E}_t$ from (9.10) was proposed in [7]. In this approach the longitudinal electric field E_z in (9.10) is written in terms of $\vec{E}_t$ using (9.12). Subsequently, we operate on the resulting equation with the vector operator $(\bar{\mu}_{tt}\cdot\hat{z}\times)$ to yield

$$\bar{\mu}_{tt}\cdot\hat{z}\times\nabla_t\times\left(\mu_{zz}^{-1}\nabla_t\times\vec{E}_t\right)-\hat{z}\times\nabla_t\left(\epsilon_{zz}^{-1}\nabla_t\cdot\bar{\epsilon}_{tt}\cdot\vec{E}_t\right)$$
$$-\omega^2\bar{\mu}_{tt}\cdot\left(\hat{z}\times\left(\bar{\epsilon}_{tt}\cdot\vec{E}_t\right)\right)=\gamma^2\hat{z}\times\vec{E}_t. \tag{9.17}$$

The development of a finite element approximation of the above equation requires an appropriate choice of basis functions for the expansion of $\vec{E}_t$. At the interface between two elements with different dielectric properties the tangential part of the transverse electric field, $\hat{n}\times\vec{E}_t$, must be continuous. Furthermore, in view of the fact that in (9.17) both the divergence and curl operators act upon $\vec{E}_t$, the choice of the basis functions must be such that both the divergence and the curl of the vector basis functions used are non-zero. This implies that the lowest 2D TV space (i.e., the popular edge elements) are not suitable as basis functions, since the edge elements only contain the zeroth-order gradient and, hence, their divergence is zero. Thus a higher-order (p-th order) polynomial vector space must be chosen, for which the curl and divergence of its basis functions are both complete to p-th order, while at the same time the basis functions exhibit tangential continuity at material interfaces. Since the construction of such a space is cumbersome, the simpler choice of expanding the two components of $\vec{E}_t$ in terms of p-order scalar basis functions was adopted in [8]. Obviously, with such a choice, special constraints must be imposed to ensure tangential field continuity at material interfaces.

One way to avoid the added complexity associated with working directly with the finite element approximation of (9.17) is to effect the cancelation of E_z during the development of the finite element approximation of the eigenvalue problem. To demonstrate how this is done, let us first multiply (9.10) by a weighting function, $\vec{w}$, and integrate the result over the waveguide's cross-section.

$$\iint_S (\nabla_t \times \vec{w}) \cdot \mu_{zz}^{-1} \left(\nabla_t \times \vec{E}_t \right) ds - \oint_{\partial S} (\hat{n} \times \vec{w}) \cdot \mu_{zz}^{-1} \left(\nabla_t \times \vec{E}_t \right) dl$$

$$- \gamma \iint_S (\hat{z} \times \vec{w}) \cdot \bar{\mu}_{tt}^{-1} \cdot (\hat{z} \times \nabla_t E_z) \, ds - \gamma^2 \iint_S (\hat{z} \times \vec{w}) \cdot \bar{\mu}_{tt}^{-1} \cdot \left(\hat{z} \times \vec{E}_t \right) ds \quad (9.18)$$

$$- \omega^2 \iint_S \vec{w} \cdot \bar{\epsilon}_{tt} \cdot \vec{E}_t ds = 0.$$

In deriving the above result use was made of the vector identity (A.6). If $\vec{w}$ is chosen to be a tangentially-continuous vector basis function with its tangential component, $\hat{n} \times \vec{w}$, on the PEC boundary zero, then the boundary integral along ∂S_1 vanishes. The boundary integral along ∂S_2 also vanishes because $\nabla_t \times \vec{E}_t$ is zero on a PMC boundary due to (9.14). Thus only the impedance boundary condition on ∂S_3 contributes to the boundary integral.

With $\vec{E}_t$ expanded in terms of tangentially-continuous vector basis functions $\vec{w}$, and E_z in terms of scalar basis functions ϕ, then testing of (9.18) with all the functions used in the expansion of $\vec{E}$ yields the following matrix equation:

$$\mathcal{A}x_{e,t} - \gamma \mathcal{C}x_{e,z} - \gamma^2 \mathcal{B}x_{e,t} = 0, \qquad (9.19)$$

where $x_{e,t}$ is the vector of the expansion coefficients for $\vec{E}_t$, and $x_{e,z}$ is the vector of the expansion coefficients for E_z. The elements of the matrices in (9.19) are calculated in terms of the following integrals:

$$\mathcal{A}_{m,n} = \iint_S (\nabla_t \times \vec{w}_m) \cdot \mu_{zz}^{-1} (\nabla_t \times \vec{w}_n) \, ds - \omega^2 \iint_S \vec{w}_m \cdot \bar{\epsilon}_{tt} \cdot \vec{w}_n ds$$

$$+ j\omega \int_{\partial S_3} (\hat{n} \times \vec{w}_m) \cdot \frac{1}{Z_s} (\hat{n} \times \vec{w}_n) \, dl$$

$$\mathcal{B}_{m,n} = \iint_S (\hat{z} \times \vec{w}_m) \cdot \bar{\mu}_{tt}^{-1} \cdot (\hat{z} \times \vec{w}_n) \, ds \qquad (9.20)$$

$$\mathcal{C}_{m,n} = \iint_S (\hat{z} \times \vec{w}_m) \cdot \bar{\mu}_{tt}^{-1} \cdot (\hat{z} \times \nabla_t \phi_n) \, ds.$$

Next, we integrate the scalar product of (9.11) by $\hat{z}\phi$ over the waveguide's cross-section

$$\iint_S (\hat{z} \times \nabla_t \phi) \cdot \bar{\mu}_{tt}^{-1} \cdot (\hat{z} \times \nabla_t E_z) \, ds + \gamma \iint_S (\hat{z} \times \nabla_t \phi) \cdot \bar{\mu}_{tt}^{-1} \cdot \left(\hat{z} \times \vec{E}_t \right) ds$$

$$+ \oint_{\partial S} (\hat{n} \times \hat{z}\phi) \cdot \bar{\mu}_{tt}^{-1} \cdot \left[\hat{z} \times \left(\nabla_t E_z + \gamma \vec{E}_t \right) \right] dl - \omega^2 \iint_S \phi \epsilon_{zz} E_z ds = 0, \qquad (9.21)$$

where use was made of (A.6) and (A.7). If ϕ is chosen to be zero on the PEC boundary, the boundary integral along ∂S_1 vanishes. The boundary integral along ∂S_2 also vanishes in view of the boundary condition (9.14) on PMC boundary because of (9.3) and (3.31) Thus only the impedance boundary condition on ∂S_3 contributes to the boundary integral.

With such a choice for ϕ, testing with all the functions in the expansion for E_z yields the following matrix equation:

$$\gamma \mathcal{C}^T x_{e,t} + \mathcal{D}x_{e,z} = 0, \qquad (9.22)$$

where the elements of the matrix $\mathcal{D}$ are calculated as follows:

$$
\mathcal{D}_{m,n} = \iint_S (\hat{z} \times \nabla_t \phi_m) \cdot \bar{\mu}_{tt}^{-1} \cdot (\hat{z} \times \nabla_t \phi_n) \, ds - \omega^2 \iint_S \phi_m \epsilon_{zz} \phi_n ds
$$
$$
+ j\omega \int_{\partial S_3} \phi_m \frac{1}{Z_s} \phi_n dl. \tag{9.23}
$$

Finally, multiplication of (9.12) by ϕ, followed by its integration over the cross-section S yields

$$
\iint_S \nabla_t \phi \cdot \bar{\epsilon}_{tt} \cdot \vec{E}_t ds + \oint_{\partial S} \left(\hat{n} \cdot \bar{\epsilon}_{tt} \cdot \vec{E}_t \right) \phi dl - \gamma \iint_S \phi \epsilon_{zz} E_z ds = 0, \tag{9.24}
$$

where use was made of the vector identity (A.5). As before, with ϕ chosen to be zero on the PEC boundary, the boundary integral along ∂S_1 vanishes. The boundary integral over ∂S_2 also vanishes since $\hat{n} \cdot \bar{\epsilon}_{tt} \cdot \vec{E}_t$ is zero on a PMC boundary. The boundary integral over ∂S_3 is more involved. To cast the integrand of the second term in (9.24) in terms of Z_s, let us begin by noting that from Ampère's law we have

$$
\nabla \times \vec{H} = j\omega \epsilon \vec{E} \quad \Rightarrow \quad \nabla_t \times \hat{z} H_z - \gamma \hat{z} \times \vec{H}_t = j\omega \bar{\epsilon}_{tt} \cdot \vec{E}_t. \tag{9.25}
$$

The scalar product of the above result with $\hat{n}$, followed by the substitution of (9.15) into the resulting equation yields

$$
-\nabla_t \cdot (\hat{n} \times \hat{z} H_z) + \gamma \hat{z} \cdot \left(\hat{n} \times \vec{H}_t \right) = j\omega \hat{n} \cdot \bar{\epsilon}_{tt} \cdot \vec{E}_t
$$
$$
\Rightarrow - \nabla_t \cdot \left(\frac{1}{Z_s} \hat{n} \times \hat{n} \times \vec{E}_t \right) - \frac{\gamma}{Z_s} E_z = j\omega \hat{n} \cdot \bar{\epsilon}_{tt} \cdot \vec{E}_t. \tag{9.26}
$$

This result allows us to recast the product of (9.24) by ω^2 in the following form:

$$
\omega^2 \iint_S \nabla_t \phi \cdot \bar{\epsilon}_{tt} \cdot \vec{E}_t ds - j\omega \int_{\partial S_3} (\hat{n} \times \nabla_t \phi) \cdot \left(\frac{1}{Z_s} \hat{n} \times \vec{E}_t \right) dl
$$
$$
- \gamma \left[\omega^2 \iint_S \phi \epsilon_{zz} E_z ds - j\omega \int_{\partial S_3} \phi \frac{1}{Z_s} E_z dl \right] = 0. \tag{9.27}
$$

Testing of the above equation by all functions in the expansion of E_z yields the following matrix equation

$$
\mathcal{E} x_{e,t} - \gamma \mathcal{F} x_{e,z} = 0, \tag{9.28}
$$

where the elements of the matrices $\mathcal{E}$ and $\mathcal{F}$ are calculated in terms of the following integrals

$$
\mathcal{E}_{m,n} = \omega^2 \iint_S \nabla_t \phi_m \cdot \bar{\epsilon}_{tt} \cdot \vec{w}_n ds - j\omega \int_{\partial S_3} (\hat{n} \times \nabla_t \phi_m) \cdot \left(\frac{1}{Z_s} \hat{n} \times \vec{w}_n \right) dl,
$$
$$
\mathcal{F}_{m,n} = \omega^2 \iint_S \phi_m \epsilon_{zz} \phi_n ds - j\omega \int_{\partial S_3} \phi_m \frac{1}{Z_s} \phi_n dl. \tag{9.29}
$$

In summary, the three matrix equations obtained from (9.10-9.11) are

$$
\begin{align}
\mathcal{A} x_{e,t} - \gamma \mathcal{C} x_{e,z} - \gamma^2 \mathcal{B} x_{e,t} &= 0 \tag{9.30}\\
\gamma \mathcal{C}^T x_{e,t} + \mathcal{D} x_{e,z} &= 0 \tag{9.31}\\
\mathcal{E} x_{e,t} - \gamma \mathcal{F} x_{e,z} &= 0. \tag{9.32}
\end{align}
$$

There are two ways to eliminate the longitudinal-component vector $x_{e,z}$ and derive a matrix eigenvalue equation in terms of the transverse-component vector $x_{e,t}$ only, with eigenvalue γ^2. One method uses (9.30) and (9.31) to cancel $x_{e,z}$,

$$\mathcal{A}x_{e,t} = \gamma^2 \left(\mathcal{B} - \mathcal{C}D^{-1}\mathcal{C}^T\right) x_{e,t}. \tag{9.33}$$

If the material tensors $\bar{\epsilon}$ and $\bar{\mu}$ are symmetric the above formulation results in a generalized symmetric matrix eigenvalue equation.

The second method combines (9.30) and (9.32) to effect the cancelation of $x_{e,z}$,

$$\left(\mathcal{A} - \mathcal{C}\mathcal{F}^{-1}\mathcal{E}\right) x_{e,t} = \gamma^2 \mathcal{B}x_{e,t}. \tag{9.34}$$

In both cases, the inverse of a matrix (either $\mathcal{D}$ or $\mathcal{F}$) is required. Consequently, the resulting matrix eigenvalue problem involves dense matrices and is computationally expensive. This is where the transverse-longitudinal-field (TLF) formulation becomes a computationally more attractive alternative.

9.3 TRANSVERSE-LONGITUDINAL-FIELD METHODS

The TLF formulation of the 2D eigenvalue problem can be developed in one of two different ways. The first one is in terms of the three components of the electric field vector. We will refer to this formulation as the field TLF formulation. The second utilizes the magnetic vector and electric scalar potentials $\vec{A}$ and V, respectively, and will be referred to as the potential TLF formulation. Both formulations are described in detail in the next two subsections. The section concludes with the development of numerical algorithms for the solution of the associated matrix eigenvalue problems.

9.3.1 Field TLF formulation

The field TLF formulation utilizes the system of (9.30) and (9.31), together with the following variable transformation:

$$x_{e,t} = \frac{x'_{e,t}}{\gamma} \tag{9.35}$$

proposed in [11] to develop the following matrix eigenvalue statement:

$$\underbrace{\begin{bmatrix} \mathcal{A} & 0 \\ 0 & 0 \end{bmatrix}}_{\begin{bmatrix} I \\ 0 \end{bmatrix} [\mathcal{A}] \begin{bmatrix} I & 0 \end{bmatrix}} \begin{bmatrix} x'_{e,t} \\ x_{e,z} \end{bmatrix} = \gamma^2 \begin{bmatrix} \mathcal{B} & \mathcal{C} \\ \mathcal{C}^T & \mathcal{D} \end{bmatrix} \begin{bmatrix} x'_{e,t} \\ x_{e,z} \end{bmatrix}, \tag{9.36}$$

where I is the identity matrix. Clearly, the advantage of this formulation is that, instead of the quadratic eigenvalue problem of the system of (9.30) and (9.31), a generalized eigenvalue problem in γ^2 is obtained. However, this comes at the cost of having to deal with the presence of the following set of spurious modes:

$$\left\{ \gamma = 0, \begin{bmatrix} x'_{e,t} \\ \hat{x}_{e,z} \end{bmatrix}, (x'_{e,t} = 0, \hat{x}_{e,z} \neq 0) \right\}. \tag{9.37}$$

Clearly, these spurious modes have zero transverse components and do not propagate. The removal of these spurious modes will be discussed during the presentation of the numerical algorithm for the solution of (9.36).

Let us define the matrices $\mathbf{A}$ and $\mathbf{B}$ as follows

$$\mathbf{A} = \begin{bmatrix} \mathcal{A} & 0 \\ 0 & 0 \end{bmatrix}, \quad \mathbf{B} = \begin{bmatrix} \mathcal{B} & \mathcal{C} \\ \mathcal{C}^T & \mathcal{D} \end{bmatrix}. \tag{9.38}$$

If the material is reciprocal, that is, if $\bar{\epsilon}_{tt}$ and $\bar{\mu}_{tt}$ are 2×2 symmetric tensors, the matrices $\mathcal{A}, \mathcal{B}$ and $\mathcal{D}$ are symmetric; hence, (9.36) constitutes a generalized symmetric eigenvalue problem. Therefore, we can derive the following two properties from Theorem I in Section (8.4.1).

Property I: The physical modes of (9.36) are $\mathbf{B}$-orthogonal with the spurious modes. This property follows immediately by noting that, in view of (9.37), it is

$$\begin{bmatrix} 0 & \hat{x}_{e,z}^T \end{bmatrix} \mathbf{B} \begin{bmatrix} \gamma x_{e,t} \\ x_{e,z} \end{bmatrix} = \begin{bmatrix} 0 & \hat{x}_{e,z}^T \end{bmatrix} \begin{bmatrix} \mathcal{B} & \mathcal{C} \\ \mathcal{C}^T & \mathcal{D} \end{bmatrix} \begin{bmatrix} \gamma x_{e,t} \\ x_{e,z} \end{bmatrix} = \hat{x}_{e,z}^T \left(\gamma \mathcal{C}^T x_{e,t} + \mathcal{D} x_{e,z} \right), \tag{9.39}$$

which is zero in view of (9.31).

Property II: The physical modes of (9.36) are $\mathbf{B}$-orthogonal with each other.

Let $(\gamma^{(1)} x_{e,t}^{(1)}, x_{e,z}^{(1)})$ and $(\gamma^{(2)} x_{e,t}^{(2)}, x_{e,z}^{(2)})$ denote two different physical modes. For the two modes to be $\mathbf{B}$-orthogonal it must be

$$\begin{bmatrix} \gamma^{(1)} x_{e,t}^{(1)^T} & x_{e,z}^{(1)^T} \end{bmatrix} \begin{bmatrix} \mathcal{B} & \mathcal{C} \\ \mathcal{C}^T & \mathcal{D} \end{bmatrix} \begin{bmatrix} \gamma^{(2)} x_{e,t}^{(2)} \\ x_{e,z}^{(2)} \end{bmatrix} = 0$$

$$\Rightarrow \quad \gamma^{(1)} \left(\gamma^{(2)} x_{e,t}^{(1)^T} \mathcal{B} x_{e,t}^{(2)} + x_{e,t}^{(1)^T} \mathcal{C} x_{e,z}^{(2)} \right) = 0, \tag{9.40}$$

where use was made of (9.31). In view of the integral expressions in (9.20), the above equation may be cast in terms of the fields corresponding to the two eigenvectors in the following form:

$$\gamma^{(2)} \iint_S \hat{z} \times \vec{E}_t^{(1)} \cdot \bar{\mu}_{tt}^{-1} \cdot \hat{z} \times \vec{E}_t^{(2)} ds + \iint_S \hat{z} \times \vec{E}_t^{(1)} \cdot \bar{\mu}_{tt}^{-1} \cdot \hat{z} \times \nabla_t E_z^{(2)} ds = 0,$$

$$\Rightarrow \quad \iint_S \hat{z} \times \vec{E}_t^{(1)} \cdot \underbrace{\bar{\mu}_{tt}^{-1} \cdot \hat{z} \times \left(\nabla_t E_z^{(2)} + \gamma^{(2)} \vec{E}_t^{(2)} \right)}_{j\omega \vec{H}_t^{(2)}} ds = 0, \tag{9.41}$$

$$\Rightarrow \quad \iint_S \hat{z} \cdot \left(\vec{E}_t^{(1)} \times \vec{H}_t^{(2)} \right) ds = 0.$$

We conclude that the $\mathbf{B}$-orthogonality of the physical modes is actually the well-known mode orthogonality for the eigenmodes in uniform, two-dimensional waveguides, which holds provided that the material tensors are symmetric (see, e.g., [24]).

9.3.2 Potential TLF formulation

The potential TLF formulation was originally proposed in [14]. The development of the mathematical statement of the associated eigenvalue problem begins by replacing in the transverse-component equation (9.10) and in Gauss' law for the electric field equation (9.12), the electric field components by their expressions in terms of the magnetic vector and electric scalar potentials, $\vec{A}$ and V, respectively. Equations (9.10) and (9.12) are repeated here for simplicity

$$\nabla_t \times \mu_{zz}^{-1} \nabla_t \times \vec{E}_t + \gamma \hat{z} \times \left(\bar{\mu}_{tt}^{-1} \cdot \hat{z} \times \nabla_t E_z \right)$$
$$+ \gamma^2 \hat{z} \times \left(\bar{\mu}_{tt}^{-1} \cdot \hat{z} \times \vec{E}_t \right) - \omega^2 \bar{\epsilon}_{tt} \cdot \vec{E}_t = 0, \qquad (9.42)$$

$$\nabla_t \cdot \bar{\epsilon}_{tt} \cdot \vec{E}_t - \gamma \epsilon_{zz} E_z = 0.$$

The defining statements for the two potentials are

$$\vec{E} = -j\omega \vec{A} - \nabla V, \quad \vec{H} = \bar{\mu}^{-1} \nabla \times \vec{A}. \qquad (9.43)$$

The magnetic vector potential is made unique by choosing $A_z = 0$ as the gauge condition. Thus, with $\nabla = \nabla_t - \gamma \hat{z}$, the above expressions for the electric and magnetic field vectors are rewritten as follows:

$$\vec{E} = \underbrace{-j\omega \vec{A}_t - \nabla_t V}_{\vec{E}_t} + \hat{z} \underbrace{\gamma V}_{E_z}, \quad \vec{H} = \underbrace{-\gamma \bar{\mu}_{tt}^{-1} \cdot \hat{z} \times \vec{A}_t}_{\vec{H}_t} + \underbrace{\mu_{zz}^{-1} \nabla_t \times \vec{A}_t}_{\hat{z} H_z}. \qquad (9.44)$$

Substitution of these expressions into (9.42) yields

$$\nabla_t \times \mu_{zz}^{-1} \nabla_t \times \vec{A}_t + \gamma^2 \hat{z} \times \left(\bar{\mu}_{tt}^{-1} \hat{z} \times \vec{A}_t \right) - \omega^2 \bar{\epsilon}_{tt} \cdot \vec{A}_t + j\omega \bar{\epsilon}_{tt} \cdot \nabla_t V = 0,$$
$$\gamma^2 \epsilon_{zz} V + j\omega \nabla_t \cdot \left(\bar{\epsilon}_{tt} \cdot \vec{A}_t \right) + \nabla_t \cdot (\bar{\epsilon}_{tt} \cdot \nabla_t V) = 0. \qquad (9.45)$$

It is noted that, in contrast to the system in (9.42), the propagation constant γ appears in (9.45) in terms of its square only. In terms of $\vec{A}_t$ and V the PEC and PMC boundary conditions assume the form,

$$\hat{n} \times \vec{E} = 0 \quad \Rightarrow \quad \begin{cases} \hat{n} \times \vec{A}_t = 0, \\ V = 0 \end{cases} \quad \text{(on } \partial S_1 \text{)},$$

$$\hat{n} \times \vec{H} = 0 \quad \Rightarrow \quad \begin{cases} \hat{n} \times \nabla_t \times \vec{A}_t = 0 \\ \hat{n} \times \left(\bar{\mu}_{tt}^{-1} \cdot \hat{z} \times \vec{A}_t \right) = 0 \end{cases} \quad \text{(on } \partial S_2 \text{)}. \qquad (9.46)$$

Finally, the impedance boundary condition (9.15) on ∂S_3 becomes

$$\frac{1}{Z_s} \hat{n} \times \hat{n} \times \vec{E} = \hat{n} \times \vec{H} \quad \Rightarrow \quad \begin{cases} \dfrac{1}{Z_s} \hat{n} \times \left(j\omega \vec{A}_t + \nabla_t V \right) = -\mu_{zz}^{-1} \nabla_t \times \vec{A}_t \\ \dfrac{1}{Z_s} \hat{z} V = \hat{n} \times \left(\bar{\mu}_{tt}^{-1} \cdot \hat{z} \times \vec{A}_t \right). \end{cases} \qquad (9.47)$$

Next, we proceed with the development of the weak forms of the equations in (9.45). Integration of the scalar product of the first equation by a vector testing function, $\vec{w}$, yields

$$\iint_S (\nabla_t \times \vec{w}) \cdot \mu_{zz}^{-1} \left(\nabla_t \times \vec{A}_t \right) ds - \oint_{\partial S} \hat{n} \times \vec{w} \cdot \mu_{zz}^{-1} \nabla_t \times \vec{A}_t dl - \omega^2 \iint_S$$
$$\vec{w} \cdot \bar{\epsilon}_{tt} \cdot \vec{A}_t ds + j\omega \iint_S \vec{w} \cdot \bar{\epsilon}_{tt} \cdot \nabla_t V ds = \gamma^2 \iint_S \hat{z} \times \vec{w} \cdot \left(\bar{\mu}_{tt}^{-1} \cdot \hat{z} \times \vec{w} \right) ds, \qquad (9.48)$$

where use was made of the vector identity (A.6) and the two-dimensional form of the divergence theorem.

If the testing function, $\vec{w}$, is chosen such that $\hat{n} \times \vec{w} = 0$ on the PEC portion of the boundary, ∂S_1, the boundary integral over ∂S_1 in the above equation vanishes. Furthermore, the boundary integral over ∂S_2 also vanishes in view of the PMC boundary condition in (9.46). Thus only the impedance boundary condition over ∂S_3 contributes to the boundary integral in (9.48). To make the presence of the surface impedance Z_s explicit in (9.48), we substitute (9.47) into the integrand of the boundary integral. This yields

$$
\int_{\partial S_3} \hat{n} \times \vec{w} \cdot \left(\mu_{zz}^{-1} \nabla_t \times \vec{A}_t \right) dl
$$
$$
= -j\omega \int_{\partial S_3} \hat{n} \times \vec{w} \cdot \left(\frac{1}{Z_s} \hat{n} \times \vec{A}_t \right) dl - \int_{\partial S_3} \hat{n} \times \vec{w} \cdot \left(\frac{1}{Z_s} \hat{n} \times \nabla_t V \right) dl.
$$
(9.49)

Its substitution into (9.48) leads to the following equation:

$$
\iint_S (\nabla_t \times \vec{w}) \cdot \mu_{zz}^{-1} \left(\nabla_t \times \vec{A}_t \right) ds + j\omega \int_{\partial S_3} \hat{n} \times \vec{w} \cdot \left(\frac{1}{Z_s} \hat{n} \times \vec{A}_t \right) dl
$$
$$
- \omega^2 \iint_S \vec{w} \cdot \bar{\epsilon}_{tt} \cdot \vec{A}_t ds + j\omega \iint_S \vec{w} \cdot \bar{\epsilon}_{tt} \cdot \nabla_t V ds
$$
$$
+ \int_{\partial S_3} \hat{n} \times \vec{w} \cdot \left(\frac{1}{Z_s} \hat{n} \times \nabla_t V \right) dl = \gamma^2 \iint_S \hat{z} \times \vec{w} \cdot \left(\bar{\mu}_{tt}^{-1} \cdot \hat{z} \times \vec{w} \right) ds.
$$
(9.50)

The weak form of the second equation of (9.45) is obtained by integrating its product with a scalar testing function ϕ over the waveguide cross-section S

$$
j\omega \iint_S \nabla_t \phi \cdot \bar{\epsilon}_{tt} \cdot \vec{A}_t ds + \iint_S \nabla_t \phi \cdot \bar{\epsilon}_{tt} \cdot \nabla_t V ds
$$
$$
- \oint_{\partial S} \phi \hat{n} \cdot \bar{\epsilon}_{tt} \cdot \left(\nabla_t V + j\omega \vec{A}_t \right) dl = \gamma^2 \iint_S \phi \epsilon_{zz} V ds.
$$
(9.51)

In obtaining the above result use was made of (A.5) and the two-dimensional form of the divergence theorem.

If we choose the testing function ϕ such that $\phi = 0$ on the PEC boundary, the boundary integral over ∂S_1 vanishes. Furthermore, the boundary integral over ∂S_2 vanishes also in view of the fact that $\hat{n} \cdot \bar{\epsilon}_{tt} \cdot \vec{E}_t = 0$ on a PMC. Thus only the impedance boundary condition on ∂S_3 contributes to the boundary integral in (9.51). Substitution of (9.44) into (9.26) yields

$$
-\nabla_t \cdot \left[\frac{1}{Z_s} \hat{n} \times \hat{n} \times \left(j\omega \vec{A}_t + \nabla_t V \right) \right] + \frac{\gamma^2}{Z_s} V = j\omega \hat{n} \cdot \bar{\epsilon}_{tt} \cdot \left(j\omega \vec{A}_t + \nabla_t V \right).
$$
(9.52)

Thus the boundary integral over ∂S_3 assumes the form

$$
\int_{\partial S_3} \phi \hat{n} \cdot \bar{\epsilon}_{tt} \cdot \left(\nabla_t V + j\omega \vec{A}_t \right) dl = - \int_{\partial S_3} \hat{n} \times \nabla_t \phi \cdot \left(\frac{1}{Z_s} \hat{n} \times \vec{A}_t \right) dl
$$
$$
- \frac{1}{j\omega} \int_{\partial S_3} \hat{n} \times \nabla_t \phi \cdot \left(\frac{1}{Z_s} \hat{n} \times \nabla_t V \right) dl + \frac{\gamma^2}{j\omega} \int_{\partial S_3} \phi \frac{1}{Z_s} V dl.
$$
(9.53)

Use of this result in (9.51) leads to the weak statement for the second equation in (9.45)

$$
j\omega \iint_S \nabla_t\phi \cdot \bar{\epsilon}_{tt} \cdot \vec{A}_t ds + \int_{\partial S_3} \hat{n} \times \nabla\phi \cdot \frac{1}{Z_s}\hat{n} \times \vec{A}_t dl + \iint_S \nabla_t\phi \cdot \bar{\epsilon}_{tt} \cdot \nabla_t V ds
$$
$$
+ \frac{1}{j\omega} \int_{\partial S_3} \hat{n} \times \nabla_t\phi \cdot \frac{1}{Z_s}\hat{n} \times \nabla_t V dl = \gamma^2 \iint_S \phi\epsilon_{zz} V ds + \frac{\gamma^2}{j\omega} \int_{\partial S_3} \phi\frac{1}{Z_s} V dl. \tag{9.54}
$$

The weak forms (9.50) and (9.54) are used for the development of the finite element approximation of the eigenvalue problem. Expansion of $\vec{A}_t$ in the TV space and V in the scalar space and use of Galerkin's testing yields the following matrix equation:

$$
\begin{bmatrix} A & C \\ C^T & D \end{bmatrix} \begin{bmatrix} x_{a,t} \\ x_v \end{bmatrix} = \gamma^2 \begin{bmatrix} B & 0 \\ 0 & E \end{bmatrix} \begin{bmatrix} x_{a,t} \\ x_v \end{bmatrix}, \tag{9.55}
$$

where the vector $x_{a,t}$ contains the expansion coefficients for $\vec{A}$, while x_v contains the expansion coefficients for V. The elements of the matrices A, B, C, D and E, are given in terms of the following integrals:

$$
A_{m,n} = \iint_S \nabla_t \times \vec{w}_m \cdot \bar{\mu}_{tt}^{-1} \cdot \nabla_t \times \vec{w}_n ds - \omega^2 \iint_S \vec{w}_m \cdot \bar{\epsilon}_{tt} \cdot \vec{w}_n ds
$$
$$
+ j\omega \int_{\partial S_3} \hat{n} \times \vec{w}_m \cdot \frac{1}{Z_s}\hat{n} \times \vec{w}_n dl
$$
$$
B_{m,n} = \iint_S \hat{z} \times \vec{w}_m \cdot \bar{\mu}_{tt}^{-1} \cdot \hat{z} \times \vec{w}_n ds
$$
$$
C_{m,n} = j\omega \iint_S \vec{w} \cdot \bar{\epsilon}_{tt} \cdot \nabla_t\phi_n ds + \int_{\partial S_3} \hat{n} \times \vec{w}_m \cdot \frac{1}{Z_s}\hat{n} \times \nabla_t\phi_n dl \tag{9.56}
$$
$$
D_{m,n} = \iint_S \nabla_t\phi_m \cdot \bar{\epsilon}_{tt} \cdot \nabla_t\phi_n ds + \frac{1}{j\omega} \int_{\partial S_3} \hat{n} \times \nabla_t\phi_m \cdot \frac{1}{Z_s}\hat{n} \times \nabla_t\phi_n dl
$$
$$
E_{m,n} = \iint_S \phi_m\epsilon_{zz}\phi_n ds + \frac{1}{j\omega} \int_{\partial S_3} \phi_m\frac{1}{Z_s}\phi_n dl.
$$

Let W_{tv} denote the TV space used for the expansion of $\vec{A}$, and W_s the scalar space used for the expansion of V. As discussed in Chapter 2, the gradient of the basis functions of W_s is a subset of W_{tv}. The specific relationship is cast in matrix form as follows:

$$
\underbrace{[\nabla\phi_1, \cdots, \nabla\phi_M]}_{\nabla_t W_s} = \underbrace{[\vec{w}_1, \cdots, \vec{w}_N]}_{W_{tv}} G. \tag{9.57}
$$

For the special case where W_s is the lowest-order scalar space, (i.e., the node-element space), and W_{tv} is the lowest-order TV space, (i.e., the edge-element space), the matrix G is the gradient matrix whose form was discussed in detail at Chapter 2.

In view of the bilinear form of the elements of the matrices in (9.55), use of (9.57) yields the following matrix relationships:

$$
C = \frac{1}{j\omega}AG, \quad C^T = \frac{1}{j\omega}G^T A, \quad D = -\frac{1}{\omega^2}G^T AG. \tag{9.58}
$$

Consequently, (9.55) may be cast in the following form:

$$\underbrace{\begin{bmatrix} I \\ \frac{1}{j\omega}G^T \end{bmatrix} A \begin{bmatrix} I & \frac{1}{j\omega}G \end{bmatrix}}_{\begin{bmatrix} A & C \\ C^T & D \end{bmatrix}} \begin{bmatrix} x_{a,t} \\ x_v \end{bmatrix} = \gamma^2 \begin{bmatrix} B & 0 \\ 0 & E \end{bmatrix} \begin{bmatrix} x_{a,t} \\ x_v \end{bmatrix},$$
(9.59)

where I is the identity matrix. This form makes the identification of the spurious modes of (9.55) straightforward. They are

$$\left\{ \gamma = 0, \begin{bmatrix} -\frac{1}{j\omega}Gx_v \\ x_v \end{bmatrix}, x_v \neq 0 \right\}.$$
(9.60)

It is also straightforward to show that these spurious modes are the same with the ones in (9.37), associated with the field TLF formulation. This is done by first deriving the relationship between the vectors $x_{a,t}$, x_v and $x_{e,t}$, $x_{e,z}$ as follows:

$$\vec{E} = \underbrace{-j\omega\vec{A}_t - \nabla_t V}_{\vec{E}_t} + \hat{z}\underbrace{\gamma V}_{E_z} \quad \Rightarrow \quad \begin{cases} x_{e,t} = -j\omega x_{a,t} - Gx_v \\ x_{e,z} = \gamma x_v. \end{cases}$$
(9.61)

Substitution of (9.60) into these equations results in $x_{e,t} = 0$, which is exactly the case for the spurious modes of (9.37).

Finally, it is noted that the potential statement of the eigenvalue problem (9.55) can be obtained from (9.30-9.32), by making use of the vector relations of (9.61).

Let us define the matrices **A** and **B** as follows

$$\mathbf{A} = \begin{bmatrix} A & C \\ C^T & D \end{bmatrix}, \quad \mathbf{B} = \begin{bmatrix} B & 0 \\ 0 & E \end{bmatrix}.$$
(9.62)

For reciprocal media the tensors $\bar{\epsilon}_{tt}$ and $\bar{\mu}_{tt}$ are symmetric; hence, (9.59) is a generalized symmetric eigenvalue problem. In this case, the following two properties hold from Theorem I in Section 8.4.1.

Property I: The physical modes of (9.59) are **B**-orthogonal to the spurious modes (9.60). This property follows immediately by noting that, in view of (9.60), it is

$$\begin{bmatrix} -\frac{1}{j\omega}\hat{x}_v^T G^T & \hat{x}_v^T \end{bmatrix} \begin{bmatrix} B & 0 \\ 0 & E \end{bmatrix} \begin{bmatrix} x_{a,t} \\ x_v \end{bmatrix} \Rightarrow \hat{x}_v^T \left(Ex_v - \frac{1}{j\omega}G^T Bx_{a,t} \right) = 0,$$
(9.63)

where $\hat{x}_v$ denotes the scalar potential vector coefficient for the spurious mode. If the vectors $x_{a,t}$, x_v are associated with a physical mode with non-zero eigenvalue, then a simple matrix manipulation of the two equations in (9.59) shows that the term inside the parentheses on the right-hand side of the above equation is zero. This concludes the proof of Property I.

Property II: The physical modes of (9.59) are **B**-orthogonal to each other.

To show this, let $(x_{a,t}^{(1)}, x_v^{(1)})$ and $(x_{a,t}^{(2)}, x_v^{(2)})$ denote two distinct physical modes, we have

$$\begin{bmatrix} x_{a,t}^{(2)T} & x_v^{(2)T} \end{bmatrix} \begin{bmatrix} B & 0 \\ 0 & E \end{bmatrix} \begin{bmatrix} x_{a,t}^{(1)} \\ x_v^{(1)} \end{bmatrix} = 0.$$
(9.64)

Making use of the fact that for a physical eigenmode of (9.63) it is

$$x_v = \frac{1}{j\omega} E^{-1} G^T B x_{a,t} \tag{9.65}$$

the B-orthogonality of the two modes $(x_{a,t}^{(1)}, x_v^{(1)})$ and $(x_{a,t}^{(2)}, x_v^{(2)})$ may be cast in the following form:

$$\begin{bmatrix} x_{a,t}^{(1)^T} & x_v^{(1)^T} \end{bmatrix} \begin{bmatrix} B & 0 \\ 0 & E \end{bmatrix} \begin{bmatrix} x_{a,t}^{(2)} \\ x_v^{(2)} \end{bmatrix} = 0 \Rightarrow x_{a,t}^{(1)^T} B x_{a,t}^{(2)} + \frac{1}{j\omega} x_v^{(1)^T} G^T B x_{a,t}^{(2)} = 0. \tag{9.66}$$

In view of the second of (9.56) and equation (9.57) the elements of the matrix $G^T B$ are of the form

$$(G^T B)_{mn} = \iint_S \hat{z} \times \nabla_t \phi_m \cdot \bar{\mu}_{tt}^{-1} \cdot \hat{z} \times \vec{w}_n \, ds. \tag{9.67}$$

Using this result along with the second of (9.56) for the elements of B, (9.66) may be cast in terms of the vector and scalar potentials corresponding to the two eigenvectors in the following manner:

$$\iint_S \hat{z} \times \vec{A}_t^{(1)} \cdot \bar{\mu}_{tt}^{-1} \cdot \hat{z} \times \vec{A}_t^{(2)} \, ds + \frac{1}{j\omega} \iint_S \hat{z} \times \nabla_t V^{(1)} \cdot \bar{\mu}_{tt}^{-1} \cdot \hat{z} \times \vec{A}_t^{(2)} \, ds = 0$$

$$\Rightarrow \quad \frac{1}{j\omega\gamma_2} \iint_S \hat{z} \times \underbrace{\left(-j\omega\vec{A}_t^{(1)} - \nabla_t V^{(1)} \right)}_{\vec{E}_t^{(1)}} \cdot \underbrace{\left(-\gamma_2 \bar{\mu}_{tt}^{-1} \cdot \hat{z} \times \vec{A}_t^{(2)} \right)}_{\vec{H}_t^{(2)}} \, ds = 0 \tag{9.68}$$

$$\Rightarrow \quad \iint_S \hat{z} \cdot \left(\vec{E}_t^{(1)} \times \vec{H}_t^{(2)} \right) \, ds = 0.$$

Hence, we conclude that the B-orthogonality of the physical modes is actually the well-known mode orthogonality for the eigenmodes in uniform, two-dimensional waveguides.

9.3.3 Computer algorithms for eigenvalue calculation

Since the matrices for the two TLF formulations, (9.36) and (9.55), are available, a complete matrix eigenvalue analysis can be performed using one of several numerical matrix eigensolver packages. However, when the dimension of the matrices is large, such an approach is computationally expensive. Since for most practical applications what is of interest is the extraction of the first few dominant modes, the algorithms discussed in Chapter 8 provide for a computationally efficient alternative to a full eigensolution. In the following, the application of these algorithms to the solution of (9.36) and (9.55) is presented. The emphasis of the presentation is on the key implementation issues, namely, the selection of the expansion point for the shift-and-invert technique, the removal of spurious modes, and the extraction of multiple eigenmodes.

The eigenvalue to be obtained is the propagation constant, which is, in general, a complex number $\gamma = \alpha + j\beta$, where α denotes the attenuation constant and β the phase constant for an eigenmode. Depicted on the left of Fig. 9.2 on the complex γ plane is a generic distribution of the eigenvalues for the case of a lossless, isotropic waveguide. Propagating modes in this case have purely imaginary eigenvalues, depicted by points on the imaginary axis, while modes below cutoff exhibit only attenuation and thus have real eigenvalues, depicted by points on the real axis. In the presence of loss the eigenvalues of the modes are shifted off the axes, since in this case propagating modes will also exhibit attenuation.

Accurate calculation of the eigenvalues of the modes of interest requires the finite element discretization to be fine enough to resolve with sufficient accuracy their spatial distribution. This requirement has already been discussed in the context of multigrid approximation in Chapter 4. A more detailed discussion can be found in [25]. The better the resolution of the spatial distribution of the eigenmode fields, the smaller the error in the approximation of the eigenvalue. For those modes for which spatial resolution of their fields by the discretization is not adequate, their numerical eigenvalues are highly inaccurate. Clearly, these modes exhibit rapid spatial variation; hence, the inability of the finite element grid to resolve them manifest itself in terms of a jagged, highly-oscillatory spatial profile, as captured for the simple case of the one-dimensional eigenvalue problem of Section 4.6.1 in Fig. 4.2.

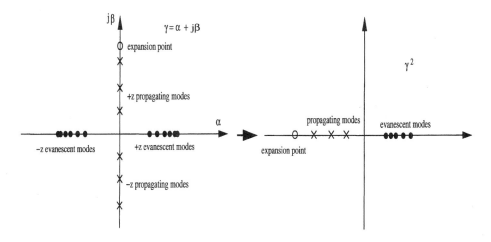

Figure 9.2   The complex γ plane for the illustration of the eigenvalue distribution of the matrix eigenvalue problem.

As discussed in Chapter 8, the Krylov-subspace-based eigensolver can extract the eigenvalues with the largest magnitude. However, in most 2D eigenvalue problems of practical interest the dominant eigenmodes are not the ones which have the largest eigenvalue magnitude. For example, consider the simple case of a homogeneously filled, parallel-plate waveguide with plate separation of 1 m. With the axis of the guide taken to be along z and the phasor of the longitudinal component of the electric field written as, $E_z(x, z) = e_z(x) \exp(\pm \gamma z)$, the eigenvalue equation for the transverse-magnetic (TM$_z$) modes is

$$\left(\frac{d^2}{dx^2} + \omega^2 \mu \epsilon \right) e_z(x) + \gamma^2 e_z(x) = 0, \quad 0 \le x \le 1 \tag{9.69}$$

with $e_z(0) = e_z(1) = 0$. The analytic result for the eigenvalues of this simple problem is well known,

$$\gamma^2 = \left[(m\pi)^2 - \omega^2 \mu \epsilon \right], \quad m = 0, 1, 2, \dots \tag{9.70}$$

with corresponding eigenmodes for $e_z(x)$,

$$e_z^{(m)}(x) = \sin(m\pi x), \quad m = 0, 1, 2, \dots \tag{9.71}$$

Clearly, for a specified value of the angular frequency ω, the eigenmodes with largest eigenvalue magnitudes are the ones corresponding to large values of m. From (9.71) these modes are recognized to be the rapidly varying modes which, with $m\pi > \omega\sqrt{\mu\epsilon}$, are non-propagating modes at the operating frequency and, hence, for most practical cases, of no interest. Furthermore, as already pointed out in the previous paragraphs and illustrated in Fig. 4.2, these modes are not well-resolved by the finite element grid. Nevertheless, these are the modes that a Krylov-subspace-based extraction process will converge to first.

As explained in Section 8.5, this can be avoided by means of the shift-and-invert transformation of the matrix eigenvalue problem into one where the magnitude of the eigenvalues of the modes of interest is large. Toward this objective, guided by (9.70), we observe that the propagating modes will exhibit propagation constant values close to $j\beta_0$, where $\beta_0 = \omega\sqrt{\mu_{max}\epsilon_{max}}$, with μ_{max}, ϵ_{max}, denoting the largest values of the magnetic permeability and electric permittivity, respectively, in the cross-section of the waveguide. Thus the point $\gamma_0 = j\beta_0$ is used as the expansion point for the shift-and-invert scheme of Section 8.5.

In this manner the eigenvalue problem of 9.36 assumes the form

$$\underbrace{\frac{1}{\gamma^2 - \gamma_0^2}}_{\lambda} \underbrace{\begin{bmatrix} \mathcal{A} - \gamma_0^2\mathcal{B} & -\gamma_0^2\mathcal{C} \\ -\gamma_0^2\mathcal{C}^T & -\gamma_0^2\mathcal{D} \end{bmatrix}}_{\mathbf{B}} \begin{bmatrix} x'_{e,t} \\ x_{e,z} \end{bmatrix} = \underbrace{\begin{bmatrix} \mathcal{B} & \mathcal{C}^T \\ \mathcal{C} & \mathcal{D} \end{bmatrix}}_{\mathbf{A}} \begin{bmatrix} x'_{e,t} \\ x_{e,z} \end{bmatrix}. \tag{9.72}$$

This is a generalized symmetric eigenvalue problem of the form (8.27) and can be solved via Algorithm (8.10). However, the presence of spurious modes (9.37) requires extra care for the development of a robust algorithm.

First, let us note that the shift-and-invert process has altered the eigenvalues of the spurious modes from 0 to the value of $-\frac{1}{\gamma_0^2}$, while the spurious eigenvectors remain unchanged; hence, the spurious eigenmodes of (9.72) are

$$\left\{ \lambda = -\frac{1}{\gamma_0^2}, \begin{bmatrix} 0 \\ I_{e,z} \end{bmatrix} \right\}, \tag{9.73}$$

where $I_{e,z}$ is an identity matrix with size equal to that of $x_{e,z}$. In view of the fact that the spurious modes are orthogonal to the physical modes, the calculation of the spurious modes can be avoided. This is achieved by forcing the Lanczos vectors in Q_m, calculated in Algorithm (8.10), to be **B**-orthogonal to the spurious modes

$$\begin{bmatrix} 0 & I_{e,z} \end{bmatrix} \underbrace{\begin{bmatrix} \mathcal{A} - \gamma_0^2\mathcal{B} & -\gamma_0^2\mathcal{C} \\ -\gamma_0^2\mathcal{C}^T & -\gamma_0^2\mathcal{D} \end{bmatrix}}_{\mathbf{B}} \begin{bmatrix} (q_i)_t \\ (q_i)_z \end{bmatrix} = 0, \tag{9.74}$$

where $(q_i)_t$ denotes the portion of the ith Lanczos vector associated with the transverse part of the electric field, while $(q_i)_z$ denotes the portion corresponding to the longitudinal component electric field. Algorithm (8.10) contains the scheme to deflate the converged eigenvectors. Incorporation of the above process for the removal of spurious modes yield the following modified version of Algorithm (8.10).

Algorithm (9.1): Restarted deflated Lanczos algorithm for n eigen-modes with spurious-mode removal (Field TLF formulation)
  1 $V_0 \leftarrow 0$;
  2 for $k = 1, 2, \cdots, n$
      2.a.1 $r_0 \leftarrow (I - \mathbf{B}V_{k-1}V_{k-1}^T)r_0$;   (Deflate converged modes)

2.a.2 $(r_0)_z \leftarrow 0;$ (Deflate spurious modes)
2.a.3 $r_0 \leftarrow \mathbf{B}^{-1}r_0;$
2.b $q_0 = 0,\ \beta_1 = \|r_0\|,\ q_1 = r_0/\beta_1;$
2.c for $m = 1, 2, \cdots, M$
   2.c.1.a $w \leftarrow (I - \mathbf{B}V_{k-1}V_{k-1}^T)\mathbf{A}q_m;$ (Deflate converged modes)
   2.c.1.b $(w)_z \leftarrow 0;$ (Deflate spurious modes)
   2.c.2 $\alpha_m \leftarrow \langle q_m, w \rangle = q_m^T w;$
   2.c.3 $\tilde{q}_{m+1} \leftarrow \mathbf{B}^{-1}w - \beta_m q_{m-1} - \alpha_m q_m;$
   2.c.4 Compute the eigen-pairs $(y_i, \tilde{\lambda}_i)$ of T_m, $(i = 1, \cdots, m);$
   2.c.5 Compute $\rho_i = \|e_m^T y_i\|\|\tilde{q}_{m+1}\|/\|\tilde{\lambda}_i q_m y_i\|,\ (i = 1, \cdots, m);$
   2.c.6 If one eigen-pair $(\tilde{\lambda}_i, \tilde{v}_i = Q_m y_i,)$ converges, goto 2.e;
   2.c.7 $\beta_{m+1} = \sqrt{\tilde{q}_{m+1}^T w},\ q_{m+1} = \tilde{q}_{m+1}/\beta_{m+1};$
2.d $r_0 \leftarrow \tilde{v}_i$ and goto 2.b;
2.e $\mathbf{B}$-normalize $\tilde{v}_i$ and $V_k \leftarrow [V_{k-1}, \tilde{v}_i].$

Steps (2.a.1–2.a.3) render the initial vector r_0 $\mathbf{B}$-orthogonal to both the converged eigenvectors and the spurious modes. Steps 2.c.1.a and 2.c.1.b explicitly force all the Lanczos vectors $\mathbf{B}$-orthogonal to both the converged eigenvectors and the spurious modes. The dominant computational cost in the algorithm is the inverse of $\mathbf{B}$.

The process for the development of an effective algorithm for the potential TLF eigenvalue problem of (9.55) is very similar. The expansion point for the shift-and-inverse statement of the problem is also taken to be $\gamma_0 = j\beta_0$. Thus the eigenvalue problem of (9.55) becomes

$$\frac{1}{\gamma^2 - \gamma_0^2}\underbrace{\begin{bmatrix} A - \gamma_0^2 B & C \\ C^T & D - \gamma_0^2 E \end{bmatrix}}_{\mathbf{B}}\begin{bmatrix} x_{a,t} \\ x_v \end{bmatrix} = \underbrace{\begin{bmatrix} B & 0 \\ 0 & E \end{bmatrix}}_{\mathbf{A}}\begin{bmatrix} x_{a,t} \\ x_v \end{bmatrix}. \qquad (9.75)$$

This is a generalized symmetric eigenvalue problem of the form (8.27) and can be solved via Algorithm (8.10), provided that extra care is taken to deal properly with the spurious modes (9.60). The shift-and-invert process has altered the eigenvalues of the spurious modes from 0 to the value of $-\frac{1}{\gamma_0^2}$, while the spurious eigenvectors remain unchanged; hence, the spurious eigenmodes of (9.75) are

$$\left\{\lambda = -\frac{1}{\gamma_0^2}, \begin{bmatrix} -\frac{1}{j\omega}G \\ I_v \end{bmatrix}\right\}, \qquad (9.76)$$

where I_v is a vector of ones of length equal to that of x_v. In view of the fact that the spurious modes are orthogonal to the physical modes, the calculation of the spurious modes can be avoided Again, this is achieved by forcing the Lanczos vectors in Q_m, calculated in Algorithm (8.10), to be $\mathbf{B}$-orthogonal to the spurious modes

$$\begin{bmatrix} -\frac{1}{j\omega}G^T & I_v \end{bmatrix}\underbrace{\begin{bmatrix} A - \gamma_0^2 B & C \\ C^T & D - \gamma_0^2 E \end{bmatrix}}_{\mathbf{B}}\begin{bmatrix} (q_i)_a \\ (q_i)_v \end{bmatrix} = 0, \qquad (9.77)$$

where $(q_i)_a$ denotes the portion of the ith Lanczos vector associated with the vector potential, while $(q_i)_v$ denotes the portion corresponding to the scalar potential. Algorithm (8.10) contains the scheme to deflate the converged eigenvectors. Incorporation of the above process for the removal of spurious mode yields the following modified version of Algorithm (8.10).

Algorithm (9.2): Restarted deflated Lanczos algorithm for n eigen-modes with spurious-mode removal (Potential TLF formulation)

```
1  V₀ ← 0;
2  for  k = 1, 2, · · · , n
   2.a.1  r₀ ← (I − BV_{k−1}V_{k−1}^T)r₀;            (Deflate converged modes)
   2.a.2  (r₀)_v ← (1/jω)G^T(r₀)_t;                  (Deflate spurious modes)
   2.a.3  r₀ ← B^{−1}r₀;
   2.b  q₀ = 0,  β₁ = ‖r₀‖,  q₁ = r₀/β₁;
   2.c  for  m = 1, 2, · · · , M
      2.c.1.a  w ← (I − BV_{k−1}V_{k−1}^T)Aq_m;       (Deflate converged modes)
      2.c.1.b  (w)_v ← (1/jω)G^T(w)_a;               (Deflate spurious modes)
      2.c.2  α_m ← ⟨q_m, w⟩ = q_m^T w;
      2.c.3  q̃_{m+1} ← B^{−1}w − β_m q_{m−1} − α_m q_m;
      2.c.4  Compute the eigenpairs (y_i, λ̃_i) of T_m, (i = 1, · · · , m);
      2.c.5  Compute ρ_i = ‖e_m^T y_i‖‖q̃_{m+1}‖/‖λ̃_i q_m y_i‖, (i = 1, · · · , m);
      2.c.6  If one eigenpair (λ̃_i, ṽ_i = Q_m y_i) has converged, goto 2.e;
      2.c.7  β_{m+1} = √(q̃_{m+1}^T w),  q_{m+1} = q̃_{m+1}/β_{m+1};
   2.d  r₀ ← ṽ_i and goto 2.b;
   2.e  B-normalize ṽ_i and V_k ← [V_{k−1}, ṽ_i].
```

9.3.4 Numerical examples

In this section we apply the TLF methodologies and algorithms developed above to the solution of typical electromagnetic eigenvalue problems, associated with the extraction of the wave transmission properties of various types of uniform electromagnetic waveguides.

Microstrip Line. The shielded microstrip line, with cross-section as depicted in the insert of Fig. 9.3 is a structure commonly encountered in multi-layered substrates used for signal distribution in high-speed/high-frequency electronic systems. The top and bottom walls, as well as the strip are assumed to be perfect electric conductors. For the purpose of finite element analysis, perfect magnetic conductors (PMC) are used as side walls to truncate the unbounded structure on either side.

The dispersion diagrams for the three modes of the structure, obtained from the finite element analysis, are compared with those published in [26]. Very good agreement is observed. The space $\mathcal{H}^1(curl)$ was used for the expansion of the transverse and longitudinal electric fields, resulting in 8168 and 3256 unknowns, respectively.

Pedestal-Supported Stripline. The cross-section of the waveguide is depicted in the insert of Fig. 9.3.4. Results from [27]) for the propagation constant of the fundamental mode of the structure, obtained using a spectral domain approach, were used for comparison with those obtained using the proposed finite element solver. As indicated in Fig. 9.3.4 the two sets of results are in very good agreement.

Dielectric Image Line. The cross-section of a rectangular dielectric image line with relative permittivity of 2.0 is depicted in the insert of Fig. 9.5. Although the fields in this image guide actually extend to infinity, the computational domain was truncated for the purposes of a finite element analysis. More specifically, a PMC wall was placed on either side and at the top of the guide to form a rectangular enclosure. The bottom wall is the physical electrically conducting wall of the structure, which was taken to be a PEC.

The computed dispersion characteristics for the first two modes are depicted in Fig. 9.5. Very good agreement between the finite element solver results and those of Ogusu, reported in [28], is observed. The number of unknowns for the expansion of the transverse and

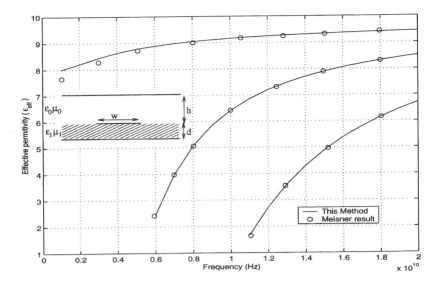

Figure 9.3   Dispersion diagram for the first three modes of a simple microstrip line ($\epsilon_1 = 9.7\epsilon_0$, $w = 9.15$ mm, $h = 1.92$ mm, $d = 0.64$ mm).

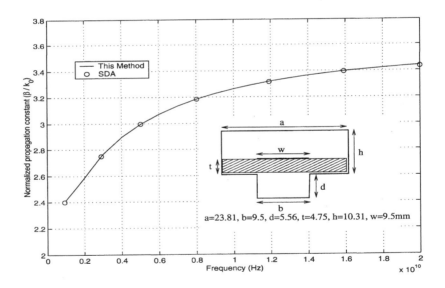

Figure 9.4   Dispersion curve for the fundamental mode of a pedestal-support stripline.

longitudinal electric fields in terms of $\mathcal{H}^1(curl)$ expansion functions were 3292 and 1330, respectively.

Three-Conductor Microstrip Waveguide.   As seen by the relative permittivity data in the caption of Fig. 9.6, the substrate for this three-conductor microstrip structure is anisotropic. Results for the propagation properties of the three (quasi-TEM) fundamental modes of this structure are given in [29]. These results were used for validation of the

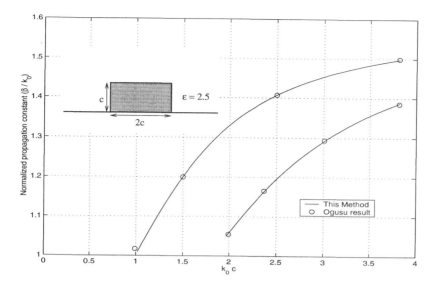

Figure 9.5 Dispersion curves for the first two modes of a dielectric image line.

accuracy of the dispersion curves for the first three modes obtained using the TL-based eigensolver. The comparison is depicted in Fig. 9.6, indicating a very good agreement between the two sets of results. The number of unknowns for the expansion of the transverse and longitudinal electric fields using $\mathcal{H}^1(curl)$ expansion functions were 9626 and 3567, respectively.

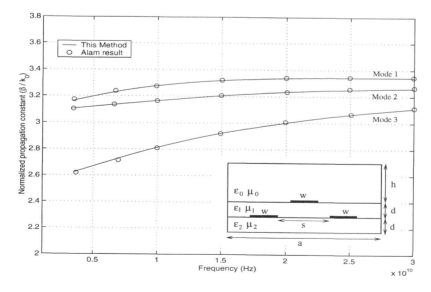

Figure 9.6 Dispersion curves for the three fundamental modes of a three-conductor microstrip structure on an anisotropic substrate ($\epsilon_{x1} = \epsilon_{x2} = 9.4\epsilon_0$, $\epsilon_{y1} = \epsilon_{y2} = 11.6\epsilon_0$, $\epsilon_{z1} = \epsilon_{z2} = 9.4\epsilon_0$ $d = w = 1.0$ mm, $h = 4.0$ mm, $a = 10.0$ mm, $s = 2.0$ mm).

9.4   TRANSVERSE-TRANSVERSE-FIELD METHOD

The TLF methods of the previous section both result in eigenvalues problems in γ^2. Thus they are suitable for the eigenvalue analysis of uniform waveguides for which the forward $(+\hat{z})$ and backward $(-\hat{z})$ modes are degenerate, which is the case only when $\bar{\epsilon}_{tz} = \bar{\mu}_{tz} = 0$ and $\bar{\epsilon}_{zt} = \bar{\mu}_{zt} = 0$ in the material tensors. For the general anisotropic case for which the above conditions are not satisfied, a quadratic eigenvalue problem results. For such cases, the transverse-transverse-field (TTF) formulation can be used for the development of a matrix approximation of the eigenvalue problem using finite elements, which is linear in the eigenvalue γ.

9.4.1   Finite element formulation

In the TTF formulation the eigen-field is define in terms of both the transverse electric field, $\vec{E}_t$, and transverse magnetic field, $\vec{H}_t$. The development of the mathematical statement of the eigenvalue problem begins with the two Maxwell curl equations cast in the following form:

$$
\begin{cases} \nabla \times \vec{E} = -j\omega\bar{\mu} \cdot \vec{H} \\ \nabla \times \vec{H} = j\omega\bar{\epsilon} \cdot \vec{E} \end{cases} \Rightarrow
$$

$$
\begin{cases} \nabla_t \times \vec{E}_t + \nabla_t \times E_z\hat{z} + \gamma\vec{E}_t \times \hat{z} = -j\omega\bar{\mu} \cdot \left(\vec{H}_t + \hat{z}H_z \right) \\ \nabla_t \times \vec{H}_t + \nabla_t \times H_z\hat{z} + \gamma\vec{H}_t \times \hat{z} = j\omega\bar{\epsilon} \cdot \left(\vec{E}_t + \hat{z}E_z \right). \end{cases} \tag{9.78}
$$

Using the above equations, the longitudinal components, E_z and H_z, can be expressed in terms of the transverse components, $\vec{E}_t$ and $\vec{H}_t$, as follows:

$$
H_z = \frac{-1}{j\omega\mu_{zz}}\hat{z} \cdot \left(\nabla_t \times \vec{E}_t + j\omega\bar{\mu}_{zt} \cdot \vec{H}_t \right),
$$

$$
E_z = \frac{1}{j\omega\epsilon_{zz}}\hat{z} \cdot \left(\nabla_t \times \vec{H}_t - j\omega\bar{\epsilon}_{zt} \cdot \vec{E}_t \right). \tag{9.79}
$$

In addition, the transverse component equations are

$$
\nabla_t \times E_z\hat{z} + \gamma\vec{E}_t \times \hat{z} = -j\omega\left(\bar{\mu}_{tt} \cdot \vec{H}_t + \bar{\mu}_{tz} \cdot H_z \right),
$$

$$
\nabla_t \times H_z\hat{z} + \gamma\vec{H}_t \times \hat{z} = j\omega\left(\bar{\epsilon}_{tt} \cdot \vec{E}_t + \bar{\epsilon}_{tz} \cdot E_z \right). \tag{9.80}
$$

Substitution of (9.79) into (9.80) to cancel E_z and H_z yields two coupled equations that involve only the transverse parts of the fields

$$
\nabla_t \times \left(\frac{\nabla_t \times \vec{H}_t}{j\omega\epsilon_{zz}} - \frac{\bar{\epsilon}_{zt}}{\epsilon_{zz}} \cdot \vec{E}_t \right) + \gamma\vec{E}_t \times \hat{z} = -j\omega\bar{\mu}_{tt} \cdot \vec{H}_t
$$

$$
+ \frac{\bar{\mu}_{tz}}{\mu_{zz}} \cdot \nabla_t \times \vec{E}_t + j\omega\frac{\bar{\mu}_{tz} \cdot \bar{\mu}_{zt}}{\mu_{zz}} \cdot \vec{H}_t,
$$

$$
\nabla_t \times \left(-\frac{\nabla_t \times \vec{E}_t}{j\omega\mu_{zz}} - \frac{\bar{\mu}_{zt}}{\mu_{zz}} \cdot \vec{H}_t \right) + \gamma\vec{H}_t \times \hat{z} = j\omega\bar{\epsilon}_{tt} \cdot \vec{E}_t \tag{9.81}
$$

$$
+ \frac{\bar{\epsilon}_{tz}}{\epsilon_{zz}} \cdot \nabla_t \times \vec{H}_t - j\omega\frac{\bar{\epsilon}_{tz} \cdot \bar{\epsilon}_{zt}}{\epsilon_{zz}} \cdot \vec{E}_t.
$$

The weak statement of the above system will be developed next. The transverse fields, $\vec{H}_t$ and $\vec{E}_t$, are expanded in terms of tangentially continuous vector basis functions. Let $\vec{w}_e$ denote the basis functions used for the expansion of the electric field, and $\vec{w}_h$ those used for the expansion of the magnetic field. Furthermore, in the spirit of Galerkin's method, $\vec{w}_e$ and $\vec{w}_h$ are also used as testing functions.

Taking the scalar product of the first equation in (9.81) with $\vec{w}_h$, and making use of the vector identity (A.6) and the two-dimensional form of the divergence theorem, we obtain

$$
-\oint_{\partial S} \frac{\hat{n} \times \vec{w}_h \cdot \left(\nabla_t \times \vec{H}_t - j\omega \bar{\epsilon}_{zt} \cdot \vec{E}_t \right)}{j\omega \epsilon_{zz}} dl + \iint_S \frac{\nabla_t \times \vec{w}_h \cdot \nabla_t \times \vec{H}_t}{j\omega \epsilon_{zz}} ds
$$
$$
- \iint_S \nabla_t \times \vec{w}_h \cdot \frac{\bar{\epsilon}_{zt}}{\epsilon_{zz}} \cdot \vec{E}_t ds + \gamma \iint_S \vec{w}_h \times \vec{E}_t \cdot \hat{z} ds \tag{9.82}
$$
$$
= -j\omega \iint_S \vec{w}_h \cdot \bar{\mu}_{tt} \cdot \vec{H}_t ds + \iint_S \vec{w}_h \cdot \left(\frac{\bar{\mu}_{tz}}{\mu_{zz}} \cdot \nabla_t \times \vec{E}_t + j\omega \frac{\bar{\mu}_{tz} \cdot \bar{\mu}_{zt}}{\mu_{zz}} \cdot \vec{H}_t \right) ds.
$$

The integral along the boundary of ∂S_2 vanishes if we choose $\vec{w}_h$ such that $\hat{n} \times \vec{w}_h = 0$ on the PMC boundary. The integral along ∂S_1 also vanishes since $E_z = 0$ on the PEC boundary. Thus only the impedance boundary contributes to the integral along the contour of the cross-section. Substitution of (9.15) into the above equation gives (after a slight reorganization of the equation along with multiplication of both sides by $-j\omega$)

$$
-j\omega\gamma \iint_S \vec{w}_h \times \vec{E}_t \cdot \hat{z} ds = \iint_S \nabla_t \times \vec{w}_h \cdot \epsilon_{zz}^{-1} \nabla_t \times \vec{H}_t ds
$$
$$
+ j\omega \int_{\partial S_3} \hat{n} \times \vec{w}_h \cdot \frac{1}{Z_s} \hat{n} \times \vec{H}_t dl - \omega^2 \iint_S \vec{w}_h \cdot \left(\bar{\mu}_{tt} - \frac{\bar{\mu}_{tz} \cdot \bar{\mu}_{zt}}{\mu_{zz}} \right) \cdot \vec{H}_t ds \tag{9.83}
$$
$$
- j\omega \iint_S \left(\nabla_t \times \vec{w}_h \cdot \frac{\bar{\epsilon}_{zt}}{\epsilon_{zz}} \cdot \vec{E}_t + \vec{w}_h \cdot \frac{\bar{\mu}_{tz}}{\mu_{zz}} \cdot \nabla_t \times \vec{E}_t \right) ds.
$$

In a similar fashion, the inner product of the second equation in (9.81) with $\vec{w}_e$ yields

$$
\oint_{\partial S} \frac{\hat{n} \times \vec{w}_e \cdot \left(\nabla_t \times \vec{E}_t + j\omega \bar{\mu}_{zt} \cdot \vec{H}_t \right)}{j\omega \mu_{zz}} dl - \iint_S \frac{\nabla_t \times \vec{w}_e \cdot \nabla_t \times \vec{E}_t}{j\omega \mu_{zz}} ds
$$
$$
- \iint_S \nabla_t \times \vec{w}_e \cdot \frac{\bar{\mu}_{zt}}{\mu_{zz}} \cdot \vec{H}_t ds + \gamma \iint_S \vec{w}_e \times \vec{H}_t \cdot \hat{z} ds \tag{9.84}
$$
$$
= j\omega \iint_S \vec{w}_e \cdot \bar{\epsilon}_{tt} \cdot \vec{E}_t ds + \iint_S \vec{w}_e \cdot \left(\frac{\bar{\epsilon}_{tz}}{\epsilon_{zz}} \cdot \nabla_t \times \vec{H}_t - j\omega \frac{\bar{\epsilon}_{tz} \cdot \bar{\epsilon}_{tz}}{\epsilon_{zz}} \cdot \vec{E}_t \right) ds.
$$

The integral along ∂S_1 vanishes if we choose the expansion functions for $\vec{E}_t$ such that $\hat{n} \times \vec{w}_e = 0$ on the PEC boundary. The integral along ∂S_2 also vanishes in view of the boundary condition (9.14). Thus only the impedance boundary contributes to the integration along the contour boundary of the cross-section. Substitution of (9.15) into the above equation and multiplication by $j\omega$ yields

$$
-j\omega\gamma \iint_S \vec{H}_t \times \vec{w}_e \cdot \hat{z} ds = \iint_S \nabla_t \times \vec{w}_e \cdot \mu_{zz}^{-1} \nabla_t \times \vec{E}_t ds
$$
$$
+ j\omega \int_{\partial S_3} \hat{n} \times \vec{w}_e \cdot \frac{1}{Z_s} \hat{n} \times \vec{E}_t dl - \omega^2 \iint_S \vec{w}_e \cdot \left(\bar{\epsilon}_{tt} - \frac{\bar{\epsilon}_{tz} \cdot \bar{\epsilon}_{zt}}{\epsilon_{zz}} \right) \cdot \vec{E}_t ds \tag{9.85}
$$
$$
+ j\omega \iint_S \left(\nabla_t \times \vec{w}_e \cdot \frac{\bar{\mu}_{zt}}{\mu_{zz}} \cdot \vec{H}_t + \vec{w}_e \cdot \frac{\bar{\epsilon}_{tz}}{\epsilon_{zz}} \cdot \nabla_t \times \vec{H}_t \right) ds.
$$

Equations (9.83) and (9.85) constitute the weak statement of the TTF formulation of the eigenvalue problem for lossy, anisotropic, inhomogeneous waveguides. The finite element matrix eigenvalue problem is obtained through testing with all functions in the expansions of $\vec{E}_t$ and $\vec{H}_t$. The result is cast in matrix form as follows:

$$-j\gamma \begin{bmatrix} 0 & D \\ D^T & 0 \end{bmatrix} \begin{bmatrix} x_{h,t} \\ x_{e,t} \end{bmatrix} = \begin{bmatrix} A & C_1 \\ C_2 & B \end{bmatrix} \begin{bmatrix} x_{h,t} \\ x_{e,t} \end{bmatrix}, \tag{9.86}$$

where, $x_{h,t}$ and $x_{e,t}$ are the vectors of the coefficients in the expansions of $\vec{H}_t$ and $\vec{E}_t$, respectively. The elements of the matrices A, B, C_1, C_2 and D are given in terms of the following integrals:

$$A_{i,j} = \iint_S \nabla_t \times \vec{w}_{h,i} \cdot \frac{1}{\epsilon_{zz}} \nabla_t \times \vec{w}_{h,j} ds - \omega^2 \iint_S \vec{w}_{h,i} \cdot \left(\bar{\mu}_{tt} - \frac{\bar{\mu}_{tz} \cdot \bar{\mu}_{zt}}{\mu_{zz}} \right) \cdot \vec{w}_{h,j} ds$$

$$+ j\omega \int_{\partial S_3} \hat{n} \times \vec{w}_{h,i} \cdot \frac{1}{Z_s} \hat{n} \times \vec{w}_{h,j} dl$$

$$B_{i,j} = \iint_S \nabla_t \times \vec{w}_{e,i} \cdot \frac{1}{\mu_{zz}} \nabla_t \times \vec{w}_{e,j} ds - \omega^2 \iint_S \vec{w}_{e,i} \cdot \left(\bar{\epsilon}_{tt} - \frac{\bar{\epsilon}_{tz} \cdot \bar{\epsilon}_{zt}}{\epsilon_{zz}} \right) \cdot \vec{w}_{e,j} ds$$

$$+ j\omega \int_{\partial S_3} \hat{n} \times \vec{w}_e \cdot \frac{1}{Z_s} \hat{n} \times \vec{E}_t dl$$

$$(C_1)_{i,j} = -j\omega \iint_S \left(\nabla_t \times \vec{w}_{h,i} \frac{\bar{\epsilon}_{zt}}{\epsilon_{zz}} \cdot \vec{w}_{e,j} + \vec{w}_{h,i} \cdot \frac{\bar{\mu}_{tz}}{\mu_{zz}} \cdot \nabla_t \times \vec{w}_{e,j} \right) ds$$

$$(C_2)_{i,j} = j\omega \iint_S \left(\nabla_t \times \vec{w}_{e,i} \cdot \frac{\bar{\mu}_{zt}}{\mu_{zz}} \cdot \vec{w}_{h,j} + \vec{w}_{e,i} \cdot \frac{\bar{\epsilon}_{tz}}{\epsilon_{zz}} \cdot \nabla_t \times \vec{w}_{h,j} \right) ds$$

$$D_{i,j} = \omega \iint_S \vec{w}_{h,i} \times \vec{w}_{e,j} \cdot \hat{z} ds. \tag{9.87}$$

In the following we examine the properties of the eigenmodes of the above system for three special types of material tensors, commonly encountered in practical applications.

Case I: Symmetric material tensors. In this case it is $\epsilon = \epsilon^T$ and $\mu = \mu^T$. Consequently, A and B are symmetric matrices, while $C_2 = -C_1^T$. Equation (9.86) assumes the form

$$-j\gamma \underbrace{\begin{bmatrix} 0 & D \\ -D^T & 0 \end{bmatrix}}_{\mathbf{B}} \begin{bmatrix} x_{h,t} \\ x_{e,t} \end{bmatrix} = \underbrace{\begin{bmatrix} A & C_1 \\ C_1^T & -B \end{bmatrix}}_{\mathbf{A}} \begin{bmatrix} x_{h,t} \\ x_{e,t} \end{bmatrix}. \tag{9.88}$$

Hence, $\mathbf{A}$ is symmetric while $\mathbf{B}$ is skew-symmetric. It follows immediately from Theorem III in Section 8.4.1 that the non-zero eigenmodes of (9.88) are paired in the sense that for each non-zero eigenvalue γ there exists another eigenvalue $-\gamma$. Clearly, these two eigenmodes constitute a pair of one forward- and one backward-propagating mode.

Next we examine whether there is a relationship between the transverse electric and magnetic field profiles of these modes over the waveguide cross-section. Cancelation of $x_{h,t}$ in (9.88) yields two eigen-matrix equations, one for the $(+)$ mode, $(x_{e,t}^+, x_{h,t}^+)$, with propagation constant $+\gamma$, and one for the $(-)$ mode, $(x_{e,t}^-, x_{h,t}^-)$, with propagation constant $-\gamma$,

$$(j\gamma D^T - C_1^T)A^{-1}(j\gamma D + C_1)x_{e,t}^+ = Bx_{e,t}^+,$$

$$(j\gamma D^T + C_1^T)A^{-1}(j\gamma D - C_1)x_{e,t}^- = Bx_{e,t}^-. \tag{9.89}$$

Clearly, the two eigen-matrix equations are different; hence, we conclude that, even though the propagation constants of the two eigenmodes satisfy the relationship $\gamma^+ = -\gamma^-$, their transverse electric and magnetic fields are different.

From Theorem III in Section 8.4.1 we also know that any two eigenmodes, $(x_{e,t}^{(1)}, x_{h,t}^{(1)})$, $(x_{e,t}^{(2)}, x_{h,t}^{(2)})$, with propagation constants γ_1 and γ_2, respectively, such that $\gamma^{(1)} + \gamma^{(2)} \neq 0$, are **B**-orthogonal; hence, it is

$$\begin{bmatrix} x_{h,t}^{(1)T} & x_{e,t}^{(1)T} \end{bmatrix} \begin{bmatrix} 0 & D \\ -D^T & 0 \end{bmatrix} \begin{bmatrix} x_{h,t}^{(2)} \\ x_{e,t}^{(2)} \end{bmatrix} = 0 \;\Rightarrow\; x_{h,t}^{(1)T} D x_{e,t}^{(2)} - x_{e,t}^{(1)T} D^T x_{h,t}^{(2)} = 0. \quad (9.90)$$

In view of the form of the elements of D, this relationship may be cast in terms of the transverse electric and magnetic fields for the two modes as follows:

$$\iint_S \left(\vec{E}_t^{(1)} \times \vec{H}_t^{(2)} - \vec{E}_t^{(2)} \times \vec{H}_t^{(1)} \right) \cdot \hat{z} \, ds = 0. \quad (9.91)$$

This result is immediately recognized as the consequence of the source-free Lorentz reciprocity theorem, which holds inside a domain with reciprocal media [23],

$$\nabla \cdot \left(\vec{E}^{(1)} \times \vec{H}^{(2)} - \vec{E}^{(2)} \times \vec{H}^{(1)} \right) = 0. \quad (9.92)$$

To show this, consider the volume integral of the above equation in a longitudinal section of the waveguide of length l, bounded on the left and on the right by planes $z = z_0$ and $z = z_0 + l$, respectively. Making use of the divergence theorem, the integration yields

$$\iint_{S_{z_0+l}} \hat{z} \cdot \left(\vec{E}^{(1)} \times \vec{H}^{(2)} - \vec{E}^{(2)} \times \vec{H}^{(1)} \right) ds - \iint_{S_{z_0}} \hat{z} \cdot \left(\vec{E}^{(1)} \times \vec{H}^{(2)} \right.$$
$$\left. - \vec{E}^{(2)} \times \vec{H}^{(1)} \right) ds + \iint_{\text{walls}} \hat{n} \cdot \left(\vec{E}^{(1)} \times \vec{H}^{(2)} - \vec{E}^{(2)} \times \vec{H}^{(1)} \right) ds = 0. \quad (9.93)$$

Since $\hat{n} \times \vec{E}$ vanishes on the PEC portion of the waveguide walls and $\hat{n} \times \vec{H}$ vanishes on the PMC portion, any contribution to the surface integral in the above equation will be associated with the impedance portion of the boundary. However, use of the impedance boundary condition (9.15) yields

$$\iint_{\text{walls}} \hat{n} \cdot \left(\vec{E}^{(1)} \times \vec{H}^{(2)} - \vec{E}^{(2)} \times \vec{H}^{(1)} \right) ds$$
$$= \iint_{\text{walls}} \left(\hat{n} \times \vec{E}^{(1)} \cdot \frac{1}{Z_s} \hat{n} \times \vec{E}^{(2)} - \hat{n} \times \vec{E}^{(1)} \cdot \frac{1}{Z_s} \hat{n} \times \vec{E}^{(2)} \right) ds = 0. \quad (9.94)$$

Thus the surface integral over the side walls of the guide is zero. Finally, substitution of (9.6) into the integrals in (9.93) over the cross-sections at $z = z_0$ and $z = z_0 + l$ yields

$$\left(e^{-(\gamma_1+\gamma_2)(z_0+l)} - e^{-(\gamma_1+\gamma_2)z_0} \right) \iint_S \left(\vec{E}_t^{(1)} \times \vec{H}_t^{(2)} - \vec{E}_t^{(2)} \times \vec{H}_t^{(1)} \right) \cdot \hat{z} \, ds = 0 \quad (9.95)$$

since the terms involving the axial components of the fields vanish upon taking their dot product with $\hat{z}$. With $\gamma_1 + \gamma_2 \neq 0$, the above result implies that the integral over the cross-section of the guide is zero, which is identical with (9.91).

Case II: Material tensors with skew-symmetric transverse-to-longitudinal parts. In this case it is, $\bar{\epsilon}_{tt} = \bar{\epsilon}_{tt}^T$, $\bar{\mu}_{tt} = \bar{\mu}_{tt}^T$, $\bar{\epsilon}_{tz}^T = -\bar{\epsilon}_{zt}$, and $\bar{\mu}_{tz}^T = -\bar{\mu}_{zt}$. Under this condition, A and B are symmetric and $C_2 = C_1^T$. Thus the eigen-matrix equation (9.86) becomes

$$-j\gamma \underbrace{\begin{bmatrix} 0 & D \\ D^T & 0 \end{bmatrix}}_{\mathbf{B}} \begin{bmatrix} x_{h,t} \\ x_{e,t} \end{bmatrix} = \underbrace{\begin{bmatrix} A & C_1 \\ C_1^T & B \end{bmatrix}}_{\mathbf{A}} \begin{bmatrix} x_{h,t} \\ x_{e,t} \end{bmatrix}. \tag{9.96}$$

Since $\mathbf{A}$ and $\mathbf{B}$ are both symmetric, Theorem I from Section 8.4.1 applies. It is immediate evident from the above equation that different eigenmodes correspond to eigenvalues γ and $-\gamma$. Furthermore, from Theorem 1, any two eigenmodes of (9.96), $(x_{e,t}^{(1)}, x_{h,t}^{(1)})$, $(x_{e,t}^{(2)}, x_{h,t}^{(2)})$, with different propagation constants γ_1 and γ_2, respectively, are $\mathbf{B}$-orthogonal; hence,

$$\begin{bmatrix} x_{h,t}^{(1)^T} & x_{e,t}^{(1)^T} \end{bmatrix} \begin{bmatrix} 0 & D \\ D^T & 0 \end{bmatrix} \begin{bmatrix} x_{h,t}^{(2)} \\ x_{e,t}^{(2)} \end{bmatrix} = 0 \;\Rightarrow\; x_{h,t}^{(1)^T} D x_{e,t}^{(2)} + x_{e,t}^{(1)^T} D^T x_{h,t}^{(2)} = 0$$

$$\Rightarrow \iint_S \left(\vec{E}_t^{(1)} \times \vec{H}_t^{(2)} + \vec{E}_t^{(1)} \times \vec{H}_t^{(2)} \right) \cdot \hat{z}\,ds = 0. \tag{9.97}$$

Case III: Uniaxial media. This case is encountered most frequently in practice. The material tensors satisfy the relations $\bar{\epsilon}_{tt} = \bar{\epsilon}_{tt}^T$, $\bar{\mu}_{tt} = \bar{\mu}_{tt}^T$, $\bar{\epsilon}_{tz} = \bar{\mu}_{tz} = 0$ and $\bar{\epsilon}_{zt} = \bar{\mu}_{zt} = 0$. Thus A and B are symmetric, while $C_1 = 0$ and $C_2 = 0$. The eigen-matrix equation assumes the form

$$-j\gamma \underbrace{\begin{bmatrix} 0 & D \\ D^T & 0 \end{bmatrix}}_{\mathbf{B}} \begin{bmatrix} x_{h,t} \\ x_{e,t} \end{bmatrix} = \underbrace{\begin{bmatrix} A & \\ & B \end{bmatrix}}_{\mathbf{A}} \begin{bmatrix} x_{h,t} \\ x_{e,t} \end{bmatrix}. \tag{9.98}$$

Furthermore, the expressions for the elements of A and B simplify as follows:

$$A_{i,j} = \iint_S \nabla_t \times \vec{w}_{h,i} \cdot \frac{1}{\epsilon_{zz}} \nabla_t \times \vec{w}_{h,j}\,ds - \omega^2 \iint_S \vec{w}_{h,i} \cdot \bar{\mu}_{tt} \cdot \vec{w}_{h,j}\,ds$$
$$+ j\omega \int_{\partial S_3} \hat{n} \times \vec{w}_{h,i} \cdot \frac{1}{Z_s} \hat{n} \times \vec{w}_{h,j}\,dl,$$
$$B_{i,j} = \iint_S \nabla_t \times \vec{w}_{e,i} \cdot \frac{1}{\mu_{zz}} \nabla_t \times \vec{w}_{e,j}\,ds - \omega^2 \iint_S \vec{w}_{e,i} \cdot \bar{\epsilon}_{tt} \cdot \vec{w}_{e,j}\,ds \tag{9.99}$$
$$+ j\omega \int_{\partial S_3} \hat{n} \times \vec{w}_e \cdot \frac{1}{Z_s} \hat{n} \times \vec{E}_t\,dl.$$

The interesting thing about this equation is that it can be considered as a special case of either (9.88) or (9.96). For example, in the way the matrix eigenvalue equation is written in (9.98), $\mathbf{A}$ and $\mathbf{B}$ are both symmetric; hence, Theorem I of Section 8.4.1 applies. However, recognizing (9.98) as a system of two matrix equations, and rewriting it after the second equation, $-j\gamma D^T x_{h,t} = B x_{e,t}$, is multiplied by -1, we obtain a matrix eigenvalue problem for which $\mathbf{A}$ is still symmetric but $\mathbf{B}$ is skew-symmetric. Thus Theorem III of Section 8.4.1 also applies to (9.98).

We conclude that both Theorem I and Theorem III can be used to derive the following properties for the eigenmodes for the case of uniaxial media.

1. For every nonzero eigenvalue γ of (9.98) there exist another eigenvalue $-\gamma$.

2. Two eigenmodes with eigenvalues γ_1 and γ_2 such that $\gamma_1 + \gamma_2 \neq 0$, satisfy the following relations

$$
\begin{cases}
\iint_S \left(\vec{E}_t^{(1)} \times \vec{H}_t^{(2)} - \vec{E}_t^{(2)} \times \vec{H}_t^{(1)} \right) \cdot \hat{z} ds = 0 \\
\iint_S \left(\vec{E}_t^{(1)} \times \vec{H}_t^{(2)} + \vec{E}_t^{(2)} \times \vec{H}_t^{(1)} \right) \cdot \hat{z} ds = 0.
\end{cases}
\tag{9.100}
$$

Adding and subtracting these equations yields

$$
\iint_S \hat{z} \cdot \left(\vec{E}_t^{(1)} \times \vec{H}_t^{(2)} \right) ds = \iint_S \hat{z} \cdot \left(\vec{E}_t^{(2)} \times \vec{H}_t^{(1)} \right) ds = 0,
\tag{9.101}
$$

which is recognized as the orthogonality relation for distinct eigenmodes in uniaxial waveguides.

3. For two eigenmodes of (9.98) with eigenvalues γ_1 and γ_2 such that $\gamma_1 + \gamma_2 = 0$, their tangential electric and magnetic fields satisfy the following relationship:

$$
\iint_S \left(\vec{E}_t^{(1)} \times \vec{H}_t^{(2)} + \vec{E}_t^{(2)} \times \vec{H}_t^{(1)} \right) \cdot \hat{z} ds = 0.
\tag{9.102}
$$

Clearly, this is the case of two modes, one forward-propagating and one backward-propagating, with eigenvalues the negative of each other. Recognizing this as a special case of (9.89), the two matrix equations for the two modes, resulting from the elimination of $x_{h,t}$, degenerate into a single equation

$$
-\gamma^2 D^T A^{-1} D x_{e,t}^{(\pm)} = B x_{e,t}^{(\pm)}.
\tag{9.103}
$$

From this it follows immediately that the transverse fields of the two modes are related as follows:

$$
\left\{ \gamma^{(+)}, \begin{bmatrix} x_{h,t}^{(+)} \\ x_{e,t}^{(+)} \end{bmatrix} \right\} = \left\{ \gamma, \begin{bmatrix} x_{h,t} \\ x_{e,t} \end{bmatrix} \right\}, \quad \left\{ \gamma^{(-)}, \begin{bmatrix} x_{h,t}^{(-)} \\ x_{e,t}^{(-)} \end{bmatrix} \right\} = \left\{ -\gamma, \begin{bmatrix} -x_{h,t} \\ x_{e,t} \end{bmatrix} \right\}.
\tag{9.104}
$$

This concludes the discussion of the properties of the matrix eigenvalue problems resulting from the TTF formulation. One final point worth mentioning concerns the case where PEC and/or PMC strips are present in the waveguide. Since the transverse magnetic field is solved for in this formulation, care must be exercised not to allow for infinitesimally thin PEC strips, since this will result in a discontinuity in the tangential magnetic field vector across the element edges coinciding with the surface of the strip, due to the induced electric current density on its surface. Such discontinuity is incompatible with the attributes of the expansion functions used for the finite element approximation of the transverse magnetic field vector. To avoid this, a finite thickness must be assigned to the strip. In a similar manner, to allow for the discontinuity in the tangential electric field across a PMC strip to be correctly accounted for in the finite element approximation of the eigenvalue problem, a finite thickness must be assigned to any PMC strips present in the waveguide.

Compared to TLF formulations, the dimension of the matrix eigenvalue problem for the TTF formulation is larger. More specifically, the matrix dimension for the TTF matrix is, roughly, twice the number of edges in the 2D mesh, while that for the TLF matrix is, roughly, equal to the sum of the numbers of edges and the number of nodes in the mesh. Since there are more edges than nodes, the dimension of the TTF matrix is larger than that of the TLF matrix. While, at first sight, this suggests that, from a computational efficiency point of view TLF is superior to TTF, two additional factors must be considered when deciding which one of the two formulations is preferable for a specific application. The first concerns the sparsity of the matrices. The TTF matrices are sparser than the TLF matrices. The second concerns the material complexity of the structure. The TTF formulation yields a linear generalized eigenvalue problem irrespective of the properties of the material tensors.

9.4.2 Algorithms

In this section we discuss the application of the Krylov-subspace-based eigensolvers of Chapter 8 for the extraction of a some of the dominant eigenvalues of (9.86).

Let us begin by recalling the expansion spaces used for the approximation of the transverse electric and transverse magnetic fields. Because of the tangential continuity of $\vec{E}_t$ and $\vec{H}_t$, assuming that a triangular mesh is used for the discretization of the cross-section of the guide, the expansion space for both should be $\mathcal{H}^p(curl)$ where $p = 0, 1, \cdots$. The two lowest subspaces are

$$\mathcal{H}^0(curl) = \{\lambda_i \nabla \lambda_j - \lambda_j \nabla \lambda_i \qquad | \text{ all edges } i < j \quad \},$$

$$\mathcal{H}^1(curl) = \left\{ \begin{array}{ll} \lambda_i \nabla \lambda_j - \lambda_j \nabla \lambda_i & | \text{ all edges } i < j \\ \lambda_i \nabla \lambda_j + \lambda_j \nabla \lambda_i & | \text{ all edges } i < j \\ 4\lambda_i(\lambda_j \nabla \lambda_k - \lambda_k \nabla \lambda_j) & | \text{ all triangles } i < j < k \end{array} \right\}. \qquad (9.105)$$

For the reasons discussed in Section 9.3.3, a shift-and-invert transformation of the matrix eigenvalue equation (9.86) is utilized to render the eigenvalues of interest the ones with largest magnitude. Let γ_0 be the necessary shift. Then the matrix eigenvalue problem becomes

$$\begin{bmatrix} 0 & D \\ D^T & 0 \end{bmatrix} \begin{bmatrix} x_{h,t} \\ x_{e,t} \end{bmatrix} = \frac{j}{\gamma - \gamma_0} \begin{bmatrix} A & C_1 + j\gamma_0 D \\ C_2 + j\gamma_0 D^T & B \end{bmatrix} \begin{bmatrix} x_{h,t} \\ x_{e,t} \end{bmatrix}. \qquad (9.106)$$

This is the eigenvalue equation to be processed by the appropriate Krylov-subspace-based eigensolvers of Chapter 8 for the extraction of the first few physically dominant modes.

For the case where the system has degenerate modes (i.e., when the tz and zt components of the material tensors are zero), it is numerically more robust to work with the reduced version of the original eigenproblem for this case, namely, (9.103). To explain, consider the case where some of the dominant evanescent modes are desired. Then, despite the shift, the convergence of the eigensolver will be slow since the shift value is equidistant from both the $+z$ and $-z$ evanescent modes.

Depicted on the right of Fig. 9.2 is the complex γ^2 plane for this case (since the eigenvalue of interest is γ^2). For a lossless waveguide, γ^2 is negative real for propagating modes and positive real for evanescent modes. With γ_0^2 denoting the appropriate shift, the transformed version of (9.103) is as follows:

$$\underbrace{D^T A^{-1} D}_{\mathbf{A}} x_{e,t} = \frac{1}{\gamma_0^2 - \gamma^2} \underbrace{\left(B + \gamma_0^2 D^T A^{-1} D \right)}_{\mathbf{B}} x_{e,t}. \qquad (9.107)$$

If the material tensors are symmetric, the above is a symmetric, generalized eigenvalue problem. With regards to computational complexity, it is clear that the application of a

Krylov-subspace-based eigensolver for its solution is dominated by the inversion of the matrices A and $(\gamma_0^2 D^T A^{-1} D + B)$. Especially, the inversion of $(\gamma_0^2 D^T A^{-1} D + B)$ is the primary issue of concern since this is a dense matrix. However, the following relationship may be used to expedite its solution:

$$(B + \gamma_0^2 D^T A^{-1} D)^{-1} = \begin{bmatrix} 0 & I \end{bmatrix} \begin{bmatrix} A & j\gamma_0 D \\ j\gamma_0 D^T & B \end{bmatrix}^{-1} \begin{bmatrix} 0 \\ I \end{bmatrix}, \qquad (9.108)$$

where I is the identity matrix. Note that the inversion of the dense matrix has been recast in terms of the inversion of a larger but sparser matrix, the factorization of which can be performed efficiently using a sparse matrix solver.

If the material tensors are not symmetric, a restarted deflated Arnoldi algorithm can be implemented to solve (9.107), recast in the form

$$(B + \gamma_0^2 D^T A^{-1} D)^{-1} D^T A^{-1} D x_{e,t} = \frac{1}{\gamma_0^2 - \gamma^2} x_{e,t}. \qquad (9.109)$$

Certainly the computational complexity of this approach stems from the calculation of the product of a vector with the matrix $(B + \gamma_0^2 D^T A^{-1} D)^{-1} D^T A^{-1} D$. However, the calculation of this matrix-vector product can be expedited by noting that

$$(B + \gamma_0^2 D^T A^{-1} D)^{-1} D^T A^{-1} D = \begin{bmatrix} 0 & I \end{bmatrix} \begin{bmatrix} A & j\gamma_0 D \\ j\gamma_0 D^T & B \end{bmatrix}^{-1} \begin{bmatrix} \frac{j}{\gamma_0} D \\ 0 \end{bmatrix}. \qquad (9.110)$$

When extracting multiple eigenmodes, in order to improve further the efficiency of the eigensolver, a *shift-value hopping process* may be utilized. The basic idea of such a process is that the shift value is updated on the fly during the iterative eigensolution process. Let $\gamma_0^{(1)}$ be the first shift value. Once the first dominant eigenvalue has been extracted, the approximate value of the next eigenvalue emerges. Then a new shift value, $\gamma_0^{(2)}$, chosen to be close to the approximate value for the second dominant eigenvalue, is used. In this manner, all the desired modes, either propagating or evanescent, can be extracted in a systematic and efficient fashion.

9.4.3 Applications

In this section results are presented from the application of the TTF 2D eigensolvers to the analysis of several types of waveguides of practical interest.

Rectangular Metallic Waveguide. The availability of analytic solutions for the eigenvalues of an air-filled, metallic waveguide of rectangular cross-section and perfectly conducting walls provide for the validation of the TTF eigensolver. The geometry of the guide is depicted in the insert of Fig. 9.7. The propagation constants for the first five dominant modes are plotted versus frequency and are compared to their analytic values. Since the width of the guide is twice its height, modes TE_{20} and TE_{01} have the same eigenvalues. The next two higher-order modes are TE_{11} and TM_{11}, both of which have the same eigenvalue. $\mathcal{H}^1(curl)$ bases were used to expand both the transverse electric and magnetic fields. The number of unknowns in the expansions for the electric and magnetic fields are 954 and 1056, respectively. The first five dominant modes are extracted. The dispersion curves for the five modes are shown in Fig. 9.7.

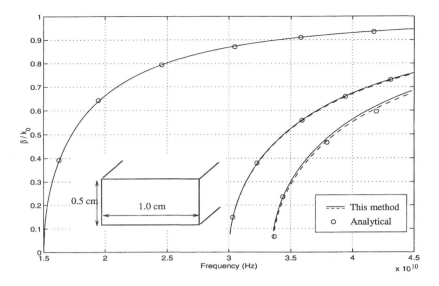

Figure 9.7   Dispersion curves for the first five dominant modes of a rectangular waveguide.

The eigenvalues obtained from the eigensolver at 40 GHz are compared to the analytic ones in Table 9.1.

Table 9.1   Normalized Eigenvalues (β/k_0) of the Five Dominant Modes of a Metallic Waveguide at $f = 40$ GHz, Extracted Using $\mathcal{H}^1(curl)$ Basis Functions

Mode	Analyt.	Numerical	Residual	Error
1	0.92712842791	0.9271356	7.815897e-16	0.0008%
2	0.66201849471	0.6620445	3.954094e-09	0.0039%
3	0.66201849471	0.6620569	1.978055e-08	0.0058%
4	0.54574317145	0.5457919	2.559151e-07	0.0089%
5	0.54574317145	0.5457521	2.716896e-09	0.0016%

As discussed in the previous section, for waveguides with uniaxial media, we can either use (9.98) or (9.103) to solve for the γ or γ^2, respectively. When using (9.98), for each eigenvalue, γ, there exits another valid eigenvalue, $-\gamma$. In the left plot in Fig. 9.8, the extracted eigenvalues using (9.98) are plotted at $f = 40GHz$. The shift points are placed at $j838.3$ and $-j838.3$, respectively. When using (9.103), the eigenvalues γ and $-\gamma$ are degenerate, represented by a single eigenvalue, γ^2. The right plot of Fig. 9.8 depicts the extracted eigenvalue using (9.103), with the shift point taken to be -838.3^2.

Ferrite-Loaded Waveguide.   The ferrite-loaded waveguide of Fig. 9.9 was analyzed using $\mathcal{H}^1(curl)$. The finite element approximation of the eigenvalue problem involved 1540 and 1660 tangential electric and magnetic field unknowns, respectively. The ferrite

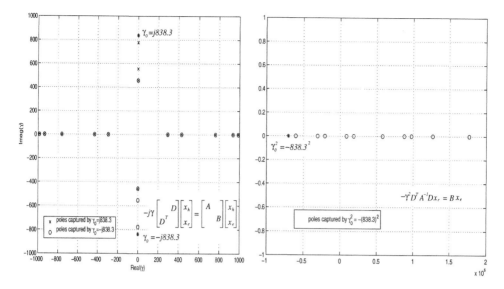

Figure 9.8   Eigenvalue distribution for the first several modes of the waveguide at $f = 40$ GHz.

slab electric permittivity and its magnetic permeability tensor are

$$\epsilon = 10\epsilon_0, \quad \bar{\mu} = \begin{bmatrix} 0.875 & 0 & -j0.375 \\ 0 & 1 & 0 \\ j0.375 & 0 & 0.875 \end{bmatrix} \mu_0. \qquad (9.111)$$

Because of the skew symmetry of the transverse part of the permeability tensor the modes propagating along $+z$ and $-z$ direction are not degenerate modes. The dispersion curves for the first six modes are shown in Fig. 9.9.

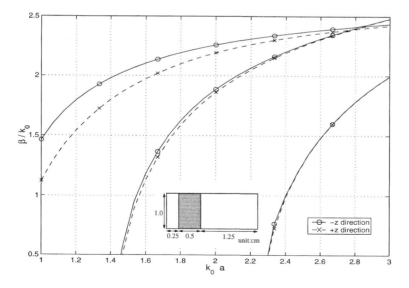

Figure 9.9   Dispersion curves for the first six modes in a ferrite-loaded rectangular waveguide.

Figure 9.10 depicts the dominant eigenvalues for $k_0 a = 2$. In order to extract modes corresponding to the wave propagating in $+z$ and $-z$ directions, the shift points $j500$ and $-j500$ were used, respectively.

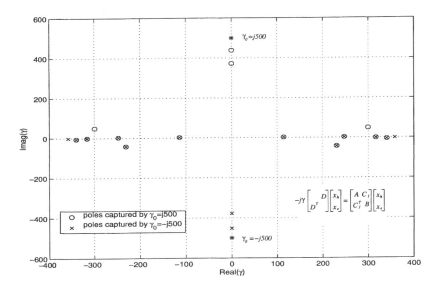

Figure 9.10   Eigenvalues of the ferrite-loaded rectangular waveguide at $k_0 a = 2$.

9.5   EQUIVALENT TRANSMISSION-LINE FORMALISM FOR PLANAR WAVEGUIDES

The last example in Section 9.3.4 considered the extraction of the propagation constants of the modes supported by a three-conductor, planar transmission line system. This is a simple example of a very important class of planar waveguiding structures used for signal distribution in high-speed/high-frequency integrated electronic systems. Because of their importance, a very rich literature is available on the subject, with [30] and [31] offering a limited sample of useful texts on the subject, discussing the electromagnetic properties of the most commonly-used types of planar waveguides, and providing analytical results in support of design of simple structures involving at most two coupled waveguides.

Recent advances in high-speed digital integrated electronic systems have steered the interest toward more complex planar waveguiding structures, involving multiple coupled wires. These structures are typically operated at the so-called *quasi transverse electric and magnetic* (quasi-TEM) regime, understood to mean that the electric and magnetic fields of the fundamental set of propagating modes are predominantly transverse. In other words, the longitudinal components, even though non-zero in general, are of magnitude much smaller than that of the transverse components. Under such conditions, the longitudinal components of the fields are assumed negligible and, thus, position-dependent (along the waveguide axis) voltage and current quantities can be defined for each conductor in terms of line integrals over the cross-section of the waveguide, with one of the conductors serving as a reference for the definition of the voltage. This, then, provides for the extension of the well-known transmission line theory, used for the quantification of one-dimensional wave propagation along the axis of a two-conductor, uniform, waveguide, to a *multiconductor transmission line* (MTL) model for the quantification of wave propagation in a multiple-conductor waveguide.

The reader is referred to [32] and [33] for the details of the theoretical development of the equivalent transmission-line model for wave propagation in a set of parallel, coupled wires. In the following, our discussion assumes that such an MTL model is applicable for the analysis of electromagnetic wave propagation in the system. The emphasis of the discussion is on the background mathematical formalism needed for the establishment of a finite element-based methodology that will be used for the extraction of the parameters needed for the quantitative description of the MTL model. In our presentation only the case of reciprocal multi-conductor systems is considered. The modifications needed for handling the general case of non-reciprocal waveguides are presented in [33].

9.5.1 Multi-conductor transmission line theory

Shown on the left in Fig. 9.11 is the MTL model for a system of $m+1$ parallel conductors of length l, with their axes taken parallel to the z axis of the reference coordinate system. The cross-sectional geometry of the system is assumed independent of z. Otherwise, conductor cross-sections are of arbitrary shapes and the properties of the insulating material filling the space between the wires are assumed arbitrary, as depicted on the right in Fig. 9.11.

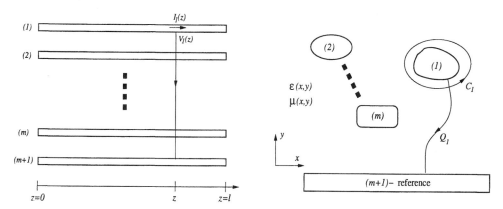

Figure 9.11   a) Multiconductor transmission line model of a uniform multi-conductor waveguide; b) Cross-sectional geometry of the multi-conductor waveguide.

Under the assumption of quasi-TEM propagation in the system, wave propagation along the axis z of the structure is governed by the generalized system of telegrapher's equations. For the purposes of our discussion, time-harmonic excitation of angular frequency ω is assumed. Hence, telegrapher's equations are written in their complex phasor form

$$\begin{cases} j\omega C(\omega)V(z,\omega) + G(\omega)V(z,\omega) + \dfrac{dI(z,\omega)}{dz} = 0 \\[2mm] j\omega L(\omega)I(z,\omega) + R)\omega)I(z,\omega) + \dfrac{dV(z,\omega)}{dz} = 0. \end{cases} \tag{9.112}$$

In the above system $V(z,\omega)$ is the (phasor) voltage vector of length m, with elements the voltages of the m conductors with respect to the $(m+1)$-st conductor, which is taken as reference. Thus the voltage $V_i(z = z_0, \omega)$ for the ith wire is calculated in terms of the line integral of the electric field on a path, Q_i, between the i-th conductor and the reference conductor, on the cross-section with coordinate $z = z_0$. The current vector, $I(z, \omega)$, is also a vector of length m, with elements the net currents on the m wires. The calculation of the current for the ith conductor for a given value of z involves the circulation of the magnetic field on a closed path, C_i, around the cross-section of the ith wire. The four matrices R, L, C, and G, are $m \times m$ matrices, referred to as the resistance, inductance, capacitance and conductance matrix, respectively, per unit length. Their values depend on the material properties and the frequency. For the case of reciprocal media, the matrices are symmetric. A methodology for their calculation will be discussed later in this section.

Let us rewrite the system of (9.112) as follows:

$$\begin{cases} Y(\omega)V(z, \omega) + \dfrac{dI(z, \omega)}{dz} = 0 \\ Z(\omega)I(z, \omega) + \dfrac{dV(z, \omega)}{dz} = 0, \end{cases} \tag{9.113}$$

where

$$Y(\omega) = G(\omega) + j\omega C(\omega), \qquad Z(\omega) = R(\omega) + j\omega L(\omega) \tag{9.114}$$

are define, respectively, as the *per-unit-length admittance* and *per-unit-length impedance* matrices for the MTL. In the following, the dependence of the phasors on ω will be suppressed for simplicity.

The two matrix equations in (9.113) can be combined to derive the following two second-order matrix differential equations for the voltage and current phasors:

$$\frac{d^2 I(z)}{dz^2} - YZI(z) = 0, \tag{9.115}$$

$$\frac{d^2 V(z)}{dz^2} - ZYV(z) = 0. \tag{9.116}$$

For wave propagation along z the solutions of interest for the voltage phasor are of the form

$$V(z) = \begin{bmatrix} \hat{V}_1 \\ \vdots \\ \hat{V}_m \end{bmatrix} e^{\pm \gamma z} = \hat{V} e^{\pm \gamma z}, \tag{9.117}$$

where γ is recognized to be the propagation constant. Substitution of (9.117) in (9.116) yields the following matrix eigenvalue equation:

$$(ZY)\hat{V} = \gamma^2 \hat{V}. \tag{9.118}$$

Thus γ^2 is actually one of the eigenvalues of ZY, and $\hat{V} = [\hat{V}_1, \cdots, \hat{V}_m]^T$ is the associated eigenvector.

In view of the above result, the eigen-decomposition of ZY has the form

$$ZY = \Gamma_V \underbrace{\begin{bmatrix} \gamma_1^2 & & \\ & \ddots & \\ & & \gamma_m^2 \end{bmatrix}}_{\Lambda^2} \Gamma_V^{-1}, \tag{9.119}$$

where γ_i is the propagation constant of the i-th mode, and

$$\Gamma_V = \begin{bmatrix} \hat{V}^{(1)} & \hat{V}^{(2)} \dots & \hat{V}^{(m)} \end{bmatrix}, \tag{9.120}$$

where the i-th column of Γ_V is the voltage eigenvector of the i-th mode.

In a similar manner, a propagating solution to (9.115) will be of the form

$$I(z) = \begin{bmatrix} \hat{I}_1 \\ \vdots \\ \hat{I}_m \end{bmatrix} e^{\pm\gamma z} = \hat{I} e^{\pm\gamma z}. \tag{9.121}$$

Its substitution into (9.115) reveals that γ^2 is actually one of the eigenvalues of the matrix YZ, with $\hat{I}$ the associated eigenvector. Hence, the following eigen-decomposition relation can be written for YZ:

$$YZ = \Gamma_I \underbrace{\begin{bmatrix} \gamma_1^2 & & \\ & \ddots & \\ & & \gamma_m^2 \end{bmatrix}}_{\Lambda^2} \Gamma_I^{-1} \tag{9.122}$$

where γ_i is the propagation constant of the i-th eigenmode, and

$$\Gamma_I = \begin{bmatrix} \hat{I}^{(1)} & \hat{I}^{(2)} & \dots & \hat{I}^{(m)} \end{bmatrix}, \tag{9.123}$$

where the i-th column of Γ_I is the associated eigenvector of the i-th mode.

If Z and Y are symmetric matrices it can be shown that the matrices YZ and ZY have the same eigenvalues. More specifically, if Y and Z are symmetric and the matrices YZ and ZY have the following eigen-decompositions:

$$YZ = \Gamma_I \Lambda_I \Gamma_I^{-1}, \quad ZY = \Gamma_V \Lambda_V \Gamma_V^{-1}, \tag{9.124}$$

then $\Lambda_I = \Lambda_V = \Lambda^2$ and $\Gamma_V^T \Gamma_I$ is a diagonal matrix.

To prove this, consider the transpose of the eigen-decomposition statement for YZ. Making use of the symmetry of Y and Z

$$(YZ)^T = ZY = \left(\Gamma_I^T\right)^{-1} \Lambda_I \Gamma_I^T. \tag{9.125}$$

A direct comparison of this result with the eigen-decomposition statement for ZY reveals the $\Lambda_V = \Lambda_I$ and that $\Gamma_V^T \Gamma_I$ is a diagonal matrix, P,

$$\Gamma_V^T \Gamma_I = \underbrace{\begin{bmatrix} p_1 & & \\ & \ddots & \\ & & p_m \end{bmatrix}}_{P}. \tag{9.126}$$

Clearly, the elements p_i, $i = 1, 2, \dots, m$, have units of power. Furthermore, the above relationship serves as the orthogonality relationship between the m quasi-TEM eigenmodes of the MTL, and is analogous to the electric and magnetic field-based relationship of (9.101).

Substitution of (9.117) and (9.121) into the system of (9.116) and (9.115) yields

$$\begin{cases} Y\Gamma_V = \Gamma_I \Lambda \\ Z\Gamma_I = \Gamma_V \Lambda \end{cases} \Rightarrow \begin{cases} Y = \Gamma_I \Lambda \Gamma_V^{-1} \\ Z = \Gamma_V \Lambda \Gamma_I^{-1}. \end{cases} \tag{9.127}$$

Furthermore, use of (9.126) allows us to re-write Y and Z in terms of the matrices Γ_V, Γ_I, P and Λ as follows,

$$Z = \Gamma_V \Lambda P^{-1} \Gamma_V^T, \quad Y = \Gamma_I \Lambda P^{-1} \Gamma_I^T. \tag{9.128}$$

It is immediately evident from these equations that if P is diagonal then Z and Y are symmetric. In other words, the orthogonality relation (9.126) implies the symmetry of Y and Z and, hence, the reciprocity of the MTL model.

We are now in a position to consider the methodology to be used for the extraction of the per-unit-length matrices Y and Z from the set of the m fundamental (quasi-TEM) eigenmodes obtained from the finite element-based eigenvalue analysis of the multi-conductor waveguide. In doing so, several physically meaningful requirements must be imposed, to ensure that the MTL model captures correctly the physical attributes of electromagnetic wave propagation in the multi-conductor waveguide.

The first requirement is that the two systems have the same dispersion characteristics. This is easily enforced by requiring that the propagation constants of the m eigenfields are also the propagation constants of the m voltage-current eigenpairs of the MTL. The second requirement is that, since the waveguide system is reciprocal, the generated MTL model is also reciprocal. As we saw above, this is ensured by enforcing the orthogonality relationship of (9.126).

It is evident from (9.126) and (9.128) that, given Λ and a process for calculating either Γ_V or Γ_I, a specific choice of the elements of the diagonal matrix P will result in the determination of Y and Z. As long as P is diagonal, the choice of its elements is understood as a particular orthonormalization of the voltage-current eigenpairs. Thus, for example, P may be chosen to be the $m \times m$ identity matrix. Such a choice results in the so-called *reciprocity-based* MTL model for the multi-conductor waveguide [33]. Alternatively, in view of the eigenfield orthogonality relation (9.101), one may calculate the diagonal elements of P in terms of the integrals,

$$p_i = \iint_S \hat{z} \cdot \left(\vec{E}_t^{(i)} \times \vec{H}_t^{(i)} \right) ds, \quad i = 1, 2, \ldots, m \tag{9.129}$$

where $\vec{E}_t^{(i)}$ and $\vec{H}_t^{(i)}$ are the transverse electric and magnetic fields for the ith mode over the cross-section of the MTL system.

As a review of the MTL literature reveals, a third requirement often imposed in the development of the MTL model is that of the conservation of complex power. The mathematical form of this requirement is

$$(\Gamma_V)_i^T (\Gamma_I)_j^* = \iint_S \hat{z} \cdot \left(\vec{E}_t^{(i)} \times \vec{H}_t^{(j)*} \right) ds, \quad i, j = 1, 2, \ldots, m \tag{9.130}$$

where $(\Gamma_V)_i$ denotes the ith column of Γ_V. In compact matrix form this expression becomes

$$\Gamma_V^T \Gamma_I^* = P_c. \tag{9.131}$$

It can be shown that P_c is diagonal only for the case of lossless, reciprocal waveguides [33]. Consequently, if the conservation of complex power (9.131) is used in place of (9.126) the matrices Y and Z of the MTL model will be symmetric only for lossless waveguides.

One way to address this difficulty is to require that conservation of complex power holds only for the individual modes of the MTL model and the multi-conductor waveguide. In other words, the following requirement is imposed [33]:

$$(\Gamma_V)_i^T (\Gamma_I)_i^* = \iint_S \hat{z} \cdot \left(\vec{E}_t^{(i)} \times \vec{H}_t^{(i)*} \right) ds = P_c^{(i)}, \quad i = 1, 2, \ldots, m. \tag{9.132}$$

This result, combined with (9.126), yields the following expressions for the elements of P:

$$p_i = \frac{P_c^{(i)}}{\sum_{j=1}^{m} (\Gamma_I)_{ji}^* (\Gamma_I^{-1})_{ij}}. \tag{9.133}$$

A similar relationship involving Γ_V can be developed by eliminating $(\Gamma_I)_i$ rather than $(\Gamma_V)_i$ from (9.126) and (9.132). In this manner, a reciprocal MTL model with some degree of complex power conservation can be constructed.

The final step in the development of the MTL model involves the calculation of the matrices Γ_V and Γ_I. If the MTL structure is such that the definition of voltages for each of the m conductors with respect to the reference conductor $m+1$ is possible in terms of line integrals on the cross-section of the MTL, then the voltage eigenvector matrix, Γ_V can be calculated using the integrals

$$(\Gamma_V)_{ij} = \int_{Q_i} \vec{E}_t^{(j)} \cdot \vec{dl}, \tag{9.134}$$

where Q_i is a directed path on the MTL cross-section from the cross-section of the ith conductor to the cross-section of the reference conductor as depicted in Fig. 9.11. Once Γ_V and P have been computed, Γ_I can be obtained from (9.126). Finally, (9.128) are utilized for the calculation of Y and Z.

The described process is suitable for MTLs for which the integrals in (9.134) are insensitive to the choice of the integration path Q_i. Clearly, this is the case for structures for which the electromagnetic fields for the fundamental set of the m eigenmodes are predominantly transverse with almost negligible longitudinal components. Microstrip- and stripline-type MTLs, for which the ground planes serve as the reference conductor, fall in the category of structures for which the aforementioned process is most suitable.

Alternatively, one can compute the current eigenvector matrix, Γ_I. The (i, j) element of Γ_I is the current flowing in the ith conductor for the jth eigenmode. Its calculation is through the integral

$$(\Gamma_I)_{ij} = \oint_{C_i} \vec{H}_t^{(j)} \cdot \vec{dl}, \tag{9.135}$$

where the closed path, C_i, encloses the cross-section of conductor i as depicted in Fig. 9.11. For example, C_i may be taken to be along the circumference of the cross-section of the ith conductor. The sense of the line integral is determined by the right-hand-screw rule, with the $+z$ direction taken as the direction of advance of the screw. With Γ_I calculated, Γ_V can be obtained from (9.126). Finally, (9.128) are utilized for the calculation of Y and Z.

The aforementioned process for the extraction of the R, L, C, G matrices for the MTL model is summarized in terms of the following algorithm.

Algorithm (9.3): RLCG extraction
```
1 Compute the first m propagating modes (propagation constants
  and transverse magnetic and electric fields) using the
  electromagnetic eigensolvers discussed in previous sections;
2 Build the propagation constant matrix Λ = diag(γ₁,···,γₘ);
3 Build the power matrix P = diag(p₁,···,pₘ) using either (9.129) or (9.133);
4 Build either the voltage eigenvector matrix Γᵥor the current
  eigenvector matrix Γᵢ;
5 Compute either Γᵢ = (Γᵥᵀ)⁻¹P or Γᵥ = (Γᵢᵀ)⁻¹P, depending on which is available
  from Step 4;
6 Compute Z and Y using (9.128);
7 Extract R and L from Z; Extract G and C from Y.
```

9.5.2 S-parameter representation of a section of an MTL

Once available, the per-unit-length matrices R, L, C and G can be used for the development of multiport network representations of the MTL, compatible with general-purpose, network analysis-oriented simulators. For transient simulators, such as the popular simulator SPICE and its derivatives, multiport representations in terms of distributed equivalent circuits are often used. A methodology for the direct synthesis of broadband, passive distributed, multiport circuit models for an MTL section, called *BestFit* has been described in [34] and [35]. For frequency-domain simulators, a multiport matrix representation of the MTL section in terms of its *scattering matrix* is used most commonly [36]. The development of the frequency-domain circuit scattering matrix, S, for an MTL section with per-unit-length admittance and impedance matrices Y and Z, respectively, and length l, is described next.

For the $m+1$-conductor MTL depicted in Fig. 9.11 and with the near- and far-end terminals of the reference conductor taken as reference, m near-end and m far-end ports are defined at $z = 0$ and $z = l$, respectively, with port voltages and port currents taken to be the $V_i(z = 0), I_i(z = 0), i = 1, 2, \ldots, m$ at the near end of the MTL and $V_i(z = l), I_i(z = l)$, $i = 1, 2, \ldots, m$ at the far end. Let $Z_{r1}^{(i)}, i = 1, 2, \ldots, m$, be reference impedances for the near-end ports, and $Z_{r2}^{(i)}, i = 1, 2, \ldots, m$, reference impedances for the far-end ports of the MTL. Then the scattering parameters, $a_1^{(i)}, b_1^{(i)}$, at the near-end port of the i-th conductor are defined in terms of the equations

$$
\begin{aligned}
V_i(z = 0) &= \sqrt{Z_{r1}^{(i)}} \left(a_1^{(i)} + b_1^{(i)} \right) \\
I_i(z = 0) &= \sqrt{Z_{r1}^{(i)}}^{-1} \left(a_1^{(i)} - b_1^{(i)} \right)
\end{aligned}
\tag{9.136}
$$

with similar expressions written for the remaining near-end and far-end ports. In matrix form it is, then,

$$
\begin{aligned}
V(z = 0) &= Z_{r1}^{1/2}(a_1 + b_1) \\
I(z = 0) &= Z_{r1}^{-1/2}(a_1 - b_1)
\end{aligned}
\tag{9.137}
$$

for the near-end ports, and

$$
\begin{aligned}
V(z = l) &= Z_{r2}^{1/2}(a_2 + b_2) \\
I(z = l) &= Z_{r2}^{-1/2}(a_2 - b_2)
\end{aligned}
\tag{9.138}
$$

for the far-end ports, where a_1, b_1 denote the vectors of length m with elements the corresponding scattering parameters for the near-end ports, while a_2, b_2 denote the vectors of length m with elements the corresponding scattering parameters for the far-end ports. The $m \times m$ diagonal matrices $Z_{r1}^{1/2}$ and $Z_{r2}^{1/2}$ have as elements the square roots of the port impedances at the near-end and far-end, respectively. A similar definition holds for the $m \times m$ diagonal matrices $Z_{r1}^{-1/2}$ and $Z_{r2}^{-1/2}$.

With these definitions, the $2m \times 2m$ scattering matrix of the MTL is defined in terms of the relation

$$
\begin{bmatrix} b_1 \\ b_2 \end{bmatrix} = \begin{bmatrix} S_{11} & S_{12} \\ S_{21} & S_{22} \end{bmatrix} \begin{bmatrix} a_1 \\ a_2 \end{bmatrix}.
\tag{9.139}
$$

In the following we derive expressions for the matrices S_{11}, S_{12}, S_{21} and S_{22} in terms of the propagation characteristics of the MTL.

To begin with we note that the vectors $V(z)$ and $I(z)$ can be expanded in terms of the voltage and current eigenvectors of the MTL as follows:

$$V(z) = \underbrace{\Gamma_V e^{-\Lambda z} A^+}_{V^+(z)} + \underbrace{\Gamma_V e^{\Lambda z} A^-}_{V^-(z)}, \tag{9.140}$$

where the vectors A^+ and A^- contain the coefficients in the expansion for the forward-propagating voltage wave vector $V^+(z)$ and the backward propagating voltage wave vector $V^-(z)$, respectively. These two vectors are determined by the boundary conditions at the two ends of the MTL. Substitution of (9.140) into the second equation of (9.113) yields the following decomposition of the current vector at z

$$I(z) = \underbrace{Z_c^{-1} V^+(z)}_{I^+(z)} - \underbrace{Z_c^{-1} V^-(z)}_{I^-(z)}, \tag{9.141}$$

where $I^+(z)$ is the forward-propagating current wave vector, $I^-(z)$ is the backward-propagating current wave vector, and the matrix Z_c, which is referred to as the *characteristic impedance matrix*, is given by

$$Z_c = \Gamma_V \Lambda^{-1} \Gamma_V^{-1} Z. \tag{9.142}$$

Application of (9.140) and (9.141) at the two ends of the MTL yields the following matrix relationships between the scattering-parameter vectors and the vectors A^+ and A^-:

$$Z_{r1}^{1/2}(a_1 + b_1) = \Gamma_V A^+ + \Gamma_V A^-$$
$$Z_{r1}^{-1/2}(a_1 - b_1) = Z_c^{-1} \Gamma_V A^+ - Z_c^{-1} \Gamma_V A^- \tag{9.143}$$

and

$$Z_{r2}^{1/2}(a_2 + b_2) = \Gamma_V e^{-\Lambda l} A^+ + \Gamma_V e^{\Lambda l} A^-$$
$$Z_{r2}^{-1/2}(a_2 - b_2) = Z_c^{-1} \Gamma_V e^{-\Lambda l} A^+ - Z_c^{-1} \Gamma_V e^{\Lambda l} A^-. \tag{9.144}$$

Elimination of the vectors A^+, A^- from (9.143) and (9.144) yields the following expressions:

$$\begin{bmatrix} b_1 \\ b_2 \end{bmatrix} = \begin{bmatrix} P_1 & Q_1 \\ P_2 e^{-\Lambda l} & Q_2 e^{\Lambda l} \end{bmatrix} \begin{bmatrix} Q_1 & P_1 \\ Q_2 e^{-\Lambda l} & P_2 e^{\Lambda l} \end{bmatrix}^{-1} \begin{bmatrix} a_1 \\ a_2 \end{bmatrix}, \tag{9.145}$$

where the following matrices have been defined

$$P_1 = Z_{r1}^{1/2} \left(Z_{r1}^{-1} - Z_c^{-1} \right) \Gamma_V, \quad Q_1 = Z_{r1}^{1/2} \left(Z_{r1}^{-1} + Z_c^{-1} \right) \Gamma_V,$$
$$P_2 = Z_{r2}^{1/2} \left(Z_{r2}^{-1} - Z_c^{-1} \right) \Gamma_V, \quad Q_2 = Z_{r2}^{1/2} \left(Z_{r2}^{-1} + Z_c^{-1} \right) \Gamma_V. \tag{9.146}$$

REFERENCES

1. S. Ahmed and P. Daly, "Finite element method for inhomogeneous waveguide," *IEE Proc.* vol. 116, pp. 1661-1664, Oct. 1969.

2. P. Daly, "Hybrid-mode analysis of microstrip by finite-element methods," *IEEE Trans. Microwave Theory Tech.*, vol. MTT-19, pp. 19-25, Jan. 1971.

3. Z. J. Cendes and P. Silvester, "Numerical solution of dielectric loaded waveguides: I. Finite element analysis," *IEEE Trans. Microwave Theory Tech.*, vol. 18, pp. 1124-1131, Dec. 1970.

4. C. Yel. K. Ha, S. B. Dong, and W. P. Brown, "Single-mode optical waveguides," *Appl. Opt.*, vol. 18, pp. 1490-1504, May 1979.

5. K. Oyamada and T. O. Okoshi, "Two-dimensional finite-element method calculation of propagation characteristics of axially nonsymmetrical optical fibers," *Radio Science*, vol 17, pp. 109-116, Jan.-Feb. 1982.

6. P. Savi, I. L. Gheorma, and R.D. Graglia, "Full-wave high-order FEM model for lossy anisotropic waveguides," *IEEE Trans. Microwave Theory Tech.*, vol. 50, pp. 495-500, Feb. 2002.

7. W. C. Chew and M. A. Nasi "A variational analysis of anisotropic, inhomogeneous dielectric waveguides," *IEEE Trans. Microwave Theory and Tech.,* vol. 37, pp. 661-668, Apr. 1989.

8. K. Radhakrishnan and W. C. Chew, "Full-wave analysis of multiconductor transmission lines on anisotropic inhomogeneous substrates," *IEEE Trans. Microwave Theory Tech.*, vol. 47, no. 9, Sep. 1999.

9. A. J. Kobelansky and J. P. Webb, "Eliminating spurious modes in finite-element waveguides problems by using divergence-free fields," *Electron. Lett.*, vol. 22, pp. 569-570, 1986.

10. J. P. Webb, "Finite element analysis of dispersion in waveguides with sharp metal edges," *IEEE Trans. Microwave Theory Tech.*, vol. 36, pp. 1719-1824, Dec. 1988

11. J. F. Lee, D. K. Sun, and Z. J. Cendes, "Full-wave analysis of dielectric waveguides using tangential vector finite elements," *IEEE Trans. Microwave Theory Tech.*, vol. 37, pp. 1262-1271, Aug. 1991.

12. J. F. Lee, "Finite element analysis of lossy dielectric waveguides," *IEEE Trans. Microwave Theory Tech.,* vol. 42, pp. 1025-1031, Jun. 1994.

13. S. V. Polstyanko and J. F. Lee, "$\mathcal{H}_1(curl)$ tangential vector finite element method for modeling anisotropic optical fibers," *J. Lightwave Technol.*, vol. 13, pp. 2290-2295, Nov. 1995.

14. S. Lee, J. Lee, and R. Lee, "Hierarchical vector finite elements for analyzing waveguiding structures," *IEEE Trans. Microwave Theory Tech.,* vol. 51, pp. 1897-1905, Aug. 2003.

15. T. Angkaes, M. Matsuhara, and N. Kumagai, "Finite-element analysis of waveguide modes: A novel approach that eliminates spurious modes," *IEEE Trans. Microwave Theory Tech.*, vol. 35, no. 2, pp. 117-125, Feb. 1987.

16. K. Hayata, K. Miura, and M. Koshiba, "Full vectorial finite element foralism for lossy anisotropic waveguides," *IEEE Trans. Microwave Theory Tech.*, vol. 37, pp. 875-883, May 1989.

17. I. Bardi and O. Biro, "An efficient finite-element forulation without spurious modes for anisotropic waveguides," *IEEE Trans. Microwave Theory Tech.*, vol. 37, pp. 1131-1139, Jul. 1991.

18. Y. Lu and F. A. Fernández, "An efficient finite element solution of inhomogeneous anisotropic and lossy dielectric waveguides," *IEEE Trans. Microwave Theory Tech.*, vol. 41, pp. 1215-1223, Jun.-Jul. 1993.

19. M. S. Alam, K. Hirayama, Y. Hayashi, and M. Koshiba, "A vector finite elemene tanalyhsis of complex modes in shielded microstrip lines," *Microwave and Optical Tech. Lett.*, vol. 6, no. 16, pp. 873-875, Dec. 1993.

20. B. M. Dillon and J. P. Webb, "A comparison of formulations for the vector finite element analysis of waveguides," *IEEE Trans. Microwave Theory Tech.*, vol. 41, pp. 308-316. Feb. 1994.

21. L. Valor and J. Zapata, "Efficient finite element analysis of waveguides with lossy inhomogeneous anisotropic materials characterized by arbitrary permittivity and permeability tensors," *IEEE Trans. Microwave Theory Tech.*, vol. 43, no. 10, pp. 2452-2552, Oct. 1995.

22. B. M. Dillon and J. P. Webb, "A comparison of formulations for the vector finite element analysis of waveguides," *IEEE Trans. Microwave Theory Tech.*, vol. 42, no. 2, pp. 308-316, Feb. 1994.

23. R. E. Collin, *Field Theory of Guided Waves*, 2nd. ed., Piscataway, NJ: IEEE Press, 1991.

24. A. D. Bresler, "Vector formulations for the field equations in anisotropic waveguides," *IRE Trans. Microwave Theory Tech.*, vol. 7, 1959.

25. A. C. Cangellaris, "Numerical error in finite element solution of electromagnetic boundary value problems," in *Finite Element Software for Microwave Engineering*, T. Itoh, G. Pelosi and P.P. Silvester, Eds., pp. 347 - 382, New York: John Wiley & Sons, 1996.

26. R. H. Jansen, "High-speed computation of single and coupled microstrip parameters including dispersion, high-order modes, loss, and finite strip thickness," *IEEE Trans. Microwave Theory Tech.*, vol. 26, pp. 75-82, Feb. 1978.

27. C. H. Chan, K. T. Ng, and A. B. Kouki, "A mixed spectral-domain approach for dispersion analysis of suspended planar transmission lines with pedestals," *IEEE Trans. Microwave Theory Tech.*, vol. 37, pp. 1716-1723, Nov. 1989.

28. K. Ogusu, "Numerical analysis of the rectangular dielectric waveguide and its modifications," *IEEE Trans. Magnetics*, vol. MAG-19, pp. 2551-2554, 1983.

29. M. S. Alam, M. Koshiba, K. Hirayama, and Y. Hayashi, "Hybrid-mode analysis of multilayered and multiconductor transmission lines," *IEEE Trans. Microwave Theory Tech.*, vol. 44, pp. 205-211, Feb. 1997.

30. K. C. Gupta, R. Garg, I. Bahl, and P. Bhartia, *Microstrip Lines and Slotlines*, 2nd edition, Norwood, MA: Artech House, 1996.

31. F. Di Paolo, *Networks and Devices Using Planar Transmissions Lines*, CRC Press, 2000.

32. C.R. Paul, *Analysis of Multiconductor Transmission Lines*, New York: Wiley-Interscience, 1994.

33. F. Olyslager, *Electromagnetic Waveguides and Transmission Lines*, Oxford: Clarendon Press, 1999.

34. T. V. Yioultsis and A. C. Cangellaris, "Optimal synthesis of finite difference models for multi-conductor transmission lines via Gaussian spectral rules," *Proc. 2002 Electronic Components and Technology Conference (ECTC)*, May 2002.

35. T. V. Yioultsis, A. Woo, and A. C. Cangellaris, "Passive synthesis of compact frequency-dependent interconnect models via quadrature spectral rules," *ICCAD-2003 Digest of Technical Papers*, pp. 827 - 834, Nov. 2003.

36. D. M. Pozar, *Microwave Engineering*, 2nd. ed., New York: John Wiley & Sons, Inc., 1998.

CHAPTER 10

THREE-DIMENSIONAL EIGENVALUE ANALYSIS OF RESONATORS

Superior geometry and material modeling versatility are the two primary reasons why the method of finite elements has become one of the most popular numerical methods for the eigen-analysis of complex three-dimensional electromagnetic cavities and resonators [1]-[17]. However, its application has been predominantly restricted to lossless electromagnetic systems, where the absence of loss yields an easier to handle, standard, linear matrix eigenvalue problems. On the other hand, lossy electromagnetic resonators, where loss is manifested either as material loss or as electromagnetic energy leakage due to the unbounded volume of the resonator, are more difficult to analyze numerically using finite elements. The reason for this is that the presence of loss leads to a nonlinear matrix generalized eigenvalue problems. To address this obstacle, methodologies based on the method of subspace iteration [10] and the transformation of the nonlinear eigenvalue problem into a linear one [12] have been proposed.

Our discussion of the development of finite element approximations of three-dimensional (3D) eigenvalue problems will be founded on the field-flux formulation of the electromagnetic system discussed in Sections 3.4.4. This formulation of the finite element approximation of the electromagnetic system was first proposed in [19] and [20] in the context of model order reduction of state-space representations of discrete approximations of electromagnetic systems. This topic will be discussed in detail in Chapter 11. At this point, it suffices to say that model order reduction is understood to mean the utilization of a set of the dominant eigenstates of the system for the approximation of the system response. Thus model order reduction and eigen-analysis of a system are closely related.

The field-flux formulation is based on the finite element discretization of the coupled system of Maxwell's curl equations; hence, both the electric field intensity, $\vec{E}$, and the magnetic flux density, $\vec{B}$, contribute to the degrees of freedom of the discrete problem. At first sight, this formulation is expected to be computationally more expensive than the popular electric-field formulation, where the vector wave equation for the electric field is discretized and hence, only the electric field contributes to the vector of unknowns. However, it will be shown that a special relationship between the tangentially continuous vector space used for the expansion of $\vec{E}$ and the normally continuous vector space used for the expansion of $\vec{B}$ can be exploited to render the computational complexity of the field-flux formulation commensurate to that of the conventional finite element formulation of the vector wave equation.

The chapter is organized as follows: First, the mathematical development of the field-flux formulation and the associated eigen-analysis algorithms are presented. Emphasis is placed on the development of effective means for the removal of spurious DC modes that are known to occur and hinder numerical convergence and algorithm robustness. This development is then followed by several example applications from the eigen-analysis of typical electromagnetic resonators.

10.1 FEM FORMULATION OF THE THREE-DIMENSIONAL ELECTROMAGNETIC EIGENVALUE PROBLEM

For our purposes, the computational domain, Ω, is assumed to be source-free and unbounded. Furthermore, the media may exhibit anisotropy, described in the mathematical statement of the problem in terms of the material tensors, $\bar{\mu}$, $\bar{\epsilon}$, and $\bar{\sigma}$. In the presence of ohmic loss in the media, the electric field satisfies the vector Helmholtz equation

$$\nabla \times \bar{\mu}^{-1} \cdot \nabla \times \vec{E} + j\omega\bar{\sigma} \cdot \vec{E} - \omega^2\bar{\epsilon} \cdot \vec{E} = 0. \tag{10.1}$$

The typical finite element approximation of (10.1) is effected through the expansion of $\vec{E}$ in terms of tangentially continuous vector basis functions, $\vec{w}_t$. A Galerkin process leads to the finite element approximation of the problem which, in matrix form, is written as follows:

$$(S + j\omega Z' - \omega^2 T)x_e = j\omega U, \tag{10.2}$$

where x_e is the vector containing the expansion coefficients of $\vec{E}$, and the elements of the matrices S, Z', and T are given by

$$S_{i,j} = \iiint_\Omega \nabla \times \vec{w}_{t,i} \cdot \bar{\mu}^{-1} \cdot \nabla \times \vec{w}_{t,j}\, dv, \quad T_{i,j} = \iiint_\Omega \vec{w}_{t,i} \cdot \bar{\epsilon} \cdot \vec{w}_{t,j}\, dv,$$

$$Z'_{i,j} = \iiint_\Omega \vec{w}_{t,i} \cdot \bar{\sigma} \cdot \vec{w}_{t,j}\, dv, \qquad\qquad U_{i,j} = \iint_{\partial\Omega} \vec{w}_{t,i} \cdot \hat{n} \times \vec{H}\, ds. \tag{10.3}$$

The term on the right-hand side of (10.2) is used for a truncation boundary condition to be imposed on a fictitious boundary, $\partial\Omega$, used to truncate the portion of the computational domain that, in reality, extends to infinity. For example, the following impedance boundary condition can be imposed:

$$\hat{n} \times \vec{H} = \frac{1}{\eta}\hat{n} \times \hat{n} \times \vec{E}, \tag{10.4}$$

where η denotes the intrinsic impedance of the medium in the unbounded region. In this case, (10.2) becomes

$$(S + sZ + s^2T)x_e = 0, \tag{10.5}$$

where $s = j\omega$, and the elements of the matrix Z are calculated as follows:

$$Z_{i,j} = Z'_{i,j} + \iint_{\partial\Omega} \hat{n} \times \vec{w}_{t,i} \cdot \frac{1}{\eta} \hat{n} \times \vec{w}_{t,j} ds. \tag{10.6}$$

For the case of lossless, bounded regions Z is zero and (10.3) reduces to a linear generalized eigenvalue problem that can be solved directly using the Lanzcos or Arnoldi algorithms discussed in Chapter 8. For the general case of a lossy and/or unbounded region, (10.5) is a nonlinear eigenvalue problem.

For the case of the field-flux formulation, both Maxwell's curl equations

$$\begin{cases} \nabla \times \vec{E} = -j\omega\vec{B} \\ \nabla \times \left(\bar{\mu}^{-1} \cdot \vec{B}\right) = j\omega\bar{\epsilon} \cdot \vec{E} + \bar{\sigma} \cdot \vec{E} \end{cases} \tag{10.7}$$

are discretized. Because of the boundary condition properties of electric field intensity and magnetic flux density at material interfaces, $\vec{E}$ is expanded in terms of tangentially continuous vector basis functions, $\vec{w}_t$, while $\vec{B}$ is expanded in terms of normally continuous vector basis functions, $\vec{w}_n$.

10.1.1 Finite element approximation: The case of symmetric material tensors

We will consider first the case where the material tensor is symmetric, a property that implies reciprocity within the domain of the electromagnetic resonator. To develop the finite element approximation of (10.7), first we multiply the first equation by $(\bar{\mu}^T)^{-1}$. With the expansions for $\vec{E}$ and $\vec{B}$ substituted in the resulting equation, the functions $\vec{w}_n$ are used for its testing. Similarly, the second equation in (10.7) is tested using the functions $\vec{w}_t$. The resulting finite element system assumes the form

$$\begin{bmatrix} 0 & D_1 \\ D_2 & -Z' \end{bmatrix} \begin{bmatrix} x_b \\ x_e \end{bmatrix} = -j\omega \begin{bmatrix} P & 0 \\ 0 & -T \end{bmatrix} \begin{bmatrix} x_b \\ x_e \end{bmatrix} - \begin{bmatrix} 0 \\ U \end{bmatrix}. \tag{10.8}$$

In the above equations x_e and x_b are the vectors containing the expansion coefficients for $\vec{E}$ and $\vec{B}$, respectively. Z' and T are the matrices appearing in (10.3). The elements of the new matrices D_1, D_2 and P are given by

$$D_{1,i,j} = \iiint_\Omega \vec{w}_{n,i} \cdot (\bar{\mu}^T)^{-1} \cdot \nabla \times \vec{w}_{t,j} dv = \iiint_\Omega \left(\bar{\mu}^{-1} \cdot \vec{w}_{n,i}\right) \cdot \nabla \times \vec{w}_{t,j} dv,$$

$$D_{2,i,j} = \iiint_\Omega \nabla \times \vec{w}_{t,i} \cdot \left(\bar{\mu}^{-1} \cdot \vec{w}_{n,j}\right) dv, \tag{10.9}$$

$$P_{i,j} = \iiint_\Omega \vec{w}_{n,i} \cdot (\bar{\mu}^T)^{-1} \cdot \vec{w}_{n,j} dv = \iiint_\Omega \left(\bar{\mu}^{-1} \cdot \vec{w}_{n,i}\right) \cdot \vec{w}_{n,j} dv.$$

In view of the assumed symmetry of the material tensors, it is $D_1 = D_2^T = D$. Also, P, Z' and T are symmetric. Imposing the impedance boundary condition (10.4) yields the final statement of the eigenvalue problem

$$\underbrace{\begin{bmatrix} 0 & D \\ D^T & -Z \end{bmatrix}}_{G} \begin{bmatrix} x_b \\ x_e \end{bmatrix} = -s \underbrace{\begin{bmatrix} P & 0 \\ 0 & -T \end{bmatrix}}_{C} \begin{bmatrix} x_b \\ x_e \end{bmatrix}. \tag{10.10}$$

Clearly, the resulting matrix eigenvalue problem is linear with respect to frequency; hence, this formulation is most suitable for the development of the finite element eigen-solution for the general case of lossy media.

Let us discuss, next, two useful properties of the eigenvectors of the field-flux formulation of (10.10).

Property I: For the case of reciprocal media, (10.10) is a generalized symmetric eigenvalue problem. Then, in view of Theorem I of Section 8.4.1, the two distinct eigenmodes $(s_i, x_{b,i}, x_{e,i})$ and $(s_j, x_{b,j}, x_{e,j})$ are **C**-orthogonal; hence, they satisfy the following orthogonality relation:

$$\begin{bmatrix} x_{b,i}^T & x_{e,i}^T \end{bmatrix} \begin{bmatrix} P & 0 \\ 0 & -T \end{bmatrix} \begin{bmatrix} x_{b,j} \\ x_{e,j} \end{bmatrix} = 0$$

$$\Rightarrow \iiint_\Omega \left(\bar{\mu}^{-1} \cdot \vec{B}_i \right) \cdot \vec{B}_j dv - \iiint_\Omega \vec{E}_i \cdot \bar{\epsilon} \cdot \vec{E}_j dv = 0, \quad (i \neq j), \tag{10.11}$$

where $(\vec{E}_i, \vec{B}_j)$ and $(\vec{E}_j, \vec{B}_j)$ are the fields corresponding to the two eigenmodes. ∎

Property II: If the material is reciprocal and lossless (hence, there is no surface impedance boundary), the matrix Z vanishes in (10.10). The matrix equation is then recast in the form

$$\underbrace{\begin{bmatrix} 0 & D \\ -D^T & 0 \end{bmatrix}}_{\hat{G}} \begin{bmatrix} x_b \\ x_e \end{bmatrix} = -s \underbrace{\begin{bmatrix} P & 0 \\ 0 & T \end{bmatrix}}_{\hat{C}} \begin{bmatrix} x_b \\ x_e \end{bmatrix}. \tag{10.12}$$

Then, in view of Theorem III in Section 8.4.1, for any two nonzero eigenvalues s_i and s_j, such that $s_i + s_j \neq 0$, their corresponding eigenvectors, $(x_{b,i}, x_{e,i})$ and $(x_{b,j}, x_{e,j})$ are $\hat{C}$-orthogonal,

$$\begin{bmatrix} x_{b,i}^T & x_{e,i}^T \end{bmatrix} \begin{bmatrix} P & 0 \\ 0 & T \end{bmatrix} \begin{bmatrix} x_{b,j} \\ x_{e,j} \end{bmatrix} = 0,$$

$$\Rightarrow \iiint_\Omega \bar{\mu}^{-1} \cdot \vec{B}_i \cdot \vec{B}_j dv + \iiint_\Omega \vec{E}_i \cdot \bar{\epsilon} \cdot \vec{E}_j dv = 0, \quad (s_i + s_j \neq 0). \tag{10.13}$$

Its addition and substraction with (10.11) (which, obviously, still holds in this case) yields

$$\begin{cases} \iiint_\Omega \vec{H}_i \cdot \vec{B}_j dv = 0 \\ \iiint_\Omega \vec{E}_i \cdot \vec{D}_j dv = 0 \end{cases}, \quad (i \neq j, \ s_i + s_j \neq 0), \tag{10.14}$$

which is the orthogonality relation for the lossless and reciprocal resonators. ∎

Let us now consider (10.8) without any constraints imposed on the terms associated with the matrix U. In other words, U may be used either for imposing a radiation (truncation) condition or as a forcing term for the electromagnetic problem. Multiplication on the left by $\begin{bmatrix} x_b^T & -x_e^T \end{bmatrix}$, followed by the interpretation of the matrix vector products in terms of surface and volume integrals of vector products of the field vectors, yields

$$\begin{bmatrix} x_b^T & -x_e^T \end{bmatrix} \left(\begin{bmatrix} 0 & D \\ D^T & -Z' \end{bmatrix} \begin{bmatrix} x_b \\ x_e \end{bmatrix} + j\omega \begin{bmatrix} P & 0 \\ 0 & -T \end{bmatrix} \begin{bmatrix} x_b \\ x_e \end{bmatrix} = -\begin{bmatrix} 0 \\ U \end{bmatrix} \right)$$

$$\Rightarrow \iint_{\partial\Omega} \vec{E} \times \vec{H} \cdot d\vec{s} + \iiint_\Omega \vec{E} \cdot \bar{\sigma} \cdot \vec{E} dv = -j\omega \iiint_\Omega \left(\vec{H} \cdot \vec{B} + \vec{E} \cdot \vec{D} \right) dv \tag{10.15}$$

or, reverting back to a time-dependent form,

$$\iint_{\partial\Omega} \vec{E} \times \vec{H} \cdot d\vec{s} + \iiint_{\Omega} \vec{E} \cdot \bar{\sigma} \cdot \vec{E} dv = -\iiint_{\Omega} \left(\vec{H} \cdot \frac{\partial \vec{B}}{\partial t} + \vec{E} \cdot \frac{\partial \vec{D}}{\partial t} \right) dv. \quad (10.16)$$

This is immediately recognized as a conservation of energy equation statement for the electromagnetic field, with the surface integral of the Poynting vector, $\vec{E} \times \vec{H}$, representing electromagnetic power flow through the domain boundary. Recalling the expression for the total energy density in the field

$$u = \frac{1}{2} \left(\vec{E} \cdot \vec{D} + \vec{B} \cdot \vec{H} \right) \quad (10.17)$$

the volume integral on the right is interpreted as the time rate of change of the electromagnetic energy inside the volume Ω. Finally, the volume integral on the left represents energy dissipation due to ohmic loss.

10.1.2 Finite element approximation: The case of Hermitian material tensors

In this case, the development of the finite element approximation begins with the multiplication of the first curl equation in (10.7) by $(\bar{\mu}^H)^{-1}$. The resulting equation is then tested by the expansion functions $\vec{w}_n$ in the approximation of $\vec{B}$. Similarly, the second curl equation is tested by the expansion functions $\vec{w}_t$, used for the approximation of $\vec{E}$. The resulting system is of the form of (10.8) where now the elements of the matrices D_1 and P are given by

$$D_{1,i,j} = \iiint_{\Omega} \vec{w}_{n,i} \cdot (\bar{\mu}^H)^{-1} \cdot \nabla \times \vec{w}_{t,j} dv = \iiint_{\Omega} ((\bar{\mu}^*)^{-1} \cdot \vec{w}_{n,i}) \cdot \nabla \times \vec{w}_{t,j} dv,$$

$$P_{i,j} = \iiint_{\Omega} \vec{w}_{n,i} \cdot (\bar{\mu}^H)^{-1} \cdot \vec{w}_{n,j} dv = \iiint_{\Omega} ((\bar{\mu}^*)^{-1} \cdot \vec{w}_{n,i}) \cdot \vec{w}_{n,j} dv,$$

$$(10.18)$$

where the superscript $*$ denotes complex conjugation. As far as the matrix D_2 is concerned, it is $D_1 = D_2^H = D$. Since the material tensors are Hermitian, i.e., $\bar{\mu} = \bar{\mu}^H$, $\bar{\epsilon} = \bar{\epsilon}^H$, and $\bar{\sigma} = \bar{\sigma}^H$, P, Z' and T are Hermitian. Finally, with the impedance boundary condition (10.4) imposed on the truncation boundary, the following generalized Hermitian eigenvalue problem results:

$$\underbrace{\begin{bmatrix} 0 & D \\ D^H & -Z \end{bmatrix}}_{\mathbf{G}} \begin{bmatrix} x_b \\ x_e \end{bmatrix} = -s \underbrace{\begin{bmatrix} P & 0 \\ 0 & -T \end{bmatrix}}_{\mathbf{C}} \begin{bmatrix} x_b \\ x_e \end{bmatrix}. \quad (10.19)$$

It is noted that for both the symmetric and the Hermitian matrix eigenvalue problem statements of (10.10) and (10.19), the following compact form will be used occasionally

$$(\mathbf{G} + s\mathbf{C})x = 0, \quad (10.20)$$

where

$$\mathbf{G} = \begin{bmatrix} 0 & D \\ D^T & -Z \end{bmatrix} \text{ or } \begin{bmatrix} 0 & D \\ D^H & -Z \end{bmatrix}, \quad \mathbf{C} = \begin{bmatrix} P & 0 \\ 0 & -T \end{bmatrix}, \quad x = \begin{bmatrix} x_b \\ x_e \end{bmatrix}. \quad (10.21)$$

Two useful properties of the eigenmodes of the Hermitian eiegenvalue problem are considered next.

Property I: For the case of a medium with Hermitian material tensors, (10.19) is a generalized Hermitian eigenvalue problem. Then, in view of Theorem II in Section 8.4.1, any two distinct eigenmodes $(s_i, x_{b,i}, x_{e,i})$ and $(s_j, x_{b,j}, x_{e,j})$ with eigenvalues $s_i \neq s_j^*$, satisfy the following C-orthogonality relation:

$$\begin{bmatrix} x_{b,i}^H & x_{e,i}^H \end{bmatrix} \begin{bmatrix} P & 0 \\ 0 & -T \end{bmatrix} \begin{bmatrix} x_{b,j} \\ x_{e,j} \end{bmatrix} = 0$$

$$\Rightarrow \iiint_\Omega \left(\bar{\mu}^{-1} \cdot \vec{B}_i \right)^* \cdot \vec{B}_j dv - \iiint_\Omega \vec{E}_i^* \cdot \epsilon \cdot \vec{E}_j dv = 0, \quad (s_i \neq s_j^*). \tag{10.22}$$

where, once again, the matrix-vector products in the finite element statement of the problem are replaced with the corresponding volume and surface integrals of vector products of the electromagnetic fields. ∎

Property II: In the absence of loss (hence, in the absence of a surface impedance boundary) and under the assumption of Hermitian material tensors, the matrix Z vanishes in (10.19). Thus the matrix eigenvalue statement may be cast in the following form:

$$\underbrace{\begin{bmatrix} 0 & D \\ -D^H & 0 \end{bmatrix}}_{\hat{G}} \begin{bmatrix} x_b \\ x_e \end{bmatrix} = -s \underbrace{\begin{bmatrix} P & 0 \\ 0 & T \end{bmatrix}}_{\hat{C}} \begin{bmatrix} x_b \\ x_e \end{bmatrix}. \tag{10.23}$$

Then, in view of Theorem IV of Section 8.4.1, we conclude that for any two nonzero eigenvalues s_i and s_j such that $s_i + s_j^* \neq 0$, their corresponding eigenvectors $(x_{b,i}, x_{e,i})$ and $(x_{b,j}, x_{e,j})$ are conjugate $\hat{C}$-orthogonal

$$\begin{bmatrix} x_{b,i}^H & x_{e,i}^H \end{bmatrix} \begin{bmatrix} P & 0 \\ 0 & T \end{bmatrix} \begin{bmatrix} x_{b,j} \\ x_{e,j} \end{bmatrix} = 0,$$

$$\Rightarrow \iiint_\Omega \left(\bar{\mu}^{-1} \cdot \vec{B}_i \right)^* \cdot \vec{B}_j dv + \iiint_\Omega \vec{E}_i^* \cdot \bar{\epsilon} \cdot \vec{E}_j dv = 0, \quad (s_i^* + s_j \neq 0). \tag{10.24}$$

Since in this case the eigenmodes of the system also satisfy Property I above, addition and substraction of this last result with (10.22) yields

$$\begin{cases} \iiint_\Omega \vec{H}_i^* \cdot \vec{B}_j dv = 0 \\ \iiint_\Omega \vec{E}_i^* \cdot \vec{D}_j dv = 0 \end{cases}, \quad (s_i - s_j^* \neq 0, \ s_i + s_j^* \neq 0) \tag{10.25}$$

which is the orthogonality relation for eigenmodes of lossless electromagnetic resonators with media characterized by Hermitian material tensors.

For the special case where the material tensors $\bar{\epsilon}$ and $\bar{\mu}$ are positive definite, the matrices P and T are positive definite also. Furthermore, for the case of lossless media, $\bar{\epsilon}$ and $\bar{\mu}$ are real, and the eigen-matrix equation is of the form of (10.12) and (10.23). Consequently, all eigenvalues are purely imaginary, occurring in complex-conjugate pairs. ∎

Let us, next, consider the development of a conservation of energy relationship for the electromagnetic fields in an electromagnetic resonator for the case of media with Hermitian material tensors. Starting with (10.8), the multiplication of the first equation by x_b^H yields

$$x_b^H \left(Dx_e + j\omega P x_b \right) = 0$$
$$\Rightarrow \iiint_\Omega \vec{H}^* \cdot \nabla \times \vec{E} dv + j\omega \iiint_\Omega \vec{H}^* \cdot \vec{B} dv = 0. \tag{10.26}$$

Multiplication of the second equation in (10.8) by x_e^H yields

$$x_e^H \left(D^H x_b - Z' x_e - j\omega T x_e \right) = -x_e^H U \quad \Rightarrow$$
$$\iiint_\Omega \nabla \times \vec{E}^* \cdot \vec{H} dv - \iiint_\Omega \vec{E}^* \cdot \bar{\sigma} \cdot \vec{E} dv - j\omega \iiint_\Omega \vec{E}^* \cdot \vec{D} dv = -\iint_{\partial \Omega} \vec{E}^* \cdot \hat{n} \times \vec{H} ds. \tag{10.27}$$

Subtracting the complex conjugate of (10.27) from (10.26) yields

$$\frac{1}{2} \iint_{\partial\Omega} \vec{E} \times \vec{H}^* \cdot d\vec{s} + \frac{1}{2} \iiint_\Omega \vec{E} \cdot \bar{\sigma}^* \cdot \vec{E}^* dv = -j\omega \frac{1}{2} \iiint_\Omega \left(\vec{H}^* \cdot \vec{B} - \vec{E} \cdot \vec{D}^* \right) dv, \tag{10.28}$$

where the factor $1/2$ was introduced to make the result consistent with the commonly used statement of electromagnetic energy conservation in conjunction with linear time-harmonic electromagnetic fields [21]. The vector $(1/2)\vec{E} \times \vec{H}^*$ is the complex Poynting vector, while the terms $(1/2)\vec{E} \cdot \vec{D}^*$ and $(1/2)\vec{H} \cdot \vec{B}^*$ are associated, respectively, with the electric and magnetic energy densities. Finally, the term $(1/2)\vec{E} \cdot \bar{\sigma}^* \cdot \vec{E}^*$ describes the per-unit-volume energy dissipation due to material loss.

In the absence of loss, the left-hand sides of (10.15) and (10.28) are zero; hence, the energy conservation statement for the case of an ideal resonator is obtained, written for the two cases considered as follows:

$$\iiint_\Omega \vec{H} \cdot \vec{B} dv = -\iiint_\Omega \vec{E} \cdot \vec{D} dv, \quad \iiint_\Omega \vec{H}^* \cdot \vec{B} dv = \iiint_\Omega \vec{E} \cdot \vec{D}^* dv. \tag{10.29}$$

The electromagnetic energy conservation statements (10.15) and (10.28) can be derived directly from Maxwell's equations [21], [22]. As our development illustrates, the finite element approximations derived using the field-flux formulation satisfy discrete energy conservation statements consistent with those satisfied by the continuous system. This is important for several reasons. Conservation of energy in the discrete system, especially for the case of modeling of a linear, passive electromagnetic device or system, is important for ensuring the passivity of the discrete system. This issue is deliberated further in Chapter 11 in the context of reduced-order modeling of discrete electromagnetic systems.

For the purposes of this chapter, the consistency of the discrete statement of electromagnetic energy conservation with the analytic one implies that the matrix-vector products in (10.15) and (10.28), associated with the calculation of stored and dissipated energy, can be used with confidence in the calculation of the quality factor, Q, of the electromagnetic resonator, defined as

$$Q = \omega_r \frac{\text{stored energy at resonance}}{\text{power loss at resonance}}, \tag{10.30}$$

where ω_r is the angular resonant frequency. At this point, it is appropriate to comment on the close similarity between the discrete system resulting from the finite element approximation of the electromagnetic system and the circuit equations describing a simple parallel lumped circuit resonator. This similarity is most useful since it allows us to extend compact results for the calculation of Q and the bandwidth of a resonator, which are readily derived for lumped circuit resonators, to various classes of three-dimensional electromagnetic resonators, both physically occurring [21], and man-made for microwave and optical applications [22], [23].

10.1.3 Lumped parallel resonant circuit

Consider the parallel resonant circuit shown in Fig.10.1. Using the inductor current, $i(t)$, and the voltage, $v(t)$, across each one of the lumped elements as the state variables of the circuit the relevant equations are written in matrix form as follows:

$$\begin{cases} l\dfrac{di}{dt} - v = 0 \\ i + gv + c\dfrac{dv}{dt} = -i_0 \end{cases} \Rightarrow \begin{bmatrix} & -1 \\ -1 & -g \end{bmatrix} \begin{bmatrix} I \\ V \end{bmatrix} = -j\omega \begin{bmatrix} l & \\ & -c \end{bmatrix} \begin{bmatrix} I \\ V \end{bmatrix} + \begin{bmatrix} 0 \\ I_0 \end{bmatrix}, \quad (10.31)$$

where upper-case symbols, V, I, are used for the complex phasor form of the voltage and current, respectively, for the special case of time-harmonic variation $\exp(j\omega t)$. A direct comparison of the above phasor form of the system with (10.8) reveals the formal similarity between the state-space form of circuit equations with nodal voltages and inductor currents as state variables with the field-flux formulation-based finite element approximation of the electromagnetic system with state variables the coefficients in the expansion of the electric field and the magnetic flux density.

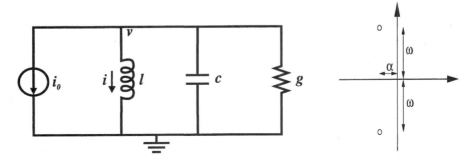

Figure 10.1   Lumped parallel resonant circuit.   **Figure 10.2**   Poles.

The equation for the complex power in the circuit is written in terms of the phasors V, I and I_0 in the following manner:

$$-\frac{1}{2}VI_0^* = \frac{1}{2}\left(g - j\omega c + j\frac{1}{\omega l}\right)VV^* \Rightarrow$$

$$\frac{1}{2}VI_0^* + \frac{1}{2}gVV^* = j\omega\frac{1}{2}\left(cVV^* - lII^*\right). \quad (10.32)$$

In the above expression the terms $(1/2)VI_0^*$ and $(1/2)gVV^*$ are identified, respectively, as the complex power provided by the source and the time-average power dissipated at the conductance g. The terms $(1/2)cVV^*$ and $(1/2)lII^*$ are associated with energy storage in the capacitor and inductor, respectively. In particular, for the case of time-harmonic excitation, the quantities

$$P_g = \frac{1}{2}gVV^*, \quad W_c = \frac{1}{4}cVV^*, \quad W_l = \frac{1}{4}lII^* = \frac{1}{4}\frac{VV^*}{\omega^2 l} \tag{10.33}$$

are often defined. Next, let us consider the following ratio:

$$\omega\frac{\text{time-average stored energy}}{\text{energy loss in one period}} = \omega\frac{(1/4)cVV^* + (1/4)(\omega^2 l)^{-1}VV^*}{(1/2)gVV^*}$$

$$= \frac{\omega c + (\omega l)^{-1}}{2g}. \tag{10.34}$$

It is straightforward to show that this ratio assumes its maximum value when $\omega = \omega_r = \frac{1}{\sqrt{lc}}$. The frequency ω_r is recognized as the resonant frequency of the parallel circuit. This maximum value is called the quality factor, Q, of the resonator. This explains the definition (10.30) for the quality factor of any electromagnetic resonator. For the case of the parallel resonant circuit the quality factor is given by

$$Q = \frac{\omega_r c}{g} = \frac{1}{g}\sqrt{\frac{c}{l}}. \tag{10.35}$$

The voltage response of the resonator is given by

$$V = -\frac{1}{c}\frac{s}{s^2 + sg/c + \omega_r^2}I_0, \tag{10.36}$$

where $s = j\omega$. The poles of the circuit are readily found to be

$$s_0 = -\frac{g}{2c} \pm j\omega_r\sqrt{1 - \left(\frac{g}{2\omega_c c}\right)^2}$$

$$= -\frac{g}{2c} \pm j\omega_r\sqrt{1 - \left(\frac{1}{2Q}\right)^2}. \tag{10.37}$$

For $g > 0$ both poles lie on the left-hand side of the complex frequency plane. For a good resonator $Q \gg 1$. Thus the two poles are approximated as

$$s_0 \approx -\frac{g}{2c} \pm j\omega_r = -\alpha \pm j\omega_r. \tag{10.38}$$

In view of this result and (10.35), Q may be approximated as follows:

$$Q \approx \frac{\omega_r}{2\alpha}. \tag{10.39}$$

This is a very useful result since, for the case of low-loss resonators, it provides for an expedient means of calculating Q for a given eigenmode in terms of the real and imaginary parts of the associated eigenvalue.

10.2 EIGENSOLVER FOR LOSSLESS MEDIA

For the case of lossless media (10.5) assumes the simple form $Sx_e = \omega^2 Tx_e$. In this case Krylov subspace eigenvalue algorithms such as Lanczos or Arnoldi successfully and accurately capture the first few dominant eigenvalues well before the generation of the entire Krylov subspace. However, as already discussed in Chapter 8, the dominant eigenvalues calculated by these algorithms are usually the ones with the largest magnitude (i.e., the highest eigenfrequencies). What is of interest, instead, are the eigenfrequencies of the lower eigenmodes, since these are the ones most commonly employed in practical applications of electromagnetic resonators. Therefore, the following spectral transformation [24] is used:

$$Sx_e = \omega^2 T x_e \;\Rightarrow\; \delta(S - \omega_0^2 T)x_e = \omega_0^2 T x_e, \quad \delta = \frac{\omega_0^2}{\omega^2 - \omega_0^2} \tag{10.40}$$

where ω_0 is chosen to be of value close to the value of the desired eigenfrequencies. Clearly, this transformation makes the desired eigenvalues to be among the dominant ones, while the corresponding eigenvectors remain unchanged. For reciprocal resonators, (10.40) is a symmetric generalized eigenvalue problem and can be solved by generalized Lanczos Algorithm (8.10).

For the case of non-reciprocal media, the matrices S and T are not symmetric. Thus the following spectral transformation is used instead:

$$Sx_e = \omega^2 T x_e \;\Rightarrow\; \delta x_e = \omega_0^2 (S - \omega_0^2 T)^{-1} T x_e, \quad \delta = \frac{\omega_0^2}{\omega^2 - \omega_0^2}. \tag{10.41}$$

This eigenvalue problem is then solved using the non-symmetric eigenvalue Arnoldi Algorithm (8.8).

In both cases, the success of the Lanczos and Arnoldi processes is strongly dependent on the elimination of spurious DC modes and, when multiple modes are solved for simultaneously, on the on-the-fly extraction of the converged modes (deflation) during the iterative solution process. Both of these issues are discussed in detail in the nest two sections.

10.2.1   Elimination of spurious DC modes

The presence of spurious DC modes in the finite element approximation of the electromagnetic system has been discussed in detail in Chapter 3. For the approximations pertinent to this discussion it suffices to recall that, even though the use of a tangentially continuous vector space instead of a scalar space for the expansion of $\vec{E}$ can eliminate spurious modes with non-zero eigenfrequencies, the presence of spurious DC modes may not be avoided [3]. This is because the tangentially continuous vector space, W_t, contains the gradient of the scalar space Φ. Thus, using the notation adopted in Chapter 2, there exists a gradient matrix G_r such that

$$\nabla \Phi = \vec{W}_t G_r \;\rightarrow\; \left[\nabla \phi_1, \cdots, \nabla \phi_{N_s}\right] = \left[w_{t,1}, \cdots, w_{t,N_t}\right] G_{r\,N_t \times N_s}. \tag{10.42}$$

The gradient matrix G_r is defined in Sections 2.4 and 2.7. From the definition of S and D and their bilinear forms it is straightforward to deduce that

$$G_r^T S = 0, \quad S G_r = 0. \tag{10.43}$$

In view of the above equations and the matrix eigenvalue statement (10.5) for the case of lossless media (i.e., when $Z = 0$), it is immediately evident that G_r contains all the DC modes (i.e., modes with eigenfrequency equal to 0). Unless these DC modes are eliminated, the Lanczos/Arnoldi process produces a large amount of eigenvalues of value close to zero or, more precisely (considering the finite precision of the numerical solution), of value several orders of magnitude less than the desired ones. The presence of these DC modes slows down convergence and wastes memory during the development of the Krylov subspace.

Since the DC modes represented by the column vectors in G_r do not satisfy the divergence-free condition $\nabla \cdot \vec{D} = 0$ at $\omega = 0$, they can be eliminated through the imposition of the divergence-free requirement on every generated Lanczos/Arnoldi vector. This is done using the technique suggested in [3] and [20]. The basic idea is to add to the electric field vector x_e an extra vector $G_r\psi$, where ψ is a vector to be determined, such that the spurious DC modes are canceled. The new vector satisfies the divergence-free condition at DC. The requirement $\nabla \cdot \vec{D} = 0$ must be imposed in the weak sense. Thus, with the new vector written as

$$x_e \leftarrow x_e + G_r\psi \tag{10.44}$$

and in view of (10.42) and the integral form of the elements of T given in (10.3), it is straightforward to show that the matrix statement of the enforcement of the divergence free condition yields the following equation for the vector ψ:

$$M\psi = -G_r^T T x_e. \tag{10.45}$$

In the above equation M is the finite element matrix for the associated electrostatic problem with elements given by

$$M_{i,j} = \iiint_\Omega \nabla\phi_i \cdot \epsilon \nabla\phi_j \, dv \tag{10.46}$$

with ϕ_i and ϕ_j scalar basis functions. Once ψ is found, (10.44) is used to eliminate the spurious DC modes.

To elaborate, use of (10.42) yields the following relationship between matrices M and T as

$$M = G_r^T T G_r. \tag{10.47}$$

Therefore, the elimination of spurious modes can be summarized formally in terms of the following statement:

$$x_e \leftarrow (I - G_r(\underbrace{G_r^T T G_r}_{M})^{-1}G_r^T T)x_e. \tag{10.48}$$

However, it should be noted that, in practice, the matrix M is not constructed from T; rather it is built using (10.46).

In the ideal case of no roundoff error, the divergence-free condition need be imposed only on the initial vectors. For example, such a divergence-free initial vector can be chosen to be the electric field induced by a loop current source. However, due to roundoff error, the selection of divergence-free initial vectors alone is not sufficient to guarantee the absence of the spurious DC modes during the Lanczos/Arnoldi eigensolution process. Therefore, the technique described above for their elimination during the solution process is necessary.

As a final remark, let us note that the error due to the presence of these modes is most severe when the shift frequency ω_0 assumes low values.

10.2.2   Extraction of multiple modes

When several of the dominant modes of an electromagnetic cavity must be computed, the deflation technique discussed in [3] can be used for the extraction of the modes one at a time.

Deflation takes advantage of the orthogonality of the eigenmodes in (10.14) for the on-the-fly removal of the converged eigenvectors during the iterative eigensolution process. For example, for the case of the Lanczos algorithm, this is effected by means of the following operation

$$x_e \quad \leftarrow \quad \left(I - VV^T T\right) x_e \tag{10.49}$$

where I is the identity matrix and the column vectors in V are the converged eigenvectors, which are T-normalized, that is, $V^T T V = I$. It is straightforward to show that the new vector x_e updated according to (10.49), is orthogonal to the converged eigenvectors, that is, $V^T T x_e = 0$. Theoretically, this operation needs to be done once only, for the initial Lanczos vector. However, because of roundoff error, its application is recommended for every new Lanczos vector generated at each iteration.

A comparison of (10.48) with (10.49) reveals the difference between the process of elimination of spurious DC modes and extraction of the converged modes. The spurious DC modes stored in G_r can be viewed as the a-priori known, converged modes. However because they are not T-normalized, the inverse of M is needed for their extraction. On the other hand, such an operation is not needed for the extraction of multiple modes since the modes are T-normalized during the iteration.

For the case of lossless media, either the Lanczos or the Arnoldi process can be applied for the solution of the matrix eigenvalue problem obtained through the $\vec{E}$-formulation. The cost of the numerical computation is dominated by the solution of the matrix equation $(S - \omega_0^2 T)$ at each iteration. Depending on the size of the problem, the solution is obtained either through LU factorization or through the application of an iterative solver. Pseudo code descriptions of the resulting algorithms are given below.

Algorithm (10.1): Deflated Lanczos algorithm for n eigenmodes of a lossless cavity.
1 $V_0 \leftarrow 0$;
2 for $k = 1, 2, \cdots, n$
  2.a $r_0 = (I - V_{k-1} V_{k-1}^T T) r_0$;
  2.b Remove spurious DC modes: $r_0 = (I - G_r M^{-1} G_r^T T) r_0$;
  2.c $q_0 = 0$, $\beta_1 = \|r_0\|$, $q_1 = r_0/\beta_1$;
  2.d for $m = 1, 2, \cdots, M$
    2.d.1 $w \leftarrow (I - T V_{k-1} V_{k-1}^T) \omega_0^2 T q_m$;
    2.d.2 $\alpha_m \leftarrow \langle q_m, w \rangle = q_m^T w$;
    2.d.3 $\tilde{q}_{m+1} \leftarrow (S - \omega_0^2 T)^{-1} w - \beta_m q_{m-1} - \alpha_m q_m$;
    2.d.4 Remove spurious DC modes: $\tilde{q}_{m+1} = (I - G_r M^{-1} G_r^T T) \tilde{q}_{m+1}$;
    2.d.5 Compute the eigenpairs $(\tilde{\lambda}_i, y_i)$ of T_m, $(i = 1, \cdots, m)$;
    2.d.6 Compute $\rho_i = \|e_m^T y_i\| \|\tilde{q}_{m+1}\| / \|\tilde{\lambda}_i Q_m y_i\|$, $(i = 1, \cdots, m)$;
    2.d.7 If one eigenpair $(\tilde{\lambda}_i, \tilde{v}_i = Q_m y_i)$ converges, goto 2.f;
    2.d.8 $\beta_{m+1} = \sqrt{\tilde{q}_{m+1}^T w}$, $q_{m+1} = \tilde{q}_{m+1}/\beta_{m+1}$;
  2.e $r_0 \leftarrow \tilde{v}_i$ and goto 2.c;
  2.f T-Normalization: $v_i = \tilde{v}_i / \sqrt{\tilde{v}_i^T T \tilde{v}_i}$;
  2.g $V_k \leftarrow [V_{k-1}, v_i]$.

Steps (2.b) and (2.d.4) ensure the Lanczos vectors q_m are free of spurious DC modes in G_r. Steps (2.a) and (2.d.1) render the next Lanczos vector T-orthogonal to the converged eigenvectors in V_k.

The Lanczos algorithm has the advantage of efficient memory usage due to its short recurrence. On the other hand, the more expensive Arnoldi process has the advantage that can prevent loss of orthogonality. The extra cost comes from the complete orthonormalization of all Arnoldi vectors at each iteration step. However, this cost may be affordable in the case where a small number of dominant eigenvectors is of interest and the algorithm converges fast.

Algorithm (10.2): Arnoldi algorithm for n eigenmodes of a lossless cavity.

```
1  V₀ ← 0;
2  for k = 1, 2, ⋯ , n
```
    2.a Deflate converged eigenmodes: $r_{0,b} \leftarrow (I - V_{k-1}V_{k-1}^T T)r_0$;

    2.b Remove Spurious Modes: $r_0 \leftarrow (I - G_r M^{-1} G_r^T T)r_0$;

    2.c $q_1 \leftarrow r_0/\|r_0\|$;

    2.d for $m = 1, 2, \cdots, M$

        2.d.1 $\tilde{q}_{m+1} \leftarrow \omega_0^2(S - \omega_0^2 T)^{-1}T q_m$;

        2.d.2 Deflate converged eigenmodes: $\tilde{q}_{m+1} \leftarrow (I - V_{k-1}V_{k-1}^T T)\tilde{q}_{m+1}$;

        2.d.3 Remove Spurious Modes: $\tilde{q}_{m+1} \leftarrow (I - G_r M^{-1} G_r^T T)\tilde{q}_{m+1}$;

        2.d.4 for $i = 1, \cdots, m$ (Normalize with the Arnoldi vectors)

            2.d.4.a $H_m(i, m) \leftarrow \langle q_i, \tilde{q}_{m+1}\rangle = q_i^T \tilde{q}_{m+1}$;

            2.d.4.b $\tilde{q}_{m+1} \leftarrow \tilde{q}_{m+1} - H_m(i, m)q_i$;

        2.d.5 Compute m eigenpairs (λ_i, y_i) of H_m $(i = 1, \cdots, m)$;

        2.d.6 Compute $\rho_i = \|e_m^T y_i\| \|\tilde{q}_{m+1}\| / \|\lambda_i Q_m y_i\|$, $(i = 1, \cdots, m)$;

        2.d.7 If one eigenpair $(\lambda_i, \tilde{v}_i = Q_m y)$ converges, then goto 2.f;

        2.d.8 $H_m(m + 1, m) = \|\tilde{q}_{m+1}\|$;

        2.c.9 $q_{m+1} = q_{m+1}/H_m(m + 1, m)$;

    2.e $r_0 \leftarrow \tilde{v}_i$ and goto 2.c;

    2.f T-Normalization: $v_i \leftarrow \tilde{v}_i / \sqrt{\tilde{v}_i^T T \tilde{v}_i}$;

    2.g $V_k \leftarrow [V_{k-1}, v_i]$.

The above algorithms are only valid for generalized symmetric eigenvalue problems, resulting when the media inside the resonators are reciprocal. For the case of non-reciprocal media, the matrices S and T are not symmetric. Thus the eigenvectors are not T-orthogonal any more. For this case the formulation of (10.41) is the appropriate one to use, and the deflated Arnoldi Algorithm (8.8) for non-symmetric eigenvalue problems must be used instead.

10.3 EIGENSOLVER FOR LOSSY MEDIA

In the presence of loss, either in the form of media loss or a surface complex impedance with a positive real part, the field-flux formulation (10.10) is most suitable. Once again, a spectral transformation (frequency shift) is used to improve convergence. Hence, the matrix statement of the discrete eigenvalue problem becomes

$$(\mathbf{G} + s\mathbf{C})x = 0 \;\Rightarrow\; \delta(\mathbf{G} + s_0\mathbf{C})x = s_0\mathbf{C}x, \quad \delta = \frac{s_0}{s_0 - s} \tag{10.50}$$

where $s_0 = j\omega_0$ is the complex frequency shift, chosen to close to the anticipated eigenvalues of the eigenmodes of interest.

For the case of reciprocal media, (10.50) is a symmetric, generalized eigenvalue problem which can be solved using the generalized Lanczos Algorithm (8.10). If, instead, the Arnoldi algorithm is to be used, the frequency-shifted matrix eigen-problem is cast in the following form:

$$(\mathbf{G} + s\mathbf{C})x = 0 \;\Rightarrow\; \delta x = s_0(\mathbf{G} + s_0\mathbf{C})^{-1}\mathbf{C}x, \quad \delta = \frac{s_0}{s_0 - s}. \tag{10.51}$$

For both algorithms the inverse of $\mathbf{G} + s_0\mathbf{C}$ must be calculated by either a direct or an iterative method. Compared to the matrix obtained from the E-field formulation, the dimension of the finite element matrix in this case is almost doubled, since both the electric field and the magnetic flux density are solved for; hence, at first appearance, the computational cost of the eigen-analysis of lossy electromagnetic problems is significantly higher than that for lossless ones. However, an algorithm is presented next that enables the solution of (10.51) at a cost only marginally higher than that for the lossless case.

In both cases, the inversion of $\mathbf{G} + s_0\mathbf{C}$ may be cast in terms of the solution of the following matrix equation:

$$(\mathbf{G} + s_0\mathbf{C})x = \mathbf{C}p. \tag{10.52}$$

With the vectors p and x split into their electric field and magnetic flux density parts,

$$p = \begin{bmatrix} p_b \\ p_e \end{bmatrix} \qquad x = \begin{bmatrix} x_b \\ x_e \end{bmatrix} \tag{10.53}$$

(10.52) becomes

$$\begin{bmatrix} s_0 P & D \\ D^T & -Z - s_0 T \end{bmatrix} \begin{bmatrix} x_b \\ x_e \end{bmatrix} = \begin{bmatrix} P p_b \\ T p_e \end{bmatrix}. \tag{10.54}$$

Through simple matrix manipulation of the resulting system of matrix equations, an equation is obtained for the calculation of x_e

$$(D^T P^{-1} D + s_0 Z + s_0^2 T)x_e = D^T p_b + s_0 T p_e. \tag{10.55}$$

The second equation is, simply, a matrix relationship for the calculation of x_b in terms of x_e

$$x_b = \frac{1}{s_0}\left(p_b - P^{-1} D x_e\right). \tag{10.56}$$

At this point it is appropriate to review the relationship between tangentially continuous and normally continuous vector spaces. From Chapter 2 we recall that the curl of the elements of the tangentially continuous vector space form a subset of the normally continuous vector space, $\nabla \times W_t \subset W_n$. Hence, a linear matrix relationship exists between the set of basis functions, W_t, used for the expansion of the electric field, and the set of the basis functions, W_n, used for the expansion of the magnetic flux density,

$$\left[\nabla \times \vec{w}_t^{(1)}, \cdots, \nabla \times \vec{w}_t^{(N_t)}\right] = \left[\vec{w}_n^{(1)}, \cdots, \vec{w}_n^{(N_n)}\right] C_r, \tag{10.57}$$

where the $N_n \times N_t$ matrix C_r is the *circulation matrix* defined in (2.126). When using the lowest-order space for W_t and W_n, N_n is the number of non-PEC facets and N_t is the number of non-PEC edges. Because of the bilinear form of the elements of the finite element matrices it is straightforward to show that

$$D^T P^{-1} D = S, \qquad D = P C_r. \tag{10.58}$$

In view of the above results, we conclude that, in the application of (10.55) and (10.56), the matrix $D^T P^{-1} D$ is already available from the finite element approximation of the vector Helmholtz equation for the electric field. Substitution of (10.58) into (10.55) and (10.56) yields the system

$$\begin{cases} (S + s_0 Z + s_0^2 T)x_e = D^T p_b + s_0 T p_e, \\ x_b = \dfrac{1}{s_0}\left(p_b - C_r x_e\right). \end{cases} \tag{10.59}$$

The matrix C_r need not be calculated explicitly. This matrix relates field to flux. The matrix-vector product associated with this matrix in the second equation is effected through a localized operation, namely, by summing up the three edge expansion coefficients associated with a facet, for each facet in the computational domain. The matrices S, Z and T are available from the field formulation. Thus the final issue to be addressed is how to avoid the generation of the matrix D.

Toward this objective, the following transformation is introduced for both vectors p and x,

$$\begin{bmatrix} p_b \\ p_e \end{bmatrix} = \begin{bmatrix} P^{-1}D & 0 \\ 0 & I \end{bmatrix} \begin{bmatrix} p_{bb} \\ p_e \end{bmatrix}, \qquad \begin{bmatrix} x_b \\ x_e \end{bmatrix} = \begin{bmatrix} P^{-1}D & 0 \\ 0 & I \end{bmatrix} \begin{bmatrix} x_{bb} \\ x_e \end{bmatrix}. \tag{10.60}$$

In view of the above transformation, the update of the next Arnoldi vector is computed as follows:

$$\begin{cases} (S + s_0 Z + s_0^2 T)x_e = Sp_{bb} + s_0 Tp_e, \\ x_{bb} = \dfrac{1}{s_0}\left(p_{bb} - x_e\right), \\ x_b = C_r x_{bb}. \end{cases} \tag{10.61}$$

The normalization of a converged eigenvector v is done in the following manner:

$$\begin{aligned} \|v\|_B^2 &= \begin{bmatrix} v_b^T & v_e^T \end{bmatrix} \begin{bmatrix} P & 0 \\ 0 & -T \end{bmatrix} \begin{bmatrix} v_b \\ v_e \end{bmatrix} \\ &= v_b^T P v_b - v_e^T T v_e \\ &= v_{bb}^T S v_{bb} - v_e^T T v_e. \end{aligned} \tag{10.62}$$

Thus all necessary matrix operations are carried out without making use of D. In the next section it will be shown that the above process helps with the elimination of spurious DC modes.

To summarize, the application of the Arnoldi process to (10.51) does not require the construction of any matrices other than the ones associated with the finite element approximation of the vector Helmholtz equation for the electric field. Furthermore, because of the splitting of the generated Arnoldi vectors into their electric field and magnetic flux density parts and the updating process of (10.61), the computational complexity of the resulting modified algorithm is only slightly higher than that for the lossless case. From the point of view of memory requirements, the auxiliary vectors Q_{bb} have to be stored in addition to the Arnoldi vectors (Q_b, Q_e). However, due to the fast convergence of the Arnoldi algorithm, the associate memory overhead is not significant since only a small number of vectors are generated.

10.3.1   Elimination of spurious DC modes

There are two types of spurious DC modes we need to be concerned about in the field-flux formulation. The first type will be referred to as *field-type* DC modes and are associated with the gradient matrix G_r. However, contrary to the case of lossless media, not every column vector in G_r corresponds to a spurious DC mode of the matrix eigenvalue problem in the presence of loss. This becomes evident from the product of the matrix on the left-hand side of (10.10) (i.e., the matrix **G** in our compact form notation) with the matrix $[0, G_r^T]^T$. Use of the results in (10.43) yields

$$\begin{bmatrix} 0 & D \\ D^T & -Z \end{bmatrix} \begin{bmatrix} 0 \\ G_r \end{bmatrix} = \begin{bmatrix} 0 \\ -ZG_r \end{bmatrix}. \tag{10.63}$$

Consequently, the DC modes are those column vectors of G_r for which the multiplication on the left by Z yields the null vector. To elaborate, each column vector in G_r corresponds to a node in the computational domain. Considering the lowest-order tangentially continuous vector space for $\vec{E}$, the nonzero entries in the column vector of G_r correspond to the edges that connect to the node associated with the specific column. Thus the spurious DC modes are associated with those nodes that are neither inside the portion of the computational domain that contains the lossy media nor on the surface of an impedance boundary. We conclude that the *field-type* DC (spurious) modes are the column vectors of the matrix G'_r, which is derived by removing the column vectors of G_r associated with the nodes in the lossy media and on the surface of the impedance boundary.

The second type of spurious DC modes is referred to as *flux-type* DC modes. Observing that $D^T = C_r^T P^T$, and in view of the relationship, $D_r C_r = 0$, between the circulation matrix C_r and divergence matrix D_r, we conclude that the flux-type DC modes are the column vectors of the matrix $(P^T)^{-1} D_r^T$. More specifically, it is

$$\begin{bmatrix} 0 & D \\ D^T & -Z \end{bmatrix} \begin{bmatrix} (P^T)^{-1} D_r^T \\ 0 \end{bmatrix} = \begin{bmatrix} 0 \\ C_r^T D_r^T \end{bmatrix} = \begin{bmatrix} 0 \\ 0 \end{bmatrix}. \tag{10.64}$$

When using the lowest-order normally continuous vector space for $\vec{B}$, the size of C_r is $N_f \times N_e$, where N_f denotes the number of non-PEC facets and N_e the number of non-PEC edges; the size of D_r is $N_t \times N_f$, where N_t denote the number of non-PEC tetrahedra. Therefore the number of *flux-type* DC modes is N_t. In summary, the two types of spurious DC modes are

$$\left\{ \begin{bmatrix} 0 \\ G'_r \end{bmatrix}, \begin{bmatrix} (P^T)^{-1} D_r^T \\ 0 \end{bmatrix}, s = 0 \right\}. \tag{10.65}$$

The removal of these two sets of spurious modes is effected by forcing the Arnoldi vectors to be orthogonal to them. More specifically, let $\begin{bmatrix} x_b \\ x_e \end{bmatrix}$ denote an Arnoldi vector. Then, for the case of the field-type modes, the following orthogonality constraint is imposed

$$\begin{bmatrix} 0 & G'^T_r \end{bmatrix} \begin{bmatrix} P & \\ & -T \end{bmatrix} \begin{bmatrix} 0 \\ x_e \end{bmatrix} = 0 \quad \Rightarrow \quad G'^T_r T x_e = 0. \tag{10.66}$$

The removal of the field-type spurious DC modes is achieved using the same technique as in the case of lossless media. More specifically, the vector $G'_r \psi$ is added to the electric field vector x_e

$$x_e \leftarrow x_e + G'_r \psi, \tag{10.67}$$

where ψ is obtained as the solution to the finite element approximation of $\nabla \cdot (\epsilon \nabla \psi) = 0$ only in the lossless region

$$M\psi = -G'^T_r T x_e. \tag{10.68}$$

In the above equation M is the finite element matrix in the lossless region, with elements

$$M_{i,j} = \iiint_{\Omega_{ll}} \nabla \phi_i \cdot \epsilon \nabla \phi_j \, dv. \tag{10.69}$$

The lossless domain, Ω_{ll}, is understood to be the region of the resonator domain, Ω, that remains after the exclusion of the lossy region and those boundaries on which a lossy surface impedance condition is imposed. In view of the integral form of the elements of M, it is rather straightforward to show that M and T satisfy the relation $M = G'^T_r T G'_r$. Thus the elimination of field-type spurious modes can be summarized as follows:

$$x_e \leftarrow (I - G'_r \underbrace{(G'^T_r T G'_r)^{-1} G'^T T}_{M}) x_e. \tag{10.70}$$

The orthogonality of the Arnoldi vector with the flux-type DC modes is given in terms of the following matrix relation

$$\begin{bmatrix} D_r P^{-1} & 0 \end{bmatrix} \begin{bmatrix} P & \\ & -T \end{bmatrix} \begin{bmatrix} x_b \\ x_e \end{bmatrix} = 0 \quad \Rightarrow \quad D_r x_b = 0. \tag{10.71}$$

Note that, in view of (10.60), the magnetic flux part of the Arnoldi vector is computed via an auxiliary vector $x_b = C_r x_{bb}$. Since $D_r C_r = 0$, (10.71) is satisfied already. Therefore, the magnetic flux part of the Arnoldi vector computed via (10.60) is free from the flux-type spurious DC modes.

In summary, the spurious DC modes can be eliminated by maintaining the generated Arnoldi vectors divergence free. This is achieved by, first, making sure that the initial vectors are divergence free. For this purpose, the fields generated by closed electric current loops should be used as initial vectors in the iteration. In addition, during the Lanczos/Arnoldi process, the solution of (10.45) is utilized through the correction operation of (10.44) to ensure the elimination from the constructed Arnoldi vectors of any spurious DC modes that may arise because of round-off error. Finally, let us point out once again that, for the case of lossy media, G_r and M are constructed only for the nodes in the lossless region of the computational domain.

10.3.2 Extraction of multiple modes

When multiple dominant modes are to be computed a block Arnoldi algorithm is applied. To extract the modes one at a time, a deflation technique is used. In a manner similar to the lossless case, the deflation process relies on the orthogonality of different eigenvectors of (10.11). Thus each Arnoldi vector generated is forced to be orthogonal to the converged eigenvectors. More specifically, assuming that $\begin{bmatrix} V_b \\ V_e \end{bmatrix}$ contains the converged eigenvectors, it is

$$\begin{bmatrix} V_b^T & V_e^T \end{bmatrix} \begin{bmatrix} P & \\ & -T \end{bmatrix} \begin{bmatrix} V_b \\ V_e \end{bmatrix} = V_{bb}^T S V_{bb} - V_e^T T V_e = I. \tag{10.72}$$

Let $\begin{bmatrix} x_b \\ x_e \end{bmatrix}$ denote the next Arnoldi vector. Then, the following operation is used to render it orthogonal to the converged eigenvectors:

$$\begin{aligned} \begin{bmatrix} x_b \\ x_e \end{bmatrix} &\leftarrow \begin{bmatrix} x_b \\ x_e \end{bmatrix} - \begin{bmatrix} V_b \\ V_e \end{bmatrix} \begin{bmatrix} V_b^T & V_e^T \end{bmatrix} \begin{bmatrix} P & \\ & -T \end{bmatrix} \begin{bmatrix} x_b \\ x_e \end{bmatrix} \\ &= \begin{bmatrix} x_b \\ x_e \end{bmatrix} - \begin{bmatrix} V_b \\ V_e \end{bmatrix} [V_{bb}^T S x_{bb} - V_e^T T x_e]. \end{aligned} \tag{10.73}$$

The techniques described above for the matrix-vector product of (10.61), the removal of the spurious DC modes of (10.70), and the simultaneous extraction of multiple modes of (10.73), are combined into an Arnoldi process, which is described below in pseudo-code form.

Algorithm (10.3): Arnoldi algorithm for n eigenmodes of a lossy, reciprocal cavity

1 $V_{0,e} \leftarrow 0$, $V_{0,b} \leftarrow 0$, $V_0 = \begin{bmatrix} V_{0,b} \\ V_{0,e} \end{bmatrix}$, and $V_{0,bb} \leftarrow 0$;

2 for $k = 1, 2, \cdots, n$

  2.a Select $r_{0,e}$, $r_{0,b} \leftarrow 0$; $r_0 = \begin{bmatrix} r_{0,b} \\ r_{0,e} \end{bmatrix}$, and $r_{0,bb} \leftarrow 0$;

  2.b Deflate-Converged-Eigenvectors(r_0, $r_{0,bb}$, V_{k-1}, $V_{k-1,bb}$);

  2.c Remove-Spurious-Modes($r_{0,e}$);

  2.d $q_1 \leftarrow r_0/\|r_0\|$, $q_{1,bb} \leftarrow r_{0,bb}/\|r_0\|$;

  2.e for $m = 1, 2, \cdots, M$

    2.e.1 $(\tilde{q}_{m+1}, \tilde{q}_{m+1,bb}) \leftarrow$ Matrix-Vector-Product(q_m, $q_{m,bb}$, V_{k-1}, $V_{k-1,bb}$);

    2.e.2 for $i = 1, \cdots, m$

      2.e.2.a $H_m(i,m) \leftarrow \langle q_i, \tilde{q}_{m+1} \rangle = q_i^T \tilde{q}_{m+1}$;

      2.e.2.b $\tilde{q}_{m+1} \leftarrow \tilde{q}_{m+1} - H_m(i,m)q_i$;

      2.e.2.c $\tilde{q}_{m+1,bb} \leftarrow \tilde{q}_{m+1,bb} - H_m(i,m)q_{i,bb}$;

    2.e.3 Compute the eigenpairs $(\tilde{\lambda}_i, y_i)$ of H_m $(i = 1, \cdots, m)$;

    2.e.4 Compute $\rho_i = \|e_m^T y_i\| \|\tilde{q}_{m+1}\|/\|\tilde{\lambda}_i Q_m y_i\|$, $(i = 1, \cdots, m)$;

    2.e.5 If one eigenpair $(\lambda_i, \tilde{v} = Q_m y, \tilde{v}_{bb} = Q_{m,bb} y)$ converges, then goto 2.g;

    2.e.6 $H_m(m+1, m) = \|\tilde{q}_{m+1}\|$;

    2.e.7 $q_{m+1} = \tilde{q}_{m+1}/H_m(m+1,m)$ and $q_{m+1,bb} = \tilde{q}_{m+1,bb}/H_m(m+1,m)$;

  2.f $r_0 \leftarrow v$, $r_{0,bb} \leftarrow v_{bb}$ and goto 2.b;

  2.g Normalize:

    1 $norm = \sqrt{\begin{bmatrix} \tilde{v}_b^T & \tilde{v}_e^T \end{bmatrix} \begin{bmatrix} P & \\ & -T \end{bmatrix} \begin{bmatrix} \tilde{v}_b \\ \tilde{v}_e \end{bmatrix}} = \sqrt{\tilde{v}_{bb}^T S \tilde{v}_{bb} - \tilde{v}_e^T T \tilde{v}_e}$;

    2 $v \leftarrow \tilde{v}/norm$ and $v_{bb} \leftarrow \tilde{v}_{bb}/norm$;

  2.h $V_k \leftarrow [V_{k-1}, v]$, $V_{k,bb} \leftarrow [V_{k-1,bb}, v_{bb}]$

Step 2.b deflates the converged eigenvectors stored in V_k from the initial vector r_0. The deflation is performed as follows:

Algorithm (10.4): Deflate-Converged-Eigenvectors $(x, x_{bb}, V_{k-1}, V_{k-1,bb})$

1 $p = \begin{bmatrix} V_{k-1,b}^T & V_{k-1,e}^T \end{bmatrix} \begin{bmatrix} P & \\ & -T \end{bmatrix} \begin{bmatrix} x_b \\ x_e \end{bmatrix} = V_{k-1,bb}^T S x_{bb} - V_{k-1,e}^T T x_e$;

2 $x \leftarrow x - V_{k-1}p$;

3 $x_{bb} \leftarrow x_{bb} - V_{k-1,bb}p$.

The process of removing of spurious DC modes involves only the electric part of the vectors, since, as explained in the previous section, the magnetic flux density vector is, by construction, free from the flux-type DC modes. The algorithm for step 2.c is as follows:

Algorithm (10.5): Remove-Spurious-Modes (x_e)

1 $x_e \leftarrow x_e - G_r'^T M^{-1} G_r'^T T x_e$

Step 2.e.1 performs the product of the matrix $(G + s_0 C^{-1})C$ with a vector q_m. The result is then forced to be free of spurious DC modes and orthogonal to the converged eigenvectors.

Algorithm (10.6): Matrix-Vector Product $(q_m, q_{m,bb}, V_{k-1}, V_{k-1,bb})$

1 $p = S q_{m,bb} + s_0 T q_{m,e}$

2 $q_{m+1,e} \leftarrow (S + s_0 Z + s_0^2 T)^{-1} p$

3 Remove-Spurious-Mode $(q_{m+1,e})$

4 $q_{m+1,bb} \leftarrow \frac{1}{s_0}(q_{m,bb} - q_{m+1,e})$

5 $q_{m+1,b} \leftarrow C_r q_{m+1,bb}$

6 $q_{m+1} \leftarrow \begin{bmatrix} q_{m+1,b} \\ q_{m+1,e} \end{bmatrix}$

7 Deflate Converged Eigenvectors(q_{m+1}, $q_{m+1,bb}$, V_{k-1}, $V_{k-1,bb}$);

8 return $(q_{m+1}, q_{m+1,bb})$.

10.4 MULTIGRID/MULTILEVEL EIGENVALUE ANALYSIS

The convergence of the proposed eigensolvers is strongly dependent on the proximity of the complex frequency shift to the desired eigenfrequencies. When extraction of multiple modes is desired, the proposed deflation techniques allow for the extraction to be performed one mode at a time. Thus the *shift-value hopping process* discussed in Section 9.4.2 can be used for each eigenfrequency.

The most time-consuming part in the eigen-solution process is the solution of the matrices $(S + s_0 Z + s_0^2 T)$ and M. This is particularly the case for large-size resonators, exhibiting significant material inhomogeneity, and thus requiring a high-resolution grid. In such cases the direct solution of the matrices may not be possible. Thus an iterative solution becomes the computationally attractive alternative, and the multigrid and multilevel preconditioning techniques of Chapter 6 and 7 can be applied to accelerate convergence.

10.5 NUMERICAL VALIDATION

In this section, some representative results are presented from the application of the eigen-analysis methodologies and associated algorithms described above to the extraction of the eigenfrequencies of typical lossless and lossy electromagnetic resonators. All calculations are performed using double precision arithmetic. The tolerance for convergence in the iterative eigen-solution was taken to be $\rho = 1.0e^{-6}$.

Dielectric-Loaded Rectangular Cavity. The first example is a rectangular cavity, with a lossy dielectric block in its center, as depicted in the insert of Fig. 10.3. The cross-sectional geometry is square of side 0.1 m. The length of the cavity is 1.0 m. The top and bottom plates are perfect electric conductors. The front and back walls are perfect electric conductors also, while the two remaining side walls (along the length of the cavity) are taken to be perfect magnetic conductors. The relative dielectric constant of the dielectric block is 2.0, and its conductivity is 0.1 S/m. The finite element approximation of the eigenvalue problem involves 104 nodes, 487 edges, 688 facets, and 304 tetrahedra. From these, there are 37 non-PEC nodes, 334 non-PEC edges, and 602 non-PEC facets. Since only the walls of the cavity are PEC, all tetrahedra are non-PEC.

The field-flux formulation is used to develop the finite element matrix statement of the numerical eigenvalue problem. First, Matlab® is used to perform the complete eigenvalue analysis. The Matlab® results will constitute the reference solution for the assessment of the accuracy of the iterative eigensolvers developed in this chapter.

Let us first consider the lossless case, where the conductivity of the dielectric block is set to zero. In this case the eigensolution can be done using either the *E-B* (field-flux) formulation, yielding the results in the left plot of Fig. 10.3, or the *E*-field formulation, yielding the results in its right plot of Fig. 10.3. The reference eigenvalues from Matlab® are indicated in the plots using the symbol ·.

As expected, for the lossless case the eigenvalues of the *E-B* formulation are on the imaginary axis of the complex *s* plane. On the other hand, the eigenvalue of the *E*-field formulation are on the positive-real axis of the complex ω plane. For each non-zero eigenvalue of the *E-B* formulation, there exists another eigenvalue symmetric to it with respect to the real axis. These two modes degenerate into a single mode in the *E*-field formulation.

Another interesting observation can be made by counting the number of zero eigenvalues for both formulations. For the E-B formulation, there are 341 zero eigenvalues, of which 37 are "field-type" and 304 are "flux-type." For the E-field formulation, there are only 37 "field-type" zero eigenvalues, and no "flux-type" ones.

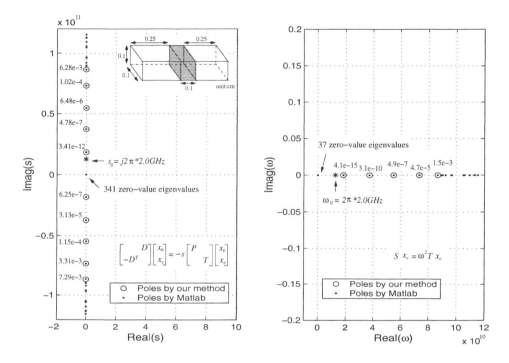

Figure 10.3   Eigenvalues of a rectangular cavity with a lossless dielectric block in the middle. Left plot depicts the eigenvalues obtained using the E-B formulation; right plot depicts the eigenvalues obtained using the E-field formulation.

The Arnoldi algorithm with spurious DC mode removal scheme is then used to calculate the dominant modes. The frequency-shift point is taken to be $f_0 = 2$ GHz which is marked by the symbol $*$ in both plots. After 11 iterations, the calculated eigenfrequencies with residual less than $1.0e^{-2}$ are plotted and indicated by the symbol $\circ$. Clearly, the eigenfrequencies close to the expansion point are calculated first by the Arnoldi eigensolvers. Furthermore, the zero eigenfrequencies, corresponding to the spurious DC modes, are successfully avoided.

We consider, next, the lossy case, with the conductivity of the dielectric block taken to be 0.1 S/m. For this case, only the E-B formulation can be used. The reference eigenvalues computed by Matlab® are indicated by the symbol $\cdot$ in Fig. 10.4. Due to the passivity of the cavity, all the eigenvalues are on the left side of the complex s plane. The finite element problem has 338 zero eigenvalues, 304 of which are flux-type. The remaining 34 field-type zero eigenvalues are three fewer than those for the lossless case. The reason for this is that three non-PEC nodes are either inside or on the surface of the dielectric block. As the conductivity of the dielectric bock increases, the three eigenvalues depart from zero, moving along the real axis in the negative direction.

For the application of the Arnoldi eigensolver the shift point was taken to be $f_0 = 2$ GHz, as indicated by the symbol $*$ in Fig. 10.4. The eigenfrequencies obtained after 11 iterations with residual less than $5.0e^{-2}$ are plotted in the figure and are indicated using the symbol $\circ$. The accompanying values are their residuals. Clearly, the eigenfrequencies close to the expansion point are calculated first by the Arnoldi eigensolvers. Furthermore, the zero eigenvalues are successfully avoided.

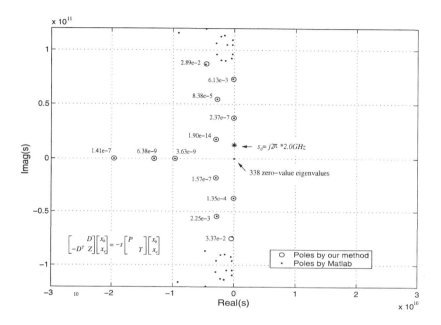

Figure 10.4 Eigenvalues of the rectangular cavity with a lossy dielectric block in the middle.

Partially Filled Circular Cylindrical Cavity. The second resonator considered is a circular cylindrical cavity, partially filled with a dielectric rod of relative permittivity $\epsilon_r = 37.6$. The cavity walls are assumed to be perfect electric conductors. The geometry is shown in the insert of Fig. 10.5. The objective is to extract the lowest eigenmodes of the cavity. Thus the expansion frequency is taken to be $f_0 = 1.2$ GHz.

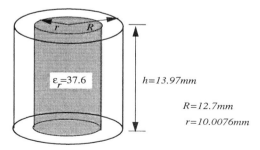

Figure 10.5 A circular cylindrical cavity, partially filled with a dielectric rod.

Since this is a lossless case, we use the E-field formulation which solves for the electric field only. The number of degrees of freedom in the finite element approximation using

$\mathcal{H}^0(curl)$ elements is 5748. The calculated resonant frequencies of the six lowest eigenmodes obtained are compared to those obtained from the analytic solution in Table 10.1. The Arnoldi process required, on the average, 9 iterations for each mode to converge. No spurious DC modes were generated. Furthermore, every calculated mode corresponds to an analytic one, indicating that no spurious AC modes occur either.

Table 10.1   Lowest Eigenvalues of a Partially Filled Circular, Cylindrical Cavity

Mode	(lossless case, $\omega/2\pi 1.0e6$) Analyt.	Numerical	(lossy case, $s/(2\pi 1.0e6)$) Numerical
1	1498	1500.0	-23.88+1499.9j
2	2435	2441.5	-23.86+2441.4j
3	2435	2442.4	-23.86+2442.3j
4	2504	2505.1	-23.56+2505.5j
5	2504	2505.6	-23.56+2505.0j
6	3029	3045.6	-23.86+3045.5j

For the case where the dielectric rod inside the cylindrical cavity is lossy, the finite element eigensolution is done using the E-B formulation, which augments the number of degrees of freedom in the electric field formulation with 10933 flux unknowns for the case of $\mathcal{H}^0(curl)$ elements. For the specific application, the conductivity of the rod was taken to be $\sigma = 0.1$ S/m. The calculated eigenvalues of the six lowest modes are shown in the last column of Table 10.1. Once again the frequency-shift point of $f_0 = 1.2$ GHz was used. Convergence was achieved, on the average, in 14 iterations for each mode.

Dielectric Half-filled Rectangular Cavity. The geometry of this cavity is depicted in the insert of Fig. 10.6. An analytic solution for this eigenproblem is possible and its results are used as the reference solution. The base of the cavity is a square of side 22.86 mm, while its height is 10.16 mm. The average grid size of the mesh is 2.49 mm, and the number of electric field unknowns is 2734 or 16274 for $\mathcal{H}^0(curl)$ elements or $\mathcal{H}^1(curl)$ elements, respectively. The number of magnetic flux density unknowns is 5403 or 16200 for $\mathcal{H}^0(curl)$ elements or $\mathcal{H}^1(curl)$ elements, respectively. The dominant eigenfrequencies (in GHz) calculated for different values of the conductivity of the dielectric are compared to their analytic counterparts in Table 10.2. Figure 10.6 depicts the impact of material loss on convergence. The higher the loss the larger the number of iterations needed for convergence.

These examples offer convincing evidence of the computational efficiency and accuracy of the three-dimensional electromagnetic eigensolvers presented in this chapter. Both lossless and lossy electromagnetic resonators can be handled by these eigensolvers with the same numerical robustness. This robustness stems primarily from the extra care taken to prevent the occurrence of spurious modes which are known to hinder the convergence and accuracy of the iterative extraction of the dominant eigenmodes. While the electromagnetic eigenvalue problem formulations used are free of spurious AC modes, the occurrence of

Table 10.2   Eigenfrequency of the Dominant Mode of a Partially Filled Rectangular
Resonator

σ(S/m)	Analytic	$\mathcal{H}^0(curl)$	$\mathcal{H}^1(curl)$
0.1	-0.354+7.379j	-0.353+7.378j	-0.354+7.386j
0.5	-1.819+7.236j	-1.813+7.236j	-1.817+7.244j
1.0	-3.864+6.579j	-3.852+6.585j	-3.861+6.589j
1.3	-5.197+5.711j	-5.182+5.724j	-5.193+5.725j

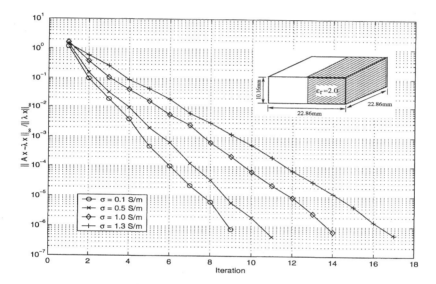

Figure 10.6   Convergence of the dominant mode of a half-filled, lossy resonator for different values
of the conductivity of the filling dielectric. (After Zhu and Cangellaris [20], ©2002, IEEE.)

spurious DC modes (i.e., modes that exhibit nonzero divergence) is possible and must be
avoided. As the numerical studies indicate, the two-prong strategy we have employed for
their avoidance, namely, the selection of divergence-free initial vectors and the explicit
elimination from the subsequently generated Lanczos/Arnoldi vectors of any spurious DC
parts caused by roundoff, is very effective and provides for the accurate calculation of the
desired subset of the dominant eigenmodes and their associated eigenvalues.

REFERENCES

1. J. F. Lee and R. Mittra, "A note on the application of edge-elements for modeling three-
dimensional inhomogeneously-filled cavities," *IEEE Trans. Microwave Theory Tech.*, vol. 40,
pp. 1767-1773, Sep. 1992.

2. J. F. Lee, G. M. Wilkins, and R. Mittra, "Finite element analysis of axisymmetric cavity resonator using hybrid edge element technique," *IEEE Trans. Microwave Theory Techn.*, vol. 41, no. 11, pp. 1981-1987, Nov. 1993.

3. S. Prepelitsa, R. Dyczij-Edlinger, and J. F. Lee, "Finite-element analysis of arbitrarily shaped cavity resonators using H1(curl) elements," *IEEE Trans. Magnetics*, vol. 33, pp. 1776-1779, Mar. 1997.

4. D. B. Davidson, "Comments on and extension of "A note on the application of edge-elements for modeling three-dimensional inhomogeneously-filled cavities," *IEEE Trans. Microwave Theory Tech.*, vol. 46, no. 9, Sep. 1998.

5. C. Liu and J. F. Lee, "Jacobi-Davidson algorithm and its application to modeling RF/Microwave detection circuits," *Comput. Methods Appl. Mech. Engrg.*, vol. 169, pp. 359-375, 1999.

6. W. P. Carpes Jr. L. Pichon, A. Razek, "Efficient analysis of resonant cavities by finite element method in the time domain," *IEE Proc. Microw. Antennas Propag.*, vol. 147, no. 1, pp. 53-57, Feb. 2000.

7. D. Schmitt and T. Weiland, "2D and 3D computations of eigenvalue problems," *IEEE Trans. Magnetics*, vol. 28, pp. 1793-1796, Mar. 1992.

8. D. Schmitt and T. Weiland, "2D and 3D computations of lossy eigenvalue problems," *IEEE Trans. Magnetics*, vol. 30, no. 5, pp. 3578-3581, Sep. 1994

9. Y. Lu, S. Zhu, and F. A. Fernandez, "The efficient solution of large sparse nonsymmetric and complex eigensystems by subspace iterations," *IEEE Trans. Magnetics*, vol. 30, no. 5, pp. 3582-3585, Sep. 1994.

10. S, Groiss, I. Bardi, O. Biro, K. Preis, and K. R. Richter, "Parameters of lossy cavity resonators calculated by the finite element method," *IEEE Trans. Magnetics*, vol. 32, pp. 894-897, May 1996.

11. J. Wang and N. Ida, "Eigenvalue analysis in EM cavities using divergence-free finite elements," in *Proc. Fourth Biennial IEEE Conf. on Electromagnetic Field Computation*, Toronto, Canada, 1990.

12. J. S. Wang and N. Ida, "Eigenvalue analysis in electromagnetic cavities using divergence free finite elements," *IEEE Trans. Magnetics*, vol. 27, pp. 3378-3381, Sep. 1991.

13. A. Konrad, "A direct three-dimensional finite element methods for the solution of electromagnetic fields in cavities," *IEEE Trans. Magnetics*, MAG-21, no. 6, pp. 2276-2279, 1985.

14. J. P. Webb, "The finite-element method for finding modes of dielectric-loaded cavities," *IEEE Trans. Microwave Theory Tech.*, MTT-33, pp. 635-639, Jul. 1985.

15. A. J. Kobelansky and J. P. Webb, " Eliminating spurious modes in finite-element waveguides problems by using divergence-free fields," *Electron. Lett.*, vol. 22, pp. 569-570, 1986.

16. J. P. Webb, " Efficient generation of divergence-free fields for the finite element analysis of 3D cavity resonators," *IEEE Trans. Magnetics*, vol. 24, no. 1, pp. 162-165, 1988.

17. K. Sakiyama, H. Kotera, and A. Ahagon, "3-D electromagnetic field mode analysis using finite-element mode by edge element," *IEEE Trans. Magnetics*, vol. 26, no 5, pp. 1759-1761, 1990.

18. D. Baillargeat, S. Verdeyme, M. Aubourg, and P. Guillon, "CAD applying the finite-element method for dielectric resonator filters," *IEEE Trans. Microwave Theory Tech.*, vol. 46, no. 1. pp. 10-17, Jan. 1998.

19. Y. Zhu and A. C. Cangellaris, "A new FEM formulation for reduced order electromagnetic modeling," *IEEE Microw. Guid. Wave Lett.*, vol. 11, pp. 211-213. May 2001.

20. Y. Zhu and A. C. Cangellaris, "Robust finite element solution of lossy and unbounded electromagnetic eigenvalue problems," *IEEE Trans. Microwave Theory Tech.*, vol. 50, pp. 2331-2338, Oct. 2002,

21. J.D. Jackson, *Classical Electrodynamics*, 3rd ed., New York: John Wiley & Sons, Inc., 1999.

22. R.E. Collin, *Field Theory of Guided Waves*, 2nd ed., Piscataway, NJ: IEEE Press, 1991.

23. R.S. Elliott, *An Introduction to Guided Waves and Microwave Circuits*, Englewood Cliffs, NJ: Prentice-Hall, Inc., 1993.

24. Y. Saad, *Numerical Methods for Large Eigenvalue Problems*, New York: John Wiley & Sons, 1992.

25. J. F. Lee, R. Lee, and A. Cangellaris, "Time-domain finite-element methods," *IEEE Trans. Antennas Propagat.*, vol. 45, pp. 430-442, Mar. 1997.

CHAPTER 11

MODEL ORDER REDUCTION OF ELECTROMAGNETIC SYSTEMS

Despite continuing advances in algorithmic sophistication, computational efficiency, and numerical robustness of both integral-equation-based and differential-equation-based numerical methods, system-level electromagnetic modeling still remains a formidable challenge. The term "system" is understood to mean any physical structure within which electromagnetic field interactions exhibit significant spatial complexity due to either one or several of the following factors. Large electrical size; material inhomogeneity; complicated geometry exhibiting large variation in characteristic feature size over the volume of the structure. Typical examples of electrically-large structures are, a) the electromagnetic field distribution inside a building associated with a wireless RF network; b) the electromagnetic scattering by an airplane at GHz frequencies; c) the electromagnetic analysis of a very large antenna array. As far as material and geometry complexity are concerned, the most striking example is the integrating substrate (be that the semiconductor die, the package or the board) used for high-density, mixed-signal, integrated electronic systems.

The computational complexity of the electromagnetic analysis of such structures is compounded by several factors, the most obvious of which is the large number of degrees of freedom required for the development of an accurate numerical model. For the case of electrically-large structures, the increase in the number of degrees of freedom stems from requirements for sufficient spatial resolution of the fields, quantified (roughly) in terms of sampling points per wavelength, to ensure accuracy in the electromagnetic response. For the case of high geometric/material complexity, it is the minimum geometry feature size that dictates complexity, since the numerical grid used to discretize the structure should be fine enough to capture all geometric and material attributes that impact the electromagnetic response.

The challenge of system-level electromagnetic modeling is not new. The computational electromagnetic community has significant experience with tackling such problems, albeit in more ad-hoc rather than systematic ways. One might argue that, to date, it is only in the case of microwave circuits and networks that a systematic methodology is available and is being successfully applied for sub-system and system-level microwave and millimeter-wave design (e.g., [1]).

For the case of microwave circuit design, the methodology is founded on a *divide-and-conquer* process, where the system is decomposed into sub-systems or components, each one of which can be analyzed independently from the others. Once the electromagnetic responses for all components have been generated, they are combined together using microwave network analysis practices to enable system-level analysis.

The success and accuracy of such a *domain decomposition* process is strongly dependent on the selection of the decomposition boundaries across which the individual domains interact. For the case of microwave circuits, these boundaries are most commonly associated with cross-sections of waveguides, which serve as the only (or, at least, the dominant) electromagnetic paths of interaction between components. For such cases, *electromagnetic ports* are readily defined in terms of the amplitudes of the electric and magnetic fields of the modes participating in the transmission of electromagnetic power through the port. This makes possible the application of well-documented microwave network analysis techniques for system-level electromagnetic simulation (e.g., [2], [3]).

It should be evident that such a domain decomposition approach to system-level electromagnetic modeling can be generalized to be made applicable to systems for which the decomposition surfaces do not coincide with waveguide ports. Based on well-known uniqueness arguments [4], the tangential components of the electric and magnetic fields on the decomposition boundaries can be used in place of physical ports, to serve as the glue for stitching together the electromagnetic responses of the sub-domains.

The most popular example of such a domain decomposition strategy in computational electromagnetics is associated with the hybrid finite element/integral equation solution of unbounded scattering and radiation problems [5]. In this case the decomposition boundary is a mathematical surface that encloses the (typically) highly-inhomogeneous scatterer. The electromagnetic problem in the interior region is discretized using a finite element model with either the tangential electric field (or the tangential magnetic) field on the enclosing boundary serving as one of the "sources" for the problem. The interior problem can then be formally solved for the tangential magnetic field (or the tangential electric field) on the boundary. A similar strategy is applied to the exterior problem, where now an integral equation statement is used to establish a relationship between the tangential electric and magnetic fields on the decomposition surface consistent with the electromagnetic properties of the exterior. The solution for the "system" is obtained by enforcing the continuity of the fields at the decomposition boundary.

The aforementioned example is a representative sample from the very active field of domain decomposition methods for the solution of partial differential equations in complex domains. A major emphasis of the most recent research activities in this area is on the development of methodologies for the exchange of information between sub-domains across decomposition boundaries to ensure both solution consistency and expedient convergence in the iterative solution of the entire domain. Of particular relevance to electromagnetic BVPs are the recent developments on decomposition methods for the Helmholtz equation and Maxwell's system [6]-[8], which have prompted exciting advances toward the establishment of robust methodologies for the solution of electromagnetic BVPs of unprecedented complexity [9].

When viewed in the context of a means for expedient iterative solution of system-level problems, domain decomposition algorithms are most commonly used for single-frequency, time-harmonic electromagnetic analysis. However, as the applications in microwave circuit/network design indicate, domain decomposition is also most useful for broadband system-level analysis. More specifically, if the broadband electromagnetic response of each sub-domain can be computed efficiently, then system-level, frequency-domain and/or time-domain analysis can be conveniently performed using general-purpose, network analysis-oriented simulation tools such as [1].

In this chapter, finite element-based methodologies are presented for the expedient computation of the broadband electromagnetic response of an electromagnetic structure. In the spirit of the aforementioned discussion, the methodologies to be presented are driven by the objective to produce a broadband matrix transfer function representation of the electromagnetic structure, that serves as a *compact*, multi-port, macromodel, compatible with network analysis-oriented simulators. The *compactness* of the generated model is understood to imply that the number of degrees of freedom and, hence, the number of poles kept in the approximate, matrix transfer function response of the system, is the minimum possible required to capture with acceptable accuracy the attributes of the system over the desirable frequency bandwidth.

Such compact macromodeling can be obtained from finite element approximations of electromagnetic systems by means of model order reduction techniques. Model order reduction of linear systems is a topic that has received significant attention by the systems community over the past thirty years. Following the successful application of reduced-order macro-modeling in the reduction of the computational complexity of very-large-scale circuit simulation (e.g., [10]-[18]), the computational electromagnetics community was prompted to explore the opportunities of model order reduction techniques for facilitating and expediting the broadband analysis of electromagnetic systems. Among the first results reported in the literature was the application of the asymptotic waveform evaluation (AWE) ideas for the fast generation of the broadband response of an electromagnetic system [19]-[22]. However, eventually, the more robust and effective model order reduction techniques based on Krylov subspace methods (e.g., [23]- [30]), took over as the methods of choice for the reduced-order macromodeling of electromagnetic systems exhibiting retardation.

It is these Krylov-subspace methods, used in the context of finite element modeling, that are discussed in depth in this chapter. Nevertheless, it is instructive to commence our presentation with brief overview of the ideas behind the so-called AWE method. This is followed by the discussion of the Krylov subspace-based methodology for model order reduction of the discrete state-space models generated by the finite element approximation of the electromagnetic system. Issues such as definition of electromagnetic ports for the construction of the reduced macromodel and the passivity of the discrete approximation of the electromagnetic system receive careful attention. The discussion concludes with some examples from the application of the proposed model order reduction methods to the generation of broadband, multi-port transfer function macromodels for a variety of electromagnetic devices and systems.

11.1   ASYMPTOTIC WAVEFORM EVALUATION

Asymptotic waveform evaluation (AWE) is a methodology first used for the expedient calculation of signal delay in digital circuits modeled in terms of large resistive/capacitive (RC) trees [10]. Its introduction led to a significant reduction in the computational effort required for RC-tree delay prediction, and motivated further research in its extension to propagation delay and noise generation and coupling in distributed (transmission line theory-based) models of complex networks of interconnects in high-speed digital circuits (e.g., [11]).

We begin the review of the AWE technique by considering its application to the expedient calculation of the response of the following single-input, single-output (SISO) linear system

$$\begin{cases} C\dfrac{dx}{dt} = -Gx + bu(t) \\ y(t) = l^T x(t), \end{cases} \tag{11.1}$$

where $C, G \in \mathbb{R}^{N \times N}$ are real matrices, $l, b \in \mathbb{R}^{N \times 1}$ are given column vectors, x denotes the vector of the state variables in the system, u is the scalar excitation (input) and y is the scalar output of interest. The transfer function of the single-input, single-output system is, obviously, a scalar function. For the purposes of this discussion, and without loss of generality, it is assumed that the initial state of the system is zero. Then, the Laplace transform of the above system leads to the following expression for its scalar transfer function:

$$H(s) = l^T (G + sC)^{-1} b, \tag{11.2}$$

where s is the complex frequency (Laplace transform variable). Expansion of $H(s)$ in the form of Taylor's series around an expansion point s_0 yields

$$\begin{aligned} H(s) &= l^T \left[I + (s - s_0)(G + s_0 C)^{-1} C \right]^{-1} (G + s_0 C)^{-1} b \\ &= l^T \left[I - (s - s_0) A \right]^{-1} r \\ &= \sum_{i=0}^{\infty} m_i (s - s_0)^i, \end{aligned} \tag{11.3}$$

where

$$A = -(G + s_0 C)^{-1} C, \quad r = (G + s_0 C)^{-1} b, \quad m_i = l^T A^i r. \tag{11.4}$$

The coefficients, m_i, in the resulting Taylor's series are often referred to as the *moments* of the expansion. The AWE proposition was to utilize a small number of the moments of the system, which, for the case of RC trees can be computed expediently, for the purposes of constructing reasonably accurate approximations to the system response. More specifically, let N be the order of poles of the system. Then the transfer function may be cast in the pole-residue form

$$H(s) = \sum_{i=1}^{N} \frac{k_i}{s - p_i}. \tag{11.5}$$

For real-world problems, with networks involving hundreds of thousands of lumped elements, the extraction of all the poles of the system is computationally intensive and often times prohibitive. Thus a lower-order approximation of $H(s)$ is sought, of order $n \ll N$,

$$H(s) \approx \tilde{H}(s) = \sum_{i=1}^{n} \frac{\hat{k}_i}{s - \hat{p}_i}. \tag{11.6}$$

The way the approximation is effected is by *matching* the $2n$ moments of (11.6) to the first $2n$ moments of $H(s)$, calculated from the last equation in (11.4). This matching provides the $2n$ equations needed for obtaining the n poles and n residues in (11.6).

The application of such an approximation process to the discrete electromagnetic system resulting from the finite element approximation of the vector Helmholtz equation for the electric field is fairly straightforward. The pertinent finite element matrix equation is recalled here for convenience,

$$(S + sZ + s^2 T)x_e = f_e, \tag{11.7}$$

where x_e is the vector of unknowns from which a desirable *response* quantity is computed. This post-processing is described in terms of the linear relationship

$$H(s) = l^T x_e, \tag{11.8}$$

where l is a known column vector.

In a manner similar to the development in the previous paragraphs, an approximation to $H(s)$ is sought, constructed in terms of the first few moments in the Taylor's series expansion of $H(s)$ at the expansion frequency s_0. Let

$$x_e = \sum_{n=0} x_{e,n}(s - s_0)^n,$$
$$f_e = \sum_{n=0} f_{e,n}(s - s_0)^n \tag{11.9}$$

be the Taylor's series expansions for the unknown and the forcing vectors. Their substitution into (11.7) yields

$$\begin{cases} (S + sZ + s^2 T) \sum_{n=0} x_{e,n}(s - s_0)^n = \sum_{n=0} f_{e,n}(s - s_0)^n, \\ H(s) = l^T \sum_{n=0} x_{e,n}(s - s_0)^n \\ \qquad = \sum_{n=0} \left(l^T x_{e,n}\right)(s - s_0)^n \\ \qquad = \sum_{n=0} m_n(s - s_0)^n. \end{cases} \tag{11.10}$$

Introducing the shift $\hat{s} = s - s_0$ and matching coefficients of same powers of $\hat{s}$ on the two sides of the resulting equation yields the following sequence of recursive matrix equations for the calculation of the moments, m_n, in the Taylor's series approximation of the response For $n = 0$:

$$(S + s_0 Z + s_0^2 T)x_{e,0} = f_{e,0},$$
$$m_0 = l^T x_{e,0}. \tag{11.11}$$

For $n = 1$:

$$(S + s_0 Z + s_0^2 T)x_{e,1} = f_{e,1} - Z x_{e,0} - 2 s_0 T x_{e,0},$$
$$m_1 = l^T x_{e,1}. \tag{11.12}$$

For $n \geq 2$:

$$(S + s_0 Z + s_0^2 T) x_{e,n} = f_{e,n} - Z x_{e,n-1} - T(x_{e,n-2} + 2s_0 x_{e,n-1}),$$
$$m_n = l^T x_{e,n}. \tag{11.13}$$

The key observation is that in the above process for the recursive computation of the moments of the response the matrix equation $(S + s_0 Z + s_0^2 T)$ has to been assembled and solved at the expansion frequency s_0 only once. Thus, assuming that the accuracy of the truncated Taylor's series expansion is acceptable over some bandwidth, ω_{BW}, around the expansion frequency s_0, this process provides for a computationally efficient alternative to solving the finite element system repeatedly, for several frequency points over the bandwidth ω_{BW}.

It should be apparent that the proposed process is valid provided that the material properties and any impedance boundary conditions in the computational domain are frequency independent so that the matrices S, Z and T are also independent of frequency. If the electromagnetic properties of the materials exhibit frequency dependence, then Taylor's series expansions for the matrices S, T and Z must be developed and taken into account in the development of the recursive process for the generation of the moments of the response.

The direct utilization of the truncated Taylor's series expansion in (11.10) as an approximation of the frequency-dependent response is hindered by its very narrow bandwidth of accuracy. To address this difficulty, a Padé expansion, which is known to exhibit broader bandwidth of accuracy, is used instead as an approximation of the response. Thus the application of the aforementioned recursive computation of the first $2n$ moments, $m_i, (i = 0, 1, \cdots, 2n - 1)$, of the response, is followed by setting these moments equal to the moments of the following Padé approximation of the original transfer function,

$$\tilde{H}(s) = \frac{a_0 + a_1(s - s_0) + a_2(s - s_0)^2 + \cdots + a_{n-1}(s - s_0)^{n-1}}{1 + b_1(s - s_0) + b_2(s - s_0)^2 + \cdots + b_n(s - s_0)^n}. \tag{11.14}$$

It can be shown that the $2n$ Padé coefficients a_i and b_i are solved through the following moment-matching equations: [11]

$$\begin{bmatrix} m_0 & m_1 & \cdots & m_{n-1} \\ m_1 & m_2 & \cdots & m_n \\ \vdots & \vdots & & \vdots \\ m_{n-1} & m_n & \cdots & m_{2n-2} \end{bmatrix} \begin{bmatrix} b_n \\ b_{n-1} \\ \vdots \\ b_1 \end{bmatrix} = - \begin{bmatrix} m_n \\ m_{n+1} \\ \vdots \\ m_{2n-1} \end{bmatrix},$$

$$\begin{cases} a_0 = m_0 \\ a_1 = m_1 + b_1 m_0 \\ \vdots \\ a_{n-1} = m_{n-1} + \sum_{i=1}^{n-1} b_i m_{n-i-1}. \end{cases} \tag{11.15}$$

In summary, the application of the AWE process to the development of a frequency-dependent approximation to the solution of the finite element model of the electromagnetic BVP involves two steps. First, the first $2n$ moments are computed at the expansion frequency s_0. Second, the Padé approximate of order n is built such that its moments are equal to the computed $2n$ moments. This process may be summarized algorithmically as follows:

Algorithm (11.1): Asymptotic Waveform Evaluation
```
1 Generate 2n moments using (11.11-11.13);
2 Calculate Padé approximate of order n to match the 2n moments
  using (11.15).
```

This process is often referred to in the computational electromagnetics literature as *fast frequency sweep*. Obviously, this term expresses the computational efficiency gain associated with the development of the Padé approximate, where the calculation of the response over the bandwidth of validity of the Padé approximate is achieved through the decomposition of the finite element system at a single frequency only.

The generated Padé approximation exhibits very good accuracy in the immediate neighborhood of the expansion frequency. However, the accuracy decreases for more distant frequencies, prompting the possibility of utilizing a higher-order Padé approximation. While, in principle, a larger order helps extend the bandwidth of accuracy of the Padé approximate, the matrix in (11.15) becomes ill-conditioned as the order n increases. Therefore, the bandwidth of accuracy around the expansion point s_0 does not consistently increase with the order n. Thus most applications of the AWE process for the development of Padé approximations around an expansion frequency s_0 tend to limit the order to 10 or 15, and provide reliable accuracy over a rather narrow bandwidth.

This shortcoming is addressed through the development of multiple Padé approximates at several frequency points over the desirable bandwidth. To most popular scheme for this purpose is the complex frequency hopping (CFH) process [19]. In this process, the binary search algorithm is applied to determine the number and location of expansion points over the frequency bandwidth of interest, $[\omega_{min}, \omega_{max}]$.

Algorithm (11.2): Binary search algorithm
```
1 Let ω₁ = ω_min and ω₂ = ω_max;
2 Apply AWE at ω₁ and ω₂ and obtain H₁(ω) and H₂(ω);
3 Choose ω_mid = (ω₂ + ω₁)/2 and calculate H₁(ω_mid) and H₂(ω_mid);
4 If ‖H₁(ω_mid) − H₂(ω_mid)‖ < ε, stop;
5 Otherwise, apply AWE at ω_mid and repeat Steps 3 and 4 for each
  frequency subinterval.
```

In the practical implementation of the algorithm, the error $\|H_1(\omega) - H_2(\omega)\|$ is not calculated at the center frequency, but over the range $(\omega_{mid} - \Delta\omega < \omega < \omega_{mid} + \Delta\omega)$.

To offer a sample from the successful application of the CFH-assisted AWE process to the generation of the broadband response of a linear electromagnetic device, the calculation of the input reflection coefficient of a microstrip patch antenna is considered (see Fig. 11.1). While the magnitude of the reflection coefficient is plotted over the frequency range 0 - 25 GHz, the finite element solutions at frequencies 8 GHz and 18 GHz were used to generate Padé approximations of order 8 at these two points. In the spirit of the binary search-based CFH process, the responses provided by the two approximations were compared in the vicinity of the mid-frequency point (13 GHz). Lack of good agreement between the two responses necessitated the generation of a third Padé approximation of order 7 at 13 GHz. Comparison of the response obtained by the new approximation to those obtained by the 8-GHz and 18-GHz approximations at frequencies 10.5 GHz and 15.5 GHz, respectively, were found in good agreement, prompting the termination of the CFH process.

It should be mentioned that the plotted values for the magnitude of the reflection coefficient outside the CFH interval, that is, over the frequency ranges of $0 - 8$ GHz and $18 - 25$ GHz, were obtained, respectively, from the calculated Padé approximations at 8.0 GHz and 18.0 GHz. The fact that the predicted values at the lower end of the $0 - 8$ GHz

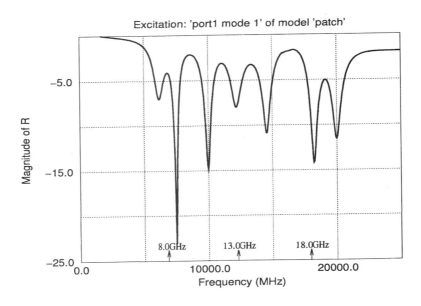

Figure 11.1   Magnitude of the reflection coefficient of a microstrip patch antenna generated by means of the AWE process.

range converge correctly to the physically consistent, capacitive behavior of the antenna in the dc limit, serves as an indicator of the potential of the AWE process to extrapolate the frequency-domain response of electromagnetic devices to frequency regions in which numerical electromagnetic solution may be hindered by lack of robustness and/or accuracy.

11.2   KRYLOV SUBSPACE-BASED MODEL ORDER REDUCTION

The lack of numerical robustness of the AWE process is intimately connected to the condition number of the matrix of moments used in (11.15) for the calculation of the coefficients in the denominator of the Padé approximation. More specifically, the condition number of the matrix worsens as the order increases. An inspection of the matrix suggests that this worsening implies a progressively more severe linear dependence of the rows of the matrix, which stems from the fact that higher-order moments are almost equal to each other. This, in turn, and in view of (11.4), must be attributed to the increasingly worsening (for larger values of n) linear dependence of the vectors

$$\left\{ r, Ar, A^2r, \cdots, A^{2n-1}r \right\} \tag{11.16}$$

generated by the AWE process. More specifically, a recollection of the power method for the iterative extraction of matrix eigenvectors [37] reveals that the above sequence quickly converges to the dominant eigenvector with eigenvalue in the vicinity of the expansion point s_0. Thus the higher-order vectors are almost linearly dependent with one another, thus resulting in almost indistinguishable higher-order moments. Consequently, the moment matrix is almost singular and the subsequent computation of the Padé approximation coefficients in (11.15) becomes numerically unreliable if not impossible.

To remedy this shortcoming of the AWE process, model order reduction methodologies based on the numerically stable construction of Krylov subspaces for the state-space system matrix have been developed. Two of these methodologies are discussed in detail next.

11.2.1 Padé Via Lanczos process

The most obvious way in which the poor conditioning of the AWE moment matrix can be improved is through the construction of an orthonormal basis for the Krylov subspace (11.16). The notation

$$\mathcal{K}_n(A, r) = \text{span} \left\{ r, Ar, A^2r, \cdots, A^n r \right\} \tag{11.17}$$

will be used in the following for referring to a Krylov subspace generated by the vector r and matrix A. If A is symmetric, the Lanczos process can be used to construct an orthonormal basis for the Krylov subspace. The process reduces the symmetric matrix A into a tri-diagonal matrix $T_{n \times n}$ that satisfies the following matrix relation [37]:

$$A V_{N \times n} = V_{N \times n} T_{n \times n} + v_{n+1} e_n^T. \tag{11.18}$$

In the above equation the matrix $V_{N \times n}$ contains the vectors that constitute the orthonormal basis for $\mathcal{K}_n(A, r)$. $T_{n \times n}$ is the symmetric tri-diagonal matrix defined in (4.70), while e_n is a vector of zeros except for its n-th entry which has a value of one.

Once the tri-diagonal matrix is generated through the Lanczos process in Algorithm (4.3), the original model of (11.1) is reduced in the following manner.

Let $\delta = s - s_0$, where s_0 is a complex frequency expansion point. Then, the Laplace-domain form of the model may be written as follows:

$$\begin{cases} (I - \delta A)x = ru \\ y = l^T x, \end{cases} \tag{11.19}$$

where the matrices A and r are defined in (11.4). Thus the transfer function of the original model is formally obtained through the following matrix operations:

$$H(s) = \frac{y(s)}{u(s)} = l^T (G + sC)^{-1} b = l^T (I - \delta A)^{-1} r. \tag{11.20}$$

Next, the orthonormal basis vectors $V_{N \times n}$ of $\mathcal{K}_n(A, r)$ are used to define a reduced vector, $\tilde{x}$, of length n, through the transformation

$$x = V_{N \times n} \tilde{x}. \tag{11.21}$$

Substitution of (11.21) into (11.19), followed by multiplication on the left by $V_{N \times n}^T$ yields the transfer function of the reduced model

$$\tilde{H}_n(s) = \frac{y(s)}{u(s)} = l^T V_{N \times n} (I - \delta T_{n \times n})^{-1} V_{N \times n}^T r. \tag{11.22}$$

In deriving this result use was made of the fact that

$$V_{N \times n}^T A V_{N \times n} = T_{n \times n}. \tag{11.23}$$

We will show, next, that the first n moments of $H(s)$ match those of $\tilde{H}_n(s)$. More specifically, we will show that the following relationship holds:

$$l^T A^i r = l^T V_{N \times n} T_{n \times n}^i V_{N \times n}^T r, \qquad (i = 0, 1, \cdots, n - 1). \tag{11.24}$$

To prove this result we need to show that it is

$$A^i r = V_{N \times n} T_{n \times n}^i V_{N \times n}^T r, \qquad (i = 0, 1, \cdots, n - 1). \tag{11.25}$$

Since the Krylov subspace is built from r, the first column vector of V is $r/\|r\|$. It is, then,

$$V_{N \times n}^T r = \begin{bmatrix} \|r\| \\ 0 \\ \vdots \\ 0 \end{bmatrix} \tag{11.26}$$

and thus (11.25) is valid for $i = 0$. Assume, next, that (11.25) is correct up to $i = k - 1$. That is,

$$A^{k-1} r = V_{N \times n} T_{n \times n}^{k-1} V_{N \times n}^T r. \tag{11.27}$$

Multiplication on the left by A and use of (11.18) yields

$$A^k r = A V_{N \times n} T_{n \times n}^{k-1} V_{N \times n}^T r = V_{N \times n} T_{n \times n}^k V_{N \times n}^T r + v_{n+1} e_n^T T_{n \times n}^{k-1} V_{N \times n}^T r. \tag{11.28}$$

In view of (11.26), the last term in the above equation becomes

$$e_n^T T_{n \times n}^{k-1} V_{N \times n}^T r = \|r\| T_{n \times n}^{k-1}(n, 1), \tag{11.29}$$

where the notation $T_{n \times n}^{k-1}(n, 1)$ is used to denote the element of $T_{n \times n}^{k-1}$ at row n and column 1. However, since $T_{n \times n}$ is a tri-diagonal matrix, $T_{n \times n}^{k-1}(n, 1)$ and thus the last term in (11.28) is zero until $k = n - 1$. This completes the proof of (11.25).

While G and C are symmetric in several cases, the matrix A, in (11.4), is not. However, if G and C are symmetric, it is possible to develop a formal statement for the transfer function of the system in terms of a symmetric matrix A. Toward this objective, consider the Cholesky factorization of $G + s_0 C$,

$$G + s_0 C = PP^T, \tag{11.30}$$

where P is a low-triangular matrix. Its substitution in (11.2) yields

$$H(s) = \left(P^{-1} l\right)^T \left(I + \delta P^{-1} C \left(P^{-1}\right)^T\right)^{-1} P^{-1} b. \tag{11.31}$$

Clearly, the matrix $A = -P^{-1} C \left(P^{-1}\right)^T$ is symmetric.

The case where the matrices G and C are not symmetric is more challenging. In order to reduce a non-symmetric matrix A to a tri-diagonal matrix, two sets of Lanczos vectors are required. We refer to these sets of vectors as the *left* and *right* Lanczos vectors. They span, respectively, the *left* and *right* Krylov subspaces, $\mathcal{K}_n(A^T, l)$ and $\mathcal{K}_n(A, r)$, defined as follows:

$$\begin{aligned} \mathcal{K}_n(A, r) &= \text{span} \left\{ r, Ar, A^2 r, \cdots, A^{n-1} r \right\}, \\ \mathcal{K}_n(A^T, l) &= \text{span} \left\{ l, A^T l, (A^T)^2 l, \cdots, (A^T)^{n-1} l \right\}. \end{aligned} \tag{11.32}$$

Let $V_{N \times n}$ be the matrix with columns the basis vectors of the right subspace $\mathcal{K}_n(A, r)$, and $W_{N \times n}$ the matrix with columns the basis vectors of the left subspace $\mathcal{K}_n(A^T, l)$. The two subspaces are constructed to be bi-orthogonal. The recurrences used for their generation can be summarized as follows:

$$\begin{cases} AV_{N \times n} = V_{N \times n} T_{n \times n} + v_{n+1} e_n^T \\ A^T W_{N \times n} = W_{N \times n} T_{n \times n}^T + w_{n+1} e_n^T \\ W_{N \times n}^T V_{N \times n} = I_{n \times n}, \end{cases} \tag{11.33}$$

where $I_{n \times n}$ is the identity matrix of dimension n, and $T_{n \times n}$ is a tri-diagonal matrix, which is not symmetric in general,

$$T_{n \times n} = \begin{bmatrix} \alpha_1 & \gamma_1 & & & \\ \beta_2 & \alpha_2 & \gamma_2 & & \\ & \beta_3 & \alpha_3 & \gamma_3 & \\ & & \ddots & \ddots & \ddots \\ & & & \beta_m & \alpha_m \end{bmatrix}. \tag{11.34}$$

Comparison of (11.33) with (11.18) reveals that for the case where A is symmetric and $l = r$ the right and left Lanczos subspace $V_{N \times n}$ and $W_{N \times n}$ are the same and $T_{n \times n}$ becomes a symmetric tri-diagonal matrix. The pertinent Lanczos algorithm for the construction of V and W is given below.

> **Algorithm (11.3): Lanczos Algorithm for a Non-Symmetric Matrix**
> 1 $v_0 \leftarrow 0$ and $w_0 \leftarrow 0$; $\quad \beta_1 = \|l^T r\|$; $\quad v_1 \leftarrow r/\beta_1$ and $w_1 \leftarrow l/\beta_1$;
> 2 for $i = 1 \cdots m$
>   2.a $\tilde{v}_{i+1} \leftarrow A v_i$;
>   2.b $\alpha_i = \langle w_i, \tilde{v}_{i+1} \rangle$;
>   2.c $\tilde{v}_{i+1} \leftarrow \tilde{v}_{i+1} - \gamma_{i-1} v_{i-1} - \alpha_i v_i$;
>   2.d $\tilde{w}_{i+1} \leftarrow A^T w_i$;
>   2.e $\tilde{w}_{i+1} \leftarrow \tilde{w}_{i+1} - \beta_i w_{i-1} - \alpha_i w_i$;
>   2.f $\beta_{i+1} = \langle \tilde{w}_{i+1}, \tilde{v}_{i+1} \rangle$; $\quad v_{i+1} = \tilde{v}_{i+1}/\beta_{i+1}$ and $w_{i+1} = \tilde{w}_{i+1}/\beta_{i+1}$;

Since in this case matrix-vector multiplications with both A and A^T are required, this algorithm is computationally twice as expensive as Algorithm (4.3). Once V and W have been generated, the transfer function for the reduced-order system may be constructed in the following manner. Substitution of (11.21) into (11.19), followed by multiplication on the left by W^T yields

$$\tilde{H}_n(s) = \frac{y(s)}{u(s)} = l^T V_{N \times n} (I - \delta T_{n \times n})^{-1} W_{N \times n}^T r, \tag{11.35}$$

where use was made of the fact that $W_{N \times n}^T A V_{N \times n} = T_{n \times n}$. This model order reduction process is referred to as *Padé Via Lanczos* (PVL) [12].

It can be shown that the first $2n$ moments of $H(s)$ match those of $\tilde{H}_n(s)$. This amounts to proving that the following relationship holds:

$$l^T A^i r = l^T V_{N \times n} T_{n \times n}^i W_{N \times n}^T r, \quad (i = 0, 1, \cdots, 2n - 1). \tag{11.36}$$

For $i < 2n - 1$, we write $i = j + k$, where both j and k are less than $n - 1$. From (11.25), we have

$$\begin{aligned} A^j r &= V_{N \times n} T_{n \times n}^j V_{N \times n}^T r, \\ (A^T)^k l &= W_{N \times n} \left(T_{n \times n}^T \right)^k W_{N \times n}^T l. \end{aligned} \tag{11.37}$$

The inner product of the transpose of the second vector with the first yields

$$l^T A^i r = l^T W_{N \times n} T^i_{n \times n} V^T_{N \times n} r, \quad (i < 2n - 1). \tag{11.38}$$

If $i = 2n - 1$, from (11.28) we have

$$A^n r = V_{N \times n} T^n_{n \times n} V^T_{N \times n} r + v_{n+1} e^T_n T^{n-1}_{n \times n} V^T_{N \times n} r,$$
$$(A^T)^{n-1} l = W_{N \times n} \left(T^T_{n \times n} \right)^{n-1} W^T_{N \times n} l. \tag{11.39}$$

Taking the inner product of the transpose of the second vector with the first one, and noting that it is $W^T_{N \times n} v_{n+1} = 0$, we obtain

$$l^T A^{2n-1} r = l^T W_{N \times n} T^{2n-1}_{n \times n} V^T_{N \times n} r. \tag{11.40}$$

This completes the proof that the first $2n$ moments of $\tilde{H}_n(s)$ match those of $H(s)$.

11.2.2 Arnoldi process: SISO system

An alternative to the PVL process for the case of non-symmetric matrices is the use of the Arnoldi Algorithm (4.1) to reduce A into an upper Hessenberg matrix $H_{n \times n}$ as follows:

$$A V_{N \times n} = V_{N \times n} H_{n \times n} + v_{n+1} e^T_n, \tag{11.41}$$

where $H_{n \times n}$ has the structure defined in (4.56) and $V_{N \times n}$ contains the orthonormal basis of the Krylov subspace (11.17). Using the above reduction of A into $H_{n \times n}$, the reduced-order approximation of the transfer function is obtained in the following manner. Use of the projection (11.21) of the original state vector to the reduced one in (11.19), followed by multiplication on the left by $V^T_{N \times n}$ yields

$$\tilde{H}_n(s) = \frac{y(s)}{u(s)} = l^T V_{N \times n} (I - \delta H_{n \times n})^{-1} V^T_{N \times n} r, \tag{11.42}$$

where use was made of the following result from (11.41) $V^T_{N \times n} A V_{N \times n} = H_{n \times n}$.

It can be shown that the first n moments of $\tilde{H}_n(s)$ match those of $H(s)$ by proving that

$$l^T A^i r = l^T V_{N \times n} H^i_{n \times n} V^T_{N \times n} r, \quad (i = 0, 1, \cdots, n - 1). \tag{11.43}$$

The proof of this follows from the proof of the following relation:

$$A^i r = V_{N \times n} H^i_{n \times n} V^T_{N \times n} r, \quad (i = 0, 1, \cdots, n - 1). \tag{11.44}$$

To prove this we note first that, since the initial vector used for the construction of the Krylov subspace is r, the first column vector of $V_{N \times n}$ is $r / \|r\|$; hence, $V^T_{N \times n} r$ has the form of (11.26). Thus (11.44) is valid for $i = 0$.

Assume, next, that (11.44) is valid up to $i = k - 1$; hence, it is

$$A^i r = V_{N \times n} H^i_{n \times n} V^T_{N \times n} r. \tag{11.45}$$

Multiplication of the above expression on the left by A, followed by substitution of (11.41) in the resulting equation yields

$$A^k r = A V_{N \times n} H^{k-1}_{n \times n} V^T_{N \times n} r$$
$$= V_{N \times n} H^k_{n \times n} V^T_{N \times n} r + v_{n+1} e^T_n H^{k-1}_{n \times n} V^T_{N \times n} r. \tag{11.46}$$

In view of (11.26) the last term is

$$e_n^T H_{n \times n}^{k-1} V_{N \times n}^T r = \|r\| H_{n \times n}^{k-1}(n, 1), \qquad (11.47)$$

where $H_{n \times n}^{k-1}(n, 1)$ denotes the n-th element in the first column of $H_{n \times n}^{k-1}$. However, since $H_{n \times n}$ is a Hessenberg matrix of lower bandwidth of one, the elements $H_{n \times n}^{k-1}(n, 1)$ are zero for $k \leq n - 1$. Hence, we conclude that (11.44) holds for $k \leq n - 1$. This completes the proof of (11.43).

11.2.3 Arnoldi process: MIMO system

Up till now our discussions have considered a single-input, single-output system; thus l and r are column vectors. For the case of multiple-input, multiple-output (MIMO) systems, matrices L and R, of dimension $N \times p_o$, and $N \times p_i$, respectively, are involved in the state-space description (11.19) of the linear system, where p_i is the number of inputs and p_o the number of outputs. For most applications pertinent to reduced-order macromodeling of electromagnetic devices and systems, the number of input and output ports are equal. Thus, in the following, we will limit ourselves to the reduced-order macromodeling of MIMO systems with $p_i = p_o = p$.

The state-space model of (MIMO) is

$$\begin{cases} C\dfrac{dx}{dt} = -Gx + Bu(t) \\ y(t) = L^T x(t), \end{cases} \qquad (11.48)$$

where $C, G \in \mathbb{R}^{N \times N}$ are real matrices and $L, B \in \mathbb{R}^{N \times p}$ are given column vectors. The vector x is the vector of the state variables in the system, u is the excitation (input) vector and y is the output of interest. Then, the Laplace transform of the above system leads to the following expression for its transfer function matrix:

$$\begin{aligned} y(s) &= L^T (G + sC)^{-1} Bu(s) = H(s)u(s) \Rightarrow \\ H(s) &= L^T (G + sC)^{-1} B = L^T (I - \delta A)^{-1} R, \end{aligned} \qquad (11.49)$$

where $A = -(G + s_0 C)^{-1} C$ and $R = (G + s_0 C)^{-1} B$.

For the case of MIMO systems, the aforementioned Lanczos and Arnoldi processes are replaced by their *block* counterparts. The block-Arnoldi process is discussed next. For the discussion of the block-Lanczos algorithm the reader is referred to [31].

Using R as the starting matrix, the Krylov subspace used in the Arnoldi process is

$$\mathcal{K}_n(A, R) = \text{span}\left\{R, AR, A^2 R, \ldots, A^{\hat{n}-1} R, A^{\hat{n}} r_0, A^{\hat{n}} r_1, \ldots A^{\hat{n}} r_{l-1}\right\}, \qquad (11.50)$$

where n is the order of the subspace, r_i is the ith column of R, $\hat{n} = \lfloor n/p \rfloor$, and $l = n - \hat{n} p$. In addition to the generation of the orthonormal basis for the Krylov subspace, an upper (block) Hessenberg matrix $H_{n \times n}$ is generated, such that

$$A V_{N \times n} = V_{N \times n} H_{n \times n} + V_{N \times l} [\, 0, \; 0 \; \ldots, I_{l \times l}]_{l \times n}. \qquad (11.51)$$

The matrix $V_{N \times n}$ acts as the projection matrix of the original matrix A of dimension N to the Hessenberg matrix $H_{n \times n}$. In matrix form, the above statement is

$$
A V_{N \times n} = V_{N \times n}
\begin{bmatrix}
h_{1,1} & h_{1,2} & \cdots & * & \cdots & * \\
\vdots & \vdots & & \vdots & & \vdots \\
h_{p+1,1} & h_{p+1,2} & \cdots & * & \cdots & * \\
& h_{p+2,2} & & & & \\
& & \ddots & \vdots & & \vdots \\
& & & h_{n,\hat{n}p} & \cdots & h_{n,n}
\end{bmatrix}
$$
$$
+
\begin{bmatrix}
0 & \cdots & 0 & v_{1,\hat{n}p+1} & \cdots & v_{1,n} \\
\vdots & & \vdots & \vdots & & \vdots \\
0 & \cdots & 0 & v_{N,\hat{n}p+1} & \cdots & v_{N,n}
\end{bmatrix} .
\tag{11.52}
$$

Except for the block form of R the block Arnoldi algorithm is similar to the one used for the SISO case. Its pseudo-code description is given below.

Algorithm (11.4): Block Arnoldi Algorithm, $\mathcal{K}_n(A, R)$

```
1 Solve (G + s₀ C) R = B for R;
2 V₀ = qr(R);                                    (QR factorization of R)
3 For k = 0, 1, ···, ⌊n/p⌋ − 1
   3.a M = CVₖ;
   3.b Solve (G + s₀ C)Vₖ₊₁ = −M for Vₖ₊₁;
   3.c For j = 0, ···, k
      3.c.1  H = VⱼᵀVₖ₊₁;
      3.c.2  Vₖ₊₁ = Vₖ₊₁ − Vⱼ H;
   3.d Vₖ₊₁ = qr(Vₖ₊₁);                         (QR factorization of Vₖ₊₁)
4 Set V = [V₀, V₁, ···, V⌊n/p⌋] and truncate V to have n columns only.
```

Once $V_{N \times n}$ and $H_{n \times n}$ are available, the reduced-order approximation of the transfer function matrix of the MIMO systems in constructed as follows: First, the reduced-order state vector is formed through the projection (11.21). Its introduction in (11.49) followed by multiplication on the left by $V_{N \times n}^T$ yields

$$
\tilde{H}_n(s) = L^T V_{N \times n} (I - \delta H_{n \times n})^{-1} V_{N \times n}^T R.
\tag{11.53}
$$

The reduced transfer function matrix, $\tilde{H}_n(s)$, is of dimension $p \times p$.

It can be shown that the first $\lfloor n/p \rfloor$ moments of $\tilde{H}_n(s)$ match those of $H(s)$ by proving that it is

$$
L^T A^i R = L^T V_{N \times n} H_{n \times n}^i V_{N \times n}^T R, \quad (i = 0, 1, \cdots, \lfloor n/p \rfloor - 1).
\tag{11.54}
$$

To prove this it suffices to show the following:

$$
A^i R = V_{N \times n} H_{n \times n}^i V_{N \times n}^T R, \quad (i = 0, 1, \cdots, \lfloor n/p \rfloor - 1).
\tag{11.55}
$$

We begin by noting that, since R is used for the construction of the initial vector in the Krylov subspace, the above equation holds for $i = 0$. Furthermore, it is

$$
V_{N \times n}^T R =
\begin{bmatrix}
U_{p \times p} \\
0 \\
\vdots \\
0
\end{bmatrix},
\tag{11.56}
$$

where $U_{p \times p}$ is an upper triangular matrix of dimension p. Assume, next, that (11.55) holds for values of i up to and equal to $k - 1$. Multiplication on the left of the equation for $i = k - 1$ by A, followed by substitution of (11.51) in the resulting expression, yields

$$A^k R = A V_{N \times n} H_{n \times n}^{k-1} V_{N \times n}^T R$$
$$= V_{N \times n} H_{n \times n}^k V_{N \times n}^T R + V_{N \times l} [\, 0,\ 0\ \ldots, I_{l \times l}]_{l,n}\ H_{n \times n}^{k-1} V_{N \times n}^T R. \tag{11.57}$$

The last term in the above expression is

$$[\, 0,\ 0\ \ldots, I_{l \times l}]_{l,n}\ H_{n \times n}^{k-1} V_{N \times n}^T R = [\, 0,\ 0\ \ldots, I_{l \times l}]_{l,n}\ H_{n \times n}^{k-1} \begin{bmatrix} U_{p \times p} \\ 0 \\ \vdots \\ 0 \end{bmatrix}. \tag{11.58}$$

Since $H_{n \times n}$ is a Hessenberg matrix with lower bandwidth of $p + 1$, the above equation is zero for values of k up to and including $\lfloor n/p \rfloor - 1$. This completes the proof of (11.55).

Krylov subspace-based, model order reduction techniques offer a numerically more stable alternative to the AWE process. In the context of large-scale circuit modeling and simulation, an important attribute of the generated reduced-order models is the compatibility of the generated reduced state-space model with the popular, SPICE-like, general-purpose, network analysis-oriented, non-linear, transient simulators. More specifically, in their state-space form, the Lanczos and Arnoldi macromodels of (11.20, 11.35, 11.42, 11.53) can be incorporated directly in such simulators. However, prior to doing so, extra care is required to ensure that the generated reduced-order models are passive.

At this point, it is appropriate to recall the necessary and sufficient condition for the transfer function, $H(s)$, of a linear system to be passive [34]:

1. $H(s^*) = H^*(s)$ for all complex s, where $*$ denotes complex conjugation.

2. $H(s)$ is a positive matrix, that is, $z^{*T}(H(s) + H^T(s^*))z \geq 0$ for all complex value of s satisfying $Re(s) > 0$ and for any complex vector z.

The passivity requirement for the generated reduced model stems from an important result in circuit theory that states that interconnections of stable systems may not necessarily be stable; however, strictly passive circuits are asymptotically stable, and arbitrary interconnections of strictly passive circuits are strictly passive, and thus asymptotically stable [32]. Thus ensuring the passivity, not just stability, of the generated macromodel is of critical importance in the development of an effective and useful model order reduction process.

From the various methodologies proposed for rendering Krylov subspace-based, reduced-order models passive, we choose to discuss the so-called PRIMA process [17]. The reader is referred to the literature for details on other passive reduced-order modeling techniques (e.g., [14], [18] and the references within).

11.3 PASSIVE REDUCED-ORDER INTERCONNECT MACROMODELING ALGORITHM (PRIMA)

As the name indicates, PRIMA was proposed in [17] in the context of reduction of transmission-line models of high-speed electronic interconnects. Subsequently, its application to the passive reduced-order macro-modeling of discrete forms of electromagnetic systems was presented in [30]. In the following, the PRIMA process is presented for the case of the MIMO system of (11.48).

The first step in PRIMA involves the construction of a set of orthonormal vectors $V_{N \times n}$ for the Krylov subspace using the block-Arnoldi Algorithm (11.4). The orthonormal basis of the generated Krylov subspace contains the eigenvectors corresponding to the dominant eigenvalues of A, which are the ones with the largest magnitude. Thus the projection of the original model of (11.19) onto the constructed Krylov subspace is expected to generate a reduced-order model that captures the dominant characteristics of the original system. More specifically, the poles of the system closest to the expansion frequency s_0 are expected to be the ones present in the reduced model.

The second step in the PRIMA process involves the use of the matrix $V_{N \times n}$ as the congruence transformation matrix in the reduction process of the original system. More specifically, the reduced-order state vector is defined through the mapping,

$$x = V_{N \times n} \tilde{x}. \tag{11.59}$$

Its substitution into (11.48), followed by multiplication on the left by $V_{N \times n}^T$, yields

$$\begin{cases} s V_{N \times n}^T C V_{N \times n} \tilde{x} = -V_{N \times n}^T G V_{N \times n} \tilde{x} + V_{N \times n}^T B, \\ y = L^T V_{N \times n} \tilde{x}. \end{cases} \tag{11.60}$$

Thus the matrices in the state-space form of the reduced model are obtained from the original ones through the following congruence transformations:

$$\tilde{C} = V_{N \times n}^T C V_{N \times n}, \quad \tilde{G} = V_{N \times n}^T G V_{N \times n}. \tag{11.61}$$

Furthermore, the reduced forms of L and B are given by

$$\tilde{B} = V_{N \times n}^T B, \quad \tilde{L} = V_{N \times n}^T L. \tag{11.62}$$

In view of the above, the reduced-order transfer function is

$$\tilde{H}_n(s) = \tilde{L}^T \left(\tilde{G} + s \tilde{C} \right)^{-1} \tilde{B}. \tag{11.63}$$

Since the size of $\tilde{G}$ and $\tilde{C}$ is typically very small, it is easy to factor $(\tilde{G} + s \tilde{C})$. An alternative form of $\tilde{H}_n(s)$ in terms of the poles of the reduced transfer function can be developed. This is discussed further in Section 11.3.4.

11.3.1 Preservation of moments in PRIMA

In this section it is shown that the transfer function of the reduced-order system (11.63) preserves the $\lfloor n/p \rfloor$ block moments of the original system (11.20). The block moments, M_i, of the original system are

$$M_i = L^T A^i R, \tag{11.64}$$

where $A = -(G + s_0 C)^{-1} C$ and $R = (G + s_0 C)^{-1} B$. The block moments of the reduced system are given by

$$\tilde{M}_i = \tilde{L}^T \tilde{A}^i \tilde{R}, \tag{11.65}$$

where $\tilde{A} = -(\tilde{G} + s_0 \tilde{C})^{-1} \tilde{C}$ and $\tilde{R} = (\tilde{G} + s_0 \tilde{C})^{-1} \tilde{B}$. To prove that the two sets of moments are equal for $i \leq \lfloor n/p \rfloor - 1$, we need to show that it is

$$A^i R = V_{N \times n} \tilde{A}^i \tilde{R}, \quad (i = 0, 1, \cdots, \lfloor n/p \rfloor - 1). \tag{11.66}$$

For $i = 0$, we have

$$
\begin{aligned}
V\tilde{R} &= V(\tilde{G} + s_0\tilde{C})^{-1}\tilde{B} \\
&= V(\tilde{G} + s_0\tilde{C})^{-1}V^T B \\
&= V(\tilde{G} + s_0\tilde{C})^{-1}V^T(G + s_0C)R \\
&= V(\tilde{G} + s_0\tilde{C})^{-1}V^T(G + s_0C)VV^T R \\
&= V(\tilde{G} + s_0\tilde{C})^{-1}(\tilde{G} + s_0\tilde{C})V^T R = R,
\end{aligned}
\tag{11.67}
$$

since $VV^T R = R$. Assume, next, that (11.66) is valid for $i \leq k - 1$. Considering the equality for $i = k - 1$ we will show that the kth moments are also equal. Multiplication on the left by A yields

$$
\begin{aligned}
& A^k R = AV\tilde{A}^{k-1}\tilde{R} \\
\Rightarrow\quad & VH_{n\times n}^k V^T R = AV\tilde{A}^{k-1}\tilde{R} \\
\Rightarrow\quad & (G + s_0C)VH_{n\times n}^k V^T R = -CV\tilde{A}^{k-1}\tilde{R} \\
\Rightarrow\quad & V^T(G + s_0C)VH_{n\times n}^k V^T R = -V^T CV\tilde{A}^{k-1}\tilde{R} \\
\Rightarrow\quad & H_{n\times n}^k V^T R = -(\tilde{G} + s_0\tilde{C})^{-1}\tilde{C}\tilde{A}^{k-1}\tilde{R} \\
\Rightarrow\quad & H_{n\times n}^k V^T R = \tilde{A}^k\tilde{R} \\
\Rightarrow\quad & V^T H_{n\times n}^k V^T R = V^T\tilde{A}^k\tilde{R} \\
\Rightarrow\quad & A^k R = V\tilde{A}^k\tilde{R},
\end{aligned}
\tag{11.68}
$$

where use was made of (11.55). This completes the proof that the block moments of the original and the reduced system are equal for $i \leq \lfloor n/p \rfloor - 1$.

The number of block moments matched in the block-Arnoldi-based PRIMA is the same with the number of block moments matched by the block Arnoldi algorithm. This fact prompts us to compare the two reduced models.

First, we note that the transfer function (11.63) generated through PRIMA may be cast in the form

$$
\tilde{H}_n(s) = \tilde{L}^T\left(I - \delta\tilde{A}\right)^{-1}\tilde{R},
\tag{11.69}
$$

where $\tilde{A} = -(\tilde{G} + s_0\tilde{C})^{-1}\tilde{C}$ and $\tilde{R} = (\tilde{G} + s_0\tilde{C})^{-1}\tilde{B}$. This form should be compared with (11.53). In view of (11.62) and the fact that, from (11.67), it is $V^T R = \tilde{R}$, the comparison of the two transfer functions boils down to the comparison of the matrix $\tilde{A}$ with the Hessenberg matrix $H_{n\times n}$. However, we have shown that the block moments in the block Arnoldi process are equal to those in PRIMA (11.68). This equality may be cast in the form

$$
\tilde{A}^i\tilde{R} = H_{n\times n}^i V^T R = H_{n\times n}^i\tilde{R}, \quad (i = 0, \cdots, \lfloor n/p \rfloor - 1).
\tag{11.70}
$$

Considering this result for $i = 1$, and in view of the matrix form of $V^T R$ in (11.56), we conclude that the first p column vectors in $\tilde{A}$ and $H_{n\times n}$ are the same. Since $H_{n\times n}$ is an upper block Hessenberg matrix with lower bandwidth of $p + 1$, the first p columns of $H_{n\times n}^i$ depend on the first ip columns of $H_{n\times n}$. Thus through the recursive evaluation of (11.70) we conclude that the matrices $\tilde{A}$ and $H_{n\times n}$ are the same up to the last p columns.

11.3.2 Preservation of passivity

Reduced-order model representation of passive electromagnetic components and networks in terms of frequency-dependent matrix transfer functions is most useful when these components constitute parts of more complex functional blocks. For the purposes of design-driven simulation at the functional block level, a network-oriented simulation approach is used. The incorporation of the reduced-order models for the electromagnetic components in a network analysis-oriented circuit simulator may be effected either through convolution, utilizing the pole-residue representations of the elements of the transfer function matrix, or through the direct incorporation of the state-space models of the reduced system in the circuit simulator [33]. In either case, the passivity of the reduced system needs to be ensured in order to avoid non-physical instabilities in the subsequent simulation of the overall circuit.

The two steps in the development of the reduced macromodel that can impact passivity are the development of the discrete model of the physical system under consideration and the subsequent model order reduction. For the state-space representations of both the original discrete system and its subsequent reduced-order model, the necessary and sufficient conditions for passivity of the transfer function $H(s) = (G + sC)^{-1}$ can be considered in terms of the properties of the matrices C and G. For example, the condition that $H(s^*) = H^*(s)$ for all complex s, is automatically satisfied if C and G are real matrices. To examine the constraints on C and G for $H(s)$ to be a positive matrix, we consider the scalar quantity $z^{*T} \left(H(s) + H^T(s^*) \right) z$, where z is a complex vector,

$$z^{*T} \left[(G + sC)^{-1} + (G^T + s^*C^T)^{-1} \right] z$$
$$= z^{*T} (G^T + s^*C^T)^{-1} \left[(G + sC) + (G^T + s^*C^T) \right] (G + sC)^{-1} z. \tag{11.71}$$

Setting $w = (G + sC)^{-1}z$ and $s = \alpha + j\omega$, in the above result we obtain that

$$z^{*T} \left(H(s) + H^T(s^*) \right) z$$

is positive provided that it is

$$w^{*T}[G + G^T + \alpha(C + C^T)]w > 0, \quad (\alpha > 0). \tag{11.72}$$

Assume, next, that the discrete model of (11.19) is passive; thus C and G satisfy (11.72). We would like to investigate the passivity of the PRIMA-generated reduced-order model

$$\tilde{H}(s) = (\tilde{G} + s\tilde{C})^{-1}, \tag{11.73}$$

where $\tilde{G} = V^T G V$ and $\tilde{C} = V^T C V$. To begin with, we note that if the projection matrix V is real, the reduced matrices $\tilde{G}$ and $\tilde{C}$ are real; hence, the first requirement for passivity is satisfied. Clearly, the projection matrix V is real if the expansion point, s_0, is taken to be real. As far as the second condition for passivity is concerned, we have

$$z^{*T} \left[(\tilde{G} + s\tilde{C})^{-1} + (\tilde{G}^T + s^*\tilde{C}^T)^{-1} \right] z$$
$$= z^{*T} (\tilde{G}^T + s^*\tilde{C}^T)^{-1} \left[(\tilde{G} + s\tilde{C}) + (\tilde{G}^T + s^*\tilde{C}^T) \right] (\tilde{G} + s\tilde{C})^{-1} z \tag{11.74}$$
$$= z^{*T} (\tilde{G} + s^*\tilde{C})^{-T} V^T \left[(G + G^T) + (sC + s^*C) \right] V (\tilde{G} + s\tilde{C})^{-1} z.$$

Setting $w = V(\tilde{G} + s\tilde{C})^{-1}z$ and $s = \alpha + j\omega$ in the above equation we obtain that $z^{*T} \left(\tilde{H}(s) + \tilde{H}^T(s^*) \right) z$ is positive provided that it is

$$w^{*T}[G + G^T + \alpha(C + C^T)]w > 0, \quad (\alpha > 0). \tag{11.75}$$

Clearly, this is nothing else but (11.72), expressing the passivity of the original discrete model. Thus we conclude that, if the original discrete model is passive, the PRIMA process with real expansion frequency point s_0 preserves passivity.

11.3.3 Error estimate in model order reduction

It is to be expected that the accuracy of the reduced-order model is strongly dependent on the order of the model. From the point of view of computational efficiency, especially when the reduced model is to be used in conjunction with other models for system-level analysis, the smaller the order the better. On the other hand, from the point of view of accuracy, and in particular frequency bandwidth of accuracy, the larger the order of the reduced model the better (assuming, of course, that the reduction process is robust enough for roundoff not to be a factor). To accommodate these competing requirements, a means to assess the accuracy of the generated approximation at each step of the iterative model order reduction process is highly desirable.

Toward this objective, and in view of the fact that the response of the original system is not known *a-priori*, a simple way in which the accuracy of the reduced model can be assessed is by monitoring the difference between successive orders of reduction. Once the difference drops below a given tolerance, the reduction process is assumed to have converged and thus is terminated. Unfortunately, a small difference between two successive reduced models may not correlate with good accuracy between the responses of the reduced model and the original one. Therefore, despite its simplicity, such an approach is unreliable. The only effective way to establish a stopping criterion for the reduction process is through the development of an error bound for the reduced-order model.

The development of an error bound for PVL was presented in [35]. In the following, we derive an error bound for the PRIMA process. Our development follows the procedure in [36], and is applicable for an arbitrary expansion frequency point s_0.

The objective is to find an estimate of the difference between the original model (11.19) and the reduced model (11.69); hence, we are interested in an estimate of the quantity

$$e(s) = \tilde{H}_n(s) - H(s) = \tilde{L}^T (I - \delta\tilde{A})\tilde{R} - L^T (I - \delta A)^{-1} R. \tag{11.76}$$

Toward this objective, we begin with the Hessenberg reduction of (11.51)

$$C V = -(G + s_0 C) V H_{n\times n} - (G + s_0 C) X_l, \tag{11.77}$$

where $X_l = V_{N\times l} \begin{bmatrix} 0 & \cdots & 0 & I_{l\times l} \end{bmatrix}$. Its multiplication on the left by V^T yields, after some rearrangement of the terms,

$$-(\tilde{G} + s_0\tilde{C})^{-1}\tilde{C} - (\tilde{G} + s_0\tilde{C})^{-1}V^T(G + s_0 C)X_l = H_{n\times n}. \tag{11.78}$$

Multiplying on the left by V and using the fact that $V H_{n\times n} = AV - X_l$ yields

$$V\tilde{A} - VP + X_l = AV, \tag{11.79}$$

where it is

$$P = (\tilde{G} + s_0\tilde{C})^{-1}V^H(G + s_0 C)X_l. \tag{11.80}$$

Multiplying (11.79) by $\delta = s - s_0$ and subtracting V from each side of the resulting equation we obtain

$$V(I - \delta\tilde{A}) + \delta(VP - X_l) = (I - \delta A)V. \tag{11.81}$$

It follows that

$$
\begin{aligned}
V(I - \delta\tilde{A})^{-1} &- (I - \delta A)^{-1}V \\
&= \delta(I - \delta A)^{-1}(VP - X_l)(I - \delta\tilde{A})^{-1}.
\end{aligned}
\tag{11.82}
$$

The desired error estimate can be obtained from the above equation through multiplication on the left by L^T and on the right by $V^T R$

$$
\begin{aligned}
e(s) &= \tilde{H}(s) - H(s) \\
&= L^T(I - \delta\tilde{A})^{-1}V^T R - L^T(I - \delta A)^{-1}R \\
&= \delta L^T(I - \delta A)^{-1}(VP - X_l)(I - \delta\tilde{A})^{-1}V^T R,
\end{aligned}
\tag{11.83}
$$

where use was made of the fact that $VV^T R = R$.

From a computational efficiency point of view, the computation of the term $(I - \delta A)^{-1}$ is undesirable (if not prohibitive). Thus the following approximation is used

$$
(I - \delta A)^{-1} \approx V(I - \delta\tilde{A})^{-1}V^T.
\tag{11.84}
$$

Use of this approximation, along with the fact that $V^T X_l = 0$ and $V^T V = I$, yields the following error estimate for the PRIMA process with arbitrary expansion frequency s_0

$$
\begin{aligned}
e(s) &= \tilde{H}(s) - H(s) \\
&= \tilde{L}(I - \delta\tilde{A})^{-1}\tilde{R} - L^T(I - \delta A)^{-1}R \\
&\approx \delta\tilde{L}(I - \delta\tilde{A})^{-1}(\tilde{G} + s_0\tilde{C})^{-1}V^T(G + s_0 C)X_l(I - \delta\tilde{A})^{-1}\tilde{R}.
\end{aligned}
\tag{11.85}
$$

Computation of this error requires the calculation and storage of $V^T(G + s_0 C)X_l$. Note that $V^T G$ and $V^T C$ are required for the calculation of $\tilde{G}$ and $\tilde{C}$. Thus these matrices are the only ones that need to be stored and then multiplied with X_l for the calculation of $V^T(G + s_0 C)X_l$. All remaining matrix products in (11.85) are inexpensive since the matrices involved are of small size.

11.3.4 Pole-residue representation of the reduced order model

In this section, we consider the attributes of reduced-order modeling of a system from the point of view of the pole-residue representation of the transfer function of the linear system. To begin with, consider the transfer function of the original system (11.1) in the form

$$
H(s) = L^T(I - \delta A)^{-1}R,
\tag{11.86}
$$

where $\delta = s - s_0$, and s_0 is the expansion frequency. The matrices A and R are defined in (11.49).

If the complete eigen-decomposition of A is available, i.e., $A = Q\Lambda Q^{-1}$, (11.86) becomes

$$
H(s) = L^T Q(I - \delta\Lambda)^{-1}Q^{-1}R.
\tag{11.87}
$$

Thus the (i, j) element of $H(s)$ can be written in the following pole-residue form:

$$
H_{i,j}(s) = \sum_{n=0}^{N-1} \frac{u_n^{(i)} v_n^{(j)}}{1 - \delta\lambda_n} = \sum_{n=0}^{N-1} \frac{\lambda_n^{-1}u_n^{(i)} v_n^{(j)}}{\lambda_n^{-1} + s_0 - s} = -\sum_{n=0}^{N-1} \frac{\lambda_n^{-1}u_n^{(i)} v_n^{(j)}}{s - \left(\lambda_n^{-1} + s_0\right)},
\tag{11.88}
$$

where $u^{(i)}$ is the i-th row of $L^T Q$ and $v^{(j)}$ is the j-th column of $Q^{-1} R$.

From (11.88), it is clear that the reciprocals of the eigenvalues of A are the poles of the transfer function matrix $H(s)$. Since the dimension of discrete approximations of electromagnetic structures of practical interest can be in the order of tens or even hundreds of thousands, only a small subset of the poles have a dominant impact on the attributes of the response of the system. Clearly, these poles are associated with the eigenvalues of those eigenmodes of the system that are most influential in the determination of the properties of the output quantities of interest that quantify the system response over the desired frequency bandwidth. It is this subset of the dominant poles that the model order reduction process should capture with sufficient accuracy to ensure the usefulness of the generated reduced-order model.

It is apparent from (11.88) that the dominant eigenvalues of A are the ones with largest magnitudes. It is well known that the Lanczos and the Arnoldi algorithms can converge to the largest eigenvalues prematurely, well before the complete tri-diagonalization or Hessenberg reduction of the original eigenvalue problem is finished [37]. This is the reason why both algorithms have been exploited aggressively for model order reduction purposes, and have provided the foundation for the most successful and robust model-order reduction processes in use today.

A pole-residue representation of the elements of the transfer function of the reduced system is obtained in the following manner: First, the transfer function of the reduced-order model is cast in the form of (11.69), which is repeated here for convenience,

$$\tilde{H}_n(s) = \tilde{L}^T(\tilde{G} + s\tilde{C})^{-1}\tilde{B} = \tilde{L}(I - \delta\tilde{A})^{-1}\tilde{R}. \tag{11.89}$$

An eigen-decomposition of $\tilde{A}$ is performed first

$$\tilde{A} = Q_n\Lambda_n Q_n^{-1}, \tag{11.90}$$

where $\Lambda_n = \text{diag}(\lambda_1, \lambda_2, \ldots, \lambda_n)$. Substitution of (11.90) into (11.89) yields

$$\tilde{H}_n(s) = \tilde{L}^T Q_n(I - \delta\Lambda_n)^{-1} Q_n^{-1}\tilde{R}. \tag{11.91}$$

Let $u^{(i)}$ denote the ith row of $\tilde{L}^H Q_n$, and $v^{(j)}$ the jth column of $Q_n^{-1}\tilde{R}$. The pole-residue representation of the (i, j) element of $\tilde{H}_n(s)$ is given by

$$\tilde{H}_{n,i,j}(s) = \sum_{m=1}^{n} \frac{u_m^{(i)} v_m^{(j)}}{1 - \delta\lambda_m}. \tag{11.92}$$

In our matrix eigen-solution discussions of Chapter 8 it was mentioned that the Lanczos and Arnoldi processes converge fastest to those modes with eigenvalues closest to the expansion frequency s_0. However, "higher-order modes" with eigenvalues that fall outside the frequency bandwidth of interest as defined by the choice of the expansion frequency s_0, do have an impact on the response of the system. A convincing example in the context of electromagnetic applications is provided by the problem of obtaining the input impedance of a coaxial probe feeding a rectangular waveguide [2]. The non-propagating (higher-order) waveguide modes are associated with the reactive energy stored in the immediate vicinity of the probe and thus have a significant impact on the reactive part of the input impedance. In addition to capturing with sufficient accuracy the dominant eigenvalues, the model order reduction process will generate an additional set of poles which provide for an approximation of the effect of the "higher-order" eigenmodes of the system on the approximated transfer function of the reduced model over the bandwidth of interest.

11.4 MODEL ORDER REDUCTION OF ELECTRIC FIELD-BASED FINITE ELEMENT MODELS

In this section we begin the discussion of the application of the aforementioned Krylov subspace model order reduction techniques to the generation of compact, reduced-order transfer function matrices for discrete models of electromagnetic devices and systems. More specifically, we are interested in the reduction of finite element approximations of electromagnetic systems. However, the methodologies to be presented are also suitable for the reduction of discrete electromagnetic models obtained through other types of finite methods.

Our discussion will consider two approaches to the problem of model order reduction of finite method-based approximations of the electromagnetic system. The first one is streamlined toward the reduction of finite element systems resulting from the approximation of the electric field vector Helmholtz equation. The second is most suitable for the reduction of discrete models generated from the discretization of Maxwell's curl equations; hence, it is applicable to discrete models obtained using the FDTD method (e.g., [38]-[44]) and other related finite integral or finite volume discretizations of Maxwell's curl equations (e.g., [45]).

This section is devoted to the discussion of the first approach. Our emphasis is on the finite element model resulting from the discretization of the electric field vector Helmholtz equation using tangentially continuous vector finite elements. The term *E-field* will be used to refer to this model.

Figure 6.19 depicts the generic geometry of a multi-port electromagnetic structure. All media are assumed to be linear, isotropic, passive, and frequency-independent. The development of the weak form of the vector Helmholtz equation for the electric field was presented in Section 3.3.2. The weak statement is repeated below, with the surface integral taken over the enclosing surface S_0 and the surfaces S_i, $i = 1, 2, \ldots, p$, associated with all physical ports through which the electromagnetic structure interacts with other electromagnetic devices. For the purposes of this discussion and without loss of generality, S_0 will be taken to be a numerical grid truncation boundary surface on which absorbing or radiation boundary conditions are imposed. For example, in the development that follows the first-order absorbing boundary condition (ABC) of (5.7) will be imposed on S_0. In a more general sense, S_0 may be though of as an impedance boundary on which a known impedance relationship between the tangential electric and magnetic fields is imposed.

In the Galerkin sense, the testing functions are the same with the expansion functions. Thus, using $\vec{w}$ to denote the tangentially continuous vector finite elements in the TV space W_t discussed in Chapter 2, the weak statement assumes the form

$$\iiint_{\Omega} \left(\nabla \times \vec{w} \cdot \frac{1}{\mu} \nabla \times \vec{E} + s\vec{w} \cdot \sigma\vec{E} + s^2\vec{w} \cdot \epsilon\vec{E} \right) dv = s \iint_{S_0 \cup \sum_{n=1}^{p} S_n} \hat{n} \times \vec{H} \cdot \vec{w}ds,$$

$$(11.93)$$

where $s = j\omega$, and $\hat{n}$ is the outward-pointing unit normal vector on the surface enclosing the computational domain Ω.

On the port surfaces, $S_1, S_2, \ldots, S_p$, the tangential magnetic field is expanded in terms of known magnetic *mode* functions, $\vec{h}_n^{(i)}$, $i = 1, 2, \ldots, p$. In other words, for the purposes of this discussion, it is assumed that each port is associated with a uniform waveguide structure for which the transverse electric and magnetic fields associated with the waveguide eigenmodes can be computed (e.g., using the techniques in Chapter 9). Furthermore, for the sake of simplicity and without loss of generality, the assumption is made that only a single (fundamental) mode is propagating in the waveguide. Thus, with the boundary S_n placed sufficiently far away from the junction of the waveguide attached at port n with the domain Ω so that higher-order, evanescent modes have attenuated sufficiently, only the fundamental mode participates in the exchange of power between the domain Ω and the waveguide. Consequently, a single mode will be used to represent the tangential magnetic field on the surface of each port, yielding the following mathematical statement for the approximation of the tangential magnetic field on the port surfaces,

$$\hat{n} \times \vec{H} = \sum_{n=1}^{p} i_n \vec{h}_n, \tag{11.94}$$

where i_n are the unknown expansion coefficients. Thus, in the following, $\vec{h}_n$ denotes the transverse magnetic field of the fundamental mode on the cross-section of the nth waveguide associated with the nth port.

Before we proceed any further, it is important to point out that the methodology to be presented for the development of reduced-order macromodels for the electromagnetic structure enclosed within Ω is not limited only to cases where the ports S_n, $n = 1, 2, \ldots, p$, are waveguide ports. The general case, where a number of the boundaries S_n, $n = 1, 2, \ldots, p$, are, simply, mathematical surfaces used to identify junctions through which the domain Ω connects to the outside world, can be handled in a very similar fashion, by using in place of the eigenmode fields a set of properly selected expansion functions to approximate the tangential magnetic field over the boundary. Thus each one of the functions $\vec{h}_n$ in (11.94) is replaced by an expansion of the form

$$\vec{h}_n = \sum_{m=1}^{M_n} i_m^{(n)} \vec{w}_{h,m}^{(n)}, \quad n = 1, 2, \ldots, p \tag{11.95}$$

where M_n is the number of expansion functions, $\vec{w}_{h,m}^{(n)}$, for port n, and $i_m^{(n)}$ are the associated expansion coefficients. Clearly, in this case, the number of additional state variables associated with the expansions (11.95) is equal to $P = \sum_{n=1}^{p} M_n$. Incidentally, the notation in (11.95) is appropriate for describing the case where one or several of the port surfaces are cross-sections of waveguides on which more than one eigenmodes are used to represent the tangential magnetic field vector. Clearly, the simple case of (11.94), which will be discussed in detail in the following, is recovered from this more general formulation by setting $M_n = 1$, $n = 1, 2, \ldots, p$ in (11.95). The expansion functions, $\vec{h}_n$, will be referred to as *magnetic mode functions*.

The finite element matrix approximation of (11.93) using Galerkin's method assumes the form

$$(S + sZ + s^2 T)x_e = sBI(s), \tag{11.96}$$

where x_e is the vector of the degrees of freedom in the approximation of $\vec{E}$ in the TV space W_t, $I(s)$ is the vector containing the expansion coefficients i_n for the tangential magnetic field on the ports,

$$I(s) = \begin{bmatrix} i_1 & i_2 & \cdots & i_p \end{bmatrix}^T \tag{11.97}$$

and the elements of the matrices S, Z, T, and B are given by

$$S_{ij} = \iiint_{\Omega} \nabla \times \vec{w}_{t,i} \cdot \frac{1}{\mu} \nabla \times \vec{w}_{t,j} \, dv$$

$$B_{ij} = \iint_{S_0} \hat{n} \times \vec{w}_{t,i} \cdot \hat{n} \times \vec{h}_j ds$$

$$T_{ij} = \iiint_{\Omega_1} \vec{w}_{t,i} \cdot \epsilon \vec{w}_{t,j} \, dv$$

$$Z_{ij} = \iiint_{\Omega_1} \vec{w}_{t,i} \cdot \sigma \vec{w}_{t,j} \, dv + \iint_{S_0} \hat{n} \times \vec{w}_{t,i} \cdot \frac{1}{\eta} \hat{n} \times \vec{w}_{t,j} \, ds,$$

(11.98)

where η is the intrinsic impedance of the unbounded medium associated with the truncation boundary S_0.

From the point of view of model order reduction, input (or excitation) and output (or observable) quantities need to be identified for use in the development of the matrix transfer function for the macro-model representation of the electromagnetic multi-port. As noted earlier, the electromagnetic multi-port interacts with the outside world through its p ports. Thus the column vector $I(s)$ is taken as the input vector, while the tangential electric fields over the port surfaces are chosen as the output quantities. More specifically, in a manner similar to the one used for the expansion of the tangential magnetic field on the ports, *electric mode functions* $\vec{e}_n$, $n = 1, 2, \ldots, p$, that is, the transverse parts of the electric field vectors for the fundamental eigenmodes in the ports, are used for the representation of the tangential electric field

$$\vec{E} = \sum_{n=1}^{p} v_n \vec{e}_n.$$

(11.99)

The expansion coefficients v_n, $n = 1, 2, \ldots, p$, constitute the output quantities that will be used for the definition of the matrix transfer function. These coefficients can be extracted from the electric filed on each port surface making use of the mode orthogonality property

$$\iint_{S_i} (\vec{e}_i \times \vec{h}_j) \cdot \hat{n} ds = \begin{cases} 1 & (i = j) \\ 0 & (i \neq j) \end{cases}$$

(11.100)

discussed in Chapter 9. Thus, with $\hat{n} \times \vec{E}$ denoting the tangential electric field on the port surface, the output coefficients are computed as follows:

$$v_i = \iint_{S_i} (\hat{n} \times \vec{E}) \cdot \vec{h}_i \, ds.$$

(11.101)

Let $V(s)$ denote the vector of the expansion coefficients for the electric field at the ports

$$V(s) = \begin{bmatrix} v_1 & v_2 & \cdots & v_p \end{bmatrix}^T.$$

(11.102)

Then, the formal solution of (11.96), followed by the computation of the coefficients v_n, $n = 1, 2, \ldots, p$, according to (11.101), yields

$$V(s) = B^T x_e = B^T (S + sZ + s^2 T)^{-1} sBI(s).$$

(11.103)

Thus the matrix transfer function of the electromagnetic multi-port is immediately identified as

$$H(s) = B^T (S + sZ + s^2 T)^{-1} sB.$$

(11.104)

Calculation of the transfer function matrix at a given frequency requires the inversion of the finite element matrix which, for practical applications, is of dimension in the order of tens or hundreds of thousands. The dimension of the finite element matrix (i.e., the number of degrees of freedom or the order of the approximation) is equal to the number of eigenvalues of the discretized structure. From this large set of discrete eigenvalues, only a small subset, corresponding to the physical eigenvalues of the electromagnetic structure governing its behavior over the frequency bandwidth of interest, are the ones that really matter in the development of the reduced-order macromodel. Consequently, it is this subset of the eigenvalues that the model-order reduction process will attempt to extract.

The presence of the quadratic dependence on frequency in (11.104) is not directly compatible with the Krylov model-order reduction techniques described in previous sections. For this reason, early applications of Krylov model order reduction to the system of (11.104) were limited to the case of bounded and/or lossless structures [25], [26]. For such structures the matrix Z is zero. Thus, letting $s' = s^2$, the transfer function may be written as $H(s) = sH'(s')$, where $H'(s') = B^T(S + s'T)^{-1}B$. The linear dependence on the auxiliary "frequency" variable s', exhibited by the matrix inside the parentheses in this latter equation, makes possible the use of Krylov subspace-based techniques for its reduction.

11.4.1 Adaptive Lanczos-Padé sweep

The adaptive Lanczos Padé sweep (ALPS) process was proposed in [23] for the fast frequency sweep of discrete approximations of electromagnetic systems exhibiting general polynomial dependence on frequency. In the context of our presentation, we consider this process in a manner suitable for use in conjunction with the waveport-based definition of the matrix transfer function for the electromagnetic multi-port.

Starting from (11.104), and writing s in the form

$$s = s_0(1 + \delta), \quad \delta = \frac{s - s_0}{s_0} \tag{11.105}$$

yields

$$H(s) = B^T(A_0 - \delta A_1 - \delta^2 A_2)^{-1}s_0(1 + \delta)B, \tag{11.106}$$

where it is

$$A_0 = S + s_0 Z + s_0^2 T, \quad A_1 = -(s_0 Z + 2s_0^2 T), \quad A_2 = -s_0^2 T. \tag{11.107}$$

Assume, next, that a projection space, of dimension n much less than the dimension, N, of the finite element matrix, has been constructed. Let $Q = [q_1, q_2, \ldots, q_n]$, be the $(N \times n)$ matrix with columns the n basis vectors that span the constructed subspace. Then, the projection $x_e = Q\tilde{x}_e$ may be used to develop a reduced-order version of (11.106) as follows:

$$\tilde{H}_n(s) = \tilde{B}^T(\tilde{A}_0 - \delta\tilde{A}_1 - \delta^2\tilde{A}_2)^{-1}s_0(1 + \delta)\tilde{B}, \tag{11.108}$$

where

$$\tilde{B} = Q^T B, \quad \tilde{A}_i = Q^T A_i Q, \quad (i = 0, 1, 2). \tag{11.109}$$

The projection matrix Q is critical for the effectiveness and accuracy of this reduction technique. As already discussed in earlier sections, Q should contain the eigenmodes of the domain Ω that influence the electromagnetic response of the system in the vicinity of the expansion frequency s_0. These eigenmodes are associated with the dominant eigenvectors of the following matrix eigenvalue problem:

$$(A_0 - \delta A_1 - \delta^2 A_2)x = 0. \tag{11.110}$$

This eigenvalue problem is difficult to solve. Therefore, an approximation of Q is obtained in terms of the vectors in the Krylov subspace generated while solving the following generalized eigenvalue problem:

$$\lambda A_0 x' = A_1 x'. \tag{11.111}$$

For the case of a lossless domain, $Z = 0$ and $A_2 = A_1/2$. Thus (11.110) becomes

$$A_0 x = \left(\delta + \frac{\delta^2}{2}\right) A_1 x. \tag{11.112}$$

Comparison of the above equation with (11.111) reveals that the two problems have the same eigenvectors while their eigenvalues are related through the equation

$$1 - \delta\lambda - \delta^2\lambda/2 = 0. \tag{11.113}$$

Thus the eigenvalues of (11.112) are obtained from those of (11.111) through the following equations

$$\delta = -1 \pm \sqrt{1 + \frac{2}{\lambda}} \quad \Rightarrow \quad s = \pm s_0 \sqrt{1 + \frac{2}{\lambda}}. \tag{11.114}$$

Clearly, the largest (dominant) eigenvalues of (11.111), which are the ones computed through one of the Krylov-based eigensolvers presented in Chapter 10, are projected through the above equations to the ones of (11.110) (for the case of a lossless domain) close to the expansion point s_0.

It should be clear that, once the projection matrix Q has been developed, (11.109) can be constructed to provide for the expedient calculation of the system response over the desired frequency bandwidth (fast frequency sweep). However, in the presence of high loss, the frequency bandwidth over which this process exhibits good accuracy may be limited. Finally, it should be pointed out that the finite element matrix equation (11.111) contains spurious modes. This can be seen by substituting the explicit forms of A_0 and A_1 to cast the equation in the form

$$\lambda(S + s_0 Z + s_0^2 T)x' = -(s_0 Z + 2s_0^2 T)x'. \tag{11.115}$$

As discussed in Chapter 3, S is a positive-indefinite matrix. Its null space is spanned by the column vectors of the gradient matrix G_r (see Section 10.3.1). Removing the column vectors of G_r associated with the nodes inside the lossy media and on boundaries on which impedance boundary conditions are imposed, yields the matrix G'_r such that $ZG'_r = 0$. Thus the spurious modes for (11.111) are $(\lambda = -2, G'_r)$. They can be removed from Q using the technique described in Section 10.3.1.

In summary, the ALPS process may be summarized in a pseudo-code form as follows [23]:

> **Algorithm (11.5): Adaptive Lanczos Padé Sweep (ALPS)**
> 1 Construct Q for (A_0, A_1) in (11.111) using Lanczos
>   1.a Remove the spurious modes from Q using (10.48);
>   1.b Estimate poles of (11.110) from converged eigenvalues

```
            of (11.111) using (11.114);
    1.c Process ends when new poles fall out of desired frequency
        range;
    2 Project H(s) onto Q using (11.108).
```

The ALPS algorithm needs one LU factorization of the finite element matrix at s_0 to generate a reduced-order model of good accuracy over a wide frequency range. The bandwidth of accuracy depends on the loss mechanisms inside the computational domain. If the impact of the loss amounts to only a slight perturbation of the solution under the assumption of lossless media, a broad frequency bandwidth of accuracy is possible.

Let $[\omega_{min}, \omega_{max}]$ denote the frequency bandwidth of interest. Then, a good choice for s_0 is at the center of the frequency band, $s_0 = j(\omega_{min} + \omega_{max})/2$. However, such a choice results in a complex projection matrix Q since (11.111) is a complex eigenvalue problem in the presence of loss. This, in turn, results in a complex reduced transfer function $\tilde{H}(s)$. While a complex $\tilde{H}(s)$ poses no problems when ALPS is used for the fast computation of the electromagnetic response of the system over the bandwidth $[\omega_{min}, \omega_{max}]$, its interfacing with general-purpose, network analysis-oriented circuit simulators (such as SPICE) is cumbersome. For such applications of the generated reduced-order macromodel, a choice of real expansion frequency is advantageous since it leads to a real projection matrix Q and, thus a real state-space form of the reduced-order model.

11.5 MAXWELL'S CURL EQUATIONS-BASED MODEL ORDER REDUCTION

The discussion in the previous section makes clear the fact that the application of Krylov subspace techniques for the reduction of finite element approximations of the vector Helmholtz equation in lossy media is hindered by the presence of both the linear and the quadratic term of the complex frequency s. This can be avoided from the start by working directly with the system of the two Maxwell's coupled curl equations. A methodology for the reduction of the finite element approximation of Maxwell's curl equations is discussed in the following.

Once again, we concern ourselves with linear, passive, time-independent media, which, without loss of generality, will be assumed to be isotropic. Thus the Laplace domain form of Maxwell's curl equations is

$$\begin{cases} \nabla \times \vec{E} = -s\,\vec{B}, \\ \nabla \times \mu^{-1}\vec{B} = s\,\epsilon\,\vec{E} + \sigma\vec{E}. \end{cases} \tag{11.116}$$

The development of the finite element approximation of this system over the domain Ω of Fig. 6.19, begins with the expansion of $\vec{E}$ and $\vec{B}$ in terms of the tangentially-continuous vector space W_t and the normally-continuous vector space W_n, respectively. As already discussed in Chapter 10, Galerkin's testing leads to the following discrete form:

$$\begin{bmatrix} 0 & D \\ -D^T & Z \end{bmatrix} \begin{bmatrix} x_b \\ x_e \end{bmatrix} = -s \begin{bmatrix} P & 0 \\ 0 & T \end{bmatrix} \begin{bmatrix} x_b \\ x_e \end{bmatrix} + \begin{bmatrix} 0 \\ B \end{bmatrix} I(s), \tag{11.117}$$

where the vectors x_b and x_e contain, respectively, the expansion coefficients in the approximation of $\vec{B}$ and $\vec{E}$, while the vector $I(s)$ contains the expansion coefficients for the tangential magnetic field on the port boundaries. The matrices $Z, T,$ and B were defined in (11.98), while the elements of P and D are given in (10.9). In compact form, the discrete approximation of (11.117) is written as

$$(\mathbf{G} + s\mathbf{C})X = B'I(s), \tag{11.118}$$

where

$$\mathbf{G} = \begin{bmatrix} 0 & D \\ -D^T & Z \end{bmatrix}, \quad \mathbf{C} = \begin{bmatrix} P & 0 \\ 0 & T \end{bmatrix}, \quad B' = \begin{bmatrix} 0 \\ B \end{bmatrix}, \quad X = \begin{bmatrix} x_b \\ x_e \end{bmatrix}. \tag{11.119}$$

For macromodel generation purposes, the output quantities of interest are the tangential electric fields at the ports. Thus, following exactly the same procedure as in Section 11.4, the vector $V(s)$ containing the relevant expansion coefficients in the approximation of the tangential electric field vector over the port boundaries, is extracted from X using mode orthogonality as follows:

$$V(s) = B'^T X = B'^T (\mathbf{G} + s\mathbf{C})^{-1} B' I(s). \tag{11.120}$$

Finally, the finite element approximation of Maxwell's curl equations leads to the following form for the transfer function matrix of the multi-port structure:

$$H(s) = B'^T (\mathbf{G} + s\mathbf{C})^{-1} B'. \tag{11.121}$$

Clearly, the element of the transfer function matrix have units of impedance. Therefore, we will refer to $H(s)$ as the *generalized impedance matrix* (GIM) of the multi-port.

The linear dependence of (11.118) on s makes it suitable for reduction via Krylov subspace based techniques. The methodology for such model order reduction will be discussed next. Prior to doing so, we must examine the passivity properties of the generated discrete model of (11.118).

11.5.1 Passivity of discrete model

The discussion about passivity of reduced-order models is meaningless if the original discrete model is not passive. Thus in this section, we shows the passivity of the discrete model for the field-flux formulation. As we discussed earlier in this chapter the passivity of (11.118) when $\mathbf{G}$ and $\mathbf{C}$ are real matrices boils down to the positive-definiteness of $(\mathbf{G} + s\mathbf{C})^{-1}$. More specifically, the following must hold for any non-zero complex vector z,

$$(z^*)^T [(\mathbf{G} + s\mathbf{C})^{-1} + (\mathbf{G}^T + s^*\mathbf{C}^T)^{-1}]z > 0 \Rightarrow$$
$$(w^*)^T [\mathbf{G} + \mathbf{G}^T + \alpha(\mathbf{C} + \mathbf{C}^T)]w > 0, \quad (\alpha > 0) \tag{11.122}$$

where $w = (\mathbf{G} + s\mathbf{C})^{-1}z$ and $s = \alpha + j\omega$. Using the fact that $\mathbf{C}$ is symmetric, the above equation becomes

$$w^{*T} [\mathbf{G} + \mathbf{G}^T + 2\alpha\mathbf{C}]w > 0, \quad (\alpha > 0). \tag{11.123}$$

From the element expressions for the matrices T and P, given, respectively, in (11.98) and (10.9), it is immediately apparent that $\mathbf{C}$ is positive definite for media with $\epsilon > 0$ and $\mu > 0$. Thus $2\alpha w^{*T}\mathbf{C}w > 0$ for $\alpha > 0$. Furthermore, using the fact that G is skew-symmetric (see (11.119)), it is straightforward to show that

$$\mathbf{G} + \mathbf{G}^T = \begin{bmatrix} 0 & 0 \\ 0 & 2Z \end{bmatrix}. \tag{11.124}$$

However, from (11.98) it is seen that Z is non-negative for lossy media ($\sigma \geq 0$) and resistive boundary conditions ($Re(\eta) \geq 0$). Thus we conclude that the product in (11.123) is positive definite for any complex vector z and for $Re(s) > 0$; hence the discrete system of (11.117) is passive.

At this point it is worth noting that the passivity of the E-field finite element model follows from this result since (11.96) is derived from (11.117) through the elimination of x_b.

Passivity of the discrete model is important since, as shown earlier in this chapter, application of the PRIMA process to a passive state-space model leads to a guaranteed passive reduced-order model. The passive reduction of (11.117) using PRIMA is discussed next.

11.5.2 Incorporation of lumped elements

The incorporation of lumped elements in differential equation-based solvers is a useful modeling capability, since it facilitates the computationally efficient representation of electrically-small devices/components within the computational domain. The electromagnetic modeling of such components would require a very fine grid, resulting in both an unnecessary increase in the number of state variables in the discrete model and a penalty in the condition number of the finite element matrices. Use of circuit lumped element representation of such components, described in terms of appropriate current-voltage ($I - V$) relationships, alleviates these difficulties.

The incorporation of lumped circuit elements in differential equation-based electromagnetic field solvers has been practiced extensively in the computational electromagnetics community (e.g., [46]-[52]). While most of the applications in these references are in conjunction with time-domain finite methods and tend to emphasize the transient electromagnetic analysis of electromagnetic devices driven by non-linear sources, the methodology used for the incorporation of the element $I - V$ relationship in the discrete electromagnetic system is applicable to both frequency-domain and time-domain solvers.

In particular, one of the methods utilizes the integral of the electric field along the edge(s) of the grid where the element is placed to calculate the voltage drop across the lumped element. The current through the element is calculated in terms of the circulation of discrete magnetic field around the edge. In the following we apply this method to the finite element model obtained from the discretization of vector Helmholtz equation and Maxwell's curl equations [53].

For simplicity, and without loss of generality, we assume that the lumped element coincides with one of the edges in the finite element mesh. Depending on the layout of the structure, the lumped element can be split, subject to appropriate circuit rules to assure that its impedance value remains intact, into several subelements and each one of them can be assigned to one finite element edge. In the following our discussion concerns the lumped element assigned to one edge.

We insert a lumped current source to model the element. With this lumped current source, the vector Helmholtz equation becomes

$$\nabla \times \frac{1}{\mu} \nabla \times \vec{E} + s\sigma\vec{E} + s^2\epsilon\vec{E} = -sI_k\hat{l}_k, \tag{11.125}$$

while the Ampère's law curl equation gives

$$\nabla \times \mu^{-1}\vec{B} = s\epsilon\vec{E} + \sigma\vec{E} + I_k\hat{l}_k. \tag{11.126}$$

In the above equations I_k is the value of the lumped current source and $\hat{l}_k$ is unit vector pointing along the kth edge. Recalling that the finite element approximation of the vector

Helmholtz equation or Maxwell's curl equations involves the following integration

$$\iiint_{\Omega} \vec{w}_{t,k} \cdot I_k \hat{l}_k dv = I_k \iint_{area} \left(\int_{edge-k} \vec{w}_{t,k} \cdot \vec{dl} \right) ds = I_k, \tag{11.127}$$

where $\vec{w}_{t,k}$ is the edge element on the kth edge and use was made of the edge-element property (2.69). The voltage across the lumped current source is obtained through the following integration of the electric field along the kth edge

$$V_k = \int_{edge-k} \vec{E} \cdot \vec{dl} = \int_{edge-k} \sum_{i=1}^{N_e} x_{e,i} \vec{w}_{t,i} \cdot \vec{dl} = x_{e,k}, \tag{11.128}$$

where V_k is the voltage along the kth edge and $x_{e,k}$ is the expansion coefficient associated with the edge. In deriving this result use was made of (2.69).

Equations (11.127) and (11.128) suggest the following procedure for the incorporation of lumped elements in the discrete model. If the lump element is a resistor, its voltage-current relation is

$$R^{-1} V_k = I_k. \tag{11.129}$$

Hence, the insertion of a lumped resistor R at edge k can be effected through the addition of the term R^{-1} to the kth diagonal term of the Z matrix. This yields

$$Z_{kk} \quad \leftarrow \quad Z_{kk} + R^{-1}. \tag{11.130}$$

Similarly, using the fact that the voltage-current relation for a capacitor C is

$$sCV_k = I_k, \tag{11.131}$$

the assignment of a lumped capacitor C at edge k, is effected through the introduction of the term C to the kth diagonal element of the T matrix. This yields

$$T_{kk} \quad \leftarrow \quad T_{kk} + C. \tag{11.132}$$

The aforementioned procedure for the introduction of lumped resistors and capacitors in the discrete electromagnetic model can be used also for the lumped inductors in the finite element approximation of the vector Helmholtz equation. The voltage-current relation for an inductor is

$$(sL)^{-1} V_k = I_k. \tag{11.133}$$

Hence, insertion of a lumped inductor L at edge k can be effected through the addition of the term L^{-1} to the kth diagonal element of the S matrix. This yields

$$S_{kk} \quad \leftarrow \quad S_{kk} + L^{-1}. \tag{11.134}$$

This form of (11.133) is not compatible with the linear in s form of the finite element approximation of Maxwell's curl equations. Its incorporation into the discrete model is through the matrix P, since $D^T P^{-1} D = S$. However, there is a simple way in which lumped inductors can be inserted in the discrete model in this case without interfering with the matrix P. The way this is done requires an understanding of the numerical process in which Krylov-based model order reduction of (11.117) is implemented. Therefore, the incorporation of lumped inductors in the discrete electromagnetic model will be discussed in the next section, in conjunction with the presentation of the Krylov subspace-based model order reduction of (11.117).

11.5.3 PRIMA-based model order reduction

The PRIMA process of Section 11.3 is directly applicable to the passive model order reduction of (11.117). For convenience, let us summarize the steps associated with the construction of the Krylov subspace and, hence, the projection matrix V used for the congruence transformations of (11.61) and the projections of (11.62).

With the transfer function matrix of the original system cast in the form

$$H(s) = B'^T (\mathbf{G} + s\mathbf{C})^{-1} B' = B'^T (I - \delta A)^{-1} R, \tag{11.135}$$

where the matrices A and R are obtained from $\mathbf{G}$, $\mathbf{C}$, B' and the expansion frequency s_0 as follows:

$$A = -(\mathbf{G} + s_0\mathbf{C})^{-1}\mathbf{C}, \quad R = (\mathbf{G} + s_0\mathbf{C})^{-1}B'. \tag{11.136}$$

These matrices are used to build the Krylov subspace of (11.17). The projection matrix V has as columns a set of orthonormal vectors that span the constructed Krylov subspace.

It is evident that the bulk of the computational cost of building the Krylov subspace is the recursive matrix-vector product of A with a vector p

$$x \leftarrow Ap \quad \Rightarrow \quad (\mathbf{G} + s_0\mathbf{C})x = -\mathbf{C}p. \tag{11.137}$$

More specifically, the computational cost is associated with the LU decomposition of the matrix $(\mathbf{G} + s_0\mathbf{C})$. Since this matrix results from the discretization of Maxwell's curl equations, its size is, roughly, twice the size of the system matrix resulting from the E-field finite element approximation. Furthermore, two additional matrices, namely, D and P, must be generated in the case of the Maxwell's curl equations-based finite element model. Thus it appears that the direct compatibility of (11.117) with Krylov model-order reduction techniques comes at a significant increase in computational overhead.

In the following we show how this apparent increase in the computational cost of the model-order reduction process, can be avoided through the implementation of a modified version of the block Arnoldi Algorithm (11.4) for the model order reduction of (11.117).

Returning to (11.137), following the scheme used in Chapter 10.3, we split the vectors p and x split into their electric field and magnetic fields parts. This yields

$$\begin{bmatrix} s_0 P & D \\ -D^T & Z + s_0 T \end{bmatrix} \begin{bmatrix} x_b \\ x_e \end{bmatrix} = - \begin{bmatrix} P p_b \\ T p_e \end{bmatrix}. \tag{11.138}$$

From the above system we can solve for the electric field and magnetic field parts of x in terms of the following two equations:

$$\begin{cases} x_e = -(S + s_0 Z + s_0^2 T)^{-1} \left[D^T p_b + s_0 T p_e \right] \\ x_b = -\dfrac{1}{s_0} (p_b + C_r x_e). \end{cases} \tag{11.139}$$

The matrix C_r is the curl or *circulation* matrix defined in (2.126). To avoid the generation of D, we will take advantage of the relations in (10.58) between matrices P, D, S and P, D, C_r. In addition, we introduce the auxiliary vectors x_{bb} and p_{bb} to accompany x_b and p_b as in (10.60). In this manner, the update for the next Arnoldi vector is performed as follows:

$$\begin{cases} x_e = -(S + s_0 Z + s_0^2 T)^{-1} (S p_{bb} + s_0 T p_e), \\ x_{bb} = -\dfrac{1}{s_0} (p_{bb} + x_e), \\ x_b = C_r x_{bb}. \end{cases} \tag{11.140}$$

Clearly, the process of the generation and orthonormalization of the vectors that span the Krylov subspace is the same with that in Section 10.3. During the construction of the Krylov subspace, the spurious DC modes must be removed from the Arnoldi vectors. The techniques used for this removal are the same with those used in the three-dimensional eigensolvers of Chapter 10. More specifically, the field-type DC modes can be removed by solving an electrostatic problem in the lossless region of the structure, and then using its solution in conjunction with (10.70) to correct x_e. As far as the removal of the flux-type DC modes is concerned, the Arnoldi vectors are already free from such modes due to the introduction of the auxiliary vectors x_{bb} and p_{bb} in (11.140).

Upon completion of the block Arnoldi process, the projection matrix V is computed and used in the following equations for the construction of the matrices of the reduced-order model:

$$\tilde{\mathbf{G}} = \begin{bmatrix} V_b^T & V_e^T \end{bmatrix} \begin{bmatrix} 0 & D \\ -D^T & Z \end{bmatrix} \begin{bmatrix} V_b \\ V_e \end{bmatrix} = V_{bb}^T S V_e - V_e^T S V_{bb} + V_e^T Z V_e,$$

$$\tilde{\mathbf{C}} = \begin{bmatrix} V_b^T & V_e^T \end{bmatrix} \begin{bmatrix} P & \\ & T \end{bmatrix} \begin{bmatrix} V_b \\ V_e \end{bmatrix} = V_{bb}^T S V_{bb} + V_e^T T V_e, \tag{11.141}$$

$$\tilde{B}' = \begin{bmatrix} V_b^T & V_e^T \end{bmatrix} \begin{bmatrix} 0 \\ B \end{bmatrix} = V_e^T B.$$

The matrix V_{bb} is recognized to be the auxiliary matrix, corresponding to the auxiliary vector x_{bb}, which is utilized in place of V_b in order to avoid the explicit construction of D.

At this point let us consider the issue of incorporating lumped inductors in the finite element approximation. We note that in the process of generating the Arnoldi vectors and the congruence transformation that yields the reduced model, the matrices P and D are not used; instead their contribution is effected indirectly through the matrix S, which, as mentioned in the earlier section, can be used for the incorporation of the lumped inductors using (11.134).

As a final comment pertinent to passivity, it should be evident from (11.130), (11.132) and (11.134), as well as our discussion of the requirements for passivity of the discrete model, that the passivity of (11.117) is maintained provided that the values of the lumped resistors, inductors and capacitors are positive.

The major savings in computer resources and CPU time resulting from the implementation of this modified block Arnoldi algorithm come from the replacement of the larger matrix $(\mathbf{G} + s_0\mathbf{C})$ with the smaller matrix $(S + s_0 Z + s_0^2 T)$. The numerical studies presented in the next section provide a more quantitative insight in the resulting savings in computational cost. Furthermore, it is immediately apparent from (11.140) and (11.141) that the proposed Krylov-based algorithm for model-order reduction of electromagnetic systems involving unbounded regions and/or lossy media, can be used in conjunction with any finite element solver that utilizes tangentially continuous edge elements for the discretization of the vector Helmholtz equation for the electric field vector.

For practical applications, the cost of the LU decomposition of $(S + s_0 Z + s_0^2 T)$ may become computationally prohibitive. However, the conjecture is that, for a thoughtfully designed electromagnetic system, undesired interactions between different components or functional blocks are weak enough for their impact on overall system performance to be negligible. If this is the case, individual components or functional blocks can be analyzed separately. Furthermore, the size of the discrete model of each block is expected to be small enough for direct LU decomposition to be possible. For those cases where the size is too large for such an option to be viable, an alternative to direct factorization will be the use of an efficient iterative solver combined with multigrid/multilevel preconditioners.

Even for those cases where an iterative solver will have to be used in the block Arnoldi algorithm, model-order reduction for fast frequency sweep is expected to be computationally more efficient than solving the finite element problem at several frequencies over the bandwidth of interest. This is due to the fact that the finite element matrix used in the block Arnoldi algorithm is fixed, calculated once at the expansion frequency, and then used $\lfloor n/p \rfloor + 1$ times for the solution of systems with p right hand sides. On the other hand, for a brute-force frequency sweep, a different finite element matrix must be solved for each of the discrete frequencies that will be used to interpolate the solution over the bandwidth of interest.

Finally, before a systematic model-order reduction has been established, the choice of the expansion frequency, s_0, needs to be decided. If the reduced-order model is to be used for fast-frequency sweep, the expansion point s_0 can be chosen at the middle of the desired frequency range. Thus, with $[\omega_{min}, \omega_{max}]$ denoting the frequency bandwidth of interest, the expansion point s_0 will be $j(\omega_{min} + \omega_{max})/2$. If, on the other hand, the generated macromodel is to be inserted in a general-purpose, transient network simulator, in addition to its passivity, it is highly desirable for the matrices of the state-space formulation of the reduced-order model to be real. This, in turn, calls for the projection matrix V to be real. To achieve this, the expansion point should be chosen on the positive real-axis.

11.6 APPLICATIONS

This section presents some examples from the application of the Krylov subspace-based model order reduction methodology discussed in this chapter to a variety of passive electromagnetic structures pertinent to microwave circuit and system design.

Microstrip Patch Antenna. The first example considered is the one-port macromodeling of a microstrip patch antenna. The dimensions of the antenna and the substrate properties are given in the insert of Fig. 11.2. First-order absorbing boundary conditions are used at the top and side truncation boundaries of the domain. The distance of the truncation boundaries from the patch are 2.4 mm at the top and 4.5 mm at the sides. The bandwidth of interest is $[0, 22]$GHz. The expansion frequency, s_0, is taken to be in the middle of the frequency band; hence, $s_0 = j2\pi \times 10^{10}$. It is at this frequency where the LU decomposition of the finite element matrix in (11.140) is performed.

The approximation of the fields on the microstrip port boundary is done in terms of the microstrip modes. These modes are also calculated at the expansion frequency of 10 GHz. At this point it is worth noting that, while the transverse field profiles of the microstrip-line modes are frequency dependent, this dependence was not taken into account in the development of the macromodel. Since the frequency dependence of the microstrip-line mode profiles is weak at frequencies for which the electrical thickness of the substrate is small, this approximation was acceptable in this case. In the general case of waveguide ports for which such approximation may not be acceptable, use of frequency-independent expansion functions for the representation of the tangential electric and magnetic fields over the port surface is recommended.

Returning to the microstrip patch antenna in the insert of Fig. 11.2, the number of edge (electric field) unknowns in the approximation is 23177, while the number of facet (magnetic flux) unknowns is 43012. Thus, instead of working with a matrix of dimension 66189, the

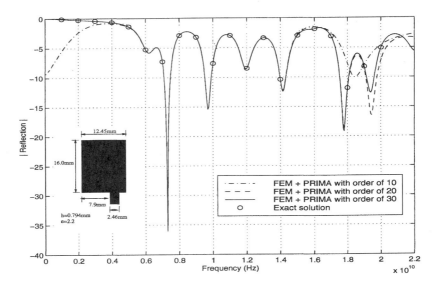

Figure 11.2   Magnitude of the input reflection coefficient of a microstrip patch antenna.

modified Arnoldi methodology of the previous section requires the LU decomposition of a matrix of size 23177. The subsequent application of the PRIMA model-order reduction algorithm yields a reduced-order model of the form of (11.121) utilizing only a small number of the eigenvalues and eigenvectors of the discrete finite element system.

Figure 11.3 depicts the real and imaginary parts of the calculated input impedance $Z_{in}(f)$ of the patch antennas. Once again, the results obtained for different values of the order of the reduced model are compared to the reference solution. As seen from both Figs. 11.2 and 11.3, a reduced-order model of order 30 provides for a highly accurate macromodel representation of the patch antenna over the entire frequency bandwidth. Furthermore, since the generated macromodel is passive, it can be safely incorporated in any general-purpose, transient circuit simulator.

At this point it is appropriate to discuss once again our choice of the expansion frequency, s_0. As mentioned in the previous section, for fast frequency sweep purposes, the expansion complex frequency, s_0, is taken to be purely imaginary, with magnitude equal to the angular frequency at the center of the frequency bandwidth of interest,

$$s_0 = (f_{max} + f_{min})\pi j. \tag{11.142}$$

A different choice of s_0 was suggested in [12], as follows:

$$s_0 = (f_{max} - f_{min})\pi + (f_{max} + f_{min})\pi j. \tag{11.143}$$

Our numerical experiments suggest that the two choices lead to responses of comparable accuracy provided that the order of the reduced model is large enough. However, for lower orders of the reduced model, (11.142) seemed to lead to reduced-order models of better accuracy. Thus this choice of the expansion frequency point was adopted for all subsequent examples.

To examine the usefulness of the approximate error bound introduced earlier in this chapter, Fig. 11.4 plots the error calculated after 10, 20, and 30 iterations in the reduction

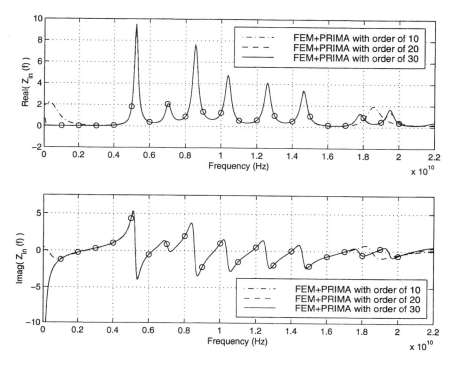

Figure 11.3   Real and imaginary parts of the input impedance of the microstrip patch antenna shown in the insert of Fig. 11.2. (After Zhu and Cangellaris, [27], ©2001 IEEE.)

process of the patch antenna model. As expected, the error is negligible at the expansion frequency (10 GHz). Furthermore, it is evident from the plot that as the order of the reduced model is increased, the bandwidth of accuracy is increased also. This suggests that the proposed error bound serves as a useful and computationally efficient means for assessing the accuracy of the generated reduced model.

Microstrip Low-pass Filter. Next we consider the reduced-order macromodeling of the low-pass microstrip filter, shown in Fig. 11.5. First-order absorbing boundary conditions were implemented at the top and side truncation boundaries of the computational domain to approximate the unbounded domain of the microstrip structure. The bandwidth of interest is $[0, 4]$GHz. Thus the expansion frequency, f_0, is taken to be in the middle of the frequency band, resulting in a value of $s_0 = j\,2\pi \times 2 \times 10^9$. It is at this frequency where the LU decomposition of $(S + s_0 Z + s_0^2 T)$ is performed. The tangential electric and magnetic fields on the microstrip ports are represented in terms of microstrip modes. These modes are calculated at 2 GHz by a two-dimensional, finite element-based eigensolver, developed according to the methodologies presented in Chapter 9. As already mentioned, the transverse field profile for the fundamental (quasi-TEM) mode of a microstrip does not exhibit strong frequency variation over a frequency bandwidth of reasonable extent. Thus the modal fields calculated by the eigensolver at 2 GHz are assumed to provide for a very accurate representation of the quasi-TEM transverse field distribution over the microstrip cross-section for the entire frequency bandwidth of 4 GHz.

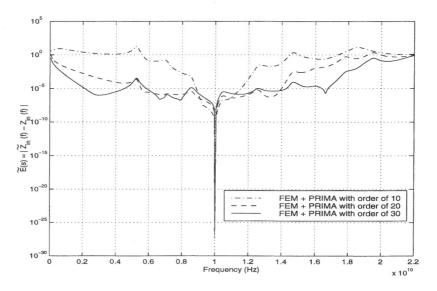

Figure 11.4   Error bound for the microstrip patch antenna macromodel, obtained after 10, 20, and 30 iterations of PRIMA.

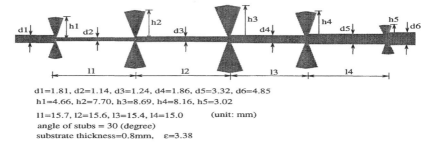

d1=1.81, d2=1.14, d3=1.24, d4=1.86, d5=3.32, d6=4.85
h1=4.66, h2=7.70, h3=8.69, h4=8.16, h5=3.02

l1=15.7, l2=15.6, l3=15.4, l4=15.0   (unit: mm)
angle of stubs = 30 (degree)
substrate thickness=0.8mm,   ε=3.38

Figure 11.5   Geometry of a low-pass microstrip filter. (After Zhu and Cangellaris [30], ©John Wiley and Sons Ltd. Reproduced with permission.)

The number of edge unknowns in the discrete approximation of the electric field is 7870, while the number of facet unknowns associated with the magnetic flux density is 14235. Thus the modified block Arnoldi algorithm requires the LU decomposition of a matrix of dimension 7870, instead of one of almost three times this dimension that would be required if the finite element matrix resulting from the discretization of the system of Maxwell's curl equations was used. The calculated magnitudes of the scattering parameters S_{11} and S_{21} of the generated reduced model of order 20 are plotted in Fig. 11.6. They are compared with the ones obtained from the brute-force solution (denoted as "exact" in the figure) of the E-field finite element approximation of the problem at several frequencies. Very good agreement is observed.

Rectangular Waveguide T-junction Circuit.   The next structure considered is the E-plane, slot-coupled, short-circuited T-junction analyzed in [54]. The geometry and dimensions of the structure are given in the insert of Fig. 7.3. The bandwidth of interest for this analysis is between 10 and 20 GHz. Thus the expansion frequency, s_0, is chosen

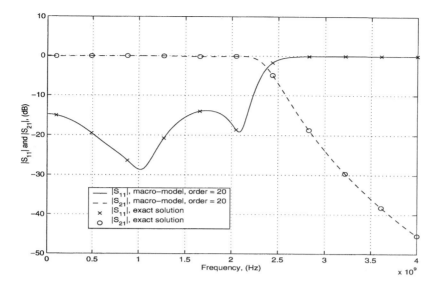

Figure 11.6   Magnitude of the scattering parameters S_{11} and S_{21} of the microstrip filter of Fig. 11.5. (After Zhu and Cangellaris [30], ©John Wiley and Sons Ltd. Reproduced with permission.)

to be $j\,2\pi \times 15 \times 10^9$. The number of edge unknowns is 21526, while the number of facet unknown is 38989. Once again, use of the modified block Arnoldi algorithm reduces by a factor of three the dimension of the matrix that has to be decomposed. Since the waveguides are homogeneous, the modal field profiles are frequency independent. However, the waveguide modal impedance is frequency dependent and is taken into account in the calculation of the scattering matrix of the two-port.

Figures 11.7 and 11.8, respectively, depict $|S_{11}|$ and $|S_{21}|$ obtained from reduced models of order 10 and 20. The results are compared to measured data provided in [54]. Also plotted in the figures, and referred to as "exact solution," are the calculated values from the brute-force solution of the problem at a set of frequencies over the desired bandwidth. It is apparent from these plots that the bandwidth of accuracy of the generated macro-model increases as the order of the reduced model increases. A reduced model of order 20 exhibits excellent agreement with the measured results.

At this point it is worth pointing out that this structure was also analyzed in [21] using a finite element-based AWE scheme. Because of the numerical stability problems of the AWE process discussed in the beginning of this chapter, the CFH technique was used in [21] to generate Padé approximations at three expansion points (13.2, 15.0, and 17.2 GHz) over the desired bandwidth. An LU decomposition of the FEM matrix was performed at each one of these frequencies, followed by the application of the AWE process for the generation of the associated Padé approximations. As can be seen from the generated plots in [21], the frequency response obtained through the CFH process is in very good agreement with the measured response only within the 12-18 GHz frequency range. In contrast, the algorithm proposed in this chapter achieves excellent accuracy over the entire bandwidth, at (roughly) the cost of only one LU decomposition of the FEM matrix.

Gap-coupled Resonance Filter. The final structure we would like to discuss is the gap-coupled resonance filter presented in [55]. The filter was realized using buried microstrip

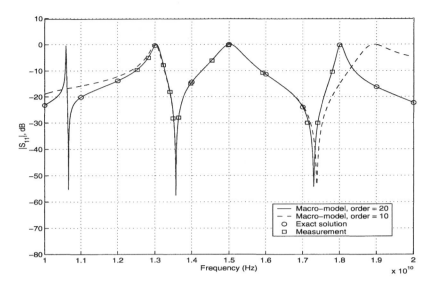

Figure 11.7 $|S_{11}|$ of the E-plane, slot-coupled, short-circuited T-junction of Fig. 7.3. (After Zhu and Cangellaris [30], ©John Wiley and Sons Ltd. Reproduced with permission.)

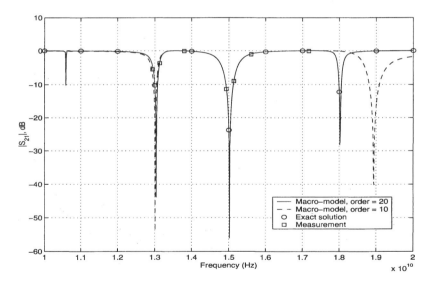

Figure 11.8 $|S_{21}|$ of the E-plane, slot-coupled, short-circuited T-junction of Fig. 7.3. (After Zhu and Cangellaris [30], ©John Wiley and Sons Ltd. Reproduced with permission.)

lines, and it is shown in Fig. 11.9. The number of edge unknowns in the finite element model for this structure is 14476, while the number of facet unknowns is 25865. For the generation of a macro-model over the 0-10 GHz frequency range, the complex expansion frequency is chosen to be $s_0 = j2\pi \times 5 \times 10^9$.

The calculated magnitudes for the reflection and transmission coefficients of the filter, obtained from the generated reduced model of order 20, are compared in Fig. 11.10 with

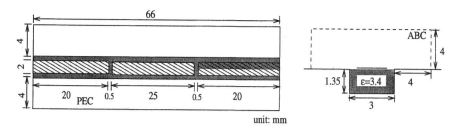

unit: mm

Figure 11.9   Geometry of a gap-coupled resonance filter. (After Zhu and Cangellaris [30], ©John Wiley and Sons Ltd. Reproduced with permission.)

the response obtained from a direct finite element solution at a set of frequencies over the bandwidth, which is indicated in the figure as the "exact solution". Excellent agreement is observed over the entire bandwidth. Furthermore, our results are in excellent agreement with those given in [55], obtained through both measurement and a Finite-Difference Time-Domain (FDTD) simulation. In fact, a careful comparison of the results generated by the proposed macro-model exhibit (at the lower end of the bandwidth) better agreement with the measured data in [55] than the FDTD results.

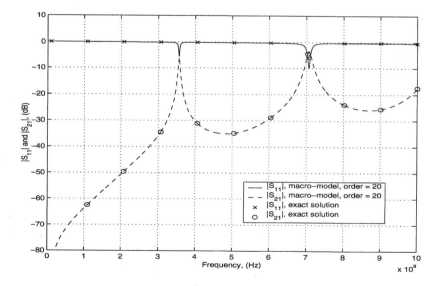

Figure 11.10   Magnitude of the scattering parameters of a gap-coupled resonance filter. (After Zhu and Cangellaris [30], ©John Wiley and Sons Ltd. Reproduced with permission.)

The examples above offer convincing evidence of the robustness and accuracy of Krylov subspace-based methods for the generation of reduced-order macromodels of multi-port, passive electromagnetic devices directly from their discrete models obtained using finite element methods. The emphasis of the presentation was on the demonstration of the accuracy with which the reduced model, obtained at the computational cost of a finite element solution at a single frequency, is able to capture the device response of a broad frequency bandwidth. Another important attribute of the model order reduction process is the fact that its output is a generalized impedance matrix representation of the electromagnetic multi-port in terms of closed-form pole-residue expressions for their elements with frequency as the parameter. These closed-form expressions are rather compact, involving only a small number of terms (i.e., a small number of poles). For the class of resonant structures considered, reduced models of order 20 were found adequate to capture the device broadband response with excellent accuracy. In particular, the frequency range of validity of the generated macromodel is specified a priori, and may span several gigahertz. In closing, let us emphasize once again what is perhaps the most impressive attribute of Krylov subspace-based model order reduction, namely, that the reduced-order macromodel is generated at a computational cost approximately equal to that required for the finite element solution of the electromagnetic problem at a single frequency.

REFERENCES

1. *Advanced Design System EDA Software*, Agilent EEsof EDA, URL: *http://eesof.tm.agilent.com/*.

2. R. E. Collin, *Foundations for Microwave Engineering*, 2nd edition, New York: McGraw-Hill Inc., 1992.

3. D. M. Pozar, *Microwave Engineering*, 2nd edition, New York: John Wiley & Sons, 1998.

4. R. F. Harrington, *Time-Harmonic Electromagnetic Fields*, 2nd edition, Piscataway, NJ: Wiley-IEEE Press, 2001.

5. J.M. Jin, *The Finite Element Method in Electromagnetics*, 2nd edition, New York: John Wiley & Sons, 2002.

6. J.-D. Benamou and B. Després, "A domain decomposition method for the Helmholtz equation and related optimal control problems," *J. Comput. Phys.*, vol. 136, pp. 68-82, 1997.

7. A. Piacentini and N. Rosa, "An improved domain decomposition method for the 3D Helmholtz equation," *Comput. Methods Appl. Mech. Engrg.*, vol. 162, pp. 113-124, 1998.

8. M. Gander, F. Magoulès, and F. Nataf, "Optimized Schwarz methods without overlap for the Helmholtz equation," *SIAM J. Sci. Comput*, vol. 24, no. 1, pp. 38-60, 2002.

9. M. N. Vouvakis and J.-F. Lee, "A fast DP-FETI like domain decomposition algorithm for the solution of large electromagnetic problems," *Proc. Eighth Copper Mountain Conference on Iterative Methods* Copper Mountain, Co, Mar. 2004.

10. L. T. Pillage and R. A. Rohrer, "Asymptotic waveform evaluation for timing analysis," *IEEE Trans. Computer-Aided Design*, vol. 33, no. 9, pp. 352-366, Apr. 1990.

11. E. Chiprout and M.S. Nakhla, *Asymptotic Waveform Evaluation and Moment Matching for Interconnect Analysis*, Norwell, MA: Kluwer Academic Publishers, 1994.

12. P. Feldmann and R. W. Freund, "Efficient linear circuit analysis by Padé approximation via the Lanczos process," *IEEE Trans. Computer-Aided Design*, vol. 34, pp. 639-649, May 1995.

13. R. Freund, "Reduced-order modeling techniques based on Krylov subspaces and their use in circuit simulation," Bell Laboratories, Murray Hill, NJ., Numerical Analysis Manuscript 98-3-02, Feb. 1998.

14. R. W. Freund, "Passive reduced-order models for interconnect simulation and their computation via Krylov-subspace algorithms," in *Proc. 1999 Design Automation Conference*, Jun. 1999.

15. D.L. Boley, "Krylov space methods on state-space control models," *Circuits, Syst. Signal Process.*, vol. 18, pp. 733-758, 1994.

16. K. Gallivan, E. Grimme, and P. Van Dooren, "Padé approximation of large-scale dynamic systems with Lanczos method," in *Proc. 33rd Conference on Decision and Control*, Buena Vista, FL., Dec. 1994, pp. 443-448.

17. A. Odabasioglu, M. Celik, and L. T. Pileggi, "PRIMA: Passive reduced-order interconnect macro-modeling algorithm," *IEEE Trans. Computer-Aided Design*, pp. 645-653, Aug. 1998.

18. M. Celik, L. Pileggi, and A. Odabasioglu, *IC Interconnect Analysis*, Norwell, MA: Kluwer Academic Publishers, 2002.

19. M. A. Kolbehdari, et al., "Simultaneous time and frequency domain solution of EM problems using finite element and CFH techniques," *IEEE Trans. Microwave Theory Tech.*, vol. 44, pp. 1526-1534, Sep. 1996.

20. J. P. Zhang and J. M. Jin, "Preliminary study of AWE method for FEM analysis of scattering problems," *Microwave and Optical Tech. Letters*, vol. 17, no. 1, pp. 7-12, Jan. 1998.

21. J. E. Bracken, D. K. Sun, and Z. J. Cendes, "S-domain methods for simultaneous time and frequency characterization of electromagnetic devices," *IEEE Trans. Microwave Theory Tech.*, vol. 42, pp. 1277-1290, Sep. 1998.

22. X. M. Zhang and J. F. Lee, "Application of the AWE method with the 3-D TVFEM to model spectral responses of passive microwave components," *IEEE Trans. Microwave Theory Tech.*, vol. 44, pp. 1735-1741, Nov. 1998.

23. D. K. Sun, J. F. Lee, and Z. Cendes "ALPS - A new fast frequency-sweep procedure for microwave devices," *IEEE Trans. Microwave Theory Tech.*, vol. 49, no. 2, pp. 398-401, Feb. 2001

24. M. Celik and A. C. Cangellaris, "Simulation of dispersive multiconductor transmission lines by Padé approximation via the Lanczos process," *IEEE Trans. Microwave Theory Tech.*, vol. 38, pp. 2525-2535, Dec. 1996.

25. M. Zunoubi, K. C. Donepudi, J. M. Jin, and W. C. Chew, "Efficient time-domain and frequency-domain finite-element solution of Maxwell's equations using spectral Lanczos algorithms," *IEEE Trans. Microwave Theory Tech.*, vol. 42, pp. 1141-1149, Aug. 1998.

26. J. Rubio, J. Garcia, and J. Zapata, "SFELP: A hybrid 3-D finite elements / segmentation method with fast frequency sweep based on the SyMPVL algorithm for the analysis of passive microwave circuits," in *Proc. of the 2000 IEEE AP-S International Symposium*, Salt Lake City, UT, July 2000.

27. Y. Zhu and A. C. Cangellaris, "A new FEM formulation for reduced order electromagnetic modeling," *IEEE Microw. Guid. Wave Lett.*, vol. 11, no. 5, May 2001, pp. 211-213. May 2001.

28. Y. Zhu and A. C. Cangellaris, "Macro-elements for efficient FEM simulation of small geometric features in waveguide components," *IEEE Trans. Microwave Theory Tech.*, vol. 48, pp. 2254-2260, Dec. 2000.

29. Y. Zhu and A. C. Cangellaris, "Macro-modeling of microwave components using finite elements and model order reduction," in *5th Workshop on Finite Element Methods for Microwave Engineering*, Boston, June 2000,

30. Y. Zhu and A. C. Cangellaris, "Finite element-based model order reduction of electromagnetic devices", *International Journal of Numerical Modeling: Electronic Networks, Devices and Fields*, vol. 15, Issue 1, pp. 73-92, 2002.

31. Z. Bai, J. Demmel, J. Dongarra, A. Ruhe, and H. van der Vorst (eds.), *Templates for the Solution of Algebraic Eigenvalue Problems – A Practical Guide*, Philadelpia: SIAM, 2000.

32. R. A. Rohrer and H. Nosrati, "Passivity considerations in stability studies of numerical integration algorithms," *IEEE Trans. Circuits Syst.*, vol. 17, pp. 857-866, 1981.

33. R. Achar and M.S. Nakhla, "Simulation of high-speed interconnects," *Proc. IEEE*, vol. 89, no. 5, pp. 693-728, May 2001.

34. A. C. Cangellaris and L. Zhao, "Passivity of discrete electromagnetic systems," in *Proc. 14th Annu. Rev. Progress Appl. Computat. Electromag.*, Monterey, CA., pp. 721-731, Mar. 1998.

35. Z. Bai, R. D. Slone, W. T. Smith, and Q. Ye, "Error bound for reduced system model by Padé approximation via the Lanczos process," *IEEE Trans. Computer-Aided Design*, vol. 48, pp. 133-141, Feb. 1999.

36. A. Odabasioglu, M. Celik, and L. T. Pileggi, "Practical considerations for RLC circuit reduction," in *IEEE/ACM Proc. of ICCAD*, San Jose, CA, Nov. 1999.

37. G. H. Golub and C. F. Van Loan, *Matrix Computation*, 3rd ed., Baltimore: Johns Hopkins University Press, 1996.

38. A. C. Cangellaris and L. Zhao, "Rapid FDTD simulation without time stepping," *IEEE Microwave and Guided Wave Letters*, vol. 9, no. 1, pp. 4-6, Jan. 1999.

39. A. C. Cangellaris, M. Celik, M. Pasha, and L. Zhao, "Electromagnetic model order reduction for system-level modeling," *IEEE Trans. Microwave Theory Tech.*, vol. 47, no. 6, pp. 840-850, Jun. 1999.

40. R. F. Remis and P. M. van den Berg, "A modified Lanczos algorithm for the computation of transient electromagnetic wavefields," *IEEE Trans. Microwave Theory Tech.*, vol. 45, no. 12, pp. 2139-2149, Dec. 1997.

41. R. F. Remis, "Low-frequency model-order reduction of electromagnetic fields without matrix factorization," *IEEE Trans. Microwave Theory Tech.*, vol. 52, no. 9, pp. 2298-2304, Sep. 2004.

42. B. Denecker, F. Olyslager, L. Knockaert, and D. De Zutter, "Generation of FDTD subcell equations by means of reduced order modeling," *IEEE Trans. Antennas Propagat.*, vol. 51, no. 8, pp. 1806-1817, Aug. 2003.

43. L. Kulas and M. Mrozowski, "Reduced order models of refined Yee's cells," *IEEE Microwave and Wireless Components Letters*, vol. 13, no. 4, pp. 164-166, Apr. 2003.

44. L. Kulas and M. Mrozowski, "Multilevel model order reduction," *IEEE Microwave and Wireless Components Letters*, vol. 14, no. 4, pp. 165-167, Apr. 2004.

45. T. Wittig, I. Munteanu, R. Schuhmann, and T. Weiland, "Two-step Lanczos algorithm for model order reduction," *IEEE Trans. Magnetics*, vol. 38, no. 2, pp. 673-676, Mar. 2002.

46. R.H. Voelker and R. J. Lomax, "A finite-difference transmission line matrix method incorporating a nonlinear device model," *IEEE Trans. Microwave Theory Tech.*, vol. 38, pp. 302-312, Mar. 1990.

47. W. Sui, D.A. Christensen, and C.H. Durney, "Extending the two-dimensional FDTD method to hybrid electromagnetic systems with active and passive lumped elements", *IEEE Trans. Microwave Theory Tech.*, vol. 40, pp. 724-730, Apr. 1992.

48. Yu-Sheng Tsuei, A.C Cangellaris, and J.L.Prince, "Rigorous electromagnetic modeling of chip-to-package (first-level) interconnections", *IEEE Trans. Comp., Hybrids, Manuf. Technol.*, vol. 16, pp. 876-883, Dec. 1993.

49. M. Piket-May, A. Taflove, and J. Baron, "FD-TD modeling of digital signal propagation in 3-D circuits with passive and active loads," *IEEE Trans. Microwave Theory Tech.*, vol. 42, pp. 1514-1523, Aug. 1994.

50. K. Guillouard, Man-Fai Wong, V.F. Hanna, and J. Citerne, "A new global finite element analysis of microwave circuits including lumped elements", *IEEE Trans Microwave Theory Tech.*, vol. 44, pp. 2587-2594, Dec. 1996.

51. M. Feliziani and F. Maradei, "Modeling of electromagnetic fields and electrical circuits with lumped and distributed elements by the WETD method," *IEEE Trans. Magnetics*, vol. 35, pp. 1666-1669, May 1999.

52. S.-H. Chang, R. Coccioli, Y. Qian, and T. Itoh, "A global finite-element time-domain analysis of active nonlinear microwave circuits," *IEEE Trans. Microwave Theory Tech.*, vol. 47, no. 12, pp. 2410-2416, Dec. 1999.

53. H. Wu and A.C. Cangellaris, "Model-order reduction of finite-element approximations of passive electromagnetic devices including lumped electrical-circuit models," *IEEE Trans. Microwave Theory Tech.,* vol. 52, no. 9, pp. 2305-2313, Sep. 2004.

54. T. Sieverding and F. Arndt, "Field theoretic CAD of open or aperture matched T-junction coupled rectangular waveguide structures," *IEEE Trans. Microwave Theory Tech.*, vol. 28, pp. 353-362, Feb. 1992.

55. T. Ishikawa and E. Yamashita, "Experimental evaluation of basic circuit components using buried microstrip lines for constructing high-density microwave integrate circuits," *IEEE Trans. Microwave Theory Tech.*, vol. 32, pp. 1074-1080, Jul. 1996.

CHAPTER 12

FINITE ELEMENT ANALYSIS OF PERIODIC STRUCTURES

Electromagnetic wave interaction with periodic structures is exploited extensively in practice for the realization of passive devices that exhibit frequency selectivity in terms of signal transmission, reflection and scattering. With applications ranging from waveguide filters, frequency-selective surfaces and phased arrays to optical gratings and photonic/electronic bandgap configurations, the design of periodic structures relies upon the availability of numerical methodologies that take advantage of the periodicity of the material and/or boundary conditions to provide for a computationally efficient solution to the problem. Often times these structures exhibit significant material and geometry complexity. For such cases the modeling versatility of the finite element method makes it the preferred method of choice for their electromagnetic analysis and computer-aided design.

The application of finite elements to the analysis of periodic structures has received significant attention in the computational electromagnetics literature over the years. In [1] and [2], a two-dimensional hybrid finite element/boundary element approach was used for the analysis of the one-dimensional infinite grating problem. The extension to the case where the structure exhibits two-dimensional periodicity, and thus required a three-dimensional, hybrid, finite-element/boundary integral equation solver was elaborated in a series of papers (e.g., [3]-[9]). In this approach, finite elements are used to discretize Maxwell's equations inside the unit cell of the periodic structure, while the fields outside the cell are calculated in terms of a boundary integral statement involving the tangential fields on the boundary surfaces of the cell parallel to the direction(s) of periodicity and the periodic Green's function for the homogeneous medium exterior to the cell [10]. The continuity of the tangential electric and magnetic fields on the boundary surfaces of the unit cell couples the two expressions for the solution, and result in a finite element sparse matrix that includes the full block that represents the integral equation-based interaction of the approximation of the tangential fields on the boundary.

As the aforementioned discussion suggests, the formalism and application of the hybrid finite element/boundary integral methodology to the analysis of periodic structures is well documented in the finite element literature and will not be repeated here. Rather, the emphasis of this chapter will be on methodologies that attempt to avoid the introduction of a full matrix sub-block in the sparse finite element matrix that results from the discretization of the unit cell. Toward this objective, two issues must be addressed. The first is the imposition of the periodic boundary condition along the direction(s) of periodicity of the structure. The second is the truncation of the unbounded domain that surrounds the structure in the remaining direction(s).

For the former, a methodology that makes use of Floquet's theorem will be elaborated. As far at the absorption of outgoing waves is concerned, two finite element grid truncation schemes suitable for periodic structures will be discussed, namely, the use of the anisotropic perfectly matched layer (PML) [11] and the concept of transfinite elements (TFE) [12]. Combined together, these methods enhance the computational efficiency of a finite element-based modeling of periodic structures [13] and make possible the analysis of structures exhibiting high geometric and material complexity within the unit cell.

12.1   FINITE ELEMENT FORMULATION OF THE SCATTERING AND RADIATION PROBLEM

The problem of time-harmonic, electromagnetic wave scattering or radiation from a periodic structure can be studied in the context of the unit cell geometry depicted in Fig. 12.1. The unit cell represents a two-dimensional periodic structure, exhibiting periodicity in material properties along the x and y axes, with period D_x and D_y, respectively. Shown in the figure are impressed electric and magnetic current densities, $\vec{J}_i$ and $\vec{M}_i$, respectively, which are also periodic along x and y, with the same period as the geometry. Without loss of generality, the media above and below the periodic structure are assumed to be unbounded and homogeneous. In addition to the impressed sources, excitation is also allowed in terms of a time-harmonic, uniform plane wave, with incident electric field vector $\vec{E}_i$, angular frequency ω, and wave vector $\vec{k}_0$. All media are assumed to be linear.

For the purposes of finite element modeling the computational domain is taken to be the unit cell. Inside the unit cell the total electric field, $\vec{E}$, satisfies the vector wave equation

$$\nabla \times \bar{\mu}_r^{-1} \cdot \nabla \times \vec{E} - k_0^2 \bar{\epsilon}_r \cdot \vec{E} = -j k_0 \eta_0 \vec{J}_i - \nabla \times \left(\bar{\mu}_r^{-1} \cdot \vec{M}_i \right), \qquad (12.1)$$

where $\bar{\epsilon}_r$ and $\bar{\mu}_r$ denote the position-dependent, relative permittivity and relative permeability tensors of the media inside the unit cell, with the permittivity and permeability of the surrounding unbounded medium taken as reference. Without loss of generality the unbounded medium is taken to be free space. Thus $k_0^2 = \omega^2 \mu_0 \epsilon_0$ and $\eta_0 = \sqrt{\mu_0/\epsilon_0}$. For any perfectly conducting regions present inside the unit cell the essential boundary condition

$$\hat{n} \times \vec{E} = 0 \qquad (12.2)$$

is imposed on their surface, with $\hat{n}$ denoting the unit normal vector on the surface.

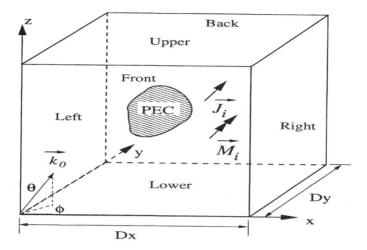

Figure 12.1   The unit cell of a two-dimensional periodic structure of infinite extent.

The weak form of (12.1) is readily obtained through its scalar multiplication by a weighting function $\vec{w}$ followed by integration over the computational domain,

$$
\iiint_\Omega \left(\nabla \times \vec{w} \cdot \bar{\mu}_r^{-1} \cdot \nabla \times \vec{E} - k_0^2 \vec{w} \cdot \bar{\epsilon}_r \cdot \vec{E} \right) dv - j\omega\mu_0 \iint_{S_{U-L}} \hat{n} \times \vec{H} \cdot \vec{w} \, ds
$$

$$
- j\omega\mu_0 \iint_{S_W} \hat{n} \times \vec{H} \cdot \vec{w} \, ds = - \iiint_\Omega \vec{w} \cdot [jk_0\eta_0 \vec{J}_i + \nabla \times (\bar{\mu}_r^{-1} \cdot \vec{M}_i)] \, dv. \tag{12.3}
$$

In deriving this result use was made of a well-known vector identity and the vector divergence theorem. S_{U-L} denotes the surface of the upper and lower boundaries of the cell, while S_W denotes the surface of the four side walls. The incident field is taken to be the field in the absence of the periodic structure; hence, it satisfies the vector wave equation in free space. A weak form for it is readily obtained over the periodic cell as follows

$$
\iiint_\Omega \left(\nabla \times \vec{w} \cdot \nabla \times \vec{E}_i - k_0^2 \vec{w} \cdot \vec{E}_i \right) dv - j\omega\mu_0 \iint_{S_{U-L}} \hat{n} \times \vec{H}_i \cdot \vec{w} \, ds
$$

$$
- j\omega\mu_0 \iint_{S_W} \hat{n} \times \vec{H}_i \cdot \vec{w} \, ds = 0. \tag{12.4}
$$

The aforementioned definition of the incident field and the linearity of the problem allows us to write the following relation between the incident and total fields:

$$
\vec{E} = \vec{E}_i + \vec{E}_s
$$

$$
\vec{H} = \vec{H}_i + \vec{H}_s, \tag{12.5}
$$

where $(\vec{E}_s, \vec{H}_s)$ is the secondary (or scattered) field due to the presence of the periodic structure. Substitution of these expressions in (12.3) and making use of (12.4) yields the following weak statement for the secondary fields:

$$\iiint_\Omega \left(\nabla \times \vec{w} \cdot \bar{\mu}_r^{-1} \cdot \nabla \times \vec{E}_s - k_0^2 \vec{w} \cdot \bar{\epsilon}_r \cdot \vec{E}_s \right) dv - j\omega\mu_0 \iint_{S_{U-L}} \hat{n} \times \vec{H}_s \cdot \vec{w}\, ds$$

$$- j\omega\mu_0 \iint_{S_W} \hat{n} \times \vec{H}_s \cdot \vec{w}\, ds = - \iiint_\Omega [\nabla \times \vec{w} \cdot (\bar{\mu}_r^{-1} - \bar{I}) \cdot \nabla \times \vec{E}_i$$

$$- k_0^2 \vec{w} \cdot (\bar{\epsilon}_r - \bar{I}) \cdot \vec{E}_i]\, dv - \iiint_\Omega \vec{w} \cdot [jk_0\eta_0 \vec{J}_i + \nabla \times (\bar{\mu}_r^{-1} \cdot \vec{M}_i)]\, dv,$$

$$(12.6)$$

where $\bar{I}$ is the identity tensor. The essential boundary condition on the PEC boundary is also written in terms of the secondary field as follows,

$$\hat{n} \times \vec{E} = 0 \quad \Rightarrow \quad \hat{n} \times \vec{E}_s = -\hat{n} \times \vec{E}_i. \qquad (12.7)$$

The linearity of the problem allows us to handle the impressed currents and the incident plane wave separately, through superposition. Thus, for example, with the impressed sources turned off and the excitation taken to be the incident plane wave, the solution of (12.6) requires the enforcement of appropriate relations between the tangential components of the electric and magnetic fields on the side walls, according to Floquet's theorem. These are

$$\begin{cases} \hat{n} \times \vec{E}_{s,Right} = -\hat{n} \times \vec{E}_{s,Left}\, e^{-j\psi_x} \\ \hat{n} \times \vec{H}_{s,Right} = -\hat{n} \times \vec{H}_{s,Left}\, e^{-j\psi_x} \\ \hat{n} \times \vec{E}_{s,Back} = -\hat{n} \times \vec{E}_{s,Front}\, e^{-j\psi_y} \\ \hat{n} \times \vec{H}_{s,Back} = -\hat{n} \times \vec{H}_{s,Front}\, e^{-j\psi_y}, \end{cases} \qquad (12.8)$$

where the unit normal $\hat{n}$ is taken to be pointing outwards. The minus sign on the right-hand side of the above equations is due to the fact that $\hat{n}$ is pointing in opposite directions on the two opposite walls. The terms ψ_x and ψ_y are the phase shifts in the electromagnetic field solution along the x and y directions, respectively, and are dictated by the wave vector (i.e., the parameters k_0, θ and ϕ) of the incident plane wave and the lengths D_x and D_y of the unit cell through the equations

$$\begin{cases} \psi_x = k_0 \sin \theta \cos \phi D_x, \\ \psi_y = k_0 \sin \theta \sin \phi D_y. \end{cases} \qquad (12.9)$$

To close the system, boundary conditions must be imposed at the top and bottom boundaries. The way the relationships (12.8) are imposed in the context of the finite element approximation of (12.6) and the boundary condition (12.7) and (12.8), as well as the types of truncation conditions used on S_{U-L}, are discussed in the following sections.

12.2   COMPUTATIONAL DOMAIN TRUNCATION SCHEMES

Our presentation of methodologies for finite element grid truncation will begin with the consideration of the top and bottom planar boundaries through which the two-dimensional periodic structure is interfaced to the unbounded domain. The secondary field in the unbounded domain will be cast in a form consistent with the periodicity of the solution. This representation of the secondary fields in terms of the so-called *space harmonics* or *Hartree harmonics*, or *Floquet modes* is discussed next.

12.2.1 Space harmonics expansion of EM fields

In the homogeneous (free-space) unbounded region exterior to the periodic structure the secondary fields will be represented in terms of a TE and TM decomposition [10]. More specifically, in view of the fact that the periodicity of the structure is taken to be along the x and y axes, a decomposition in terms of TE_z and TM_z waves will be used. The E_z and H_z components of the secondary field both satisfy the homogeneous scalar Helmholtz equation

$$\begin{cases} \dfrac{\partial^2 \Phi}{\partial^2 x} + \dfrac{\partial^2 \Phi}{\partial^2 y} + (\omega^2 \epsilon_0 \mu_0 + \gamma^2)\Phi = 0 \\ \Phi(x + D_x, y, z) = e^{-j\psi_x}\Phi(x,y,z) \\ \Phi(x, y + D_y, z) = e^{-j\psi_y}\Phi(x,y,z), \end{cases} \tag{12.10}$$

where $\Phi = H_z(x,y)\exp(-\gamma z)$ for TE_z part; and $\Phi = E_z(x,y)\exp(-\gamma z)$ for TM_z part. Furthermore, as a consequence of Floquet's theorem, the solution of (12.10) is of the form

$$\Phi(x,y,z) = \Phi_p(x,y)e^{-j(k_0 \sin\theta\cos\phi x + k_0 \sin\theta\sin\phi y)-\gamma z}, \tag{12.11}$$

where $\Phi_p(x,y)$ is periodic along x and y with periods D_x and D_y, respectively. Solutions to (12.10) exhibiting such periodicity are easily seen to be of the form

$$E_z(x,y,z) = E_0 e^{-j(\alpha_m x + \alpha_n y)}e^{-\gamma_{mn} z}$$
$$H_z(x,y,z) = H_0 e^{-j(\alpha_m x + \alpha_n y)}e^{-\gamma_{mn} z}, \tag{12.12}$$

where

$$\alpha_m = k_0 \sin\theta\cos\phi + \frac{2\pi m}{D_x},$$

$$\alpha_n = k_0 \sin\theta\sin\phi + \frac{2\pi n}{D_y}, \tag{12.13}$$

$$\gamma_{mn} = \pm j\sqrt{k_0^2 - \alpha_m^2 - \alpha_n^2},$$

where the plus and minus sign in the definition of γ represent, respectively, wave propagation in the $+z$ and $-z$ directions.

Knowledge of the general form of E_z and H_z suffices for the definition of the general form of the remaining transverse parts, $\vec{E}_t$ and $\vec{H}_t$, of the electric and magnetic fields through a manipulation of Maxwell's curl equations. More specifically, with $\vec{E}$ and $\vec{H}$ written in the split form $\vec{E} = E_z\hat{z} + \vec{E}_t$, $\vec{H} = H_z\hat{z} + \vec{H}_t$, and making use of the fact that the $\exp(-\gamma_{mn}z)$ dependence of the fields in the z direction allows us to represent the ∇ operator in the form

$$\nabla = \left(\hat{x}\frac{\partial}{\partial x} + \hat{y}\frac{\partial}{\partial y}\right) - \gamma_{mn}\hat{z} = \nabla_t - \gamma_{mn}\hat{z}. \tag{12.14}$$

Maxwell's curl equations become

$$\nabla_t \times E_z\hat{z} + \gamma_{mn}\vec{E}_t \times \hat{z} = -j\omega\mu_0\vec{H}_t,$$
$$\nabla_t \times H_z\hat{z} + \gamma_{mn}\vec{H}_t \times \hat{z} = j\omega\epsilon_0\vec{E}_t. \tag{12.15}$$

Elimination of $\vec{E}_t$ in the above system yields

$$\vec{H}_t = \frac{1}{\omega^2\epsilon_0\mu_0 + \gamma_{mn}^2}\left(-\gamma_{mn}\nabla_t H_z - j\omega\epsilon_0\hat{z} \times \nabla_t E_z\right)$$

$$= \frac{1}{\alpha_m^2 + \alpha_n^2}\left(-\gamma_{mn}\nabla_t H_z - j\omega\epsilon_0\hat{z} \times \nabla_t E_z\right), \tag{12.16}$$

where use was made of the vector identity (A.7). In a similar manner, elimination of $\vec{H}_t$ in (12.15) yields

$$
\begin{aligned}
\vec{E}_t &= \frac{1}{\omega^2 \epsilon_0 \mu_0 + \gamma_{mn}^2} \left(-\gamma_{mn} \nabla_t E_z + j\omega\mu_0 \hat{z} \times \nabla_t H_z \right) \\
&= \frac{1}{\alpha_m^2 + \alpha_n^2} \left(-\gamma_{mn} \nabla_t E_z + j\omega\mu_0 \hat{z} \times \nabla_t H_z \right).
\end{aligned}
\tag{12.17}
$$

The general expressions (12.16) and (12.17) allow us to write the general forms of the transverse components of the electromagnetic fields for TE_z and TM_z waves in a straightforward fashion. In particular, for the case of TE_z waves, use of H_z from (12.12), normalized as follows:

$$
H_z = \frac{\sqrt{\alpha_m^2 + \alpha_n^2}}{j\gamma_{mn}\sqrt{A}} e^{-j(\alpha_m x + \alpha_n y)} e^{-\gamma_{mn} z}
\tag{12.18}
$$

into (12.16) and (12.17), leads to the following general expressions for the TE_z fields:

$$
\begin{aligned}
\vec{E}_{mn}^{TE} &= \frac{j\omega\mu_0/\gamma_{mn}}{\sqrt{A(\alpha_m^2 + \alpha_n^2)}} (\alpha_n \hat{x} - \alpha_m \hat{y}) e^{-j(\alpha_m x + \alpha_n y)} e^{-\gamma_{mn} z} \\
\vec{H}_{mn}^{TE} &= \frac{\alpha_m \hat{x} + \alpha_n \hat{y} - j(\alpha_m^2 + \alpha_n^2)/\gamma_{mn}\hat{z}}{\sqrt{A(\alpha_m^2 + \alpha_n^2)}} e^{-j(\alpha_m x + \alpha_n y)} e^{-\gamma_{mn} z},
\end{aligned}
\tag{12.19}
$$

where $A = D_x D_y$ is the area of the $x - y$ cross-section of the unit cell of the periodic structure. It is straightforward to show that the TE_z waves satisfy the following orthogonality relation over the unit cell cross-section:

$$
\iint_{S_U} \hat{z} \cdot \vec{E}_{mn}^{TE} \times \left(\vec{H}_{m'n'}^{TE} \right)^* ds = \begin{cases} \dfrac{j\omega\mu_0}{\gamma_{mn}} & (m = m', \quad n = n') \\ 0 & (m \neq m', \quad n \neq n'), \end{cases}
\tag{12.20}
$$

where the superscript $*$ denotes complex conjugation. In the above expression, the surface of the upper boundary, S_U, of the unit cell of the periodic structure was taken, without loss of generality, to represent the $x - y$ cross-section over which the orthogonality of the TE_z modes holds.

In a similar fashion, substitution of E_z from (12.12), normalized as follows:

$$
E_z = \frac{\sqrt{\alpha_m^2 + \alpha_n^2}}{j\gamma_{mn}\sqrt{A}} e^{-j(\alpha_m x + \alpha_n y)} e^{-\gamma_{mn} z}
\tag{12.21}
$$

into (12.16) and (12.17) yields the following general expressions for the TM_z fields:

$$
\begin{aligned}
\vec{E}_{mn}^{TM} &= \frac{\alpha_m \hat{x} + \alpha_n \hat{y} - j(\alpha_m^2 + \alpha_n^2)/\gamma_{mn}\hat{z}}{\sqrt{A(\alpha_m^2 + \alpha_n^2)}} e^{-j(\alpha_m x + \alpha_n y)} e^{-\gamma_{mn} z}, \\
\vec{H}_{mn}^{TM} &= -\frac{j\omega\epsilon_0/\gamma_{mn}}{\sqrt{A(\alpha_m^2 + \alpha_n^2)}} (\alpha_n \hat{x} - \alpha_m \hat{y}) e^{-j(\alpha_m x + \alpha_n y)} e^{-\gamma_{mn} z}.
\end{aligned}
\tag{12.22}
$$

The TM_z modes satisfy the following orthogonality relation:

$$
\iint_{S_U} \hat{z} \cdot \vec{E}_{mn}^{TM} \times \left(\vec{H}_{m'n'}^{TM} \right)^* ds = \begin{cases} -\dfrac{j\omega\epsilon_0}{\gamma_{mn}^*} & (m = m', \quad n = n'), \\ 0 & (m \neq m', \quad n \neq n'). \end{cases}
\tag{12.23}
$$

Furthermore, it is easily verified that the TE_z and TM_z modes are mutually orthogonal

$$\iint_{S_U} \hat{z} \cdot \vec{E}_{mn}^{TE} \times \left(\vec{H}_{m'n'}^{TM}\right)^* ds = \iint_{S_U} \hat{z} \cdot \vec{E}_{mn}^{TM} \times \left(\vec{H}_{m'n'}^{TE}\right)^* ds = 0. \qquad (12.24)$$

The aforementioned results suggest the following representation for the scattered electric field in the unbounded, homogeneous regions above and below the two-dimensional, infinite periodic structure:

$$\vec{E}_s = \sum_{m=-\infty}^{+\infty} \sum_{n=-\infty}^{+\infty} \left[c_{mn} \vec{E}_{mn}^{TE} + d_{mn} \vec{E}_{mn}^{TM} \right], \qquad (12.25)$$

where the expansion coefficients, c_{mn} and d_{mn}, constitute the unknowns to be determined from the solution of the electromagnetic boundary value problem for a given excitation of the periodic structure. The physical interpretation of this result is worth noting. Contrary to the problem of plane wave scattering by a single target where scattering occurs in all directions in space, the electromagnetic fields scattered by a periodic array of similar targets are propagating in discrete directions in space, controlled by the periodicity of the ensemble. The TE_z and TM_z waves (modes) describing the electromagnetic field scattering in these discrete directions will be referred to as Floquet modes, or space harmonics, or Hartree harmonics.

Whether a space harmonic is a propagating or an evanescent wave is dependent on its order (m, n) and the operating frequency ω. Clearly, in view of the expression for γ_{mn} in (12.13), the (m, n) space harmonic is propagating provided that it is

$$\underbrace{\left(k_0 \sin\theta \cos\phi + \frac{2\pi m}{D_x} \right)^2}_{\alpha_m^2} + \underbrace{\left(k_0 \sin\theta \sin\phi + \frac{2\pi n}{D_y} \right)^2}_{\alpha_n^2} < k_0^2. \qquad (12.26)$$

For example, for the case where the excitation is provided by an incident field, and the unit cell dimensions are such that $D_x < \lambda_0/2$ and $D_y < \lambda_0/2$, where λ_0 is the free-space wavelength at the operating frequency, only the TE_{00} and TM_{00} modes are propagating modes.

When $D_x > \lambda_0/2$ and/or $D_y > \lambda_0/2$, more modes besides the $(0,0)$ mode may be propagating. Clearly, in most cases of practical interest, a finite number of propagating modes will be involved in the representation of the scattered field in the unbounded regions at a distance sufficiently far from the volume occupied by the periodic structure. This fact is exploited for the truncation of the finite element grid in the context of the transfinite element method as described next.

12.2.2   Grid truncation using transfinite flements

The finite number of the propagating space harmonics in the representation of the scattered fields from a periodic structure, combined with their known analytical form and their orthogonality relations, makes them most suitable for an effective grid truncation scheme at the top and bottom boundaries of the unit cell of the periodic structure. More specifically, a grid truncation scheme based on the transfinite element method (TFE) can be used.

The TFE method is discussed in detail in Section 6.3.2 in the context of finite element grid truncation at waveguide ports in microwave components. With the truncation surface (which, in the case of the two-dimensional periodic structure of interest is either one of the top and bottom boundaries S_U and S_L, respectively) placed at a sufficient distance from the inhomogeneous region of the unit cell, the total field on the truncation boundary is written in terms of the incident field, $\vec{E}_i$, (if any) and the truncated sum of only propagating space harmonics. Thus the electric field vector approximation at the top boundary S_U has the form

$$\vec{E}_{top} = \sum_{i=1}^{N_U} \alpha_i \vec{e}_i, \qquad (12.27)$$

where N_U is the number of Floquet modes used for the expansion of the field at S_U, while $\vec{e}_i$ represents the normalized ith Floquet mode and α_i is the corresponding coefficient. Clearly, both TE_z and TM_z modes must be included in this expansion in the general case. Furthermore, if the excitation is an incident plane wave, it can be taken into account in the above expression through the terms associated with the $(0,0)$ Floquet modes. A similar expression holds for the electric field at the bottom boundary S_L. The rest of the procedure follows closely the development of Section 6.3.2.

12.2.3 Grid truncation using anisotropic perfectly matched layers

The upper and bottom boundaries of the unit cell may also be truncated by making use of the so-called anisotropic perfectly matched layer (PML) concept [13]. The PML technique relies upon the utilization of a properly constructed anisotropic medium that provides for absorption of outgoing waves while minimizing impedance mismatch (and, hence, reflection) at its boundary with the interior of the computational domain. The PML idea was first introduced by Berenger in conjunction with grid truncation for the finite difference-time domain (FDTD) method [14]-[16]. While in Berenger's implementation the special properties of the PML medium were effected through a particular splitting of the electromagnetic fields, an alternative formulation by Chew and Weedon utilized a coordinate stretching idea to achieve the same behavior and thus render the interface between the PML region and free space reflectionless, regardless of frequency, polarization, and angle of incidence [17]. An alternative approach, where instead of coordinate stretching the absorption and reflectionless attributes of the PML were effected through the introduction of specific anisotropy in the medium, was proposed in [11]. This anisotropic formalism of the PML medium is most suitable for use in conjunction with finite element approximations of the electromagnetic problem. In this section, the anisotropic PML of [11] will be used for the truncation of the upper and lower boundaries S_U and S_L of the unit cell of a two-dimensional, infinite periodic structure.

The anisotropic PML medium has permeability and permittivity tensors of the form

$$\bar{\epsilon} = \epsilon_0 \bar{\Lambda}, \quad \bar{\mu} = \mu_0 \bar{\Lambda}, \quad \bar{\Lambda} = \begin{bmatrix} a & & \\ & b & \\ & & c \end{bmatrix}. \qquad (12.28)$$

Thus Maxwell's equations in the medium are written as follows:

$$\begin{cases} \nabla \cdot \bar{\Lambda} \cdot \vec{E} = 0 \\ \nabla \cdot \bar{\Lambda} \cdot \vec{H} = 0 \\ \nabla \times \vec{E} = -j\omega\mu_0 \bar{\Lambda} \cdot \vec{H} \\ \nabla \times \vec{H} = j\omega\epsilon_0 \bar{\Lambda} \cdot \vec{E}. \end{cases} \qquad (12.29)$$

It can be shown that the following plane waves are the eigen-solutions of the above system of Maxwell's equations:

$$\vec{E} = \vec{\mathcal{E}}e^{-j\vec{k}\cdot\vec{r}}, \quad \vec{H} = \vec{\mathcal{H}}e^{-j\vec{k}\cdot\vec{r}}, \tag{12.30}$$

where $\vec{\mathcal{E}}$ and $\vec{\mathcal{H}}$ are two constant vectors and

$$\vec{k} = k_x\hat{x} + k_y\hat{y} + k_z\hat{z}, \quad \begin{cases} k_x = k_0\sqrt{bc}\sin\theta\cos\phi \\ k_y = k_0\sqrt{ac}\sin\theta\sin\phi \ . \\ k_z = k_0\sqrt{ab}\cos\theta \end{cases} \tag{12.31}$$

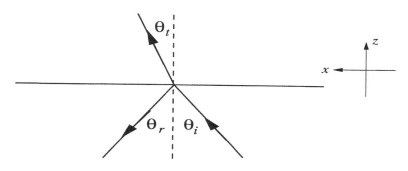

Figure 12.2 Planar $(x - y)$ interface between free-space and an anisotropic PML.

Consider, next, the planar interface between free-space in the lower region and the anisotropic PML medium in the upper region shown in Fig. 12.2. Without loss of generality, let us assume that the wave vector of the incident plane wave is on the $x - z$ plane; hence, $k_y = 0$. The incident plane wave is decomposed into its perpendicularly- and parallel-polarized parts, where, respectively, the electric field vector and the magnetic field vector of the incident field are perpendicular to the plane of incidence ($x - z$ plane). These parts will be referred to, respectively, as the TE_y and TM_y parts of the incident wave.

For the TM_y fields, the expressions for the electric fields of the incident, reflected and transmitted waves are given by

$$\begin{cases} \vec{E}_i = \hat{y}e^{-jk_0(\sin\theta_i x + \cos\theta_i z)} \\ \vec{E}_r = R^{TM}\hat{y}e^{-jk_0(\sin\theta_r x - \cos\theta_r z)} \\ \vec{E}_t = T^{TM}\hat{y}e^{-jk_0(\sqrt{bc}\sin\theta_t x + \sqrt{ab}\cos\theta_t z)}, \end{cases} \tag{12.32}$$

where R^{TM} and T^{TM} are, respectively, the reflection and transmission coefficients for TM_y waves. The corresponding magnetic field vectors are deduced from Maxwell's equations, with (12.29) used for the case of the transmitted field,

$$\begin{cases} \vec{H}_i = \sqrt{\dfrac{\epsilon_0}{\mu_0}}\left(-\cos\theta_i\hat{x} + \sin\theta_i\hat{z}\right)e^{-jk_0(\sin\theta_i x + \cos\theta_i z)} \\[2mm] \vec{H}_r = R^{TM}\sqrt{\dfrac{\epsilon_0}{\mu_0}}\left(\cos\theta_r\hat{x} + \sin\theta_r\hat{z}\right)e^{-jk_0(\sin\theta_r x - \cos\theta_r z)} \\[2mm] \vec{H}_t = T^{TM}\sqrt{\dfrac{\epsilon_0}{\mu_0}}\left(-\sqrt{\dfrac{b}{a}}\cos\theta_t\hat{x} + \sqrt{\dfrac{b}{c}}\sin\theta_t\hat{z}\right)e^{-jk_0(\sqrt{bc}\sin\theta_t x + \sqrt{ab}\cos\theta_t z)}. \end{cases}$$

$$\tag{12.33}$$

Enforcement of the continuity of the tangential electric (E_y) and magnetic (H_x) fields for the two media at the planar interface yields the following two sets of relations. First, from phase matching we obtain

$$\begin{cases} \sin \theta_i = \sin \theta_r \\ \sqrt{bc} \sin \theta_t = \sin \theta_i. \end{cases} \quad (12.34)$$

Second, enforcing phase matching the continuity of the field amplitudes yields

$$\begin{cases} 1 + R^{TM} = T^{TM} \\ \cos \theta_i - R^{TM} \cos \theta_r = T^{TM} \sqrt{\dfrac{b}{a}} \cos \theta_t \end{cases} \Rightarrow R^{TM} = \dfrac{\cos \theta_i - \sqrt{\dfrac{b}{a}} \cos \theta_t}{\cos \theta_i + \sqrt{\dfrac{b}{a}} \cos \theta_t}. \quad (12.35)$$

The calculation of the reflection and transmission coefficients, R^{TE} and T^{TE}, respectively, for the TE_y part of the incident field proceeds in exactly the same manner. The expressions for the incident, reflected and transmitted magnetic fields are

$$\begin{cases} \vec{H}_i = \hat{y} e^{-jk_0(\sin \theta_i x + \cos \theta_i z)} \\ \vec{H}_r = R^{TE} \hat{y} e^{-jk_0(\sin \theta_r x - \cos \theta_r z)} \\ \vec{H}_t = T^{TE} \hat{y} e^{-jk_0(\sqrt{bc} \sin \theta_t x + \sqrt{ab} \cos \theta_t z)}. \end{cases} \quad (12.36)$$

The corresponding electric fields are

$$\begin{cases} \vec{E}_i = \sqrt{\dfrac{\mu_0}{\epsilon_0}} \left(\cos \theta_i \hat{x} - \sin \theta_i \hat{z} \right) e^{-jk_0(\sin \theta_i x + \cos \theta_i z)} \\ \vec{E}_r = R^{TE} \sqrt{\dfrac{\mu_0}{\epsilon_0}} \left(-\cos \theta_r \hat{x} - \sin \theta_r \hat{z} \right) e^{-jk_0(\sin \theta_r x - \cos \theta_r z)} \\ \vec{E}_t = T^{TE} \sqrt{\dfrac{\mu_0}{\epsilon_0}} \left(\sqrt{\dfrac{b}{a}} \cos \theta_t \hat{x} - \sqrt{\dfrac{b}{c}} \sin \theta_t \hat{z} \right) e^{-jk_0(\sqrt{bc} \sin \theta_t x + \sqrt{ab} \cos \theta_t z)}. \end{cases} \quad (12.37)$$

Continuity of the tangential electric field components (E_x) and the tangential magnetic field component (H_y) of the total fields in the two media at the planar interface yields the phase matching condition of (12.34) and the following system of equations for R^{TE} and T^{TE}:

$$\begin{cases} 1 + R^{TE} = T^{TE} \\ \cos \theta_i - R^{TE} \cos \theta_r = T^{TE} \sqrt{\dfrac{b}{a}} \cos \theta_t \end{cases} \Rightarrow R^{TE} = \dfrac{\cos \theta_i - \sqrt{\dfrac{b}{a}} \cos \theta_t}{\cos \theta_i + \sqrt{\dfrac{b}{a}} \cos \theta_t}. \quad (12.38)$$

In view of (12.34), (12.35) and (12.38) it follows immediately that the choice

$$a = b = \frac{1}{c} \quad (12.39)$$

for the elements of the PML tensor results in the desirable result for a reflectionless interface, namely, the angle of transmission to be equal to the angle of incidence and both reflection coefficients, R^{TE} and R^{TM} to be zero.

$$a = b = \frac{1}{c} \quad \Rightarrow \quad \theta_i = \theta_t \text{ and } R^{TE} = R^{TM} = 0 \quad (12.40)$$

Furthermore, allowing a to be complex, $a = \alpha - j\beta$, the transmitted wave in the PML medium is attenuated in the direction perpendicular to the interface

$$\vec{E}_t = \vec{\mathcal{E}}e^{-k_0\beta\cos\theta_t z}e^{-jk_0(\sin\theta_t x + \alpha\cos\theta_t z)} \tag{12.41}$$

at a rate controlled by β and the angle of incidence. The apparent wavelength in the direction of attenuation is *scaled* by α.

Next, we consider these properties of the PML medium in the context of finite element grid truncation. In particular, we are interested in the design of an anisotropic PML layer to serve as a reflectionless absorber for the top and bottom sides of the unit cell of a two-dimensional, infinite periodic structure. The waves to be absorbed are any outward propagating space harmonics. The placement of the PML layer is guided by the observation that, in view of (12.41), the PML layer is most effective as an absorber for propagating waves. Thus a buffer layer is introduced between the end of the inhomogeneous topography inside the unit cell and the PML layer, to prevent the interaction of the PML with the evanescent space harmonics. The thickness of this buffer layer is taken to be a fraction of the free-space wavelength at the operating frequency. For example, a choice of $0.3\lambda_0$ for the thickness of the buffer layer has been found to provide for good accuracy in the analysis of periodic antenna arrays.

The design of the PML layer involves the determination of its thickness and the values α and β. Toward this objective, let us assume that a PEC wall will be placed at the end of the PML layer. Then, the reflection coefficient of a PEC-backed anisotropic PML layer of thickness t is easily found to be

$$|R(\theta_i)| = e^{-2\beta k_0 t\cos\theta_i} \quad \Rightarrow \quad |R(\theta_i)| = -17.372\beta k_0 t\cos\theta_i \text{ (dB)}. \tag{12.42}$$

In view of this result, a choice for t is made which is then combined with a desirable attenuation at normal incidence ($\theta_i = 0$) to deduce the appropriate value for β. While, in theory, the choice of α has no impact on the reflectionless attenuation attribute of the PML, its choice, as noted above, scales the apparent wavelength in the direction of attenuation inside the PML region from $\lambda_z = \lambda_0/\cos\theta_t$ in free space to

$$\hat{\lambda}_z = \frac{1}{\alpha}\frac{\lambda_0}{\cos\theta_t}. \tag{12.43}$$

This, then, implies, that in the discrete model extra care should be taken to avoid choices for α which reduce the wavelength to the point that its resolution by the finite element grid raises concerns about undesirable numerical dispersion. However, the PML being a "lossy" medium, the penetration depth

$$d_p = \frac{1}{\beta}\frac{\lambda_0}{2\pi\cos\theta_t} \tag{12.44}$$

is another characteristic length that must be resolved sufficiently well to ensure good accuracy. A direct comparison of (12.43) and (12.44) indicates that the choice $\alpha = \beta$, with β chosen according to (12.42) to provide for the desired attenuation and subject to the requirement that d_p is resolved sufficiently well by the finite element grid, provides for a fine resolution of $\hat{\lambda}_z$ as well. Thus this choice is very common in the practical implementation of the anisotropic PML [18] [19].

In the more general context of computational domain truncation for the finite element-based modeling of three-dimensional electromagnetic scattering and radiation problems, anisotropic PMLs may be utilized on all six surfaces of a rectangular domain enclosing the scatterer/radiator. While the design of the anisotropic PML used on the six faces of the box is done according to the considerations in the previous paragraphs, the handling of edges and corners requires extra care.

To elaborate, consider the drawing in Fig. 12.3, depicting one of the eight corners of the computational domain. Also shown in the figure are the tensors $\bar{\Lambda}^x_{side}$, $\bar{\Lambda}^y_{side}$, and $\bar{\Lambda}^z_{side}$, representing the anisotropic PMLs used as absorbers on the x, y and z faces. In the notation used in the figure, s_x, s_y, and s_z are used in place of a to specify the properties of each of the three tensors. With regards to the choice of the material properties of the PML for an edge region, one approach is to choose its properties such that it is perfectly matched to the two adjoint side layers [19] (assuming the interfaces of the edge with the two side layers are infinite). This approach results in choices for the tensors along the edges and at the corner as indicated in Fig. 12.3.

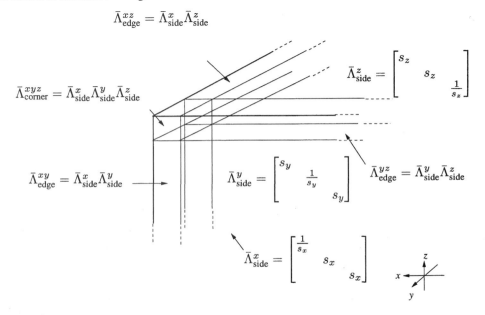

Figure 12.3 Portion of a rectangular PML box enclosing the computational domain.

Let us consider the PML used along the yz edge. To explain the choice of its parameters, let us extend the result derived earlier for a reflectionless free-space/PML interface to the case where the interface is between two anisotropic PMLs with tensors

$$\bar{\Lambda} = \begin{bmatrix} a & & \\ & b & \\ & & c \end{bmatrix}, \quad \text{and} \quad \bar{\Lambda}' = \begin{bmatrix} a' & & \\ & b' & \\ & & c' \end{bmatrix}. \tag{12.45}$$

Enforcing the interface reflectionless yields the following constraints on the elements of the two tensors:

$$\begin{cases} \dfrac{a}{a'} = \dfrac{b}{b'} = \dfrac{c'}{c} & \text{(for an } x - y \text{ interface)} \\[2mm] \dfrac{a}{a'} = \dfrac{b'}{b} = \dfrac{c}{c'} & \text{(for an } x - z \text{ interface)} \\[2mm] \dfrac{a'}{a} = \dfrac{b}{b'} = \dfrac{c}{c'} & \text{(for an } y - z \text{ interface)}. \end{cases} \tag{12.46}$$

The relation (12.39) can be viewed as the special case of the above result when one of the media is free space and material interface is on $x - y$ plane. Based on this result it can be shown that in order to match the yz edge region with the two layers used on the y and z faces the following tensor, $\bar{\Lambda}^{yz}_{\text{edge}}$, should be used

$$\bar{\Lambda}^{yz}_{\text{edge}} = \bar{\Lambda}^{y}_{\text{side}} \bar{\Lambda}^{z}_{\text{side}} = \begin{bmatrix} s_y s_z & & \\ & s_z/s_y & \\ & & s_y/s_z \end{bmatrix}. \tag{12.47}$$

Similar expressions are derived for the tensors of the PMLs to be used at the other edges; they are shown in Fig. 12.3. Finally, the PML tensor for the corner regions is given by

$$\bar{\Lambda}^{xyz}_{\text{corner}} = \bar{\Lambda}^{x}_{\text{side}} \bar{\Lambda}^{y}_{\text{side}} \bar{\Lambda}^{z}_{\text{side}}. \tag{12.48}$$

12.3 FINITE ELEMENT APPROXIMATION INSIDE THE UNIT CELL

At this point we return to the development of the finite element approximation from the weak statement of the electromagnetic boundary value problem inside the unit cell. The computational domain Ω is discretized using a finite element grid of tetrahedra. The approximations for the electric field inside the computational domain and on the cell boundary and for the magnetic field on the cell boundary are given by

$$\vec{E} = \sum_{n=0}^{N} \vec{w}_n x_{e_n}, \quad \vec{H} = \sum_{n=0}^{N_w^h} \vec{w}_n x_{h_n} \tag{12.49}$$

where $\vec{w}_n$ are the tangentially continuous finite elements discussed in Chapter 2, x_{e_n} and x_{h_n} are the unknown expansion coefficients, N is the number of electric field unknowns, and N_w^h is the number of magnetic unknowns on the side walls (i.e., periodic boundaries).

For ease of implementation of the Floquet boundary conditions, the surface meshes on opposing walls are made identical; however, the tetrahedra associated with corresponding identical elements on the two walls need not be the same. In addition to the two surface meshes being identical, the ordering of the nodes for corresponding edges on opposing walls should be the same. Since the definition of edge-element basis functions is determined by the ordering of the nodes, the enforcement of the same ordering for corresponding edges ensures a one-to-one correspondence between the associated basis functions.

For ease of formulation the vector x_e, containing the coefficients in the electric field expansion, is split into two sub-vectors as follows:

$$x_e = \begin{bmatrix} x_e^{in} \\ x_e^{w} \end{bmatrix}. \tag{12.50}$$

In the above expression x_e^{in} contains the coefficients for the elements in the interior of the computational domain while x_e^{w} contains the coefficients for the edges on the cell walls. With this splitting of the vector of unknowns, application of Galerkin's testing to (12.6) yields the following finite element system matrix:

$$\begin{bmatrix} S^{in,in} & S^{in,w} \\ S^{w,in} & S^{w,w} \end{bmatrix} \begin{bmatrix} x_e^{in} \\ x_e^{w} \end{bmatrix} - j\omega\mu_0 \begin{bmatrix} 0 \\ P \end{bmatrix} x_h^{w} = \begin{bmatrix} f_e^{in} \\ f_e^{w} \end{bmatrix}. \tag{12.51}$$

The elements of the matrices and vectors in the above system are given in terms of the following integrals:

$$[S]_{i,j} = \iiint_{\Omega} \left(\nabla \times \vec{w}_i \cdot \bar{\mu}_r^{-1} \cdot \nabla \times \vec{w}_j - k_0^2 \vec{w}_i \cdot \bar{\epsilon}_r \cdot \vec{w}_j \right) dv,$$

$$[P]_{i,j} = \iint_{S_W} \hat{n} \times \vec{w}_j \cdot \vec{w}_i \, ds,$$

$$[f_e]_i = -\iiint_{\Omega} [\nabla \times \vec{w} \cdot (\bar{\mu}_r^{-1} - \bar{I}) \cdot \nabla \times \vec{E}^{inc} - k_0^2 \vec{w} \cdot (\bar{\epsilon}_r - \bar{I}) \cdot \vec{E}^{inc}] \, dv$$

$$- \iiint_{\Omega} \vec{w} \cdot [j k_0 \eta_0 \vec{J}_i + \nabla \times (\bar{\mu}^{-1} \cdot \vec{M}_i)] \, dv.$$

(12.52)

It is noted that in (12.51) there is no contribution from the surface integral on the upper and lower boundaries, S_U and S_L, since it is assumed that PEC-backed anisotropic PMLs are used for grid truncation. Also, for the sake of generality, the elements of the source vectors include contributions from both an incident field and impressed electric and magnetic current densities inside the unit cell.

If a TFE truncation scheme is used, instead, for the top and bottom surfaces, then the system of (12.51) must be augmented by matrices associated with the additional coefficients used in the space harmonic expansion of the scattered fields. The augmented finite element matrix has the form

$$\begin{bmatrix} S^{in,in} & S^{in,w} & S^{in,UL}B \\ S^{w,in} & S^{w,w} & S^{w,UL}B \\ B^H S^{UL,in} & B^H S^{UL,w} & B^H S^{UL,UL}B + j\omega\mu_0 I \end{bmatrix} \begin{bmatrix} x_e^{in} \\ x_e^w \\ x_\alpha \end{bmatrix}$$

$$- j\omega\mu_0 \begin{bmatrix} 0 \\ P \\ 0 \end{bmatrix} x_h^w = \begin{bmatrix} f_e^{in} \\ f_e^w \\ 0 \end{bmatrix}.$$

(12.53)

In the above expression x_α is the vector containing the unknown expansion coefficients for the Floquet modes used in the approximation of the scattered field at the upper and lower boundaries. The matrices $S^{in,UL}$ and $S^{UL,in}$ describe the coupling between the finite element basis functions inside the domain with those on the upper and lower boundaries. Similarly, the matrices $S^{w,UL}$ and $S^{UL,w}$ describe the coupling of the basis functions on the four side walls with those on the upper and lower boundaries. The matrix $S^{UL,UL}$ describes the coupling between basis functions associated with edges on the upper and lower boundaries. Finally, the matrix B is the projection matrix between the finite element basis functions on the upper and lower boundaries and the space harmonics used on the two boundaries for the expansion of the scattered fields (see (6.139)).

In both (12.51) and (12.53) the number of unknowns exceeds the number of equations by N_w^h. In order to close the system, the periodic boundary conditions, according to Floquet's theorem, must be imposed. The way this is done is discussed next.

12.4   PERIODIC BOUNDARY CONDITION

The choice of identical surface meshes on opposing walls, combined with the same ordering of the corresponding edges as discussed in the previous section ensures that the associated basis functions are the same. According to Floquet's theorem, the periodic boundary condition (12.8) requires that the electric and magnetic fields at each point on a side wall are the same with those at the corresponding point on the opposing wall except for a known (for the case of a driven problem) phase shift. Thus, to impose the periodic boundary condition in the context of an edge-element based finite element approximation of the fields, one only needs to relate the expansion coefficients associated with corresponding edges on the two walls through the appropriate phase shift.

Pertinent to the imposition of the periodic boundary condition is the matrix P with elements

$$P_{i,j} = \iint_{S_W} \hat{n} \times \vec{w}_j \cdot \vec{w}_i \, ds. \tag{12.54}$$

This matrix relates the electric field unknowns on the walls to the magnetic field unknowns on the wall. Hence, with N_w^e denoting the number of electric field unknowns on the four side walls and N_w^h the number of magnetic field unknowns on the walls, the dimension of P is $N_w^e \times N_w^h$. The following two properties of the elements of P follow immediately from the properties of tangentially-continuous vector finite element spaces.

Property I. If the expansion coefficients associated with $\vec{w}_i$ and $\vec{w}_j$ are not both on the boundary then

$$\iint_s \hat{n} \times \vec{w}_j \cdot \vec{w}_i \, ds = 0. \tag{12.55}$$

This is depicted in Fig. 12.4.

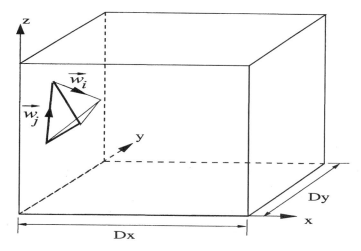

Figure 12.4   Graphical interpretation of Property I of the $P_{i,j} = \iint_s \hat{n} \times \vec{w}_j \cdot \vec{w}_i \, ds$.

Property II. For a corresponding pair of elements on the two opposing walls it is

$$\iint_s \hat{n} \times \vec{w}_j \cdot \vec{w}_i \, ds = - \iint_{s'} \hat{n} \times \vec{w}_{j'} \cdot \vec{w}_{i'} \, ds. \tag{12.56}$$

This is depicted in Fig. (12.5). The minus sign is due to the opposite direction of the outward pointing unit normal vector, $\hat{n}$, on the two opposing walls.

These two properties of (12.54) are used in the following for the imposition of the periodic boundary condition and, thus, the completion of the finite element approximation of the electromagnetic boundary value problem.

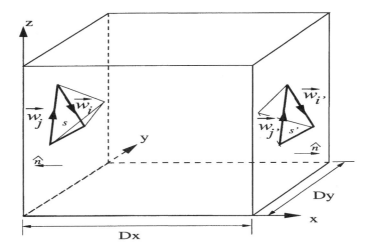

Figure 12.5   Illustration of Property II of the integral $P_{i,j} = \iint_s \hat{n} \times \vec{w}_j \cdot \vec{w}_i \, ds$.

Let us consider, first, the simpler case of a one-dimensional periodic structure, exhibiting periodicity along x. The Floquet boundary condition relates the unknown fields on the left boundary wall along x to those on the right boundary wall. The electric and magnetic field unknowns on the walls are arranged as follows:

$$x_e^w = \begin{bmatrix} x_e^l \\ x_e^r \end{bmatrix}, \qquad x_h^w = \begin{bmatrix} x_h^l \\ x_h^r \end{bmatrix} \tag{12.57}$$

where the superscripts l and r denote, respectively, the unknowns on the left and right walls. Furthermore, since the surface meshes on the two walls are identical, the unknowns on the left and right walls are taken in the same order, that is, the elements $[x_e^l]_i$ and $[x_e^r]_i$ represent two corresponding electric field unknowns, while $[x_h^l]_i$ and $[x_h^r]_i$ represent two corresponding magnetic field unknowns. Thus, in view of the aforementioned properties of the matrix P, the following relationship holds:

$$P x_h^w = \begin{bmatrix} A \\ & -A \end{bmatrix} \begin{bmatrix} x_h^l \\ x_h^r \end{bmatrix}, \tag{12.58}$$

where A represents a known matrix. The exact values of its elements are not relevant to the discussion that follows. Making use of the periodic boundary condition (12.8), we have

$$\begin{bmatrix} x_e^l \\ x_e^r \end{bmatrix} = T^e \begin{bmatrix} x_e^l \end{bmatrix}, \qquad \begin{bmatrix} x_h^l \\ x_h^r \end{bmatrix} = T^h \begin{bmatrix} x_h^l \end{bmatrix} \tag{12.59}$$

where the matrices T^e and T^h have the forms

$$T^e = \begin{bmatrix} I_{m_e} \\ c_x I_{m_e} \end{bmatrix}, \quad T^h = \begin{bmatrix} I_{m_h} \\ c_x I_{m_h} \end{bmatrix}. \tag{12.60}$$

In the above expressions m_e and m_h denote, respectively, the number of electric field and magnetic field unknowns on each wall. In general, m_e is not equal to m_h. The matrix I_{m_i}, $i \in \{e, m\}$, is the identity matrix of dimension m_i. Finally, $c_x = \exp(-j\psi_x)$ is the phase shift along x. Substitution of (12.59) into (12.58), followed by multiplication on the left by $(T^e)^H$ yields

$$(T^e)^H P T^h = 0, \tag{12.61}$$

where $(T^e)^H$ is the complex transpose of T^e. In deriving this result use was made of the special form of P shown in (12.58).

The result in (12.61) suggests a way to eliminate the extra unknowns x_h^w from the finite element system. More specifically, making use of the first of (12.59) and multiplying the second matrix equation in (12.51) and (12.53) on the left with $(T^e)^H$ we obtain

$$\underbrace{\begin{bmatrix} S^{in,in} & S^{in,w}T^e \\ (T^e)^H S^{w,in} & (T^e)^H S^{w,w}T^e \end{bmatrix}}_{S'} \begin{bmatrix} x_e^{in} \\ x_e^l \end{bmatrix} = \underbrace{\begin{bmatrix} f_e^{in} \\ (T^e)^H f_e^w \end{bmatrix}}_{f_e'} \tag{12.62}$$

for (12.51), and

$$\begin{bmatrix} S^{in,in} & S^{in,w}T^e & S^{in,UL}B \\ (T^e)^H S^{w,in} & (T^e)^H S^{w,w}T^e & (T^e)^H S^{w,UL}B \\ B^H S^{UL,in} & B^H S^{UL,w}T^e & B^H S^{UL,UL}B + j\omega\mu_0 I \end{bmatrix} \begin{bmatrix} x_e^{in} \\ x_e^l \\ x_\alpha \end{bmatrix} = \begin{bmatrix} f_e^{in} \\ (T^e)^H f_e^w \\ 0 \end{bmatrix} \tag{12.63}$$

for (12.53). In addition to eliminating the magnetic field unknowns on the side walls, electric field unknowns on only one of the two walls (the left wall) are kept in the finite element system.

Next, we consider the extension of this process to the case of a two-dimensional periodic structure. The periodicity is along the x and y directions, with unit cell as depicted in Fig. 12.1. The four side walls involved in the application of Floquet boundary conditions will be referred to as left and right for the walls along x and front and back for the walls along y. The electric and magnetic field unknowns on the walls are arranged as follows:

$$x_e^w = \begin{bmatrix} x_e^l \\ x_e^{lf} \\ x_e^{lb} \\ x_e^f \\ x_e^r \\ x_e^{rf} \\ x_e^{rb} \\ x_e^b \end{bmatrix}, \quad x_h^w = \begin{bmatrix} x_h^l \\ x_h^{lf} \\ x_h^{lb} \\ x_h^f \\ x_h^r \\ x_h^{rf} \\ x_h^{rb} \\ x_h^b \end{bmatrix}. \tag{12.64}$$

In the above expressions the superscripts l, r, f, and b indicate the corresponding vectors contain the unknowns in the interior of the left, front, right, and back wall, respectively, excluding any unknowns on wall edges. The unknowns on wall edges are incorporated in the vectors with superscripts lf, lb, rf, and fb, indicating the unknowns on the left-front, left-back, right-front, and right-back edge, respectively. Furthermore, like in the one-dimensional case described above, the same ordering is used for unknowns on opposite walls.

In view of the properties of (12.54), the matrix-vector product Px_h^w has the following form

$$Px_h^w =$$

$$
\begin{bmatrix}
A & X_1 & X_2 & 0 & 0 & 0 & 0 & 0 \\
X_3 & Y_1 + Y_2 & 0 & X_5 & 0 & 0 & 0 & 0 \\
X_4 & 0 & -Y_2 + Y_3 & 0 & 0 & 0 & 0 & -X_5 \\
0 & X_6 & 0 & B & 0 & X_7 & 0 & 0 \\
0 & 0 & 0 & 0 & -A & -X_1 & -X_2 & 0 \\
0 & 0 & 0 & X_8 & -X_3 & -Y_1 - Y_4 & 0 & 0 \\
0 & 0 & 0 & 0 & -X_4 & 0 & Y_4 - Y_3 & -X_8 \\
0 & 0 & -X_3 & 0 & 0 & 0 & -X_7 & -B
\end{bmatrix}
\begin{bmatrix}
x_h^l \\
x_h^{lf} \\
x_h^{lb} \\
x_h^f \\
x_h^r \\
x_h^{rf} \\
x_h^{rb} \\
x_h^b
\end{bmatrix},
\quad (12.65)
$$

where the exact values of the elements of each submatrix is irrelevant to the discussion that follows. To explain this form, let us consider the off-diagonal submatrices first. More specifically, let us consider submatrices X_1 and $-X_1$. X_1 describes the coupling between the elements on the left wall with those on the left-front edge, while $-X_1$ describes the coupling between elements on the right wall with those on the right-front edge. In view of Property II of the integral (12.54), the two submatrices are the negative of each other. A similar explanation holds for all remaining pairs of off-diagonal submatrices. Next we consider the diagonal submatrices. The fact that the submatrices associated with x_h^l and x_h^r are the negative of each other follows immediately from Property II of (12.54). Similarly, the submatrices associated with x_h^f and x_h^b are the negative of each other. Finally, the remaining diagonal submatrices are associated with the self contributions of the unknown coefficients on the four edges. As indicated in Fig. 12.6, contributions from both walls associated with each edge are involved in the calculation of the elements of these matrices. In doing so, Property II is again exploited to write these matrices in terms of only four submatrices, Y_1, Y_2, Y_3 and Y_4.

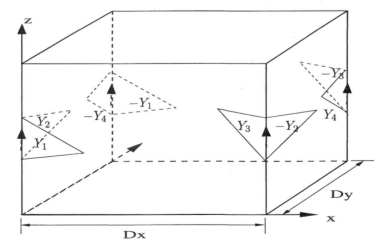

Figure 12.6 Edge-elements on the four side edges of the unit cell.

In view of the properties of the tangentially-continuous vector finite elements and the same ordering of the unknowns on opposing walls, the periodic boundary condition (12.8) is cast in discrete form as follows:

$$
\begin{bmatrix} x_e^r \\ x_e^{rf} \\ x_e^{rb} \\ x_e^b \end{bmatrix} = \begin{bmatrix} c_x I_{m_{e,1}} & & & \\ & c_x I_{m_{e,2}} & & \\ & & c_x I^{m_{e,2}} & \\ & & & c_y I^{m_{e,3}} \end{bmatrix} \begin{bmatrix} x_e^l \\ x_e^{lf} \\ x_e^{lb} \\ x_e^f \end{bmatrix} \tag{12.66}
$$

and

$$
x_e^{lb} = \begin{bmatrix} c_y I_{m_{e,2}} \end{bmatrix} x_e^{lf}, \tag{12.67}
$$

where $c_x = \exp(-j\psi_x)$ and $c_y = \exp(-j\psi_y)$, and $m_{e,1}$, $m_{e,2}$ and $m_{e,3}$ are, respectively, the number of electric fields unknowns on the left/right wall, each of the four side edges, and the front/back wall. Similarly, the following relationship holds between the magnetic field unknowns on opposing walls:

$$
\begin{bmatrix} x_h^r \\ x_h^{rf} \\ x_h^{rb} \\ x_h^b \end{bmatrix} = \begin{bmatrix} c_x I_{m_{h,1}} & & & \\ & c_x I_{m_{h,2}} & & \\ & & c_x I_{m_{h,2}} & \\ & & & c_y I_{m_{h,3}} \end{bmatrix} \begin{bmatrix} x_h^l \\ x_h^{lf} \\ x_h^{lb} \\ x_h^f \end{bmatrix} \tag{12.68}
$$

and

$$
x_h^{lb} = \begin{bmatrix} c_y I_{m_{h,2}} \end{bmatrix} x_h^{lf}, \tag{12.69}
$$

where $m_{h,1}$, $m_{h,2}$ and $m_{h,3}$ are, respectively, the number of magnetic field unknowns on the left/right wall, each of the four side edges, and the front/back wall. These numbers are, in general, different than the corresponding numbers for the electric field unknowns. From the surface coefficient relationships (12.66) and (12.67), we obtain

$$
x_e^w = T_1^e T_2^e \begin{bmatrix} x_e^l \\ x_e^{lf} \\ x_e^f \end{bmatrix}, \quad x_h^w = T_1^h T_2^h \begin{bmatrix} x_h^l \\ x_h^{lf} \\ x_h^f \end{bmatrix}, \tag{12.70}
$$

where the matrices T_1^s, T_2^s, $s \in \{e, h\}$ have the form

$$
T_1^s = \begin{bmatrix} I_{m_{s,1}} & & & \\ & I_{m_{s,2}} & & \\ & & I_{m_{s,2}} & \\ & & & I_{m_{s,3}} \\ c_x I_{m_{s,1}} & & & \\ & c_x I_{m_{s,2}} & & \\ & & c_x I_{m_{s,2}} & \\ & & & c_y I_{m_{s,3}} \end{bmatrix},
$$

$$
T_2^s = \begin{bmatrix} I_{m_{s,1}} & & \\ & I_{m_{s,2}} & \\ & c_y I_{m_{s,2}} & \\ & & I_{m_{s,3}} \end{bmatrix}. \tag{12.71}
$$

Making use of the special form of P, as shown in (12.65), and the above matrices, it is straightforward to show the following identity:

$$
(T_2^e)^H (T_1^e)^H P T_1^h T_2^h = 0. \tag{12.72}
$$

Substitution of (12.70) into the system matrix (12.51) and making using of (12.72) allows us to reduce the finite element matrix for the two-dimensional periodic structure in the following form:

$$
\begin{bmatrix} S^{in,in} & S^{in,w}T_1^e T_2^e \\ (T_1^e T_2^e)^H S^{w,in} & (T_1^e T_2^e)^H S^{w,w}T_1^e T_2^e \end{bmatrix} \begin{bmatrix} x_e^{in} \\ x_e^{w'} \end{bmatrix} = \begin{bmatrix} f_e^{in} \\ (T_1^e T_2^e)^H f_e^w \end{bmatrix},
\tag{12.73}
$$

where the vector $x_e^{w'}$ is given by

$$
x_e^{w'} = \begin{bmatrix} x_e^l \\ x_e^{lf} \\ x_e^f \end{bmatrix}.
\tag{12.74}
$$

In a similar fashion the finite element system of (12.53) is reduced as follows:

$$
\begin{bmatrix} S^{in,in} & S^{in,w}T_1^e T_2^e & S^{in,UL}B \\ (T_1^e T_2^e)^H S^{w,in} & (T_1^e T_2^e)^H S^{w,w}T_1^e T_2^e & (T_1^e T_2^e)^H S^{w,UL}B \\ B^H S^{UL,in} & B^H S^{UL,w}T_1^e T_2^e & B^H S^{UL,UL}B + j\omega\mu_0 I \end{bmatrix}
$$
$$
\begin{bmatrix} x_e^{in} \\ x_e^{w'} \\ x_\alpha \end{bmatrix} = \begin{bmatrix} f_e^{in} \\ (T_1^e T_2^e)^H f_e^w \\ 0 \end{bmatrix}.
\tag{12.75}
$$

This completes the process of imposing the Floquet periodic boundary conditions on the walls of the unit cell and, thus, closing the finite element system for the approximation of the periodic electromagnetic boundary value problem. It should be evident from the above discussion that, once the matrices T_1^e and T_2^e have been constructed, the imposition of the Floquet boundary condition amounts to straightforward left and right multiplications of the original system matrices (12.51) and (12.53), aimed at eliminating unknowns which are redundant due to the periodicity of the fields. While our development was for structures exhibiting periodicity along one or two directions, it should be apparent that the general case of structures exhibiting periodicity in all three directions can be handled through a straightforward extension of the presented methodology.

12.5 MULTILEVEL/MULTIGRID PRECONDITIONER

In most classical applications of practical interest (e.g., antenna arrays of infinite extent and periodic waveguiding structures) the dimensions of the unit cell along the directions of periodicity are of the order of the wavelength at the application frequency. However, both material and geometric complexity inside the unit cell may be significant. For example, this is typically the case with planar antenna arrays, where a multi-layered substrate of significant complexity may be associated with each element in the periodic cell. A second example is the case of frequency selective surfaces (FSS), where multiple planar FSS layers are stacked together, thus resulting in a multi-layered unit cell of high complexity. For such cases, an iterative solution may become necessary. In this section we present the way a multilevel/multigrid preconditioner can be develop in support of a robust iterative finite element solver for periodic structures of high unit cell complexity.

The main two ingredients of a multielvel/multigrid preconditioner are the formulation of the electromagnetic finite element system using potentials and the inter-grid transfer operators. Let us first consider the potential formulation for the case of an electromagnetic boundary problem with Floquet-type periodicity. The potential formulation involves the representation of the electric field vector in terms of the magnetic vector potential, $\vec{A}$, and the scalar electric potential, V, as follows:

$$\vec{E} = \vec{A} + \nabla V. \tag{12.76}$$

The way the weak statement and, subsequently, the finite element approximation of the boundary value problem are developed has been discussed in detail in Chapter 6 and will not be repeated here. As s summary of the process we note that substitution of the above equation for the electric field, with appropriate finite element expansions for the two potentials, into (12.3), followed by Galerkin's testing, yields one of the two matrix equations involved in the finite element approximation of the electromagnetic problem. The second equation is obtained by enforcing the equation $\nabla \cdot \vec{D} = 0$. The resulting finite element system has the form

$$\begin{bmatrix} I \\ G^T \end{bmatrix} [S] \begin{bmatrix} I & G \end{bmatrix} \begin{bmatrix} x_a \\ x_v \end{bmatrix} = \begin{bmatrix} I \\ G^T \end{bmatrix} [f_e]. \tag{12.77}$$

In the above equation G is the gradient matrix, S is defined in (12.52), and x_a and x_v are the vectors containing the expansion coefficients in the finite element approximations of A and V, respectively. In the above system the contribution from the surface integrals over the boundaries in the directions of periodicity of the structures are omitted since, as we saw in the previous section, they are eventually eliminated from the final system.

Without loss of generality, let us assume that the structure is periodic along x only; hence, Floquet boundary conditions are used to relate the electromagnetic fields on the left and right walls of the unit cell depicted in Fig. 12.1. Splitting the unknowns into those in the interior of the unit cell, x_a^{in}, x_v^{in}, and those on the two side walls, x_a^w, x_v^w, and recognizing that, through the enforcement of Floquet periodicity, only the unknowns on the left wall will remain in the reduced system, we have

$$\begin{aligned} x_a &= \begin{bmatrix} x_a^{in} \\ x_a^w \end{bmatrix} = \begin{bmatrix} I & \\ & T^a \end{bmatrix} \begin{bmatrix} x_a^{in} \\ x_a^l \end{bmatrix}, \\ x_v &= \begin{bmatrix} x_v^{in} \\ x_v^w \end{bmatrix} = \begin{bmatrix} I & \\ & T^v \end{bmatrix} \begin{bmatrix} x_v^{in} \\ x_v^l \end{bmatrix}, \end{aligned} \tag{12.78}$$

where the matrix T^a is the same as T^e defined in (12.60), while T^v is similar to T^e except for the dimension of the identity matrix, I, which in this case is equal to the number of scalar potential unknowns on the left/right wall. Substitution of (12.78) into (12.77), along with the definition of a matrix G' through the relationship

$$\begin{bmatrix} I & \\ & T^a \end{bmatrix} G' = G \begin{bmatrix} I & \\ & T^v \end{bmatrix} \tag{12.79}$$

allows us to recast (12.77) in the following form:

$$\begin{bmatrix} I \\ G'^T \end{bmatrix} [S'] \begin{bmatrix} I & G' \end{bmatrix} \begin{bmatrix} x_a' \\ x_v' \end{bmatrix} = \begin{bmatrix} I \\ G'^T \end{bmatrix} [f_e'], \tag{12.80}$$

where S' and f_e' are, respectively, the system matrix and the right-hand side vector in (12.62), while x_a' and x_v' denote the reduced unknown vectors,

$$x_a' = \begin{bmatrix} x_a^{in} \\ x_a^l \end{bmatrix}, \quad x_v' = \begin{bmatrix} x_v^{in} \\ x_v^l \end{bmatrix}. \tag{12.81}$$

Equation (12.80) is the potential formulation for the finite element analysis of periodic structures. Making use of the relation

$$(T^a)^H T^a = 2I, \tag{12.82}$$

where I is identity matrix, the matrix G' may be written in terms of the following matrix product:

$$G' = \begin{bmatrix} I \\ & \frac{1}{2}(T^a)^H \end{bmatrix} G \begin{bmatrix} I \\ & T^v \end{bmatrix}. \tag{12.83}$$

Thus, rather than building and storing G', the above product may be used during the iterative solution of (12.80).

Next we consider the construction of the inter-grid operators between two nested meshes. The system matrices of two nested meshes, prior to the imposition of the Floquet periodic boundary condition, satisfy the following relation:

$$Q_h^{2h} S_h Q_{2h}^h x_{e,2h} = Q_h^{2h} f_{e,h}, \tag{12.84}$$

where Q_h^{2h} and Q_{2h}^h are the inter-grid transfer operators, with the superscripts $2h$ and h denoting the coarser and finer meshes, respectively. S_h and $f_{e,h}$ are, respectively, the matrix and the right-hand side vector in (12.52) for the case of the dense mesh. As discussed in Chapter 6, the two inter-grid operators satisfy the equation $Q_h^{2h} = Q_{2h}^{h^T}$.

Splitting the unknowns $x_{e,2h}$ in the coarser mesh into those in the interior of the unit cell and those on the left and right walls in the direction of periodicity, and enforcing the Floquet periodic boundary condition, we obtain

$$x_{e,2h} = \begin{bmatrix} x_{e,2h}^{in} \\ x_{e,2h}^w \end{bmatrix} = \begin{bmatrix} I \\ & T_{2h}^e \end{bmatrix} \begin{bmatrix} x_{e,2h}^{in} \\ x_{e,2h}^l \end{bmatrix}. \tag{12.85}$$

Substituting the above equation in (12.84) and defining the matrix Q'^h_{2h} through the relation

$$\begin{bmatrix} I \\ & T_h^e \end{bmatrix} Q'^h_{2h} = Q_{2h}^h \begin{bmatrix} I \\ & T_{2h}^e \end{bmatrix} \tag{12.86}$$

allows us to cast (12.84) in the following form:

$$Q'^{2h}_h S'_h Q'^h_{2h} x'_{e,2h} = Q'^{2h}_h f'_{e,h}. \tag{12.87}$$

In the above equation S'_h and $f'_{e,h}$ are, respectively, the system matrix and the right-hand side vector defined in (12.62), while the reduced vector of unknowns, $x'_{e,2h}$, is defined as follows:

$$\begin{bmatrix} x'_{e,2h} \end{bmatrix} = \begin{bmatrix} x_{e,2h}^{in} \\ x_{e,2h}^l \end{bmatrix}. \tag{12.88}$$

Equation (12.87) is the matrix equation resulting after the imposition of the Floquet periodic boundary condition. Making use of (12.82) and noting that it is $T^a = T^e$, the new inter-grid operators are cast in the form

$$Q'^h_{2h} = \begin{bmatrix} I \\ & \frac{1}{2}(T_h^e)^H \end{bmatrix} Q_{2h}^h \begin{bmatrix} I \\ & T_{2h}^e \end{bmatrix},$$

$$Q'^{2h}_h = \left(Q'^h_{2h} \right)^H. \tag{12.89}$$

We note that the intergrid operators need not be built explicitly, Instead, the above matrix products may be used in their place during the iterative solution of the finite element system.

12.6 APPLICATIONS

The aforementioned methodologies are used in this section for the analysis of various periodic structures. These numerical studies serve as demonstration of the types of structures that can be handled by the method of finite elements in the context of electromagnetic radiation and/or scattering by periodic structures.

Radiation by an Infinite Array of Rectangular Waveguides. This structure was considered in [20] and [21]. The two-dimensional waveguide array is depicted in the inserts of Fig. 12.7 and Fig. 12.8. For the case of the array in Fig. 12.7 the walls of the waveguides are taken to be infinitesimally thin and perfectly conducting. For the case of the array in Fig. 12.8 the walls are of finite thickness and perfectly conducting.

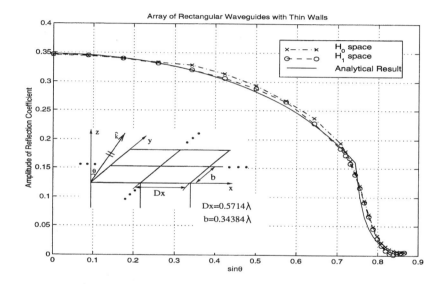

Figure 12.7 Magnitude of the reflection coefficient for an array of rectangular waveguides with infinitesimally thin PEC walls. (After Zhu and Lee [13], ©2000 EMW. Reproduced with permission.)

Each waveguide in the array is excited by the TE_{10} mode. The scan angle of the resulting radiated field is controlled by the phase-shift between the elements in the array. In order to truncate the finite element grid inside the waveguide, a TE_{10}-mode absorbing boundary condition is imposed (in the context of the transfinite element method). The truncation boundary is taken to be sufficiently far away from the waveguide opening to ensure that any higher-order (non-propagating) modes have underwent sufficient attenuation; thus only the reflected TE_{10} mode needs to be absorbed at the truncation boundary.

Plotted in the two figures is the magnitude of the reflection coefficient R for the H-plane scan for values of the scan angle θ ranging from $0°$ to $60°$. For both cases, a PML of thickness $0.23\lambda_0$ was placed $0.3\lambda_0$ above the plane of the guides. The average edge length in the finite element grid used is $0.057\lambda_0$. From 12.26, we know that for a scan angle such that $0 \leq \theta < \sin^{-1}(\lambda_0/D_x)$, only the $(0,0)$ mode is propagating. For larger values of θ, the attenuation of the $(-1,0)$ mode in the z direction (perpendicular to the plane of the array) decreases. Then, when $\theta > \sin^{-1}(\lambda_0/D_x)$, the $(-1,0)$ mode becomes a propagating mode. Its direction of propagation depends on the specific value of θ. The value of β that controls

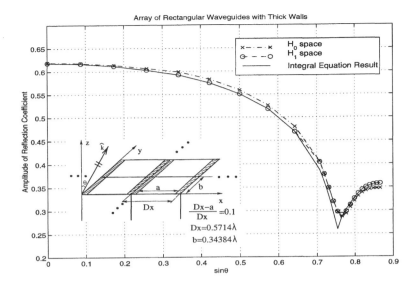

Figure 12.8   Magnitude of the reflection coefficient for an array of rectangular waveguides with PEC walls of finite thickness. (After Zhu and Lee [13], ©2000 EMW. Reproduced with permission.)

the attenuation inside the PML was assigned different values for different ranges of the scan angle θ. More specifically, $\beta = 2.0$ for $0^o \leq \theta < 20^o$; $\beta = 2.5$ for $20^o \leq \theta < 30^o$; $\beta = 3.0$ for $30^o \leq \theta < 45^o$ and $\beta = 3.7$ for $\theta \geq 45^o$.

As can be seen from Figs. 12.7 and 12.8, the results obtained using the methodology of the previous sections are in very good agreement with those obtained in [20] and [21] using an integral equation technique. Since for values of θ just below the grating lobe angle the $(-1, 0)$ mode exhibits slow decay along z and, then, for values just above the grating lobe angle it propagates at the grazing angle, its absorption by the PML for this range of values of θ is not very good. The way this inaccuracy manifests itself in the plots of the reflection coefficient amplitude in Figs. 12.7 and 12.8 is through the smoothing of the discontinuity in the slope (derivative) of $|R|$ at the grating lobe angle. This smoothing, compared to the integral equation solution, is particularly noticeable in the finite element result for the case of infinitesimally thin walls in Fig. 12.7.

At this point it is appropriate to point out that when PMLs are used for grid trunca-tion above and/or below the infinite periodic structure, the calculation of reflection and/or transmission coefficients, which are most commonly used for the quantification of the elec-tromagnetic properties of the structure, can be obtained making use of the orthogonality of the space harmonics. To elaborate, consider two planes, one above and one below the volume occupied by the periodic structure, which are also placed outside the PML regions. Assuming that the excitation of the periodic structure is through a plane wave incident from the top, let $\vec{E}^{ref}$ and $\vec{E}^{tr}$ denote, respectively, the calculated scattered (reflected and trans-mitted) electric fields at the top and bottom planes. The two-dimensional expansion of these distributions over the unit cell area in terms of Floquet modes yields directly the reflection coefficient (from $\vec{E}^{ref}$) and the transmission coefficient (from $\vec{E}^{tr}$) as the coefficients for the (0,0) Floquet mode in the two expansions.

Scattering by a Periodic Dielectric Slab. The planar dielectric slab of thickness h considered in this example was analyzed in [22] and [23]. The geometry of the slab is depicted in the insert of Fig. 12.9. The dielectric constant is changing periodically as indicated in the figure. In the same figure plots are given for the magnitude of the reflection coefficient calculated for frequencies in the range 248 to 300MHz; hence, in this range $k_0 h$ varies from 5.2 to 6.28. The average length of the edges in the finite element mesh used for this problem was 0.05m. PMLs of thickness 0.2m were placed 0.3m above and below the dielectric slab to truncate the computational domain. The attenuation constant in the PML was set by choosing $\beta = 3.0$. The results obtained using the finite element methodology outlined in the previous sections are in very good agreement with those in [22].

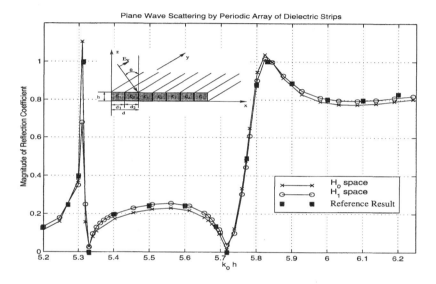

Figure 12.9   Reflection coefficient for a dielectric slab with periodically varying permittivity. $\epsilon_1 = 2.56$; $\epsilon_2 = 1.44$. $d_1 = d_2 = d/2$; $h = 1$m; $h/d = 1.713$. The incident plane wave has a $\hat{y}$-polarized electric field and impinges onto the slab at $\theta^{inc} = 45°$. (After Zhu and Lee [13], ©2000 EMW. Reproduced with permission.)

Another useful means of evaluating the accuracy of the model is by examining the power balance in the incident, reflected and transmitted fields. For example, for the case of this example, with the excitation conditions such that only the (0,0) space harmonics are propagating, the sum of the squares of the magnitudes of the scattered and reflected fields should equal 1. This sum is plotted in Fig. 12.10, for the two cases of finite element basis functions used, $\mathcal{H}^0(curl)$ and $\mathcal{H}^1(curl)$. It is apparent from the plot that the solution obtained using $\mathcal{H}^1(curl)$ elements exhibits superior accuracy especially in the neighborhood of the two peaks. Both solutions appear to diverge from the value of 1 as $k_0 h$ approaches 6.3. This error is due to the fact that in this frequency region the attenuation in the z direction of the next Floquet mode is becoming weaker as the mode approaches its "cutoff" frequency (beyond which it becomes a propagating mode). The interference of this mode with the PMLs becomes the source of the observed numerical error.

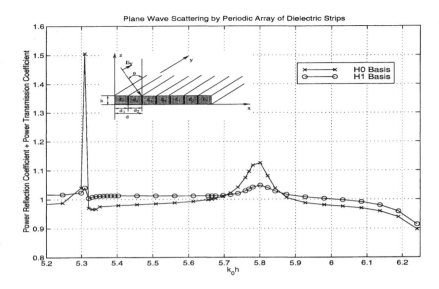

Figure 12.10 Sum of the squares of the magnitude of the reflection and transmission coefficients for the scattering problem of Fig. 12.9. (After Zhu and Lee [13], ©2000 EMW. Reproduced with permission.)

Scattering by PEC Metal Meshes. The two-dimensional periodic structure under consideration in this study is a PEC mesh formed by a periodic array of square holes in an infinite, planar PEC sheet. The side, g, of the square unit cell of the periodic array is 1 m. The side of the square hole is 0.9 m. The thickness, t, of the sheet is the variable in the study. Results will be presented for values of t of 0.00 m, 0.10 m, and 0.25 m. The excitation is taken to be a plane wave, incident normally onto the mesh, with its frequency varying from 150 to 300 MHz. Over this frequency range the electrical size, g/λ_0, of the unit cell varies from 0.5 to 1.0.

Since the frequency band of interest is wide, it is divided into three sub-bands, and a different finite element grid is used in each sub-band. Over Sub-band 1 the electrical length of the periodic cell varies from 0.5 to 0.63. The average length of the edges in the $\mathcal{H}^0(curl)$ finite element mesh used for this sub-band is 0.1 m. Over Sub-band 2 the electrical length of the periodic cell varies from 0.63 to 0.77. The average length of the edges in the $\mathcal{H}^0(curl)$ finite element mesh used for this sub-band is 0.07 m. Finally, over Sub-band 3 the electrical length of the periodic cell varies from 0.77 to 1.0. The average length of the edges in the $\mathcal{H}^0(curl)$ finite element mesh used for this sub-band is 0.05 m.

For the case of $\mathcal{H}^1(curl)$ elements two different finite element grids were used, one for the frequency band for which $0.5 \le g/\lambda_0 < 0.67$ with average edge length is 0.1m, and one for the frequency band for which $0.67 \le g/\lambda_0 \le 1.0$ with average edge length of 0.07m. The thickness of the PMLs was taken to be equal to four times the average edge length in the grid and were placed a distance of six average edge lengths away from the conducting meshes. Also, attenuation in the PML was set by choosing $\beta = 2.0$. Figures 12.11 and 12.12 plot the square of the magnitude of the transmission coefficient (power transmission coefficient) for the two types of finite elements used, $\mathcal{H}^0(curl)$ and $\mathcal{H}^1(curl)$, respectively.

In all cases, a reference solution obtained in [24] through an integral equation-based method of moments (MoM) solution is provided in the figures for reference. Very good agreement between the two sets of results is observed.

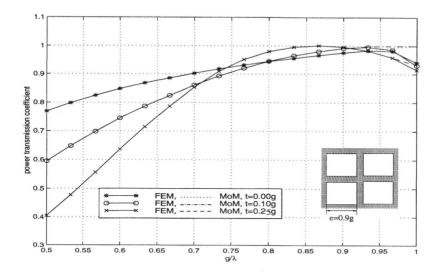

Figure 12.11   Power transmission coefficient for a plane wave incident normally onto a PEC mesh. g is the period along x and y; t is the thickness of the mesh; e is the length of the side of the square hole. Finite element solution is obtained using $\mathcal{H}^0(curl)$ elements. (After Zhu and Lee [13], ©2000 EMW. Reproduced with permission.)

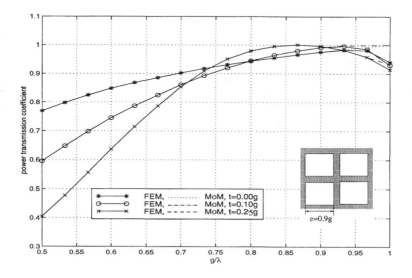

Figure 12.12   Power transmission coefficient for a plane wave incident normally onto a PEC mesh. g is the period along x and y; t is the thickness of the mesh; e is the length of the side of the square hole. Finite element solution is obtained using $\mathcal{H}^1(curl)$ elements. (After Zhu and Lee [13], ©2000 EMW. Reproduced with permission.)

Scattering by a Patch Antenna Array. The two-dimensional, infinite periodic patch antenna array of unit cell depicted in the insert of Fig. 12.13 is illuminated by a uniform plane wave at normal incidence.

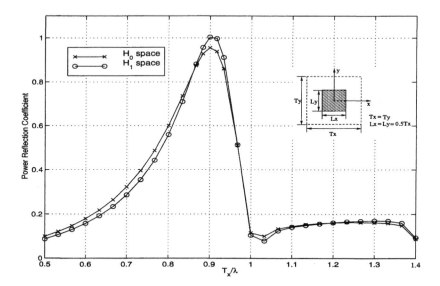

Figure 12.13 Power reflection coefficient for a uniform plane wave incident normally onto an infinite, two-dimensional planar patch antenna array. The electric field is polarized along y. (After Zhu and Lee [13], ©2000 EMW. Reproduced with permission.)

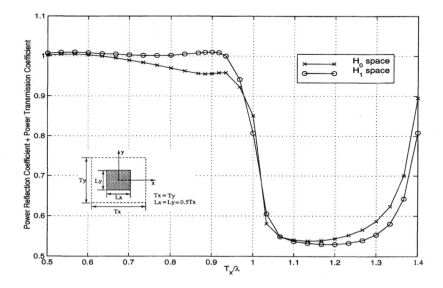

Figure 12.14 Power balance for the (0,0) space harmonics for a plane wave incident normally onto a planar patch antenna array. The electric field is polarized along y. (After Zhu and Lee [13], ©2000 EMW. Reproduced with permission.)

The patches are perfectly conducting squares of side 0.5 m. The unit cell is also square of side 1 m. For the finite element mesh with $\mathcal{H}^0(curl)$ elements the average edge length is 0.05 m. For the case of $\mathcal{H}^1(curl)$ elements, the average edge length is 0.1 m. The PMLs are 0.3 m thick and placed 0.4 m away from the array. The choice of $\beta = 2.0$ was made for setting the attenuation constant in the PML.

For frequencies higher than 300 MHz (where the free-space wavelength becomes larger than the period of the periodic array) a grating lobe appears. In order to absorb the different beams which are propagating in the different directions, β is adjusted to a value of 3.5 in this case. Figures 12.13 and 12.14 depict, respectively, the power reflection coefficient for the (0,0) space harmonic and the assessment of its power balance. Clearly, for frequencies above 300 MHz the power balance diverges from the value of 1 due to scattering into the higher-order space harmonic $(-1, 0)$.

12.7   FINITE ELEMENT MODELING OF PERIODIC WAVEGUIDES

In the previous sections we considered the finite element-based analysis of electromagnetic radiation and/or scattering from periodic structures. Another important class of practical applications involving periodic structures is that of periodic waveguides. The presence of a periodic perturbation in the material properties and/or the geometry of an otherwise uniform waveguide leads to a frequency selectivity in the transmission properties of the guides. Thus pass-bands and stop-bands can be effected in the transmission properties of the guide by controlling the periodicity of the structure. In the following, we present a brief overview of the mathematical formalism pertinent to the electromagnetic analysis of periodic waveguides. For a more detailed presentation the reader is referred to [10].

Let us consider the infinitely long, isotropic, periodic waveguide geometry shown in Fig. 12.15, with its axis taken to be along the z axis of the reference coordinate system. Let D denote the period of the material and/or geometry variation in the waveguide, with S_l and S_r denoting the left and right boundaries defining one unit cell.

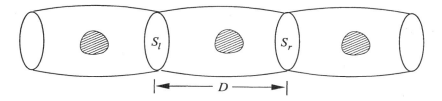

Figure 12.15   Generic geometry of an infinitely long periodic waveguide.

The electromagnetic solutions of interest inside the waveguide of Fig. 12.15 must be guided waves, exhibiting a dependence of the form $\exp(\pm\gamma z)$ along the waveguide axis. Thus any component, u, of the electromagnetic field for a wave propagating along z must satisfy the relation

$$u(z + D) = u(z)e^{-\gamma D}, \tag{12.90}$$

where γ is propagation constant, which, in general, is complex, and thus will be written as $\gamma = \alpha + j\beta$. It is noted that u is, in general, a function of x and y. However, for the sake of simplicity in the mathematical notation only the dependence of the fields on z will be shown explicitly in the equations.

Consider, next, the function, $u_p(z)$, defined as follows:

$$u_p(z) = u(z)e^{\gamma z}. \tag{12.91}$$

Since it is $u_p(z + D) = u(z + D)e^{\gamma(z+D)} = u(p)e^{\gamma z} = u_p(z)$, we conclude that $u_p(z)$ is periodic with period D. Hence, $u_p(z)$ may be expanded in a Fourier series,

$$u_p(z) = \sum_{m=-\infty}^{\infty} a_m e^{-j\frac{2\pi m}{D}z}. \tag{12.92}$$

Thus the following expression is obtained for the field component u:

$$u(z) = \sum_{m=-\infty}^{\infty} a_m e^{-\left(\gamma + j\frac{2\pi m}{D}\right)z}. \tag{12.93}$$

Clearly, the above expression can be interpreted as a superposition of plane waves (space harmonics), each one exhibiting a different phase velocity, given by

$$v_{p,m} = \frac{\omega}{\beta + \frac{2\pi m}{D}}. \tag{12.94}$$

However, the group velocity is the same for all harmonics,

$$v_{g,m} = \frac{1}{d\left(\beta + \frac{2\pi m}{D}\right)/d\omega} = \frac{1}{d\beta/d\omega} = v_{g,0}. \tag{12.95}$$

It should be apparent that the electromagnetic analysis of a periodic waveguide concerns the computation of the propagation constant γ versus frequency. In this manner, the stop-band and pass-bands of the periodic structure are obtained.

The frequency range over which γ is real defines the stop-band for the periodic structure. The pass-band is defined by the frequency range over which γ is purely imaginary, $\gamma = j\beta$. Clearly, $-j\beta$ and $j(\beta + 2\pi m/D)$ are also valid solutions for (12.90); hence, ω is a periodic function of β with period $2\pi/D$. As expected, the attributes of wave propagation are the same in the positive and negative direction along the axis of an infinite periodic structure.

The finite element modeling of periodic waveguides involves the development of a matrix eigenvalue problem for the propagation constant γ [25]. The development begins with (12.51). The vectors of unknown electric and magnetic fields on the walls, x_e^w and x_h^w, respectively, are each split into two sub-vectors, one containing the unknowns on the left wall, x_e^l and x_h^l, and one containing the unknowns on the right wall, x_e^r and x_h^r. Making use of the structure of P in (12.58) yields

$$\begin{bmatrix} S^{in,in} & S^{in,l} & S^{in,r} \\ S^{l,in} & S^{l,l} & 0 \\ S^{r,in} & 0 & S^{r,r} \end{bmatrix} \begin{bmatrix} x_e^{in} \\ x_e^l \\ x_e^r \end{bmatrix} = j\omega\mu_0 \begin{bmatrix} A & \\ & -A \end{bmatrix} \begin{bmatrix} 0 \\ x_h^l \\ x_h^r \end{bmatrix}, \tag{12.96}$$

where the matrices coupling the unknowns x_e^l and x_e^r are zero, since the two surface meshes are not related through tetrahedron elements. Elimination of the electric field unknowns, x_e^{in}, in the interior of the domain yields the reduced system

$$\begin{bmatrix} S'^{l,l} & S'^{l,r} \\ S'^{r,l} & S'^{r,r} \end{bmatrix} \begin{bmatrix} x_e^l \\ x_e^r \end{bmatrix} = j\omega\mu_0 \begin{bmatrix} A & \\ & -A \end{bmatrix} \begin{bmatrix} x_h^l \\ x_h^r \end{bmatrix}, \tag{12.97}$$

where

$$\begin{bmatrix} S'^{l,l} & S'^{l,r} \\ S'^{r,l} & S'^{r,r} \end{bmatrix} = \begin{bmatrix} S^{l,l} & \\ & S^{r,r} \end{bmatrix} - \begin{bmatrix} S^{l,in} \\ S^{r,in} \end{bmatrix} [S^{in,in}]^{-1} \begin{bmatrix} S^{in,l} & S^{in,r} \end{bmatrix}. \qquad (12.98)$$

Imposition of the Floquet periodic boundary condition links the electric and magnetic fields on the left and right walls through the equations

$$x_h^r = e^{-\gamma D} x_h^l, \quad x_e^r = e^{-\gamma D} x_e^l. \qquad (12.99)$$

Rearranging (12.97) we obtain the following generalized linear eigenvalue problem for $e^{\gamma D}$:

$$e^{\gamma D} \begin{bmatrix} S'^{l,l} & -jk_0\eta_0 A \\ S'^{r,l} & 0 \end{bmatrix} \begin{bmatrix} x_e^l \\ x_h^l \end{bmatrix} = \begin{bmatrix} S'^{l,r} & 0 \\ S'^{r,r} & -jk_0\eta_0 A \end{bmatrix} \begin{bmatrix} x_e^l \\ x_h^l \end{bmatrix}. \qquad (12.100)$$

An alternative form of the matrix eigenvalue problem for the periodic wave guide can be obtained in the following manner. First, (12.99) is substituted into (12.97). Multiplying the second equation in (12.97) with $e^{\gamma D}$ and adding it to the first equation yields

$$\left[S'^{r,l} e^{2\gamma D} + \left(S'^{l,l} + S'^{r,r} \right) e^{\gamma D} + S'^{l,r} \right] x_e^l = 0. \qquad (12.101)$$

Compared to (12.100), the dimension of the above eigenvalue problem is reduced at the expense of having to solve a nonlinear eigenvalue problem for $e^{\gamma D}$. In both cases, the matrices are dense because of the elimination of the interior unknowns through (12.98). The eigenvectors obtained in both cases provide the electric field values on the periodic walls. The corresponding electric field values in the interior of the unit cell are obtained by solving the first equation in (12.96)

$$x_e^{in} = -S^{in,in-1} \left(S^{in,l} x_e^l + S^{in,r} x_e^r \right). \qquad (12.102)$$

To avoid the expensive Schur-complement elimination of the interior unknowns and, thus, the dense matrices in the eigenvalue problems (12.100) and (12.101), an eigenvalue problem can be formulated with unknown eigenvectors involving both the interior and periodic boundary fields. Starting with (12.96) and making use of the periodic boundary condition (12.99), we obtain

$$\begin{bmatrix} S^{in,in} & S^{in,l} & S^{in,r} \\ S^{l,in} & S^{l,l} & 0 \\ S^{r,in} & 0 & S^{r,r} \end{bmatrix} \begin{bmatrix} x_e^{in} \\ x_e^l \\ e^{-\gamma D} x_e^l \end{bmatrix} = j\omega\mu_0 \begin{bmatrix} A & \\ & -A \end{bmatrix} \begin{bmatrix} 0 \\ x_h^l \\ e^{-\gamma D} x_h^l \end{bmatrix}. \qquad (12.103)$$

Rearranging the terms in the above equation, a generalized linear matrix eigenvalue problem can be obtained. Instead, one may first eliminate the magnetic field unknowns on the periodic walls, thus reducing the dimension of the matrix eigenvalue problem that must be solved for $\exp(\gamma D)$. To do this we multiply the third row of (12.103) with $e^{\gamma D}$ and add it to the second row. This yields

$$\begin{bmatrix} S^{in,in} & S^{in,l} + S^{in,r} e^{-\gamma D} \\ S^{l,in} + S^{r,in} e^{\gamma D} & S^{l,l} + S^{r,r} \end{bmatrix} \begin{bmatrix} x_e^{in} \\ x_e^l \end{bmatrix} = 0. \qquad (12.104)$$

Multiplication of the first row of the above equation by $e^{\gamma D}$, followed by the splitting of the matrix into two parts, yields the final result for the generalized linear sparse matrix eigenvalue problem for the periodic waveguide

$$-e^{\gamma D} \begin{bmatrix} S^{in,in} & S^{in,l} \\ S^{r,in} & 0 \end{bmatrix} \begin{bmatrix} x_e^{in} \\ x_e^l \end{bmatrix} = \begin{bmatrix} 0 & S^{in,r} \\ S^{l,in} & S^{l,l} + S^{r,r} \end{bmatrix} \begin{bmatrix} x_e^{in} \\ x_e^l \end{bmatrix}. \qquad (12.105)$$

This completes our development of the finite element approximation of the electromagnetic eigenvalue problem associated with wave guidance in periodic waveguides. Our development focused on periodic structures where wave propagation is supposed along one of the three direction in the reference cartesian coordinate system. Such waveguiding structures are most commonly used in conjunction with microwave and millimeter wave passive components, as well as passive components for integrated optics applications. However, the methodology presented for the development of the finite element matrix eigenvalue problem for these structures is easily extended to the general case of two- and three-dimensional periodic waveguides. Such types of waveguides are of interest to the engineering of electronic band-gap and photonic band-gap structures (e.g., [26]), with numerous applications to artificial media with extra-ordinary properties for use in conjunction with *smart* substrates for the integration of electronic and photonic functional blocks.

12.8 APPLICATION

To demonstrate the way a finite element matrix eigenvalue solver is used to compute the pass-band and stop-band characteristics of a periodic waveguide, the simple, air-filled, corrugated waveguide depicted in Fig. 12.16 was analyzed. The basic waveguide is a parallel-plate waveguide of infinite extent along y. The corrugations are also formed by conducting plates of infinite extent along y, shorted at one end as depicted in the figure. All walls are assumed to be perfect electric conductors. Thus, with no field variation along y, a two-dimensional electromagnetic eigenvalue problem must be solved. The structure supports both TE_z and TM_z waves. For the purposes of this example, we are interested in the guidance of TM_z waves only.

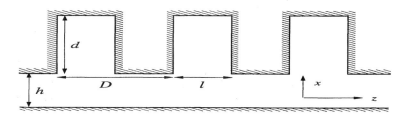

Figure 12.16   Two-dimensional corrugated waveguide.

Figure 12.17 depicts the dispersion curves for the corrugated waveguide. The pertinent $\omega - \beta$ diagram is cast in terms of the corresponding $(k_0 D) - (\beta D)$ diagram, where $k_0 = \omega\sqrt{\epsilon_0\mu_0}$. The frequency is swept from $k_0 D = 0$ to $k_0 D = 3.5$. For each frequency, the matrix eigenvalue problem of (12.105) is solved for $e^{\gamma D}$.

For propagating modes the acceptable eigenvalues are those on the unit circle of the complex λ-plane with $\lambda = e^{\gamma D}$. This is illustrated in Fig. 12.18. As already mentioned, the propagation attributes of the periodic structure are the same for both directions along its axis. Hence, a conjugate pair of eigenvalues is obtained for each propagating mode. Since $k_0 D$ is a periodic function of βD, the curves outside $\beta D \in [-\pi, \pi]$ are obtained by shifting them by the period of 2π. Once the dispersion diagram has been constructed the passbands and stopbands for the waveguiding structure can be identified as indicated in Fig. 12.17.

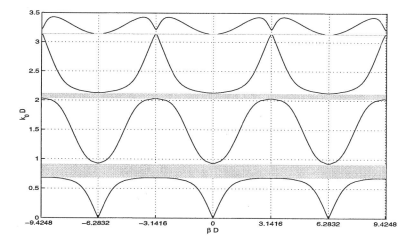

Figure 12.17   Dispersion curves for the waveguide in Fig. 12.16, $d = 2D$, $h = D$, $l = 0.5D$.

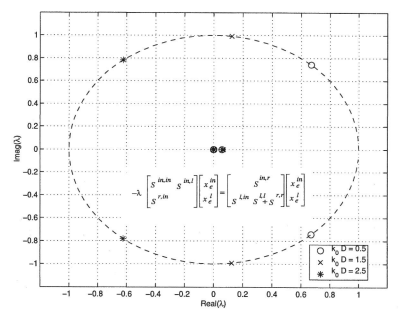

Figure 12.18   Calculated eigenvalues $\lambda = e^{\gamma D}$ for the periodic waveguide of Fig. 12.16 at selected frequencies.

REFERENCES

1. S. D. Gedney and R. Mittra, "Analysis of the electromagnetic scattering by thick gratings using a combined FEM/MM solution," *IEEE Trans. Antennas Propagat.*, vol. 39, pp. 1605-1614, Nov. 1991.

2. S. D. Gedney, J. F. Lee, and R. Mittra, "A combined FEM/MoM approach to analyze the planewave diffraction by arbitrary gratings," *IEEE Trans. Microwave Theory Tech.*, vol. 40, pp. 363-370, Feb. 1992.

3. G. Pelosi, A. Freni, and R. Coccioli, "Hybrid technique for analyzing scattering from periodic structures," *Microwaves, Antennas and Propagation, IEE Proceedings.*, vol. 140, pp. 65-70, Apr. 1993.

4. J. Jin and J. L. Volakis, "Scattering and radiation analysis of three-dimensional cavity arrays via a hybrid finite-element method," *IEEE Trans. Antennas Propagat.*, vol. 41, no. 11, pp. 1580-1585, Nov. 1993.

5. D. T. McGrath and V. P. Pyati, "Phased array antenna analysis with the hybrid finite element method," *IEEE Trans. Antennas Propagat.*, vol. 42, pp. 1625-1630, Dec. 1994.

6. E. W. Lucas and T. P. Fontana, "A 3-D hybrid finite element/boundary element method for the unified radiation and scattering analysis of general infinite period arrays," *IEEE Trans. Antennas Propagat.*, vol. 43, pp. 145-153, Feb. 1995.

7. J. D. Angelo and I. Mayergoyz, "Phased array antenna analysis," in *Finite Element Software for Microwave Engineering*, G. Pelosi, P. Silvester and T. Itoh, Eds., New York: John Wiley & Sons, Inc., 1996.

8. T.F. Eibert, J.L. Volakis, D.R. Wilton, and D.R. Jackson, "Hybrid FE/BI modeling of 3-D doubly periodic structures utilizing triangular prismatic elements and an MPIE formulation accelerated by the Ewald Transformation,", *IEEE Trans. Antennas Propagat.*, vol. 42, no. 5. pp. 843-851, May 1999.

9. J. L. Volakis, T. F. Eibert, D. S. Filipovic, Y. E. Erdemli, and E. Topsakal, "Hybrid Finite Element Methods for Array and FSS Analysis Using Multiresolution Elements and Fast Integral Techniques," *Electromagn.*, vol. 22, no. 4, pp. 297-313, 2002.

10. R.E. Collin, "Field Theory of Guided Waves," 2nd ed., Piscataway, NJ: IEEE Press, 1991.

11. D. S. Sacks, D. M. Kingsland, R. Lee, and J. F. Lee, "A perfectly matched anisotropic absorber for use asan absorbing boundary condition," *IEEE Trans. Antennas Propagat.*, vol. 43, pp. 1460-1463, Dec. 1995.

12. Z.J. Cendes and J.-F. Lee, "The transfinite element method for modeling MMIC devices," *IEEE Trans. Microwave Theory Tech.*, vol. 36, no. 12, pp. 1639-1649, Dec. 1988.

13. Y. Zhu and R. Lee, "TVFEM Analysis of Periodic Structures for Radiation and Scattering," *Progress in Electromagnetic Research Series (PIER-25)*, pp. 1-22, ISSN 1070 4698, EMW, 2000.

14. J. P. Berenger, "A perfectly matched layer for the absorption of electromagnetic waves," *J. Computational Phys.*, vol. 114, no. 2, pp. 185-200, Oct. 1994.

15. D. C. Katz, E. T. Thiele, and A. Talove, "Validation and extension to three dimensions of the Berenger absorbing boundary condition for FDTD meshes," *IEEE Microwave Guided Wave Lett.*, vol. 4, no. 3, pp. 268-270, Aug. 1994.

16. R. Mittra and U. Pekel, "A new look at the perfectly matched layer (PML) concept for the reflectionless absorption of electromagnetic waves," *IEEE Microwave Guided Wave Lett.*, vol, 5, no. 4, pp. 84-87, Mar. 1995.

17. W. C. Chew and W. H. Weedon, "A 3-D perfectly matched medium from modified Maxwell's equation with stretched coordinates," *Microwave Optical Technol. Lett.*, vol. 7, no. 13, pp. 599-604, Sep. 1994.

18. D. M. Kingsland, J. Gong, J. L. Volakis, and J. F.Lee, "Performance of an anisotropic artificial absorber for truncating finite-element meshes," *IEEE Trans. Antennas Propagat.*, vol. 44, pp. 975-981, Jul. 1996.

19. J. Y. Wu, D. M. Kingsland, J. F. Lee, and R. Lee, "A comparison of anisotropic PML to Berenger's PML and its application to the finite-element method for EM scattering," *IEEE Trans. Antennas Propagat.*, vol. 45, pp. 40-50, Jan. 1997.

20. C. P. Wu and V. Galindo, "Properties of a phased array of rectangular waveguides with thin walls," *IEEE Trans. Antennas Propagat.*, vol. 14, pp. 163-173, Mar. 1966.

21. V. Galindo and C. P. Wu, "Numerical solutions for an infinite phased array of rectangular waveguides with thick walls," *IEEE Trans. Antennas Propagat.*, vol. 14, pp. 149-158, Mar. 1966.

22. H. L. Bertoni, L. S. Cheo, and T. Tamir, "Frequency-selective reflection and transmission by a periodic dielectric layer," *IEEE Trans. Antennas Propagat.*, vol. 37, pp. 78-83, Jan. 1989.

23. W. P. Pinello, R. Lee, and A. C. Cangellaris, "Finite element modeling of electromagnetic wave interactions with periodic dielectric structures," *IEEE Trans. Microwave Theory Tech.*, vol. 33, pp. 1083-1088, Oct. 1985.

24. R. C. Compton and D. B. Rutledge, "Approximation techniques for planar periodic structures," *IEEE Trans. Microwave Theory Tech.*, vol. 33, pp. 1083-1099, Oct. 1985.

25. M. Hofer, R. Lerch, N. Finger, G. Kovacs, J. Schoberl, S. Zaglmayr, and U. Langer, "Finite element calculation of wave propagation and excitation in periodic piezoelectric systems," *Fifth World Congress on Computational Mechanics*, 7-12, Jul. 2002.

26. J. D. Joannopoulos, R. D. Meade, and J. N. Winn, *Photonic Crystals - Molding the Flow of Light*, Princeton, NJ: Princeton University Press, 1995.

APPENDIX A

IDENTITIES AND THEOREMS FROM

VECTOR CALCULUS

$$\nabla(\Phi + \Psi) = \nabla\Phi + \nabla\Psi \tag{A.1}$$

$$\nabla \cdot (\vec{A} + \vec{B}) = \nabla \cdot \vec{A} + \nabla \cdot \vec{B} \tag{A.2}$$

$$\nabla \times (\vec{A} + \vec{B}) = \nabla \times \vec{A} + \nabla \times \vec{B} \tag{A.3}$$

$$\nabla(\Phi\Psi) = \Phi\nabla\Psi + \Psi\nabla\Phi \tag{A.4}$$

$$\nabla \cdot (\Psi\vec{A}) = \vec{A} \cdot \nabla\Psi + \Psi\nabla \cdot \vec{A} \tag{A.5}$$

$$\nabla \cdot \left(\vec{A} \times \vec{B}\right) = \vec{B} \cdot \nabla \times \vec{A} - \vec{A} \cdot \nabla \times \vec{B} \tag{A.6}$$

$$\nabla \times (\Phi\vec{A}) = \nabla\Phi \times \vec{A} + \Phi\nabla \times \vec{A} \tag{A.7}$$

$$\nabla \times (\vec{A} \times \vec{B}) = \vec{A}\nabla \cdot \vec{B} - \vec{B}\nabla \cdot \vec{A} + (\vec{B} \cdot \nabla)\vec{A} - (\vec{A} \cdot \nabla)\vec{B} \tag{A.8}$$

$$\nabla(\vec{A} \cdot \vec{B}) = (\vec{A} \cdot \nabla)\vec{B} + (\vec{B} \cdot \nabla)\vec{A} + \vec{A} \times (\nabla \times \vec{B}) + \vec{B} \times (\nabla \times \vec{A}) \tag{A.9}$$

$$\nabla \cdot \nabla\Phi = \nabla^2\Phi \tag{A.10}$$

$$\nabla \cdot \nabla \times \vec{A} = 0 \tag{A.11}$$

$$\nabla \times \nabla\Phi = 0 \tag{A.12}$$

$$\nabla \times \nabla \times \vec{A} = \nabla(\nabla \cdot \vec{A}) - \nabla^2 \vec{A} \tag{A.13}$$

In the following equations, the scalar and vector fields are assumed to be well-behaved. The closed surface Γ encloses the volume Ω. The unit normal, $\hat{n}$, is pointing outwards from the volume V.

Divergence theorem:

$$\iiint_\Omega \nabla \cdot \vec{A} \, dv = \oiint_\Gamma \hat{n} \cdot \vec{A} \, ds \tag{A.14}$$

$$\iiint_\Omega \nabla \Phi \, dv = \oiint_\Gamma \Phi \hat{n} \, ds \tag{A.15}$$

$$\iiint_\Omega \nabla \times \vec{A} \, dv = \oiint_\Gamma \hat{n} \times \vec{A} \, ds \tag{A.16}$$

Green's first identity:

$$\iiint_\Omega \left(\nabla \Phi \cdot \nabla \Psi + \Phi \nabla^2 \Psi \right) dv = \oiint_\Gamma \Phi \hat{n} \cdot \nabla \Psi \, ds \tag{A.17}$$

Green's theorem:

$$\iiint_\Omega \left(\Phi \nabla^2 \Psi - \Psi \nabla^2 \Phi \right) dv = \oiint_\Gamma \hat{n} \cdot \left(\Phi \nabla \Psi - \Psi \nabla \Phi \right) ds \tag{A.18}$$

$$\iiint_\Omega \left(\nabla \times \vec{A} \cdot \nabla \times \vec{B} - \vec{A} \cdot \nabla \times \nabla \times \vec{B} \right) dv$$
$$= \iiint_\Omega \nabla \cdot \left(\vec{A} \times \nabla \times \vec{B} \right) dv \tag{A.19}$$

$$\iiint_\Omega \left(\vec{B} \cdot \nabla \times \nabla \times \vec{A} - \vec{A} \cdot \nabla \times \nabla \times \vec{B} \right) dv$$
$$= \oiint_\Gamma \hat{n} \cdot \left(\vec{A} \times \nabla \times \vec{B} - \vec{B} \times \nabla \times \vec{A} \right) ds \tag{A.20}$$

In the following equations, the open surface Γ is bounded by the contour C. The direction of the unit normal, $\hat{n}$, on Γ, is defined, in relation to the sense of line integration around C, by the right-hand-screw rule.

$$\iint_\Gamma \hat{n} \times \nabla \Phi \, ds = \oint_C \Phi \, \vec{dl} \tag{A.21}$$

$$\iint_\Gamma \nabla \times \vec{A} \cdot d\vec{s} = \oint_C \vec{A} \cdot \vec{dl} \quad \text{(Stokes' theorem)} \tag{A.22}$$

INDEX